Spectroscopy of Light and Heavy Quarks

ETTORE MAJORANA
INTERNATIONAL SCIENCE SERIES
Series Editor:
Antonino Zichichi
European Physical Society
Geneva, Switzerland

(PHYSICAL SCIENCES)

Recent volumes in the series:

A Continuation Order Plan is available for this series. A continuation order will bring delivery of
each new volume immediately upon publication. Volumes are billed only upon actual shipment.
For further information please contact the publisher.

Spectroscopy of Light and Heavy Quarks

Edited by

Ugo Gastaldi and
Robert Klapisch
CERN
Geneva, Switzerland

and

Frank Close
Rutherford Appleton Laboratory
Didcot, United Kingdom

Plenum Press ● New York and London

Library of Congress Cataloging in Publication Data

International School of Physics with Low Energy Antiprotons on Spectroscopy of
Light and Heavy Quarks (2nd: 1987: Erice, Sicily)
  Spectroscopy of light and heavy quarks / edited by Ugo Gastaldi and Robert
Klapisch and Frank Close.
    p.   cm.—(Ettore Majorana international science series, Physical sciences; v.
37)
  "Proceedings of the second course of the International School of Physics with Low
Energy Antiprotons on Spectroscopy of Light and Heavy Quarks, held May 23–31,
1987, in Erice, Sicily, Italy"—T.p. verso.
  Bibliography: p.
  Includes index.
  ISBN-13: 978-1-4612-8070-5   e-ISBN-13: 978-1-4613-0763-1
  DOI: 10.1007/978-1-4613-0763-1

  1. Quarks—Spectra—Congresses. 2. Meson—Spectra—Congresses. 3. Antiprotons—Congresses. I. Gastaldi, Ugo, 1947–   . II. Klapisch, Robert. III. Close, F.
E. IV. Title. V. Series.
QC793.5.Q2527I57 1987   88-39213
539.7'216—dc19   CIP

Proceedings of the second course of the International School of
Physics with Low Energy Antiprotons on Spectroscopy of Light and
Heavy Quarks, held May 23–31, 1987, in Erice, Italy

© 1989 Plenum Press, New York
Softcover reprint of the hardcover 1st edition 1989

A Division of Plenum Publishing Corporation
233 Spring Street, New York, N.Y. 10013

PREFACE

The second course of the International School on Physics with Low Energy Antiprotons was held in Erice, Sicily at the Ettore Majorana Centre for Scientific Culture, from May 20 to May 31, 1987.

The School is dedicated to physics accessible to experiments using low energy antiprotons, especially in view of operation of the LEAR facility at CERN with the upgraded antiproton source AAC (Antiproton Accumulator AA and Antiproton Collector ACOL). The first course in 1986 covered topics related to fundamental symmetries. This book contains the Proceedings of the second course which focused on spectroscopy of light and heavy quarks.

These Proceedings contain both the tutorial lectures and contributions presented by participants during the School.

The papers are organized in four sections:

The first section includes theoretical reviews.

Section II contains experimental reviews and covers the results in meson spectroscopy from DM2, MARK III, GAMS and Ω-WA76.

Section III presents the new meson spectroscopy experiments in preparation at CERN and Fermilab: Crystal Barrel, OBELIX, Jetset and E760.

Section IV is dedicated to LEAR and to future facilities where meson spectroscopy would be a principal component of the physics programme.

We should like to thank Dr. Alberto Gabriele and the staff of the Ettore Majorana Centre who provided for a smooth running of the School and a very pleasant stay. We are particularly grateful to Mrs. Anne Marie Bugge for her crucial help during the preparation and running of the School and for the editing of these Proceedings.

F. Close
U. Gastaldi
R. Klapisch

CONTENTS

WHAT'S NEW IN THE OLD SPECTROSCOPY?

Nathan Isgur[†]

Department of Theoretical Physics
1 Keble Road, Oxford, U.K. OX1 3NP

ABSTRACT

In the first two of these three lectures, I describe some recent work attempting to place the quark potential model on a firmer foundation: we will discuss taking first the "naive" and then the "non−relativistic" out of the quark model. In the last lecture I will give my views on the outstanding theoretical and experimental issues in "the old spectroscopy".

TAKING THE "NAIVE" OUT OF THE QUARK MODEL

I have been using the constituent quark model for many years now as a tool for understanding the spectrum and properties of the low−lying mesons and baryons. For most of this time I have been painfully aware of the difficulty of understanding from first principles (i.e., from QCD) why such a model should work. I have nevertheless had faith that we would eventually be able to justify the use of this model simply because it is such a good representation of the physics. In this lecture I will describe some recent progress in reducing ––– but not altogether eliminating –––– the gap between QCD and the quark model. The ideas I will discuss have at least convinced me that we can now take the "naive" out of the description of this model[1,2].

ADIABATIC SURFACES

The idea of adiabatic surfaces [1,2] is central to our argument for the quark model approximation to QCD. Consider first QCD without dynamical fermions in the presence of fixed $q_1\bar{q}_2$ or $q_1q_2q_3$ sources. The ground state of QCD with these sources in place will be modified, as will be its excitation spectrum. For excitation energies below those required to produce a glueball, this spectrum will presumably be discrete and continuous as, for example, shown in Figure 1 as a function of the $q_1\bar{q}_2$ spatial separation $\vec{r}$. There will be analogous spectra for $q_1q_2q_3$ which are functions of the two relative coordinates $\vec{\rho} = \sqrt{\frac{1}{2}}(\vec{r}_1 - \vec{r}_2)$ and $\vec{\lambda} = \sqrt{1/6}\,(\vec{r}_1 + \vec{r}_2 - 2\vec{r}_3)$. We call the energy surface traced out by a given level of excitation as the positions of the sources are varied an adiabatic surface.

[†] permanent address: Department of Physics, University of Toronto, Toronto, Canada, M5S 1A7

Let us now define the "quark model limit": the quark model limit obtains when the quark sources move along the lowest adiabatic surface in such a way that they are isolated from the effects of other (excited) surfaces.

Before trying to argue that this definition is relevant, let me first simply note that it has several appealing characteristics:

1) One of the great "mysteries" of the quark model is that it describes the mesons and baryons in terms of a wavefunction which only gives the amplitude for the valence quark variables, even though in QCD the general state vector must also refer to the glue fields. Indeed, in QCD for fixed q_1 and q_2, for example, there are an infinite number of possible states of the glue so that it is certainly not sufficient to simply specify the state of the quarks. In the "quark model limit", however, although there are an infinite number of possible glue states, for any fixed r there is one

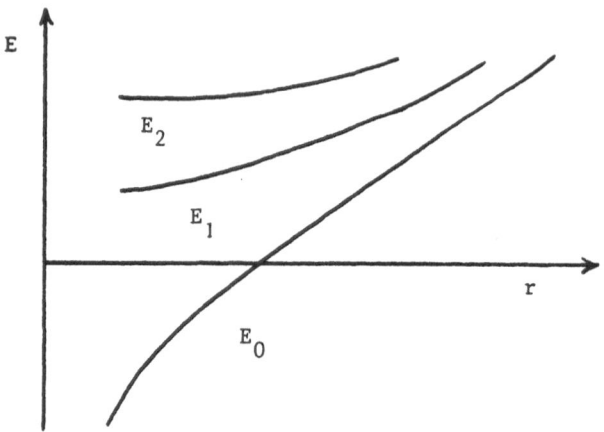

Figure 1. Schematic of the low-lying adiabatic surfaces of $q_1\bar{q}_2$ at relative separation r. $E_0(r)$ corresponds to the gluonic ground state, $E_1(r)$ to the first excited state, etc.

lowest–lying one. Moreover, although this lowest–lying state changes as r changes, it is completely determined by the quark coordinates. Thus we see the possibility that the quark model wavefunction had a "secret suppressed subscript" describing the state of the glue: $\psi_0(r)$. We will argue below that there should be analogous (but as yet undiscovered) worlds $\psi_n(r)$ for n > 0 corresponding to hybrid mesons.

2) The "quark model limit" can easily be seen to be inapplicable to any systems more complicated than $q_1\bar{q}_2$ and $q_1q_2q_3$: such systems will always have adiabatic surfaces which cross so that the condition of isolation cannot be satisfied [2]. Figure 2 gives a simple illustration of this phenomenon in the $q_1q_2\bar{q}_3\bar{q}_4$ sector. We will return to this important distinction between the familiar mesons and baryons and multiquark systems below. For now we just note that the above definition allows us to begin to see why $q_1\bar{q}_2$ and $q_1q_2q_3$ may have a special status in QCD: only in these two cases is it possible that the state of the glue is (approximately) determined by the quark coordinates (see the point x=y in Figure 2).

With these attractions for motivation, we now proceed with the argument for the relevance of this definition [1]. We first recall a simple molecular physics analogy to this proposed approximation. Diatomic molecular spectra can be obtained in an adiabatic approximation by holding the two relevant atomic nuclei at fixed separation r and then solving the Schrödinger problem for the (mutually interacting) electrons moving in the static electric field of the nuclei. The electrons will, for fixed r, have a ground state and excited states which will eventually become a continuum above energies required to ionize the molecule. The resulting adiabatic surfaces then serve as effective internuclear potentials on which vibration–rotation spectra can be built. Molecular transitions can then take place within states built on a given surface or

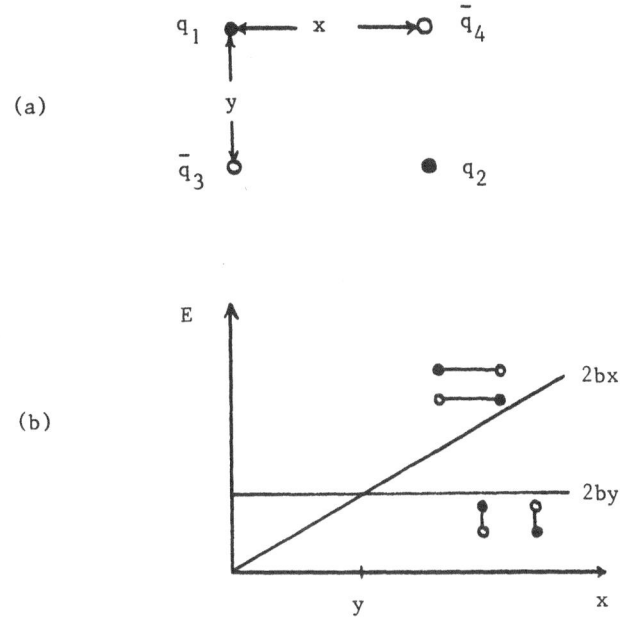

Figure 2. (a) a simple rectangular $q_1 q_2 \bar{q}_3 \bar{q}_4$ configuration with $x \equiv |\vec{r}_1 - \vec{r}_4| = |\vec{r}_3 - \vec{r}_2|$ and $y \equiv |\vec{r}_1 - \vec{r}_3| = |\vec{r}_4 - \vec{r}_2|$; (b) the two adiabatic sufaces corresponding to the "colour configurations" $(q_1 \bar{q}_3)(q_2 \bar{q}_4)$ and $(q_1 \bar{q}_4)(q_2 \bar{q}_3)$ as shown for fixed y as a function of x for x and y in the linear potential region; b is the $q \bar{q}$ string tension; the surfaces are labelled by their corresponding flux tube model states.

between surfaces. In the "quark model limit" the quark sources play the rôle of the nuclei, and the glue plays the rôle of the electrons. From this point of view we can see clearly that conventional meson and baryon spectroscopy has only scratched the surface of even $q_1 q_2$ and $q_1 q_2 q_3$ spectroscopy: so far we have only studied the vibration–rotation bands built on the lowest adiabatic surface corresponding to the gluonic ground state. We should expect to be able to build other "hadronic worlds" on the surfaces associated with excited gluonic states [2,3]: these states correspond to the hybrids first discussed in the bag model (in terms which from this point of view are inappropriate) as $q_1 \bar{q}_2 g$ and $q_1 q_2 q_3 g$ states [4].

On the basis of this analogy it seems clear that the quark model limit will apply when all of the quarks in a meson or baryon are heavy. This certainly corresponds well with the established success of the "naive" quark model in the $c\bar{c}$ and $b\bar{b}$ sectors. Although the applicability of our definition thus seems assured in this limit, it is nevertheless useful to consider the corrections to the adiabatic approximation. These considerations will not only allow us to understand how to make more accurate predictions, but will also help us to understand why the quark model limit has a much wider range of validity than might have been expected. Of course, the simple argument for the applicability of the adiabatic approximation is just that the dynamics of the heavy quark systems $Q_1\bar{Q}_2$ and $Q_1Q_2Q_3$ are completely controlled by the asymptotically free colour Coulomb interaction and so have quark velocities of the order of $\alpha_s \to 0$ as $m_Q \to \infty$. The details are more interesting. We first note (using $Q\bar{Q}$ as our prototype) that as $m_Q \to \infty$ the low–lying states have radii $r_{Q\bar{Q}} \sim (m_Q\alpha_s)^{-1} \to 0$ and frequencies $\omega_{Q\bar{Q}} \sim m_Q\alpha_s^2 \to \infty$. Note that, although as previously stated $v \sim \omega_r \sim \alpha_s \to 0$, the analogue of the condition usually quoted for the applicability of the adiabatic approximation in molecules ($\omega_{\text{nuclei}} \ll \omega_{\text{electrons}}$) does not apply. This is because $\omega_{Q\bar{Q}} \to \infty$ while ω_{glue}, corresponding to the gap between $E_0(r)$ and the first gluonic excitation, is never greater than the energy required to excite the glueball continuum. Even if we were to discount such continua, we would have to consider the other discrete surfaces which must lie no higher than about 1 GeV above $2m_Q$ corresponding to $\omega_{\text{glue}} \sim m_Q\alpha_s^2 \sim \omega_{\text{quark}}$. The physics of the decoupling of motion on the lowest adiabatic surface from excited surfaces is thus rather different than in the molecular case. It depends (in the language of second order perturbation theory) primarily on small matrix elements rather than on large energy denominators: in $Q\bar{Q}$ the decoupling occurs because the quarks produce very small oscillating colour dipole moments.

Now consider light quark systems. As the quark mass is decreased, the oscillations gradually approach a size governed by the QCD scale $b^{-\frac{1}{2}}$ (where b is the string tension, $b \sim \Lambda^2_{QCD}$). On the other hand, the quark frequencies are decreasing. For light quarks ω_{quark} also approaches the QCD scale, and we have $\omega_{\text{quark}} < \omega_{\text{glue}}$. (That ω_{quark} is somewhat less than ω_{glue} is indicated by the fact that the orbital excitations of the quark model are well known, but hybrids have not yet been discovered!) Nevertheless, given the scale of the perturbation in this case, I suspect that there is no simple argument for the validity of the quark model limit for light quark systems: they could go either way, and the issue can only be decided by explicit calculations. Since such explicit calculations necessarily involve non–perturbative gluon dynamics, any such discussion is at this time bound to be model dependent.

Having accepted that our conclusions will be model dependent, I will now introduce the flux tube model [2] for gluon dynamics. We will then be in a position to study the validity of the adiabatic approximation for all values of the quark mass within the model.

THE FLUX TUBE MODEL

Even after successful numerical calculations within QCD are possible, it will still be useful, and in complex situations essential, to have models which summarise the very complex structure of this theory. The flux tube model [2] is a model for QCD in the non–perturbative regime which emerges from considerations of Hamiltonian lattice QCD.

Hamiltonian Lattice

In the Hamiltonian version of lattice QCD, space (but not time as in most numerical studies) is discretized. In this formulation the lattice spacing "a", without reference to a perturbative expansion, plays the rôle of the regulator mass M. Latticizing the theory also has another advantage: it allows us to set up a strong coupling perturbation expansion in which the expansion parameter for lattice QCD is $1/g$ instead of g. We may expect to be able to learn more about the strongly coupled regime of the theory in terms of such an expansion, and indeed this seems to be the case: for example, confinement is an automatic property of the $g\rightarrow\infty$ limit of lattice QCD. Moreover, the natural degrees of freedom of the strong coupling regime are not quarks and gluons, but rather quarks and flux tubes, the latter being more in accord with various qualitative ideas on the nature of confinement in QCD. Of course, space is not coarse-grained (at least not on the scale of 10^{-15} metres), so that to relate lattice QCD to real QCD we must consider the limit $a\rightarrow 0$. In this limit $g\rightarrow 0$ so that a strong coupling expansion must fail; this is just the other side of the failure of the weak coupling expansion for small Q^2. If, however, it can be shown that the two regimes "match" around $g=1$, thereby proving that lattice QCD as $a\rightarrow 0$ is QCD, then one would nevertheless expect the strong coupling expansion to be useful in many situations where large scales dominate, just as the weak coupling expansion is useful for short distance physics.

A simple analogy may be useful. Consider approximating a continuous one dimensional harmonic oscillator by a particle hopping along a one dimensional lattice of points $x=na(n=\ldots, -2, -1, 0, 1, 2 \ldots)$ with lattice spacing "a". The lattice Hamiltonian can be chosen to be

$$H_{mn} = \left[\frac{1}{ma^2} + \tfrac{1}{2} ka^2 n^2\right]\delta_{mn} - \frac{1}{2ma^2}\left[\delta_{m,n+1} + \delta_{m,n-1}\right] \qquad (1)$$

since then the Schrödinger equation

$$i\,\frac{\partial\psi_m}{\partial t}(t) = H_{mn}\psi_n(t) \qquad (2)$$

becomes

$$i\,\frac{\partial\psi(x,t)}{\partial t} \quad \left[-\frac{1}{2m}\frac{\partial^2}{\partial x^2} + \tfrac{1}{2}kx^2\right]\psi(x,t) \qquad (3)$$

as $a\rightarrow 0$. Now for $a\rightarrow\infty$ with k and m fixed, the potential energy term $\tfrac{1}{2}ka^2 n^2\delta_{mn}$ dominates and the eigenstates correspond to the particle sitting on single lattice sites; corrections to this limit are of relative order $\chi=1/kma^4$ and one can systematically proceed to do perturbation theory in this hopping strength. Since the characteristic scale of the harmonic oscillator is $\alpha^{-1}=(km)^{-\frac{1}{4}}$, one will not get realistic wave functions or eigenenergies for the harmonic oscillator for $a \gg \alpha^{-1}$ where lowest order perturbation theory applies, but for $\chi\sim 1$ one will begin to get good approximations to the solutions of the continuum problem if one works to sufficiently high order in χ. By contrast, starting with free particle solutions to the continuum Hamiltonian and treating $\tfrac{1}{2}kx^2$ as a perturbation is hopeless. (The difference, of course, is that the hopping parameter expansion for the ground state, for example, will be accurate if a matrix of dimension of order $1/a\alpha$ is diagonalized.)

We now turn to the formulation of QCD on a (cubic) spatial lattice. In this formulation the quark degrees of freedom of the theory "live" on the lattice sites while the gluonic degrees of freedom "live" on the links between these sites (see Figure 3).

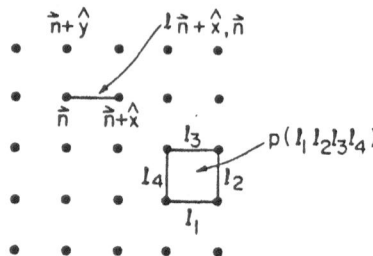

Figure 3. A two dimensional (x,y) slice of the lattice showing a typical lattice point $\vec{n}=(n_x,\ n_y,\ n_z)$, a typical link $\ell_{\vec{n}+\hat{x},\ \vec{n}}$ from $\vec{n}$ to $\vec{n}+\hat{x}$, and a typical plaquette $p(\ell_1\ell_2\ell_3\ell_4)$

Let's consider first the theory without quarks: we describe this theory in terms of link variables U_ℓ which (before quantization) are 3×3 SU(3) group elements. The pure gauge field Hamiltonian is then the sum of two parts, one involving only the U's and one which has non-trivial commutation relations with the U's:

$$H_{glue} = \frac{g^2}{2a} \sum_\ell C_\ell^2 + \frac{1}{ag^2} \sum_p Tr[\, 2-(U_{\ell_4}U_{\ell_3}U_{\ell_2}U_{\ell_1}+h.c.)\,]\tag{4}$$

with lattice spacing "a" and g the corresponding coupling constant. Here C_ℓ^2 is defined in terms of the eight generators $E^a_{\ell\pm}$ of SU(3) transformations of U_ℓ at the beginning $(-)$ or the end $(+)$ of the link ℓ

$$[\,E^a_{\ell+},\ U_\ell\,] = -\frac{\lambda^a}{2}\,U_\ell\tag{5}$$

$$[\,E^a_{\ell-},\ U_\ell\,] = +\,U_\ell\frac{\lambda^a}{2}\tag{6}$$

by $C_\ell^2=\sum_a (E^a_{\ell+})^2 = \sum_a(E^a_{\ell-})^2$. In the second term the product of the U's is taken in order around the plaquette p. To complete lattice QCD one simply adds to (10) a lattice Hamiltonian H_{quark} for the quarks interacting with the glue. With the quark fields as site variables we have

$$H_{quark} = \sum_{flavour} m_q \sum_{\vec{n}} q^\dagger_{\vec{n}}q_{\vec{n}} + \frac{1}{a}\sum_{flavour}\sum_{\ell_{ji}} q^\dagger_j U_{\ell_{ji}}\alpha_{\ell_{ji}}q_i\tag{7}$$

where $\alpha_{\ell ji}$ is the Dirac matrix in the direction of the link ℓ_{ji}. Our complete Hamiltonian $H^{lattice}_{QCD} = H_{glue}+H_{quark}$ has H_{QCD} in $A^\circ=0$ gauge as its naive continuum limit; it is, furthermore, invariant under arbitrary gauge transformations at the lattice sites. Gauss' law takes the form of a constraint in the theory that the only physically relevant states are those which are gauge invariant.

We are ready to consider the properties of $H^{lattice}_{QCD}$. We note first that in the strong coupling limit where "a" (and as we shall see, therefore g) is large

$$H^{lattice}_{QCD} \to H_{sc} = \frac{g^2}{2a}\sum_\ell C_\ell^2 + \sum_{flavour} m_q q^\dagger_{\vec{n}}q_{\vec{n}}\tag{8}$$

The eigenvalues of C_ℓ^2 are just those of the square Casimir of SU(3): 0 for the singlet, 4/3 for 3 or $\bar{3}$, 10/3 for 6 or $\bar{6}$, 3 for the octet, etc. The quark part of H_{sc} is, on the other hand, diagonalized by an arbitrary number of quarks and antiquarks at arbitrary lattice sites. Since, however, the only physically relevant eigenstates are those which are gauge invariant, the strong coupling eigenstates may be classified as follows:

1) the strong coupled vacuum: In this case all links are unoccupied ($C_{\ell}^{\bar{\ell}}=0$) and there are no fermions: the total energy is zero.

2) the pure glue sector: There are still no quarks, but links are excited in such a way that gauge invariant states are produced.   The simplest such pure glue states ("glueloops") have a closed path of links in the 3 (or $\bar{3}$) representation.   These have energy $(2g^2L)/(3a^2)$ where L is the length of the path; the simplest such state just has the links around the perimeter of an elementary plaquette excited: $\mathrm{Tr}[U_{\varrho_4}U_{\varrho_3}U_{\varrho_2}U_{\varrho_1}]|0>$, where $|0>$ is the vacuum.   Of course, more complicated configurations are allowed, including those with non–triplet flux and those with more complicated topologies.   For example, three flux links can emerge from a single lattice site since a gauge invariant combination can be formed with the ϵ_{ijk} invariant tensor. See Figure 4.

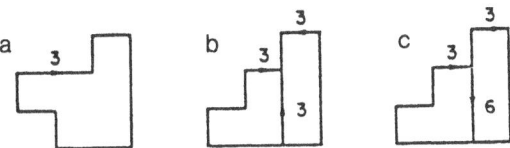

Figure 4. some primitive pure glue states.

3) the meson sector: The simplest quark–containing state consists of a quark and antiquark on the lattice joined by a path of flux lines (for gauge invariance).   These will have energy $m_q+m_{\bar{q}}+(2g^2L)/(3a^2)$ so that we automatically have quark confinement in strong coupling.   See Figure 5.

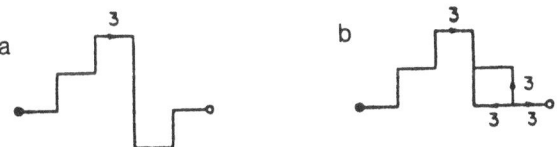

Figure 5. some primitive meson states.

4) the baryon sector: The next simplest quark–containing state consists of three quarks connected by an $\epsilon_{ijk}-$ type flux junction.   Such quarks will also be confined.   See Figure 6.

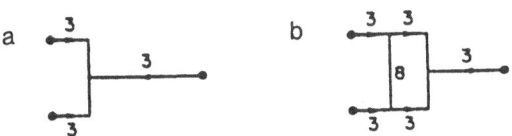

Figure 6. some primitive baryon states.

5) multi–quark sectors: When there are more quarks than those required for a meson or baryon, then in general the system will not be completely confined.   The simplest

such system consists of two quarks and two antiquarks. See Figure 7.

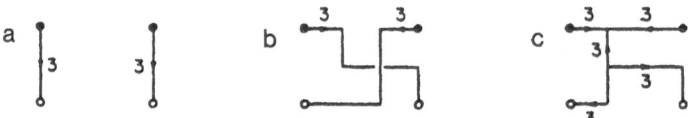

Figure 7. some primitive $q q \bar{q} \bar{q}$ states

With these examples, the general structure of the eigenstates of the strong coupling limit is clear: it consists of "frozen" gauge invariant configurations of quarks and flux lines. Of course, these are not the eigenstates of QCD, but they do form a complete basis (in the limit a→0) for the expansion of the true strong interaction eigenstates.

The full eigenstates of QCD can be found (in principle!) by considering corrections to the strong coupling limit from the terms we have neglected so far. These terms can induce a variety of effects. Consider first of all the $q^{+}U\alpha q$ term. It can, among other things,

1)   annihilate a quark at one point and recreate it at a neighbouring point with an appropriate flux link (Figure 8a);

2)   break a 3−flux line and create a pair (Figure 8b).

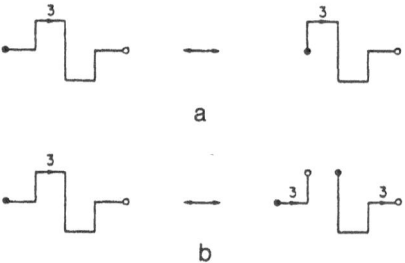

Figure 8. some effects of $q^{+}U\alpha q$: (a)quark hopping, (b)flux breaking pair creation.

This term thus plays a rôle analogous to both the usual quark kinetic energy term and quark−gluon coupling term of the weak coupled theory. Next consider the $\text{Tr}[\,2-U_{\varrho_4}U_{\varrho_3}U_{\varrho_2}U_{\varrho_1}+\text{h.c.}\,]$ term. It can, among other things,

1) allow flux to hop across plaquettes (Figure 9a)
2) change flux topology (Figure 9b).

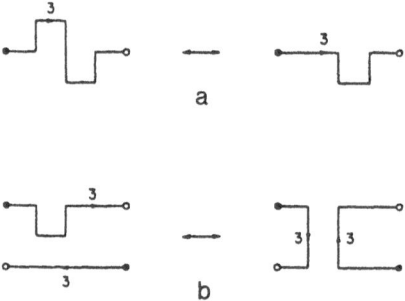

Figure 9. some effects of $\text{Tr}[\,2-(UUUU+\text{h.c.})\,]$: (a) flux tube hopping, (b) flux tube topological mixing by rearrangement.

The full diagonalization of this Hamiltonian problem as a→0 would constitute an exact solution of QCD. Unfortunately, this diagonalization represents a numerical problem which is of a magnitude well beyond presently available computing capacity.

Extracting the Flux Tube Model

To illustrate the flux tube model in the simplest possible context, and to make direct contact with the previous discussion, we consider the sector of QCD which contains $Q\bar{Q}$. The complete set of gluonic base states formed by the eigenstates of the strong coupling limit for this system can be classified by their topology [*], and as a first step in diagonalising this problem we imagine organizing the Hamiltonian into blocks of fixed topology. Within each block the Hamiltonian then consists only of kinetic energy terms corresponding to Figures 8(a) and 9(a). Other perturbations either create $q\bar{q}$ pairs or new flux topologies and so are off-block-diagonal. Within each topological block QCD has thus been reduced to a (discrete) quantum string problem (in the generalized sense: these strings may have internal loops, various local string tensions depending on the colour representation excited, extra $q\bar{q}$ pairs, etc.). Since for strong coupling topological mixing is suppressed, the flux tube model assumes that the long-wavelength properties of QCD can be approximated by treating mixing between topological blocks as a perturbation. Thus in zeroth order the flux tube model treats QCD as a theory of non-interacting, discrete, multitopological strings. The lowest-lying eigenstates in this approximation will therefore be those associated with Q and $\bar{Q}$ connected by a single string of 3-flux.

The Quark Model Limit of the Flux Tube Model

The flux tube model thus suggests a simple model for the adiabatic surfaces of QCD: the glue between Q and $\bar{Q}$ behaves like a discrete quantum string. The lowest adiabatic surface then corresponds to the ground state of this string, while the low-lying excitations correspond (at least for large r) to the excitation of "phonons" in the string. (At higher mass "topologically excited" strings will create other families of adiabatic surfaces).

With this explicit dynamical model for the long-wavelength properties of the glue, one can directly study [5] the validity of the "quark model limit" for quark masses from $m_Q >> b^{\frac{1}{2}}$ down to $m_q \simeq b^{\frac{1}{2}}$. Such a calculation shows the behaviour one would hope to see: for the low-lying spectrum of the lowest adiabatic surface, non-adiabatic corrections are essentially negligible for $m_Q = 1.5$ GeV, and they remain perturbations (~ 50 MeV) even when m_q is as small as 0.3 GeV.

Some Loose Ends

There are two important loose ends in the case we have been making for the quark potential model. The first is that since the flux tube model is at best a representation of the long-wavelength properties of QCD, we cannot be sure that there are not important adiabatic corrections in light quark systems at short distances. It nevertheless seems plausible that variants of the decoupling arguments sketched above for heavy quark systems will apply.

A more fundamental issue is related to our assumption of constituent quark masses and the dominance of the valence structure of hadrons. (Recall that all of our discussion so far has been about QCD in the absence of dynamical fermions!).

---

[*] By "topology" we mean the obvious classification into classes of flux excitations in which all the members of a given class could (in the absence of the lattice) be continuously transformed into one another.

9

Of course, our segregation of QCD into topological sectors guarantees this aspect of the quark model limit as well, but as we cut off QCD at ever smaller scales a, topological mixing both to more complicated string states and to multiquark configurations will become increasingly important as the coupling $1/g(a)$ becomes larger. As $a \to 0$ the effective quark masses $m_i(a)$ will also evolve; in particular the light quark masses will go to their current quark values. These considerations underline a fact which is well-known from deep inelastic scattering: the apparent composition of a hadron depends on the scale at which it is resolved. They also make it clear that the constituent quark model can only hope to provide a description of this composition valid above some minimum distance scale a_0 corresponding roughly to the scale where $g(a_0)=1$, i.e., a_0 of the order of 0.1 fm. Only with such a cut-off can we hope to simultaneously have a field theory with $m_{light}(a_0) \simeq b^{\frac{1}{2}}$ and small topological mixing. For the moment the belief in the existence of such a quark model scale must remain an act of faith bolstered by the phenomenological evidence: 1) light quark effective masses by any reasonable measure are about 300 MeV, 2) the valence structure of hadrons dominates spectroscopy as well as deep inelastic scattering for $Q^2 \lesssim 5$ GeV^2, and 3) the narrow resonance approximation (suggesting once again that light quark pair creation is weak) is very successful. There is also some evidence that at least this is an internally consistent possibility: constituent quark pair creation by flux tube breaking with a scale a_0 of the order of 0.1 GeV can simultaneously explain meson widths [6] while giving only modest shifts to the hadron mass spectrum and small non-valence components to the hadron Fock space expansions.

CONCLUSIONS: LECTURE ONE

I believe that it is now possible to drop the qualifier "naive" from our description of the quark potential model. In this lecture I have argued that there are good reasons for viewing this old phenomenological model as a satisfactory starting point for the description of low-lying mesons and baryons. The resulting picture at least now stands in a clearer relationship to QCD, and the requirements for establishing its pedigree are now more evident. This discussion also has important implications for multiquark physics like nuclear physics: it shows that naive quark model ideas cannot be directly applied since in multiquark systems gluonic degrees of freedom are necessarily active. Thus we reach the rather ironic conclusion that gluon dynamics are more relevant to "old-fashioned" nuclear physics than they are to the physics of (conventional) mesons and baryons.

TAKING THE "NON-RELATIVISTIC" OUT OF THE QUARK MODEL

Now that we have given reasons to consider the quark potential model as a worthwhile approximation to QCD for describing the low-lying mesons and baryons, we turn to address the other derogatory adjective usually attached to the model: "non-relativistic". I will describe some work which, by "relativizing" the quark model, suggests that the usual nonrelativistic approximations are an inessential defect, the correction of which cures many old problems.

Figure 10 provides a useful background to the issues we will address in this lecture. It lines up the $1\,^3S_1$ levels of the $b\bar{b}$, $c\bar{c}$, $s\bar{s}$, $u\bar{s}$, and ud families in order to show the evolution versus mass of the $P - S$ orbital excitation energy (corresponding to the Lyman line of the hydrogen spectrum) and the $^3S_1 - {}^1S_0$ Breit-Fermi relativistic correction due to the spin-spin interactions (corresponding to the 21 centimetre line of hydrogen). This figure certainly suggests that analogous structures occur in heavy and light quark systems and that the physics of light quark systems might be understood by extrapolation from the relatively well–understood heavy quark systems. I believe this is to be expected: in light quark systems p/m ~ Λ_{QCD} / Λ_{QCD} ~ 1 not $\gg$1 as it would be for an ultrarelativistic system. (Note, e.g., that if p = m, $(p^2+m^2)^{\frac{1}{2}}$ = $\sqrt{2}$ m $\simeq$ m + $p^2/2m$ = 3m/2). This point of view certainly has some support from experience: to use the non–relativistic quark model one finds that it is necessary to relax the quantitative relationships that would be expected in the p/m → 0 limit between different flavour sectors, between different interactions within a given sector, and between mesons and baryons. When compared to other models of hadrons, the remarkable success of the "naive, non–relativistic" quark model seems to indicate that "it is better to have the right degrees of freedom moving at the wrong speed than the wrong degrees of freedom moving at the right speed"[7].

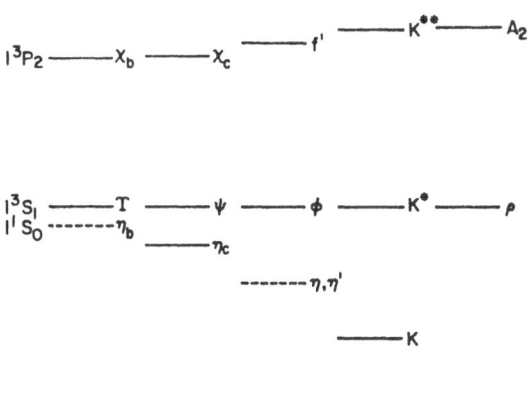

Fig. 10. a graphic illustration of the universality of meson dynamics from the π to the Υ, showing the splittings of $^3P_2$ and $^1S_0$ from $^3S_1$ in the $b\bar{b}$, $c\bar{c}$, $s\bar{s}$, $u\bar{s}$, and $u\bar{d}$ families

Of course it would be even better, presumably, if they were moving at the right speed. In an attempt to study this issue, some years ago we[8] set out to make a simple test of the idea of extrapolation suggested by Figure 10. Our goal was to:

1)   treat all mesons from the π to the Υ in the same framework

2)   incorporate all of the main dynamical features expected from QCD: a linear (scalar) confining potential, one gluon exchange with a running coupling constant, etc.

3)   use relativistic kinematics as well as relativistic potentials obtained by parameterizing (small) departures from the leading p/m .behaviour.

4)   solve the resulting equations with precision to eliminate dubious perturbative approximations.

It should be emphasized that we understand that such a programme is far from fundamental: we intended only to demonstrate that a sensible "relativization" of the quark model exists.

As an illustration of the content of our model, consider the crucial $\vec{S}_i \cdot \vec{S}_j$ interaction. In the Breit–Fermi limit it is

$$V^{ss}_{standard} - \frac{32\pi \, \alpha_s}{9m_i \, m_j} \; \vec{S}_i \cdot \vec{S}_j \; \delta^3 \left[\vec{r}_{ij}\right]$$

(9)

This interaction has several problems, all associated with its being valid only to lowest order in p/m: the argument of α_s is unclear, the $1/m_i m_j$ behaviour is unrealistic for $m \sim p$, and for $<\vec{S}_i \cdot \vec{S}_j> \, < \, 0$ the δ^3 (r_{ij}) makes V^{ss} standard an illegal operator in the Schödinger equation. Our relativization of this operator takes

$$V^{ss} - \left[\frac{m_i m_j}{E_i E_j}\right]^{\frac{1}{2} + \epsilon_{ss}} \frac{32\pi\alpha_s(r_{ij})}{9m_i m_j} \; \vec{S}_i \cdot \vec{S}_j \; \tilde{\delta}^3 \, (r_{ij}) \left[\frac{m_i m_j}{E_i E_j}\right]^{\frac{1}{2} + \epsilon_{ss}}$$

in which $\tilde{\delta}^3$ is a delta–function smeared over a quark size $r \sim m^{-1}$ and ϵ_{ss} is a free parameter which along with the m/E factors allows for the $m \leftrightarrow E$ ambiguity of the non–relativistic limit as well as relativistic non–locality in V^{ss}.

In addition to being ad hoc, this method is costly: we introduced six non–physics parameters into our quark model to describe the extrapolation to p/m $\sim$ 1. Nevertheless, the results of this exercise are quite interesting: we obtained a unified description of all mesons (i.e., hundreds of states) within a single framework with good spectroscopy (average error $\sim$ 30 MeV) and could obtain a reasonable description within the model of observed decay widths, static properties, etcetera. Figure 11 shows the two extreme spectra: $b\bar{b}$ and $u\bar{d}$, while Table I compares some of the computed decay amplitudes of the $^3P_2$ states of Figure 10 with experiment. Clearly the resulting description of meson spectroscopy is at least a useful guide!

While eliminating the usual sector–by–sector treatment of mesons in the non–relativistic quark model and explaining (at least in part) the appearance of the various pieces of the Breit–Fermi interaction with unexpected relative strengths, this work left unaddressed the problem of the relationship of mesons and baryons.

We have now also finished a study[9] of the more complex baryon system in an attempt to complete this unification of the quark model. The calculation uses the Hamiltonian of Ref. 8 for mesons with:

1) $<F_i.F_j> \, = \, 4/3 \rightarrow 2/3$ in the one–gluon exchange potentials as is appropriate for the transition from mesons to baryons

2) the linear string potential changed to the Y – string potential appropriate to the lowest adiabatic surface of the three quark system

3) most parameters taken over exactly from mesons, and all taken to be within 20% of their meson values.

The results are very interesting: baryon spectroscopy can be predicted from meson spectroscopy by this programme! Among the notable features of the calculation are that spin orbit forces, normally set to zero by hand[10], are automatically small. Indeed, the resulting states are very similar in character to those of the standard non–relativistic model of baryons[10] although some old spectroscopic puzzles are also solved (like the $\Delta 3/2^- - \Delta 1/2^-$ splitting in the $[70,1^-]$ supermultiplet). Figures 12 illustrate some of these spectroscopic features. The analog of Table I would be too lengthy to show here, but is also satisfactory.

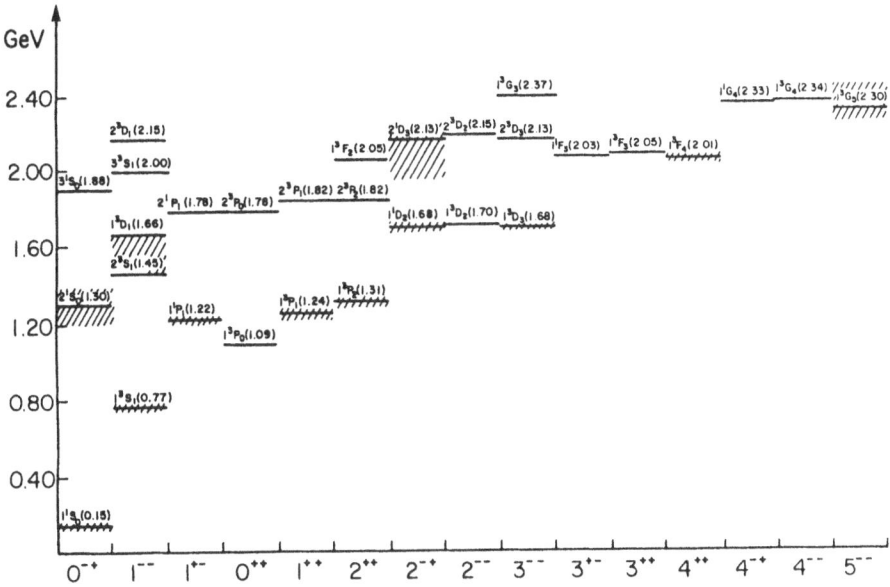

Fig. 11(a) The isovector mesons [$-u\bar{d}$, $\Lambda/2(u\bar{u}-d\bar{d})$, $d\bar{u}$]. The dominant spectral composition and predicted masses of states in GeV are shown near solid bars representing their masses. Shaded areas correspond to the experimental masses and their uncertainties, normally taken from the Particle Data Group (1984). For comparison of the 1−− and 0++ sectors with experiment see the discussion in Ref. 8.

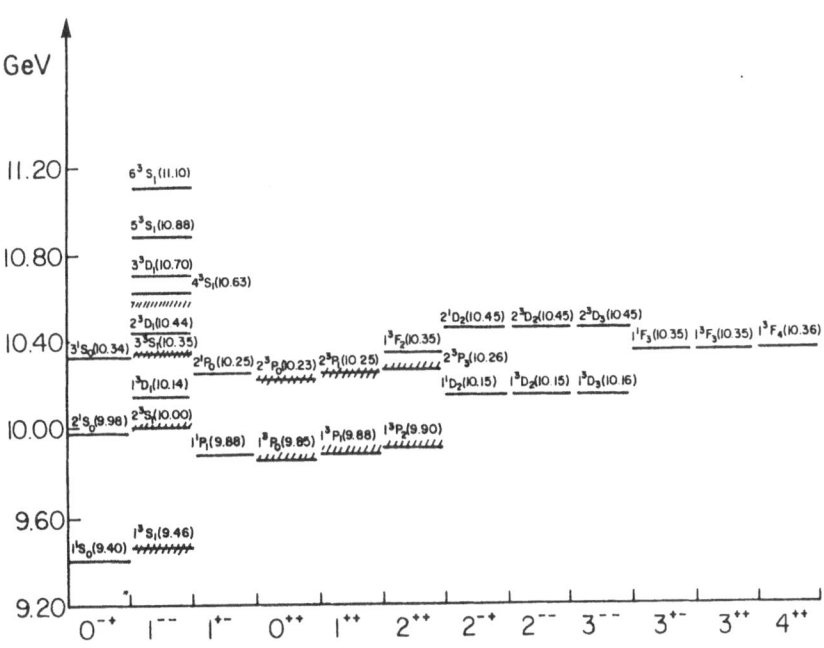

Fig. 11(b) The b-quarkonia ($b\bar{b}$). The legend is as for Fig 11 (a)

13

Table I. the five $^3P_2$ states of the "graphic illustration" of Figure 10

State	Property		Theory	Experiment
$u\bar{d}$ 1^3P_2 (a_2)	mass	(GeV)	1.31	1.32
	amp ($a_2 \to \rho\pi$)	(MeV$^{\frac{1}{2}}$)	-7.1	± 8.8±0.3
	amp ($a_2 \to \eta\pi$)	(MeV$^{\frac{1}{2}}$)	+4.8	± 4.0±0.1
	amp ($a_2 \to K\bar{K}$)	(MeV$^{\frac{1}{2}}$)	-2.7	± 2.3±0.1
	amp ($a_2 \to \eta'\pi$)	(MeV$^{\frac{1}{2}}$)	+1.1	<±1.5
	amp ($a_2 \to \pi^+\gamma$)	(MeV$^{\frac{1}{2}}$)	+0.55	± 0.55±.09
	amp ($a_2 \to \gamma\gamma$)	(keV$^{\frac{1}{2}}$)	-1.2	± 0.9±0.1
$u\bar{s}$ 1^3P_2 (K_2^*)	mass	(GeV)	1.43	1.43
	amp ($K_2^* \to K\pi$)	(MeV$^{\frac{1}{2}}$)	+7.2	± 6.7±0.5
	amp ($K_2^* \to K*\pi$)	(MeV$^{\frac{1}{2}}$)	-4.0	± 5.0±0.5
	amp ($K_2^* \to \rho K$)	(MeV$^{\frac{1}{2}}$)	-2.2	± 3.0±0.4
	amp ($K_2^* \to \omega K$)	(MeV$^{\frac{1}{2}}$)	+1.2	± 2.0±0.4
	amp ($K_2^* \to K\eta$)	(MeV$^{\frac{1}{2}}$)	-0.5	± 2.2±1.7
	amp ($K_2^* \to K^+\gamma$)	(MeV$^{\frac{1}{2}}$)	+0.48	± 0.50±.05
$s\bar{s}$ 1^3P_2 (f_2')	mass	(GeV)	1.53	1.53
	amp ($f_2' \to \pi\pi$)	(MeV$^{\frac{1}{2}}$)	+1.5	± 0.8±0.4
	amp ($f_2' \to K\bar{K}$)	(MeV$^{\frac{1}{2}}$)	-6.5	± 8±2
	amp ($f_2' \to \eta\eta$)	(MeV$^{\frac{1}{2}}$)	+2.5	< ±6
	amp ($f_2' \to \gamma\gamma$)	(keV$^{\frac{1}{2}}$)	-0.25	± 0.3±0.1
$c\bar{c}1^3P_2$ (χ_{2c})	mass	(GeV)	3.55	3.56
	amp ($\chi_{2c} \to$all)	(MeV$^{\frac{1}{2}}$)	+1.5	± 1.6±0.3
	amp ($\chi_{2c} \to \psi\gamma$)	(MeV$^{\frac{1}{2}}$)	+0.50	± 0.58± .18
	amp ($\psi' \to \chi_{2c}\gamma$)	(MeV$^{\frac{1}{2}}$)	+0.14	± 0.13± .02
$b\bar{b}$ 1^3P_2 (χ_{2b})	mass	(GeV)	9.90	9.91
	amp ($\chi_{2b} \to$all)	(MeV$^{\frac{1}{2}}$)	-0.60	-
	amp ($\chi_{2b} \to \gamma\gamma$)	(MeV$^{\frac{1}{2}}$)	-0.18	-
	amp ($\Upsilon' \to \chi_{2b}\gamma$)	(MeV$^{\frac{1}{2}}$)	-0.040	± 0.043±. 008

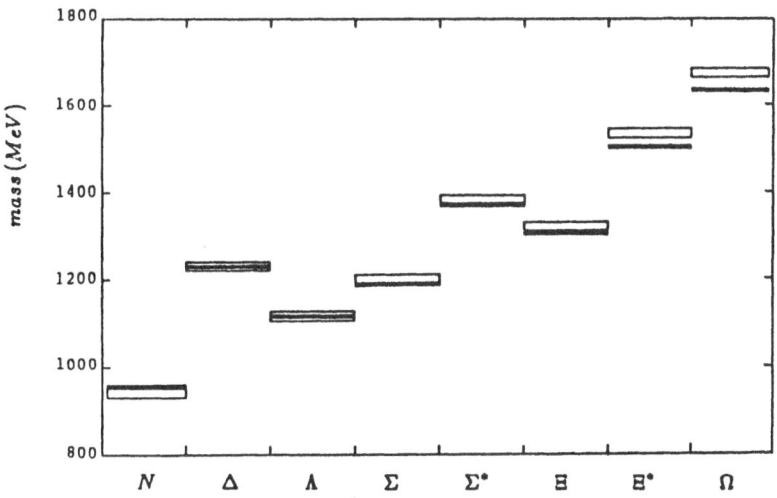

Figure 12(a). predictions (bars) for the light quark ground state baryons compared to experiment (boxes)

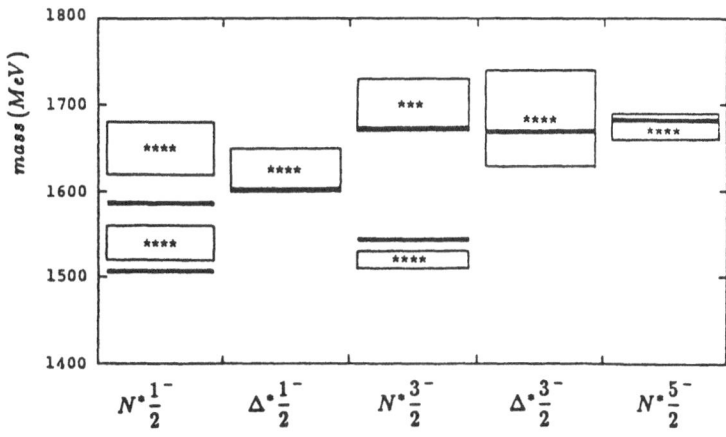

Figure 12(b). predictions (bars) for the low–lying negative parity S = 0 baryons compared to experiment (***** is a well–established state, etc.)

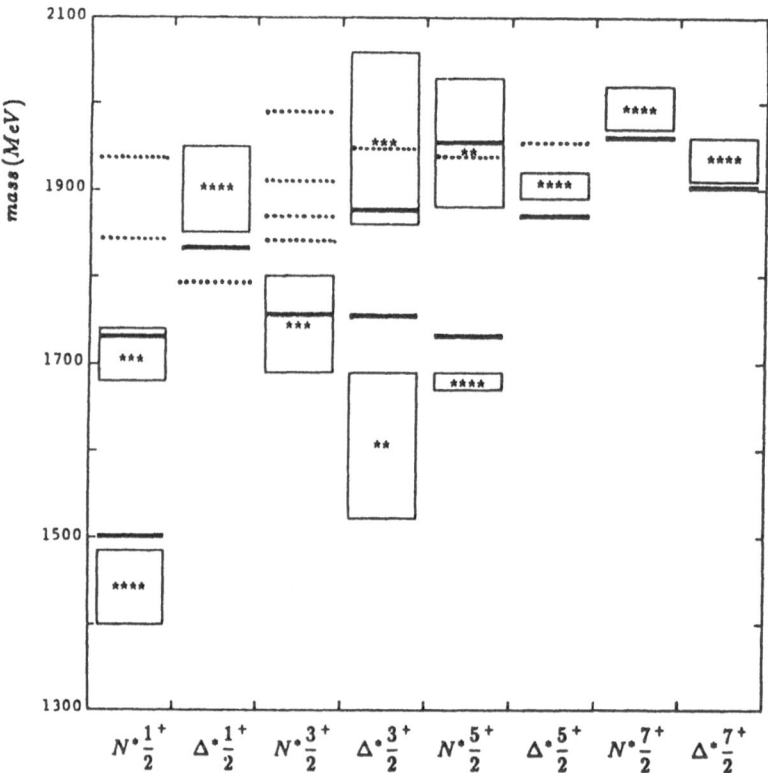

Figure 12(c). predictions for the S=0 positive parity baryons compared to experiment; the legend is: ▬▬▬▬ predicted to be seen; •••••••••••• predicted, but decouples from πN; experiment gives mass indicated by the box containing the resonance's "star rating".

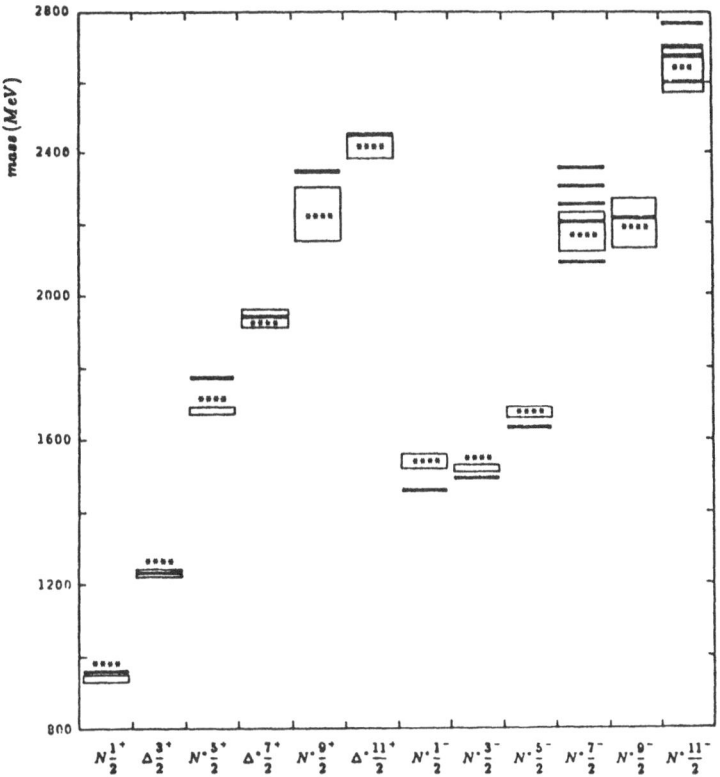

Figure 12(d). prediction (bar) for the lowest-lying S=0 baryon of each J^P up to 11/2 ±, compared to experiment (box with "star rating").

CONCLUSIONS: LECTURE TWO

Although the "relativization" of the quark potential model discussed here is crude and ad hoc, it incorporates many of the qualitative features expected from relativistic effects, including relativistic kinematics and non-local, momentum-dependent potentials. As such it allows us to explore the impact of such effects and suggests that many of the flaws of the usual models were due to the extreme non-relativistic approximation. It further suggests that Figure 10 is not misleading: there are closely analogous structures (and physics) in operation in both heavy and light quark systems, which allow an extrapolation of the known physics of the former into the kinematic regime of the latter. This relativization not only allows a unification of meson spectroscopy, but also of mesons with baryons. Indeed, this later unification is perhaps one of the strongest pieces of evidence that the good results of this programme are not illusory. For example, if the $\rho-\pi$ splitting is not due to hyperfine interactions, then it is difficult to understand how it is possible using exactly the same paramters to predict the $\Delta-N$ splitting. (Of course the $\rho-\pi$ splitting is also related to the nature of the π as a pseudo-Goldstone boson; we are arguing that from the point of view of the constituent quark model it may be viewed as the hyperfine partner of the ρ just as the η_c is the partner of the ψ.) One also sees in this approach the deep connection between such splittings as $N^*(1520)\frac{1}{2}^-$ − $N(940)\frac{1}{2}^+$ and a_2 (1320) − $\rho(770)$ and between $\Sigma-\Lambda$ /$\Delta-N$ and $K^*-K/\rho-\pi$. Thus the quark model not only survives relativization, but is remarkably improved by it.

SOME PROBLEMS IN THEORY AND EXPERIMENT

I would like to conclude with an idiosyncratic view of the outstanding problems we face today in the "old spectroscopy". I will divide my remarks into four sections: theoretical theory, phenomenological theory, baryon spectroscopy, and meson spectroscopy. Given the nature of this meeting, I will spend the most time on the last of these topics.

Theoretical Theory

Eventually the lattice will calculate everything for us. However, it is important to realize that we will be well into the next millenium before computers will exist that are powerful enough to reliably predict the basic features of the low–lying meson and baryon spectra. Even after this becomes possible we will want to have models of this physics both so that we can understand the raw numerical output of these calculations and so that we can extend our understanding to systems more complicated than those we can even then compute exactly (like, e.g., the deuteron!). At the present time we need models for both "simple" and complicated systems.

I accordingly believe that one of the most fruitful areas of strong interaction research in the future will be in using the lattice not to produce the "end products" of a hadron calculation, but rather some of the "components": we should be using the lattice to check some of the basic ingredients of our hadronic models. For example, the lattice can be used to study the adiabatic potentials and the physics behind them:

1) $V_0(r)$, the lowest $q_1 \bar{q}_2$ adiabatic potential, has long been a focus of lattice studies. Recent work[11] has lent support to the standard quark model assumption of an "effective scalar exchange" structure for the linear potential, of a short range for the $\vec{S}_i \cdot \vec{S}_j$ interaction, etc. There is also preliminary evidence [11] that $V_1(r)$, the hybrid potential, looks as expected in the flux tube model[2].

2) The extraction of $V_0 (\vec{p}, \vec{r})$, the baryon potential, has only just begun, but recent evidence[12] favours the Y–string potential.

3) The $\vec{E}$ and $\vec{B}$ fields associated with the long–range potentials can be studied. Preliminary indications[11,13] are in agreement with expectations from a dynamical string model.

4) There is also recent work[14] on the potentials in multiquark systems like $q_1 q_2 \bar{q}_3 \bar{q}_4$ and $(q_1 q_2 q_3)(q_4 q_5 q_6)$ which could eventually have important ramifications for nuclear physics.

This list is related mainly to adiabatic potentials; many other fundemental calculations could be done which should eventually lead to more and more realistic hadronic models. There are other kinds of "theoretical theory" issues to be addressed, of course. To me the most urgent is the clarification of the relationship of the constituent quark model to spontaneous chiral symmetry breaking.

Phenomenological Theory

Recent confusing results in meson spectroscopy have emphasized to me the need to develop both the physics and techniques required for the study of broad resonances, expecially when they are found in the neighbourhood of substantial backgrounds, other broad resonances, or thresholds of new strong decay channels. I suspect that the apparent need for such developments is simply a result of a selection process: most of the narrow resonances below 2 GeV have already been found! An essential ingredient of being able to extract the properties of the underlying spectrum (by which I mean the spectrum corresponding to the stationary states that one would have in the zero width approximation) will be to develop models for the coupling of the spectrum to (real and virtual) decay channels.

Another simplifying assumption of the past that may need reconsideration is the assumption that multihadron states (like the ones which lead to decay of the narrow resonances) do not produce rapidly varying amplitudes. While it is now clear that, for example, $q_1 q_2 \bar{q}_3 \bar{q}_4$ does not have a rich (nearly) discrete low mass spectrum, it may be that there are strong hadron–hadron potentials in some channels which might even occasionally produce weakly bound "molecules"[15]). Such threshold effects might lead to confusion if they are ignored or parameterized as Breit–Wigner resonances. I accordingly believe the general study of low–energy hadron–hadron potentials deserves more attention.

Baryon Spectroscopy

Although $\bar{p}p$ interactions are normally more incisive for meson than baryon spectroscopy, it should be noted that many issues in this latter area are intimately linked to those previously mentioned. For example, the "$\Lambda(1405)$ problem" may be related to the properties of the $\bar{K}N$ interhadron potential.

There is the possibility of a more direct impact. One of the central issues in baryon spectroscopy is that of the "missing resonances": the rather large numbers of states predicted by the quark potential model which have not been observed. While a plausible explanation for their absence has been advanced[16], their discovery and the confirmation of their predicted properties is a critical challenge to experimental hadron physics. This situation is relevent to $\bar{p}p$ physics because, according to Ref. 16, the missing states are absent from πN partial wave analyses because their πN couplings are very small. (In the SU(6) limit such decouplings would be very rare, but one gluon exchange induces large mixings between SU(6) states). Such resonances will either have to be detected in very high statistics (and very well understood) πN formation experiments, in γN formation experiments, or in production experiments where they are detected in channels like ηN, $\pi\Delta$, ρN, ωN, etc. It may be that $\bar{p}p$ experiments would have some advantages over other production techniques. I at least can't see why such $\bar{p}p \to B^* X$ baryon (B^*) experiments are much more difficult than their meson (M^*) counterparts $\bar{p}p \to M^* X$.

Meson Spectroscopy

Even though my comments here are constrained by my assigned title to "lie on the lowest abiabatic surface", there are many problems to be discussed[18]).

Let me begin by addressing some of the problems associated with radial excitations. Crucial to our understanding of the glueball candidate η (1440) (also known as the ι) is the status of the $2\,^1S_0$ nonet. It is in fact in rather poor shape. The Particle Data Group say that even the $\pi(1300)$ and K(1400) "backbone" of this nonet need confirmation, and although the $\eta(1275)$ effect has recently been seen in a second experiment[18], there is still some danger that it may be an artifact of the parameterization of the $f_1(1285)$ with which it is degenerate. (In particular, reconstruction of the $\eta\pi\pi$ final state in $\delta\pi$ and $\eta\epsilon$ isobars may be tricky: neither the δ nor the ϵ is very well understood at the moment (see below)). If we take the π (1300) at face value, then there should be a mainly $s\bar{s}$ $2\,^1S_0$ state at about 1650 MeV, but there is no evidence for such a state. Adding to the evidence that we don't understand this radially excited nonet are two puzzles: 1) we see $\gamma\gamma \to 1\,^1S_0$ at about the predicted levels, but no $\gamma\gamma \to 2\,^1S_0$ transitions have been seen at a level that is beginning to be difficult to excuse, and 2) in ψ radiative decay we "see" $gg \to 1\,^1S_0$ in the form $\psi \to \gamma\pi,\ \gamma\eta,\ \gamma\eta'$ at about the expected levels, but not $gg \to 2\,^1S_0$ (unless the η (1440) is such a state). It seems to me that there is little chance of clarifying the nature of the η (1440) until its environment is cleaned up!

The situation for the vector radial excitations is considerably more susceptible to an optimistic prognosis[8]. The identification of light quark radial excitations is quite different phenomenologically from the heavy quarkonium case where $e+e-$ can filter out the $^3D_1$ vector mesons: in the heavy systems the non–relativistic rule forbidding $e+e- \to \,^3D_1$ is reasonably well obeyed. In light quark systems with $v/c \sim 1$, this filtering will be much less effective and one should expect to see two nonets of vector particles in the 1.6 GeV mass region. Recent results seem to indicate that this expectation is being realized:

1) A recent experimental study[17] has confirmed the existence of two vector K^* particles, one consistent with being mainly $2\,^3S_1$ at low mass and the other, near to the known 3^- D–wave state, consistent with being mainly the $^3D_1$ state.

2) A recent analysis[20] of data on the reactions $\gamma p \to XN$ and $e^+e^- \to X$ in appropriate channels requires the existence of two ρ–like and two ω–like states in the expected mass range.

It is very important to find the two corresponding ϕ–like states. Since their behaviour is more like that of heavy quarkonium, they provide an essential bridge to the known $2\,^2S_1 - 1\,^3D_1$ states in the analogous mass range in charmonium.

It is to be expected that after the $2\,^1S_0$ and $2\,^3S_1$ states, the $2\,^3P_2$ radial excitations will be the next most easily studied nonet. Even without explicit calculations which suggest that the ρ - and ω–like members of this nonet should be found at around 1.8 GeV, given that $1\,^3P_2 - 1\,^3S_1$ in $b\bar{b}$ is 0.45 GeV versus 0.55 GeV in $u\bar{d}$ and that $2\,^3S_1 - 1\,^3S_1$ in $b\bar{b}$ is 0.56 GeV versus roughly[20] 0.68 GeV in $u\bar{d}$, knowing that $2\,^3P_2 - 2\,^3S_1$ in $b\bar{b}$ is 0.25 GeV would lead one to guess a very similar value. Indeed, there are several pieces of evidence already emerging for the existence of this nonet[21]. One of the reasons this nonet is of special interest is its ramifications for our understanding of the $\theta(1750)$ (i.e., f_2 (1750)) seen as a major channel in ψ radiative decay[17]. If this state cannot be associated with this nonet, then the possibility that it is a glueball will be much strengthened. As with all the radial excitations, one should exercise considerable caution in using an SU(3) analysis of their decays as a tool for identifying their underlying quark structure: the nodes in their radial wavefunctions can create zeroes in decay amplitudes. Such rapidly varying functions of the decay momentum could create large differences between, e.g., $\pi\pi$ and $K\overline{K}$ over and above simple phase space considerations[22].

It is sobering to realize that after all of these years we still don't understand the simple P–wave $q\bar{q}$ states! The latest news is that the $^3P_1$ nonet with $J^{PC}_n = 1^{++}$ is acting up again[18]. There is, first of all, now controrersy[23] over the mass of the a_1. The properties of the f_1 (1285) are difficult to predict because it can have some $s\bar{s}$ mixed into it. The strange meson of this nonet is strongly mixed with that of the $^1P_1$ nonet. Finally, and most distressing, is that the f_1 (1420) is not behaving as it should so that its pedigree as the $s\bar{s}$ member of this nonet is being called into question[18]. There is the associated possibility that the f_1 (1530) claimed by one group is the real $s\bar{s}$ $^3P_1$ state[24]. I have "no comment" on this mess, but I will ask a question: has anyone ever seen a 1^{++} $K^*\overline{K}$ resonant phase at 1420 MeV or is the evidence for this state based only on bumps and Dalitz plot analyses?

The P–wave problems don't end here. The scalar mesons have been notoriously difficult to understand as $q\bar{q}$ $^3P_0$ states.[18,25] It is possible that some of this confusion has been generated by the existence of $K\overline{K}$ molecules in this channel[15]. A recent analysis[25]) has indeed found a very complicated structure near $K\bar{K}$ threshold. It has been proposed[25]) that one of the scalar mesons found in that study may be a scalar glueball.

CONCLUSIONS: LECTURE THREE

The old spectroscopy, especially in the meson sector, is in ridiculously poor shape. New results in recent years have just as often created confusion as clarity, and much remains to be done in both experiment and phenomenology.

CONCLUSIONS

The spectroscopies of $q\bar{q}$, $Q\bar{Q}$, qqq, hybrids, and glueballs are really indivisible subjects: they are governed by the same underlying theory, our study of them requires the same experiments, and we must demand that our models simultaneously describe them all.

The spectroscopy of mesons ($q\bar{q}$, hydrids, and glueballs) is surprisingly poorly understood. To improve upon the present situation will require more sophisticated analysis of a wide range of data from more advanced experimental apparatus on high intensity machines. Figure 13 makes it obvious at a glance

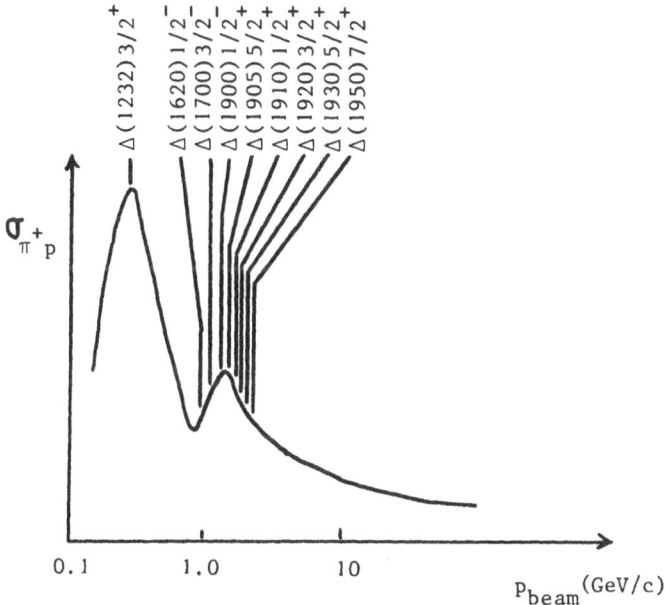

Figure 13.   one of the lessons of baryon spectroscopy

how far our experimental knowledge of baryon spectroscopy would be advanced if we had analyzed experiments in terms of Dalitz plots of bumps in cross sections. Making these advances in our understanding will not be easy, but I hope these lectures may have led you to believe that the required effort is worthwhile. It may be an old spectroscopy, but it is far from dead.

REFERENCES

1. S. Capstick, S. Godfrey, N. Isgur and J. Paton, 'Taking the "Naive" and "Non-relativistic" Out of the Quark Potential Model', Phys. Lett. B175, 457 (1986).

2. N. Isgur and J. Paton, 'A Flux-tube Model for Hadrons in QCD', Phys. Rev. D31, 2910 (1985); Phys. Lett. 124B, 247 (1983).

3. N. Isgur, R. Kokoski and J. Paton, 'Gluonic Excitations of Mesons: Why They Are Missing and How to Find Them', Phys. Rev. Lett. 54, 869 (1985).

4. F.E. Barnes, California Institute of Technology, Ph.D., thesis, 1977 (unpublished); Z. Phys. C10, 275 (1981); D. Horn and J. Mandula, Phys. Rev. D17, 898 (1978); P. Hasenfratz, R.R. Horgan, J. Kuti and J.M. Richard, Phys. Lett. 95B, 299 (1980); F.E. Barnes, F.E. Close and S. Monaghan, Nucl. Phys. B198, 380 (1982); M. Chanowitz and S. Sharpe, Nucl. Phys. B222, 211 (1983); E. Golowich, E. Haqq and G. Karl, Phys. Rev. D28, 160 (1983).

5. J. Merlin and J. Paton, J. Phys. G11, 439 (1985).

6. R. Kokoski and N. Isgur, Phys. Rev. D35, 907 (1987).

7. G. Karl, private communication.

8. S. Godfrey and N. Isgur, "Mesons in a Relativized Quark Model with Chromodynamics", Phys. Rev. D32, 189 (1985).

9. S. Capstick and N. Isgur, "Baryons in a Relativized Quark Model with Chromodynamics", Phys. Rev. D34, 2809 (1986).

10. N. Isgur and G. Karl, Phys. Lett. 72B, 109 (1977); 74B, 353 (1978); Phys. Rev. D18, 4187 (1978); D19, 2653 (1979); D21, 3175 (1980).

11. See, for example, C. Michael in "Lattice Gauge Theory: A Challenge in Large-Scale Computing" (New York, Plenum, 1986), p. 227; A. Huntley and C. Michael, Nucl. Phys. B286, 211 (1987).

12. J. Flower, Caltech preprints CALT-68-1377, 1378.

13. M. Lüscher, K. Symanzik, and P. Weisz, Nucl. Phys. B173, 365 (1980); M. Lüscher, G. Münster, and P. Weisz, Nucl. Phys. B180, 1 (1981); M. Lüscher, Nucl. Phys. B180, 317 (1981); J. Amjorn, P. Olesen, and C. Peterson, Phys. Lett. B142, 410 (1984); Nucl. Phys. B244, 262 (1984); P. Olesen, Phys. Lett. B160, 144 (1985).

14. H. Markum et al., Phys. Rev. D31, 2029 (1985); S. Ohta, M. Fukugita, and A. Ukawa, Phys. Lett. B173, 15(1986).

15. J. Weinstein and N. Isgur, Phys. Rev. Lett. 48, 659 (1982); Phys. Rev. D27, 588 (1983); J. Weinstein, University of Toronto Ph. D. Thesis, 1986 (unpublished). For a review of the status of this picture see F. E. Barnes in these proceedings. The $K\bar{K}$ molecule picture is an outgrowth of earlier work on $qq\bar{q}\bar{q}$ states in the bag model: see R. L. Jaffe, Phys. Rev. D15, 267, 281 (1977), D17, 1444 (1978); R. L. Jaffe and K. Johnson, Phys. Lett. 60B, 201 (1976).

16. R. Koniuk and N. Isgur, Phys. Rev. Lett. 44, 485 (1980); Phys. Rev. D21, 1868 (1980).

17. See the contributions by F. E. Close, G. Eigen, and F. Couchot in these proceedings.

18. For a review of the experimental situation see L. Montanet in these proceedings.

19. D. Aston et al., Nucl. Phys. B247, 261 (1984); B292, 693 (1987).

20. A. Donnachie and H. Mirzaie, Z. Phys. C33, 407 (1987); A. Donnachie and A. B. Clegg, Z. Phys. C34, 257 (1987).

21. N. M. Cason et al., Phys. Rev. Lett. 48, 1316 (1982) find evidence for the isoscalar member. See also Ref. 19 in which evidence is seen for the strange members of this nonet at 1.97 GeV.

22. M. Böhm, H. Joos, and M. Krammer, Nucl. Phys. B69, 349 (1974); W. B. Kaufmann and R. J. Jacob, Phys. Rev. D10, 1051 (1974); A. LeYaouanc et al., Phys. Lett. 71B, 397 (1977); 76B, 484 (1978); A. Bradley, J. Phys. G4, 1517 (1978); A. Bradley and D. Robson, Z. Phys. C6, 57 (1980); M. Chaichan and R. Kogerler, Ann. Phys. (N.Y.) 124, 61 (1980). E. Eichten et al., Phys. Rev. D21, 203 (1980); S. B. Gerasimov and A. B. Govorkov, Z. Phys. C13, 43 (1982); R. Kokoski and N. Isgur, Phys. Rev. D35, 907 (1987) discuss all of the 2P multiplets in this light.

23. See, for example, M. G. Bowler, Phys. Lett. B182, 400 (1986).

24. Ph. Gavillet et al., Z. Phys. C16, 119 (1982). Since giving these lectures I have learned that the LASS group also sees such a state: D. Aston et al., SLAC preprint SLAC-PUB-4340.

25. K. L. Au, D. Morgan, and M. R. Pennington, Phys. Lett. 167B, 229 (1986) and Rutherford Appleton Laboratory preprint RAL-86-076, 1986.

EXOTIC STATES BEYOND THE CONVENTIONAL QUARK MODEL :
MESONIC NUCLEI AND INTERHADRON POTENTIALS

T. Barnes

Department of Physics
University of Toronto
Toronto, Ontario, Canada

Abstract

In this review article I discuss the current theoretical and experimental status of the $q^2\bar{q}^2$ sector of the quark model (q=u,d,s). Whereas this system once was thought to accommodate many "baryonium" resonances, recent calculations suggest that $(q\bar{q})$ clustering is the dominant effect in the $q^2\bar{q}^2$ sector of Hilbert space. This implies that the spectrum of $q^2\bar{q}^2$ states consists of deuteron–like states such as the $f_0(975)$ and $a_0(980)$ kaon–antikaon nuclei "$K\bar{K}$ molecules" and two–meson continuua interacting through short–range interhadron potentials. Predictions for and experimental tests of interhadron potentials in other two–hadron systems (NN and $\pi\pi$) are also reviewed. Finally, experimental contributions to the study of mesonic nuclei and intermeson potentials which might be feasible at LEAR are discussed.

I. INTRODUCTION TO $q^2\bar{q}^2$: THE ABSENCE OF BARYONIA

In the early days of QCD and the quark model it was thought that the $q^2\bar{q}^2$ sector of the light quark system (q=u,d,s) would possess many hadronic resonances. These resonances were referred to as "baryonia", because it was expected that they would be produced copiously in baryon–antibaryon ($P\bar{P}$) annihilation.

More recent calculations of the properties of the light $q^2\bar{q}^2$ system suggest that these early predictions of $q^2\bar{q}^2$ hadrons were incorrect. The various errors which led to the expectation of such states have been reviewed elsewhere [1,2], so I will not discuss them in detail here. In brief, the problems arose from theoretical or model assumptions which imposed confinement on the entire $q^2\bar{q}^2$ system *a priori* , so that $q^2\bar{q}^2$ resonances were an unavoidable result of the initial assumptions. Although such an initial assumption of confinement is plausible for the familiar $q\bar{q}$, qqq and $\overline{qqq}$ hadrons, it is suspect for the $q^2\bar{q}^2$ system, which has the option of rearranging itself into separate color–singlet $q\bar{q}$ states, $q^2\bar{q}^2 \rightarrow (q\bar{q})(q\bar{q})$. Such a rearrangement, which is called a "fall–apart decay" [3], would imply that baryonia do not exist as resonances if there is no potential–energy or kinematic barrier to suppress the rearrangment. This possibility was evidently realized by Jaffe, who qualified his detailed bag model studies [3,4] of $q^2\bar{q}^2$ and q^6 with the observation that "Most of the 'states' listed in the table probably do not correspond to particles or resonances." [4]

As we shall see, it now appears likely that *none* of the light $q^2\bar{q}^2$ baryonia previously expected by theorists actually exists. There may be such single–hadron states in the two–heavy two–light systems (for example $Q^2\bar{q}^2$, where Q is c or b) [5,6], but light–quark $q^2\bar{q}^2$ mesons which are single four–quark clusters now appear untenable. This result follows from the variational calculations of Weinstein and Isgur [7–10], which we shall discuss in the following section.

Although the $q^2\bar{q}^2$ sector of QCD does not possess the rich spectrum of resonances once predicted by theorists, it is nonetheless proving to accommodate more interesting physics than one might have anticipated. The $f_0(975)$ and $a_0(980)$ resonances appear to be the precursors of a new branch of nuclear physics, the study of nuclei having meson constituents. The $q^2\bar{q}^2$ system is also of interest as the simplest sector of Hilbert space in which hadron–hadron scattering takes place, and the study of the associated intermeson potentials is an area in which hadron experiments at accelerators such as LEAR can make important contributions.

In the remainder of this paper we review the relatively new subjects of mesonic nuclei and interhadron potentials, and the discussion is organized as follows: in section II we describe the quark–model calculation which led Weinstein and Isgur to the suggestion that the $f_0(975)$ and $a_0(980)$ were deuteron–like kaon–antikaon bound states, and section III reviews the experimental evidence regarding this identification. Section IV is a summary of calculations of interhadron potentials in two other systems, nucleon–nucleon and $\pi\pi$. Section V suggests possible experiments relating to intermeson forces which might be possible at LEAR, and sections VI and VII contain our acknowledgements and references.

II. AN IMPROVED CALCULATION OF $q^2\bar{q}^2$ STATES

The various predictions of $q^2\bar{q}^2$ "baryonium" resonances by theorists in the 1970s were frequently artifacts of the *a priori* assumption that the entire $q^2\bar{q}^2$ system was confined to a single hadron. As a $q^2\bar{q}^2$ color–singlet state has a projection onto two separate color–singlet $q\bar{q}$ pairs, $q^2\bar{q}^2 \rightarrow (q\bar{q})(q\bar{q})$, any model in which a $q^2\bar{q}^2$ state is required to exist as a single hadron without a two–meson component has imposed an unnecessarily restrictive and perhaps unphysical assumption. A more realistic study of this system, for example as a variational calculation for the four–quark wave function in a potential model, should evidently employ an Ansatz which can interpolate between a single four–quark "baryonium" hadron and a state consisting of two separate color–singlet $q\bar{q}$ mesons.

A series of such variational calculations incorporating various refinements has been carried out by Weinstein and Isgur [7–10]. The most recent of these calculations employed a nonrelativistic potential model with a Coulomb plus linear potential and a spin–spin contact term; similar models give an accurate description of the spectrum and decays of S–wave $q\bar{q}$ mesons. The variational Ansatz for the $q^2\bar{q}^2$ wavefunction was chosen to have a great deal of freedom, so that one could search for "metastable" configurations analogous to α particles inside α–emitting nuclei, which are metastable by virtue of a potential barrier. The existence of such configurations might have allowed baryonium resonances, even if the ground state consists of two separate $q\bar{q}$ pseudoscalars.

The result found by Weinstein and Isgur is that the $J^{PC_n} = 0^{++}$ ground state of the $q^2\bar{q}^2$ system is almost always two separate $q\bar{q}$ pseudoscalar mesons, and that there are no metastable configuations. The exception to this "no go" result for baryonium occurs for quark masses intermediate between nonstrange and strange, corresponding to a kaon–antikaon system. In that special case, a weakly–bound system was found with a binding energy of $E(q^2\bar{q}^2) - 2m_K \sim -10$ Mev. A detailed study of these states revealed that they consisted of two largely unperturbed kaons with a slight overlap, rather like the PN deuteron, albeit with no repulsive core. Weinstein and Isgur suggested that these "$K\overline{K}$ molecules" were the $S^*(975)$ and $\delta(980)$ mesons, now known as the $f_0(975)$ and $a_0(980)$ respectively. Similar assignments were previously suggested by Astier *et al* [11] and by Wicklund *et al* [12]. We shall review the experimental evidence supporting the $K\overline{K}$ molecule picture in the next section.

An intuitive picture of the physics underlying these $K\overline{K}$ molecules was developed by Weinstein and Isgur, who projected their four–quark wavefunction onto the wave functions of two non–interacting kaons, thus obtaining an approximate wavefunction for the amplitude $\Psi(r_{K\overline{K}})$ to find the kaon and antikaon at a radius $r_{K\overline{K}}$. This wavefunction they substituted into the two–body Schrödinger equation, which gave an equivalent interkaon potential $V(r_{K\overline{K}})$. (This procedure assumes that the kaon wavefunctions are largely undistorted, so that an interkaon potential can be defined without much ambiguity.) The resulting $K\overline{K}$ potential was found to be approximately Gaussian in shape, with a depth and range of about $V(0) = -400$ Mev and $r_0 = 0.4$ fm, where these parameters are defined by

$$V(r_{K\overline{K}}) = V(0)\ \exp\left(-\tfrac{1}{2}(r_{K\overline{K}}/r_0)^2\right) . \tag{1}$$

The $I = 0$ interpion and kaon–antikaon potentials originally found by Weinstein and Isgur [8] are shown in figure 1.

Potentials in this region of parameter space can indeed bind kaons with typical nuclear–physics binding energies of $E_B \sim 10$ Mev. The existence of the $f_0(975)$ and $a_0(980)$ is thus something of an accident, as the potential appears to be just deep enough to form kaon–antikaon bound states.

Similar calculations have been carried out for the $\pi\pi$ system, with interesting results for the isospin dependence of interpion potentials; this will be discussed in section IV.

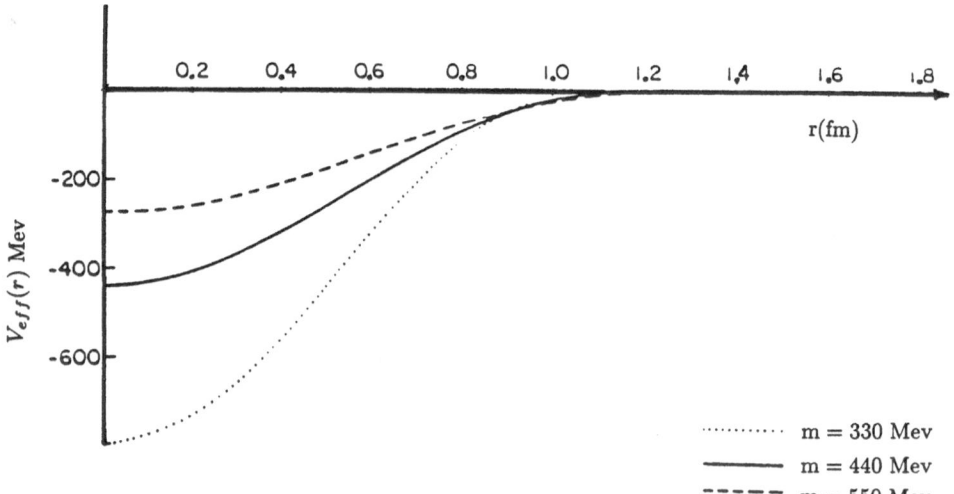

Fig. 1 I = 0 interpion (m = 330 MeV) and kaon-antikaon (m = 440 MeV) potentials found by Weinstein and Isgur [8].

III. TWO "$q^2\bar{q}^2$" ($K\overline{K}$) MESONS: THE $f_0(975)$ AND $a_0(980)$

The identification of the $f_0(975)$ and $a_0(980)$ mesons with the $K\overline{K}$ molecule states found in the variational calculation immediately explains a number of otherwise puzzling features of these states, which previously had been considered as candidates for $\ell = 1$ $q\bar{q}$ mesons or $q^2\bar{q}^2$ "baryonium" hadrons.

First, their masses appear quite natural if they are twice a kaon mass minus approximately 10 Mev of nuclear binding energy. The energy of two free kaons at rest is

$$m(K^+K^-) = 987.3 \text{ Mev} \tag{2}$$

and

$$m(K^0\bar{K}^0) = 995.4 \text{ Mev} , \tag{3}$$

so that the f_0 and a_0 masses, given by the Particle Data Group [14] as

$$m(f_0) = 975 \pm 4 \text{ Mev} \tag{4}$$

and

$$m(a_0) = 983 \pm 2 \text{ Mev} , \tag{5}$$

seem plausible in the $K\overline{K}$ molecule picture as twice a kaon mass minus nuclear binding energy. If these were $q\bar{q}$ or single $q^2\bar{q}^2$ hadrons, their proximity to $2m_K$ would be accidental.

Second, the strong decay widths of these states are too small by an order of magnitude for them to be $^3P_0$ $q\bar{q}$ states, whereas their widths are consistent with expectations for $K\overline{K}$ molecules. $q\bar{q}$ mesons with these masses would have widths (in two quark model calculations [15,16]) of

$$\Gamma(f_0(q\bar{q}) \to \pi\pi) = \begin{cases} 400 \text{ Mev} & [15] \\ 400 \text{ Mev} & [16] \end{cases} \tag{6}$$

and

$$\Gamma(a_0(q\bar{q}) \to \eta\pi) = \begin{cases} 500 \text{ Mev} & [15] \\ 225 \text{ Mev} & [16] . \end{cases} \tag{7}$$

In the $K\overline{K}$ molecule picture, assuming that these states decay by virtual annihilation through the actual, very broad 0^{++} $q\bar{q}$ states ($K\overline{K} \to (0^{++}\ q\bar{q}) \to \eta\pi$ or $\pi\pi$), the predicted strong widths [8] are

$$\Gamma(f_0(K\overline{K}) \to \pi\pi) = 15 \text{ Mev} \tag{8}$$

and

$$\Gamma(a_0(K\overline{K}) \to \eta\pi) = 40 \text{ Mev} . \tag{9}$$

The Particle Data Group [14] values for the total widths of the f_0 and a_0, which should be essentially equal to these partial widths, are

$$\Gamma_{expt}(f_0 \to \pi\pi) = 33 \pm 6 \text{ Mev} \tag{10}$$

and

$$\Gamma_{expt}(a_0 \to \eta\pi) = 54 \pm 7 \text{ Mev} . \tag{11}$$

29

The measured widths evidently are in much better agreement with the predictions of the $K\overline{K}$ molecule assignment than with the $\ell = 1$ $q\overline{q}$ assignment.

A third measurement which supports the $K\overline{K}$ molecule assignment is the partial widths of the f_0 and a_0 to two photons. These partial widths may be calculated for $K\overline{K}$ and $\ell = 1$ $q\overline{q}$ assignments [17,18], with the results

$$\Gamma(f_0(q\overline{q}) \to \gamma\gamma) = \frac{15}{4}\left(\frac{0.975}{1.274}\right)^3 \Gamma(f_2 \to \gamma\gamma) \approx 4.5 \text{ Kev}, \tag{12}$$

$$\Gamma(a_0(q\overline{q}) \to \gamma\gamma) = \frac{9}{25}\left(\frac{0.983}{0.975}\right)^3 \Gamma(f_0(q\overline{q}) \to \gamma\gamma) \approx 1.5 \text{ Kev}; \tag{13}$$

$$\Gamma(f_0(K\overline{K}) \to \gamma\gamma) = \frac{\pi\alpha^2}{m_K^2}|\Psi_{K\overline{K}}(0)|^2 \approx 0.6 \text{ Kev}, \tag{14}$$

$$\Gamma(a_0(K\overline{K}) \to \gamma\gamma) = \Gamma(f_0(K\overline{K}) \to \gamma\gamma) \approx 0.6 \text{ Kev}. \tag{15}$$

Experimentally, there is a Crystal Ball measurement of $\Gamma(a_0 \to \gamma\gamma) \cdot B(a_0 \to \eta\pi)$ [19] (the a_0 appears as a bump in $\gamma\gamma \to \eta\pi^0$, shown in figure 2). The branching fraction $B(a_0 \to \eta\pi)$ is near unity, contrary to the statement in the Review of Particle Properties [13,14], which misquotes a theoretical review by Achasov et al [20]. (See [18] for a clarification of this issue.) Taking $B(a_0 \to \eta\pi)$ to be unity and $B(f_0 \to \pi\pi) \approx 0.8$, the Crystal Ball measurement and the Crystal Ball [21] and Jade [22] limits on $\Gamma(f_0 \to \gamma\gamma) \cdot B(f_0 \to \pi\pi)$ imply

$$\Gamma_{expt}(f_0 \to \gamma\gamma) < \begin{cases} 1.0 & \text{Kev} \quad 95\% \text{ cl [21]} \\ 0.8 & \text{Kev} \qquad\qquad [22] \end{cases} \tag{16}$$

and

$$\Gamma_{expt}(a_0 \to \gamma\gamma) = 0.19 \pm 0.07 {}^{+0.10}_{-0.07} \text{ Kev}. \tag{17}$$

Both experimental results are an order of magnitude smaller than the $\gamma\gamma$ widths expected for $q\overline{q}$ states (12,13). In contrast, the $\gamma\gamma$ widths expected for $K\overline{K}$ molecules (14,15) are consistent with experiment. Clearly, improved measurements of the $\gamma\gamma$ couplings of these states would be a very interesting test of the $K\overline{K}$ molecule picture.

Finally, there is qualitative support for the $K\overline{K}$ molecule assignment in the strong $f_0(975)$ signal seen by Mark II [23], DM2 [24] and Mark III [25] in the decay $\psi \to \phi\pi^+\pi^-$ (see also the review article by Seiden [26]). Two–body hadronic ψ decays of this type are believed to be accurate tags of the flow of flavor [25]. For example, in $\psi \to \phi X$, the ψ purportedly annihilates into gluons, which then produce two $q\overline{q}$ pairs. Assuming hairpin diagrams may be neglected, the recoiling ϕ implies that both $q\overline{q}$ pairs in the final state are $s\overline{s}$, so that the final state X was made from an $s\overline{s}$ source. Comparison of $\psi \to \phi X$ with $\psi \to \omega X$ then tells us how easily one may make the state X starting from an $s\overline{s}$ versus non–strange $q\overline{q}$ pair. In $\psi \to \phi X$, $X \to \pi^+\pi^-$, there is a very prominent $f_0(975)$ signal, suggesting the presence of strange quarks in this state, as expected for a $K\overline{K}$ molecule. In contrast, the $f_2(1270)$, which is a well–established nonstrange ${}^3P_2$ $q\overline{q}$ state, is clearly observed recoiling against the ω in $\psi \to \omega\pi^+\pi^-$ but appears only very weakly in $\psi \to \phi\pi^+\pi^-$. (These decays in the DM2 data [24] are shown in figure 3.) Seiden [26] notes that quantitative tests of the $q\overline{q}$ and $K\overline{K}$ assignments should be possible given measurements of related ψ hadronic decays involving the f_0 and a_0.

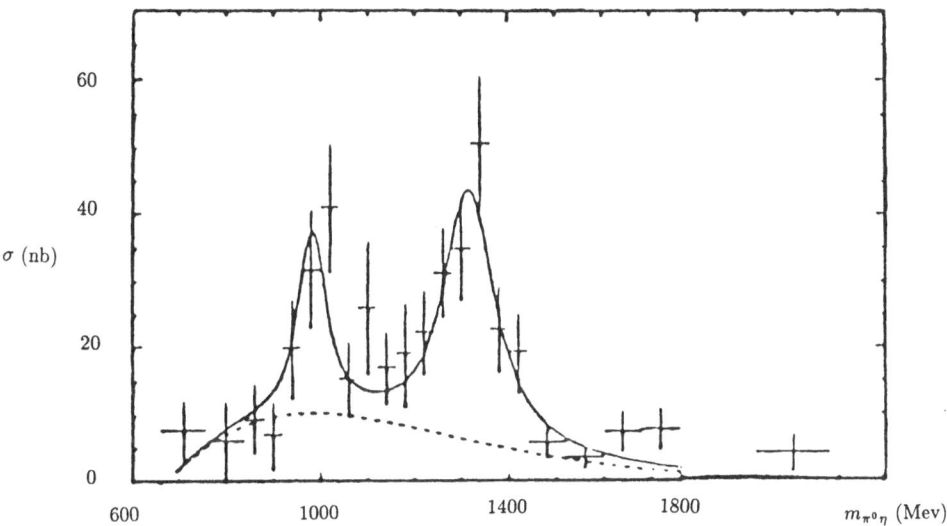

Fig. 2   The a_0(983) in $\gamma\gamma \rightarrow \eta\pi^0$
         (Crystall Ball collaboration, Antreasyan *et al* [19]).

IV. OTHER INTERHADRON POTENTIALS: NN AND $\pi\pi$

The variational calculations of Weinstein and Isgur led to interkaon potentials which were sufficiently strong to explain the $f_0(975)$ and $a_0(980)$ as weakly–bound $K\overline{K}$ states, thus solving a long–standing problem in hadron spectroscopy. One might naturally inquire about the predictions of similar variational calculations for other pairs of hadrons and whether or not there is experimental support for these predictions. Two such systems which have been studied using similar techniques are NN (nucleon–nucleon) and $\pi\pi$. As the general features of the NN potential are well understood (a repulsive core with an intermediate–range attraction), a study of this system using a nonrelativistic potential model and a variational Ansatz for the six–quark wavefunction is an important check of the reliability of calculations of this type.

The possibility of calculating the nucleon–nucleon interaction at the quark level has long been of interest to both elementary particle and nuclear physicists, and there is an extensive literature on this subject (see for example [1,28] and references cited therein). A popular starting point has been the MIT bag model, although these calculations have been only moderately successful; for example, the appearance of a repulsive core appears to be strongly dependent on approximations made in the calculation. As the bag model has problems with spurious center–of–mass excitations and with the *a priori* imposition of confinement, it is a dubious starting point for a study of the two–nucleon system.

In a study of the NN system which incorporated the essential features of the earlier work on $q^2\overline{q}^2$, Maltman and Isgur [27,28] carried out a variational calculation of the q^6 system. A nonrelativistic potential model with a spin–spin interaction was again used to model the dynamics, and a many–parameter wavefunction which interpolated between a single q^6 hadron and two separate q^3 nucleons was used as a variational Ansatz. The q^6 wavefunction thus obtained was projected onto two unperturbed nucleon wavefunctions in order to obtain an approximate two–nucleon wavefunction $\Psi(r_{NN})$. On substitution into the two–nucleon Schrödinger equation this gave an effective nucleon–nucleon potential $V(r_{NN})$. This calculation successfully predicted both a short–distance repulsive core and an intermediate–range attraction, the latter being due to an induced color electric dipole–dipole interaction. Vector meson exchange effects were found to be unimportant, contrary to two decades of nuclear physics folklore. The only meson exchange effect which was found to have important consequences was π exchange, which was required to reproduce the observed level of $^3S_1$–$^3D_1$ mixing in the deuteron.

The $\pi\pi$ system has also been the subject of a variational calculation by Weinstein and Isgur [8,9], as it is quite similar to the $K\overline{K}$ system and may be treated using the same model. The isospin channels accessible to two pions in an S–wave are $I = 0$ and $I = 2$, and each of these channels has an associated interpion potential. The variational calculation [8] finds an attractive interpion potential in $I = 0$ with an approximately Gaussian shape, with a depth of $V_{\pi\pi}^{(I=0)}(0) \approx -800$ Mev and a range of $r_0 \approx 0.4$ fm, defined by

$$V_{\pi\pi}^{(I=0)}(r_{\pi\pi}) = V_{\pi\pi}^{(I=0)}(0) \ \exp\left(-\tfrac{1}{2}(r_{\pi\pi}/r_0)^2\right) . \tag{18}$$

The $I = 2$ potential was found to be repulsive and somewhat larger in magnitude. Assuming that these potentials arise from one gluon exchange followed by rearrangement into color singlet mesons, one may show that the $I = 0$ and $I = 2$ potentials are related by

$$V_{\pi\pi}^{(I=2)}(r_{\pi\pi}) = -2 \cdot V_{\pi\pi}^{(I=0)}(r_{\pi\pi}) . \tag{19}$$

This relation is approximately consistent with the numerical results produced by Weinstein and Isgur's variational program.

As the $\pi\pi$ system has been the subject of many experimental studies, it should be possible to find tests of the purported attractive $I = 0$ and repulsive $I = 2$ interpion potentials (18) and (19). There is indeed evidence for the existence of such interpion potentials, although some of the experimental results and their interpretation are controversial at present, in particular $\gamma\gamma \to \pi\pi$.

In the $I = 0$ channel there have long been indications of a low–energy enhancement in $\pi\pi$ production, the earliest being the "ABC effect" in the reactions $PD \to {}^3H \ \pi\pi$ and $PD \to {}^3He \ \pi\pi$ [29–31]. More recently there have been reports of low–energy $\pi\pi$ enhancements in the processes $\gamma\gamma \to \pi^+\pi^-$ (which is dominantly $I = 0$) [32–36] and $\psi \to \omega\pi^+\pi^-$ [24,25] (see figures 3 and 4). In the latter case we note that $\psi \to \phi\pi^+\pi^-$ and $\psi \to \gamma\pi^+\pi^-$ show no such enhancement, which may be due to strong $m_{\pi\pi}$ dependence of the production amplitudes; clearly it is preferable to study the better–understood electromagnetic processes where data is available. Finally, the $\pi\pi$ phase shifts near threshold are positive for $I = 0$ and negative for $I = 2$ [13,37,38], as expected for attractive and repulsive potentials respectively. The $I = 2$ phase shift and the $I = 0$ one (below the range of the broad $f_0(1300)$ meson) are numerically well described by potential scattering in the non–relativistic Schrödinger equation, with (18) and (19) taken as the interpion potentials. Numerical results for the $I = 2$ phase shift with $V_{\pi\pi}^{(I=2)}(0) = 1.6$ Gev (fixed) and $r_0 = 0.37$ fm (fitted, but still close to the original Weinstein–Isgur result) are shown in figure 5, together with the data of Prukop et al [37].

The reactions $\gamma\gamma \to \pi^+\pi^-$ and $\gamma\gamma \to \pi^0\pi^0$ are especially interesting, and as the experimental results and their theoretical interpretation have recently been the subjects of controversy we shall discuss them in some detail. There are numerous measurements of $\sigma(\gamma\gamma \to \pi^+\pi^-)$ from $m_{\pi\pi} \approx 500$ Mev to above the $f_2(1270)$ [39–45], and the non–f_2 contribution is approximately consistent with the Born cross section for production of charged pointlike pions. Below $m_{\pi\pi} \approx 500$ Mev, however, the DM1, DM2 and PLUTO groups [32–36] report a $\pi^+\pi^-$ excess over the theoretical Born result. These experiments have a large background from the process $\gamma\gamma \to \mu^+\mu^-$ which must be subtracted to give the charged $\pi\pi$ cross section. Au, Morgan and Pennington [46–49] suggest that essentially all of the reported $\pi\pi$ excess in $\gamma\gamma \to \pi^+\pi^-$ consists

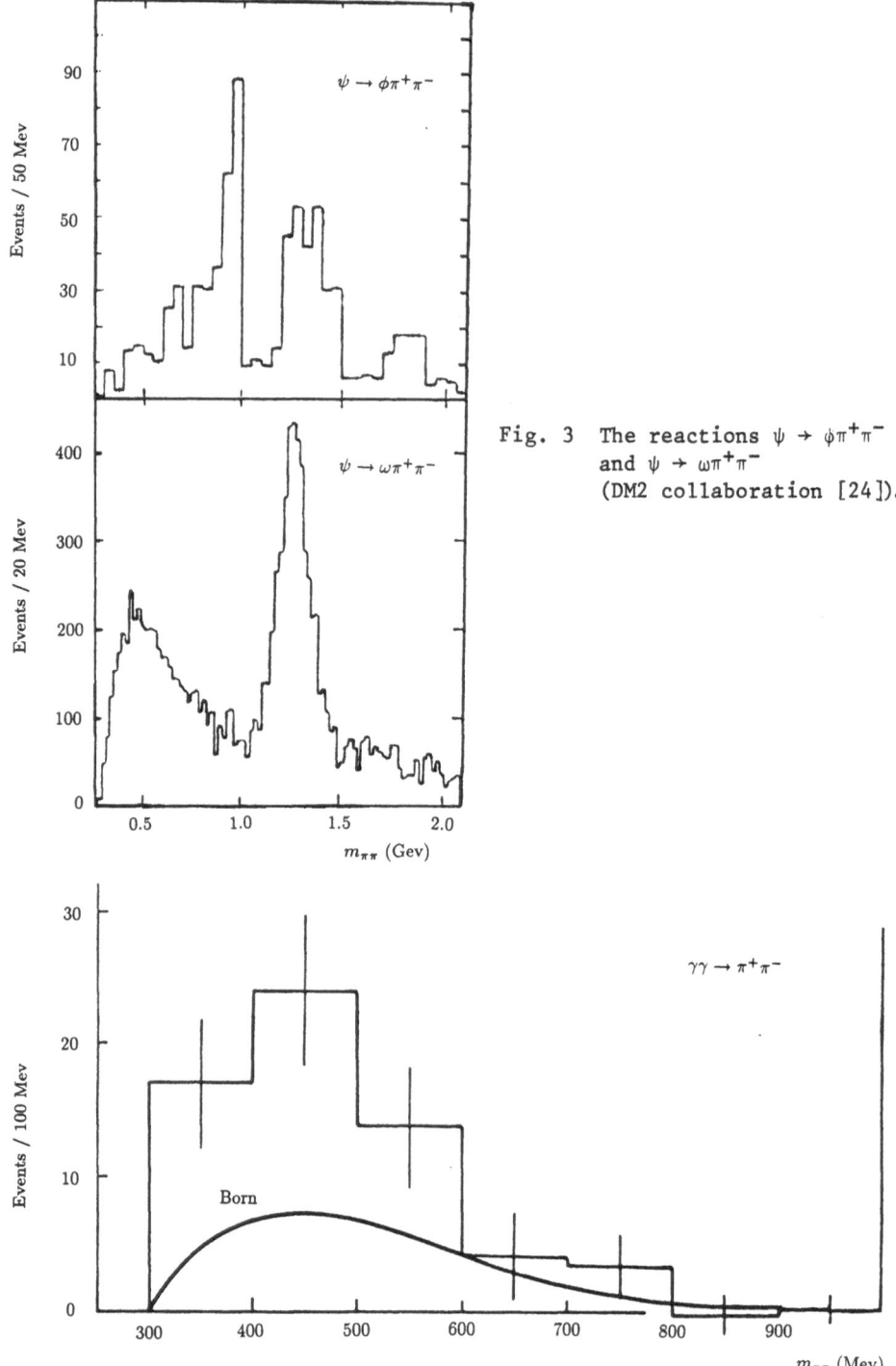

Fig. 3 The reactions $\psi \rightarrow \phi\pi^+\pi^-$ and $\psi \rightarrow \omega\pi^+\pi^-$ (DM2 collaboration [24]).

Fig. 4 Combined DM1 and DM2 results for the reaction $\gamma\gamma \rightarrow \pi^+\pi^-$ versus the Born approximation [36].

of misidentified lepton pairs. Their technique for relating final–state interactions to $\pi\pi$ scattering, which predicts little enhancement of $\gamma\gamma \to \pi^+\pi^-$, does indeed agree with the recent Crystal Ball measurement of the related process $\gamma\gamma \to \pi^0\pi^0$ [51,52]. A measurement of the $\gamma\gamma \to \pi^+\pi^-$ cross section near threshold should certainly be repeated with better statistics and improved background subtraction before a final conclusion is reached on the status of this effect.

Final–state interaction theory with the interpion potentials (18) and (19) predicts an enhancement of the process $\gamma\gamma \to \pi^+\pi^-$, although the size of this enhancement is rather sensitive to the potential parameters chosen. This is due in part to the fact that the initially–produced $\pi^+\pi^-$ state is a superposition of $I = 0$ and $I = 2$, so there is an interplay between the enhanced $I = 0$ component and the suppressed $I = 2$ one. At present it appears that the calculation of Barnes, Dooley and Isgur [50] overestimated the size of the net $\gamma\gamma \to \pi^+\pi^-$ enhancement because they were unaware of the relation (19) between $I = 0$ and $I = 2$ potentials. There may be additional complications, as we shall now see in our discussion of $\gamma\gamma \to \pi^0\pi^0$.

The cross section for $\gamma\gamma \to \pi^0\pi^0$ is zero in Born approximation. However, when the $\pi^+\pi^-$ state in $\gamma\gamma \to \pi^+\pi^-$ has its $I = 0$ part enhanced and its $I = 2$ part suppressed by the interpion potentials, it rotates slightly into a $\pi^0\pi^0$ state. Thus, unequal $I = 0$ and $I = 2$ $\pi\pi$ potentials imply a non–zero cross section for $\gamma\gamma \to \pi^+\pi^- \to \pi^0\pi^0$. A typical result for this "feedthrough" cross section is shown by Morgan and Pennington [49] (figure 6), who refer to this as the "BDI effect". $\pi\pi$ potential parameters comparable to those found by Weinstein and Isgur [8] lead to a cross section for $\gamma\gamma \to \pi^+\pi^- \to \pi^0\pi^0$ which rises to a maximum of $\sim 50 - 100$ nb near $m_{\pi\pi} = 300$ Mev and then drops quickly, passing through ~ 10 nb at 500 Mev. It is now possible to test this interesting prediction as a result of the recent Crystal Ball measurement of $\gamma\gamma \to \pi^0\pi^0$ [51,52] (also shown in figure 6), and the prediction is *not* in good agreement with experiment. The measured cross section rises quickly from threshold to a plateau of $\sim 15 - 20$ nb near $m_{\pi\pi} = 350$ Mev, and then declines to a minimum of ~ 10 nb near 750 Mev before rising to a peak near the $f_2(1270)$.

The small cross section for $\gamma\gamma \to \pi^0\pi^0$ is a serious problem for the subject of intermeson potentials. We are currently investigating the possibility that other processes such as $\gamma\gamma \to f_0(1300) \to \pi^0\pi^0$ and $\gamma\gamma \to \pi^+\pi^- \to f_0(1300) \to \pi^0\pi^0$ might interfere destructively with $\gamma\gamma \to \pi^+\pi^- \to \pi^0\pi^0$ feedthrough, thus giving a smaller cross section for $\gamma\gamma \to \pi^0\pi^0$ than naively expected. Another possible reason for the discrepancy is that conventional final–state interaction theory, which assumes that the outgoing pions are produced at contact, may be inadequate for a reaction which involves t–channel pion exchange. These issues will be discussed in detail in a future publication [53].

One interesting feature of the reaction $\gamma\gamma \to \pi^0\pi^0$ is that direct $f_0(1300)$ production $(\gamma\gamma \to f_0(1300) \to \pi^0\pi^0)$ may give an important contribution to the cross section for $\gamma\gamma \to \pi^0\pi^0$ in the 1 Gev region (see however [40]). If so, it may finally be possible to determine the mass, width and $\gamma\gamma$ coupling of the elusive $f_0(1300)$ $q\bar{q}$ meson from the Crystal Ball measurement of $\sigma(\gamma\gamma \to \pi^0\pi^0)$.

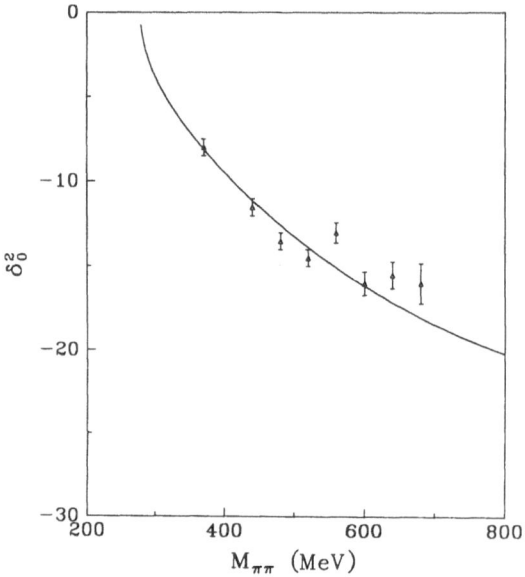

Fig. 5   The I = 2 S-wave phase shift for $V_{\pi\pi}^{I=2}(0) = 1.6$ GeV
and $r_0 = 0.37$ fm and the data of Prukop et al [37].

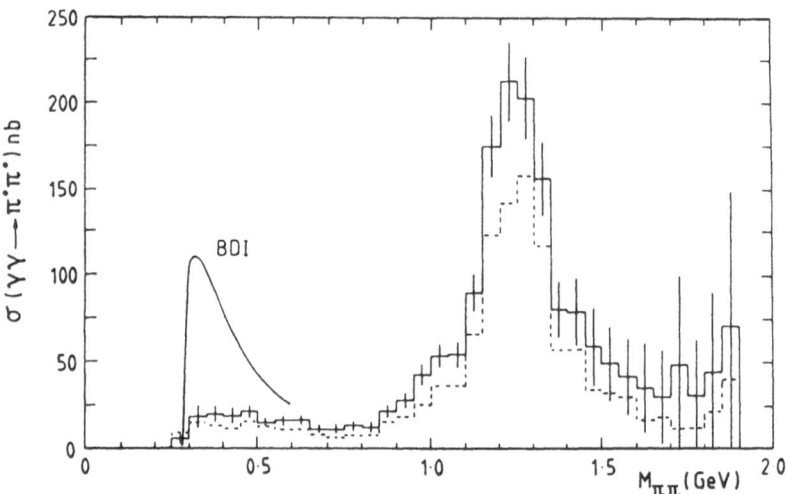

Fig. 6 The Crystal Ball measurement of $\sigma(\gamma\gamma \rightarrow \pi^0\pi^0)$ [51,52]
and the BDI effect [49].

V. POSSIBLE EXPERIMENTAL CONTRIBUTIONS AT LEAR

In the previous sections we discussed the subjects of mesonic nuclei and inter-hadron potentials and reviewed the theoretical and experimental work which has been done in this area. At present only the pseudoscalar–pseudoscalar sector of the meson–meson system has been subject to careful theoretical study, and comparison with experiment has involved measurements of properties of $K\overline{K}$ and $\pi\pi$ states. The theoretical studies suggest that the $K\overline{K}$ potential is sufficiently strong to form two nuclei, the $f_0(975)$ and $a_0(980)$, and that the $\pi\pi$ potential is attractive in $I = 0$ and repulsive in $I = 2$. These results may be qualitatively extended to other pseudoscalar–pseudoscalar two–meson states using arguments based on approximate SU(3) symmetry.

Existing experimental data appears to support the identification of the f_0 and a_0 with $K\overline{K}$ bound states, although the evidence for the closely related $\pi\pi$ potentials is more problematical; purely hadronic processes such as the $\pi\pi$ phase shifts support the calculated potentials, but there are difficulties with the recent measurement of $\sigma(\gamma\gamma \rightarrow \pi^0\pi^0)$.

There are several experimental measurements which would be particularly relevant to the study of mesonic nuclei and intermeson potentials. It is convenient to distinguish between experiments on bound states such as the $f_0(975)$ and $a_0(980)$ and experiments on scattering states such as $\pi\pi$.

a) Experiments on Mesonic Nuclei

There is clear evidence for only two mesonic nuclei, the $K\overline{K}$ bound states $f_0(975)$ and $a_0(980)$. Unless new candidates are discovered, any experimental study of mesonic nuclei must be a measurement of properties of these two resonances.

The masses and widths of the f_0 and a_0 are already well established, and studies of the branching fractions to strongly allowed states ($K\overline{K}$ versus $\pi\pi$ or $\eta\pi$) are complicated by the difficulty of distinguishing $K\overline{K}$ pairs that actually came from the resonance from those which were merely produced near threshold. A more useful and less ambiguous set of measurements would be the electromagnetic partial widths of these states, such as $\Gamma(f_0(975) \rightarrow \rho\gamma)$. The E1 decay rates of $\ell = 1$ $q\bar{q}$ states are easily calculated and could be directly compared with such measurements, thus adding to the tests of the relative merits of the $K\overline{K}$ and $q\bar{q}$ assignments. Assuming the $K\overline{K}$ assignment, these numbers would allow important tests of the $K\overline{K}$ molecule wavefunction and its decay mechanisms; for example, does $f_0(975) \rightarrow \rho\gamma$ proceed through $f_0(975) \rightarrow f_0(0^{++} q\bar{q}) \rightarrow \rho\gamma$? Measurements of all the $K\overline{K} \rightarrow (1^{--} q\bar{q})\gamma$ partial widths would be very useful for comparison with model calculations. Close [54] notes that measurements of the related transition rates for $\phi \rightarrow f_0(975)\gamma$ and $\phi \rightarrow a_0(980)\gamma$ might also be feasible, although the branching fraction will be quite small, perhaps 10^{-5}. The partial widths to $\gamma\gamma$ are also very interesting: as we have seen in section III, existing $\gamma\gamma$ measurements already make the f_0 and a_0 implausible as candidate $q\bar{q}$ states. Unfortunately, the backgrounds may be too great for $\gamma\gamma$ rates to be measured at hadron machines like LEAR.

b) Experiments on Intermeson Potentials

The most direct comparison of intermeson potentials with experiment is in measurements of two–meson scattering phase shifts and inelasticities, such as in the $I = 2$

$\pi\pi$ channel (figure 5). Without π or K beams this data is not easily accessible, so this type of experiment is not relevant to LEAR. The area in which LEAR can most easily contribute is in the measurement of intermeson potentials through the observation of final–state interactions.

Many "resonances" which are just above S–wave thresholds, such as the $J^{PC} = 1^{++}$ $f_1(1420)$ ($m_{f_1} = m_{K^*} + m_{\bar{K}} + 30$ Mev), could be threshold enhancements due to attractive intermeson potentials rather than actual $q\bar{q}$ resonances. An example is the low–mass I = 0 $\pi\pi$ bump in $\psi \to \omega\pi\pi$ (figure 3), which we believe to be the result of a final–state enhancement due to the attractive I = 0 $\pi\pi$ potential. There are many such two–meson channels which can be studied near threshold to extract the S–wave intermeson potential. In the pseudoscalar–pseudoscalar sector these include $\eta\eta$ (expected to be repulsive) and I = 3/2 $K\pi$ (analogous to I = 2 $\pi\pi$, hence also expected to be repulsive). The vector–pseudoscalar and vector–vector channels have not been carefully investigated by theorists, so experiment can provide useful constraints on future theoretical work in this area.

The technique used to extract an intermeson potential from a cross section or decay measurement near threshold is fairly straightforward. For example, suppose we are studying the S–wave I = 1 $K^*\bar{K}$ interaction, and that we have data for $\sigma(P\bar{P} \to K^{*+}\bar{K}^0\pi^-)$ as a function of $K^*\bar{K}$ invariant mass. We would assume a simple Ansatz for an I = 1 $K^*\bar{K}$ potential such as the Gaussian form used for $V_{K\bar{K}}$ in (1), with the depth $V(0)$ and range r_0 left as parameters which are to be fitted to the data. (In our examples they have typically taken on values of $\sim \pm 500$ Mev and ~ 0.5 fm.) We would then fit the cross section to a phase space distribution times the S–wave enhancement factor, where the latter is the squared ratio of the scattering–state wavefunction $\Psi(0)$ in the presence of $V(r)$ to $\Psi(0)$ with $V = 0$. It is important to restrict the fit to values of $m_{K^*\bar{K}}$ sufficiently close to threshold so that the initial $K^{*+}\bar{K}^0\pi^-$ production amplitude is approximately constant and is dominated by the $K^*\bar{K}$ S–wave. One must also be careful to avoid rapid amplitude variations due to $q\bar{q}$ resonance production. In $\psi \to \omega\pi\pi$, for example, one might attempt to fit the $m_{\pi\pi}$ dependence of the decay between threshold and 750 Mev, thus avoiding the $f_2(1270)$ region. An anti–$b_1(1235)$ cut would also be advisable in this case. A fit to $\psi \to \omega\pi\pi$ using similar methods with a square well potential has been described by Dooley [55]. This technique for calculating final–state interaction effects was originally suggested by Fermi [56] and is discussed in detail by Gillespie [57].

VI. ACKNOWLEDGEMENTS

It is a pleasure to thank the organisers, in particular F.E.Close and U.Gastaldi, for their kind invitation to the Erice school and for the opportunity to present this material and to discuss physics with my fellow participants. I am indebted to many people for discussions and information relating to the material presented in this review, and I would particularly like to thank Z.Ajaltouni, S.Cooper, A.Falvard, H.Marsiske, D.Morgan and M.R.Pennington for their assistance. I would also like to thank my colleagues K.Dooley and N.Isgur (at the University of Toronto) and J.Weinstein (at the University of Guelph) for the pleasure of collaborating with them on various aspects of this work. The financial support of the Universities of Guelph and Toronto, the Natural Sciences and Engineering Research Council of Canada and the Ettore Majorana Center for Scientific Culture is gratefully acknowledged.

VII. REFERENCES

[1] N.Isgur, Acta Physica Austriaca, Suppl. XXVII, 177 (1985).

[2] T.Barnes, Lecture Notes in Physics 234, 124 (1985).

[3] R.L.Jaffe, Phys. Rev. D15, 267 (1977).

[4] R.L.Jaffe, Phys. Rev. Lett. 38, 195, 617E (1977).

[5] J.P.Ader, J.M.Richard and P.Taxil, Phys. Rev. D25, 2370 (1982).

[6] J.L.Ballot and J.M.Richard, Phys. Lett. 123B, 449 (1983).

[7] J.Weinstein and N.Isgur, Phys. Rev. Lett. 48, 659 (1982).

[8] J.Weinstein and N.Isgur, Phys. Rev. D15, 588 (1983).

[9] J.Weinstein, University of Toronto Ph.D. thesis (1986).

[10] J.Weinstein, Toronto report UTPT–87–21.

[11] A.Astier et al , Phys. Lett. 25B, 294 (1967).

[12] A.B.Wicklund et al , Phys. Rev. Lett. 45, 1469 (1980).

[13] Review of Particle Properties, Rev. Mod. Phys. 56 (April 1984).

[14] Review of Particle Properties, Phys. Lett. 170B, 1 (April 1986).

[15] S.Godfrey and N.Isgur, Phys. Rev. D32, 189 (1985).

[16] R.Kokoski and N.Isgur, Phys. Rev. D35, 907 (1987).

[17] T.Barnes, Phys. Lett. 165B, 434 (1985).

[18] T.Barnes, in Proc. VII Internatl. Workshop on Photon–Photon
    Collisions, 25 (1986).

[19] D.Antreasyan et al (Crystal Ball collaboration), Phys. Rev. D33, 1847 (1986).

[20] N.N.Achasov, S.A.Devyanin and G.N.Shestakov, Sov. J. Nucl. Phys. 32,
    566 (1980).

[21] C.Edwards et al (Crystal Ball collaboration), Phys. Lett. 110B, 82 (1982).

[22] J.E.Olsson, in Proc. V Internatl. Workshop on Photon–Photon
    Collisions, (1983).

[23] G.Gidal et al (Mark II collaboration), Phys. Lett. 107B, 153 (1981).

[24] A.Falvard, in Proc. VII Internatl. Workshop on Photon–Photon
    Collisions, 435 (1986).

[25] U.Mallik, SLAC–PUB–4238 (February 1987).

[26] A.Seiden, in Proc. VII Internatl. Workshop on Photon–Photon
    Collisions, 193 (1986).

[27] K.Maltman and N.Isgur, Phys. Rev. Lett. 50, 1827 (1983).

[28] K.Maltman and N.Isgur, Phys. Rev. D29, 952 (1984).

[29] A.Abashian, N.E.Booth and K.M.Crowe, Phys. Rev. Lett. 5, 258 (1960).

[30] N.E.Booth, A.Abashian and K.M.Crowe, Phys. Rev. Lett. 7, 35 (1961).

[31] N.E.Booth and A.Abashian, Phys. Rev. 132, 2314 (1963).

[32] A.Courau *et al* (DM1 collaboration), Phys. Lett. 96B, 402 (1980).

[33] A.Courau *et al* (DM1 collaboration), Nucl. Phys. B271, 1 (1986).

[34] Ch.Berger *et al* (Pluto collaboration), Z. Phys. C26, 199 (1984).

[35] Z.Ajaltouni, in Proc. VII Internatl. Workshop on Photon–Photon Collisions, 475 (1986).

[36] Z.Ajaltouni *et al* , Orsay preprint LAL 87–15 (March 1987).

[37] J.P.Prukop *et al* , Phys. Rev. D10 2055 (1974).

[38] G.Grayer *et al* , Nucl. Phys. B75, 189 (1974).

[39] Ch.Berger *et al* (Pluto collaboration), Phys. Lett. 94B, 254 (1980).

[40] R.Brandelik *et al* (Tasso collaboration), Z. Phys. C10, 117 (1981).

[41] A.Roussaire *et al* (Mark II collaboration), Phys. Lett. 105B, 304 (1981).

[42] H.J.Behrend *et al* (Delco collaboration), Z. Phys. C23, 223 (1984).

[43] A.Courau *et al* (Delco collaboration), Phys. Lett. 147B, 227 (1984).

[44] J.R.Smith *et al* (Mark II collaboration), Phys. Rev. D30, 851 (1984).

[45] H.Aihara *et al* (TPC collaboration), Phys. Rev. Lett. 57, 404 (1986).

[46] K.L.Au, D.Morgan and M.R.Pennington, Phys. Rev. D35, 1633 (1987).

[47] D.Morgan and M.R.Pennington, Phys. Lett. 192B, 207 (1987).

[48] D.Morgan and M.R.Pennington, Rutherford/Durham report RAL–87–048/ DTP 87/20.

[49] D.Morgan and M.R.Pennington, Durham report DTP 87/22 and Proc. EPS Internatl. Conf. High Energy Physics, Uppsala (1987).

[50] T.Barnes, K.Dooley and N.Isgur, Phys. Lett. 183B, 210 (1987).

[51] H.Marsiske, in Proc. XXII Recontre de Moriond (March 1987).

[52] S.Cooper, MIT preprint MIT–LNS–161 (June 1987) and Proc. II Internatl. Conf. on Hadron Spectroscopy (KEK, April 1987).

[53] T.Barnes, K.Dooley, N.Isgur and J.Weinstein, work in progress.

[54] F.E.Close, personal communication.

[55] K.Dooley, University of Toronto M.Sc. thesis (1986).

[56] E.Fermi, "Elementary Particles" (Yale University Press, 1951) 58–64.

[57] J.Gillespie, "Final State Interactions" (Holden–Day, 1964).

PROSPECTS FOR GLUONIC HADRONS AT CERN

Frank Close

Rutherford Appleton Laboratory
Chilton, Didcot, Oxon OX11 0QX

Quantum chromodynamics is a theory of quarks and gluons. The successful quark model is a model of hadron spectroscopy based on quarks. So where is the glue?

According to QCD quarks (q) are in the $\underset{\sim}{3}$, antiquarks ($\bar{q}$) in $\bar{3}$ and gluons (g) in $\underset{\sim}{8}$ dimensional representations of the $SU(\widetilde{3})_c$ group. The observed particles in nature are colour $\underset{\sim}{1}$.

Combining quarks and antiquarks we have the colour representations as follows

$$q \qquad \underset{\sim}{3}$$

$$q\bar{q} \qquad 3 \times \bar{3} = 8 + 1$$

$$q q \qquad 3 \times 3 = 6 + \bar{3} \tag{1}$$

$$qqq \qquad 3 \times 3 \times 3 = 10 + 8 + 8 + 1$$

and the only colour $\underset{\sim}{1}$ are the familiar $q\bar{q}$ (mesons) and q^3 (baryons).

The simplest colour $\underset{\sim}{1}$ combinations of gluons are

$$gg \quad 8 \times 8 \qquad = 1 + 8 + \ldots$$
$$ggg \quad (8 \times 8)_8 \times 8 = 1 + \ldots \tag{2}$$

The gg states have charge conjugation C = +; the ggg can have C = ±. These states are known as glueballs or gluonia and the prejudice is that C = + glueballs lie lowest in the spectrum insofar as they are "simpler" to make than the C = -.

If C = - glueballs exist then their internal structure contains gluons combined to colour $\underset{\sim}{8}$, neutralised by another gluon:

$$\left[(gg)_8 + g_8 \right]_1 . \tag{3}$$

In this case one expects that similar possibilities should occur in the hadron (quark) spectrum. In eqs (1) we see $(q\bar{q})_8$ and $(qqq)_8$ appear. These can neutralise their colour, by analogy with eq (3) and form hybrid mesons $\left[(q\bar{q})_8 g \right]_1$ or hybrid baryons $\left[(qqq)_8 g \right]_1$.

What are the signatures for such states and the chance of producing and recognising them?

The gg and $gq\bar{q}$ have integer spins, like $q\bar{q}$ mesons, and may mix with one another, or be mixed into the wavefunctions of well known mesons (e.g. η'). This makes their easy identification hard, not least because our understanding of $q\bar{q}$ meson spectroscopy is still poor in the 1-3 GeV mass region. There is one ray of hope. There are J^{PC} combinations such as

$$J^{PC} = 0^{--} ; 0^{+-}, 1^{-+}, 2^{+-} \ldots \tag{4}$$

that are forbidden to $q\bar{q}$. Discovery of a state with such J^{PC} would show that one had a state that is "beyond" $q\bar{q}$. However it could be $q^2\bar{q}^2$ (but see also Isgur's lectures in this regard).

Hybrid baryons may offer a hope. Baryon spectroscopy is rather well understood, even above 2 GeV in some cases. However, here again there is the possibility of confusion with $q^4\bar{q}$ states.

It is amusing to contemplate the significance of discovery an exotic J^{PC} as in eq (4). Had this occurred in 1967 say it would have probably caused us to throw out the quark model - it was the absence of such exotics that gave such confidence in the $q\bar{q}$ model. However, discovery of an exotic today would be taken as confirmation of the quark model and colour and probably of the existence of gluonic hadrons. Quite a change in psychology!

In searching for gluonic hadrons there are three particular topics of concern: How do you produce gluonic states, what is their spectroscopy expected to be and how do they decay? There is rather general agreement on the first question and qualitative concensus on the second but predictions for the third are so varied as to be no useful guide at all. I will look at each of these in turn.

1. Glueball Dynamics: Production

Decades of experiments involving conventional hadron beams and targets ("quark hadrons") have produced many resonances composed of quarks. No clear gluonic candidates have emerged from all these data. This is consistent with the expectation that if gluonic states are made of valence gluons then they will appear in channels containing hard gluons, and not be copious in processes where most momentum goes into quarks.

Possible channels include proton–antiproton annihilation (where the momentum of annihilating quarks may transfer to gluons in the 1–2 GeV mass range) and the decays of heavy mesons such as J/ψ. Another strategy, which has created some recent interest, is double diffraction of protons at very high energies where gluons surrounding each proton fuse to make new mesons.

Of these three, the cleanest and most productive has been the study of radiative decays $J/\psi \to \gamma + X$. This requires the c and $\bar{c}$ quarks to annihilate. Perturbative QCD predicts the radiative branching ratio to be[1]

$$\frac{\Gamma\,(\psi \to \gamma gg)}{\Gamma\,(\psi \to ggg)} = \frac{16\alpha}{5\alpha_s} \left[1 + (2.2\pm0.6)\frac{\alpha_s}{\pi} \right] \qquad (3.3.1)$$

where α_s, the running coupling of QCD, is taken to be $\simeq 0.2$. This is consistent with data[2] and supports the belief that $\psi \to \gamma gg$ dominates the radiative decay of the ψ. Consequently this should be an excellent channel to search for glueballs with positive C-parity. It is worth noting that the ι(1440) and θ(1690) were new prominent signals seen in this chanel and, for this reason, have been prima facie candidates for gluonic hadrons, (see also Eigen's lectures).

In the hadronic decay of ψ one might also expect to produce gluonic states. If they are light enough one could produce $\psi \rightarrow ggg \rightarrow G_+ G_-$ (where the subscripts denote the C-parity). A problem is that the prejudices from spectroscopy imply that G_- states are relatively heavy, probably exceeding 2 GeV, and hence $G_+ G_-$ is probably kinematically forbidden for ψ decay. This could be important for the decay of heavier states, such as T (9.46).

Even though pairs of glueballs may be kinematically suppressed, there is still the possibility of producing a glueball recoiling against a well known quark-hadron, or of producing states that have gluonic components in their wavefunctions (such as the η or η'). Studies of $\psi \rightarrow PV$, P = pseudoscalar π, η, η', ι; V=vector ρ, ω, φ) have been made and claims made that these data show that there is glue in the η'[3].

If, as commonly believed G_+ states are lightest, then it may be possible to pair produce such states in decays of η_c or χ states ($J^{PC}=0^{-+}; 0^{++}, 1^{++}, 2^{++}$). Data on χ decays are sparse and could profitably be improved.

(i)   A danger in ψ → γ gg

The successful application of QCD perturbation theory to the radiative branching ratio has made people treat perturbative QCD with less caution then it deserves. It may work well for inclusive processes but less well to exclusive channels where final state interactions are necessarily important. Of particular concern is the popular prejudice that $\psi \nrightarrow \gamma 1^{++}$.

If ψ → γ gg → γ R, for some resonance R, proceeds via two physical on-shell gluons, then it is indeed true that $R \neq 1^{++}$. This is a consequence of Yang's theorem (essentially a consequence of the gluons transverse polarisations and Bose symmetry; see e.g. p 359 in ref 4. However if the gluons go off-shell then this suppression no longer operates. Moreover, the 1^{++} turn on may be rapid if γγ data are any guide. ($\gamma\gamma \nrightarrow 1^{++}$ for two real photons, but when one of them is slightly virtual, $Q^2 \simeq$ GeV^2, the axial production is seen[5]. This may be important in view of a signal in $\psi \to \gamma\eta\pi\pi$ where m($\eta\pi\pi$) $\simeq$ 1280 GeV, $\Gamma <$ 50 MeV. These may be consistent with $\psi \to \gamma B$ (1285)) (or f_1(1285) in modern notation) whose width $\Gamma = 25 \pm 3$ MeV. There has been much confusion about $\psi \to \gamma\eta\pi\pi$ in connection with the interpretation of ι(1440) (seen in $K\bar{K}\pi$ but not clear in $\eta\pi\pi$) and the possibility of a 0^{-+} state η(1275) in this vicinity. If 1^{++} states, such as B, can be produced, then it is important to allow for them in disentangling the data.

There are other data that support our suspicion of over zealous reliance on $\psi \to \gamma gg$ perturbation theory. These concern its consequences for $\psi \to \gamma 0^+$, 0^-, 2^+, and in particular the helicity structure (and angular distribution) of the latter.

In $\psi \to \gamma 2^{++}$, the 2^{++} can be produced with helicity $\lambda = 2,1,0$. Perturbative QCD with quasi-real gluons implies[6] that there is significant production of λ_2 ($y \equiv \lambda_2/\lambda_o = 0.54$) whereas the data for f and f' suggest that λ_2 vanishes[7]. In the case of the θ all three helicity amplitudes appear to have the same magnitude but different phases; this also is quite unlike the theoretical expectation.

These results suggest that the restriction to quasi-real gluons is unjustified. These predictions evaporate the moment that off-shell gluons (in particular longitudinal) are included.

In particular, axial mesons (1^{++} or 1^{-+}), can be produced.

(ii) Proton - antiproton annihilation at LEAR

The LEAR facility at CERN enables one to make detailed study of meson production in low energy proton-antiproton annihilation. In addition to learning more about "conventional" $q\bar{q}$ mesons, there is the possibility of discovering gluonic hadrons either in formation or production.

The essential idea is that one or more quark and antiquark in the beams annihilate. They carry colour and so there is a chance of coloured gluons being produced in the final state. The proton and antiproton are extended objects; gluons may be produced from $q\bar{q}$

annihilation while other quarks and antiquarks survive - thus a mixture
of quarks, antiquarks and hard gluons may be present. In these
circumstances there is the likelihood that hybrid mesons are formed, ($p\bar{p}$
$\rightarrow q\bar{q} + q\bar{q}g$). The hope that gluonic states are produced is raised by the
fact that the gluonium candidate ι(1440) may have been first seen in $\bar{p}p$
annihilation near threshold.

Proton-antiproton annihilation at rest occurs in S or P wave. This
limits the possible J^{PC} of the final state, which may be useful in
helping to identify the quantum numbers of new mesons. This gives $\bar{p}p$
annihilation some advantages over $e^+e^- \rightarrow \psi \rightarrow$ hadrons, though it is less
certain to be "glue-friendly".

A question of immediate concern is "what's new?" Proton-antiproton
annihilation has been studied in bubble chambers rather extensively, so
why do we hope that LEAR will provide essential new information? The
answers involve both the production mechanism and the detection.

Bubble chambers containing <u>liquid</u> hydrogen saw no sign of $\bar{p}p \rightarrow K_s K_s$
in the annihilation at rest. This was taken to imply that $\bar{p}p$
annihilation at rest in liquid is dominantly S-wave (since $K_s K_s$ is
forbidden from 0^+ or 1^-, i.e. from S-wave $\bar{p}p$). When <u>gaseous</u> hydrogen
is the target there is substantial P-wave annihilation (comparable to
S-wave). The Crystal Barrel collaboration and the Obelix collaboration
will detect low energy X-rays and thereby tag P-wave annihilations.

The ability to separate S and P wave annihilations may prove
important in isolating resonances which otherwise may have escaped
detection because of large background contributions. First, selection
rules impose marked differences for annihilations from S and P states.
Second, angular momentum barriers can suppress high mass states with
moderate spin in the S-wave case. This is already important for L=2
where $\bar{p}p \rightarrow \pi^0 f$ is much suppressed in liquid ($^1S_0$ $\bar{p}p$ with L=2) and
clearly visible in gas ($^3P_1$ or $^3P_2$ with L=1).

We can generalise this empirical observation to the following:
1986). Consider $\bar{p}p \rightarrow \pi^0 \chi^0$ (channel A), $\pi^+\pi^-\chi^0$ (B).

(a) <u>S-waves</u>

If χ^0 (channel A) has high mass and if L=2 is suppressed, then $\chi^0 =$

0^{++} or the exotic 1^{-+} in annihilation from $^1S_0$. From $^3S_1$ 1^{--}, 1^{+-} and 2^{--} are accesible as well as the exotic 0^{--}.

In channel B the $\pi^+\pi^-$ pair limits the phase space to < 1500 MeV. For high X^0 masses the dipion has no angular momentum and the J^{PC} is restricted to 0^{-+} and 1^{--}.

(b)   P-waves

If X^0 (channel A) is produced with no angular momentum then J^{PC} = 0^{-+}, 1^{--}, 2^{-+} or the exotic 1^{-+}. With one unit of angular momentum all states can be produced.

As we have noted, data with different S-wave and P-wave contributions show marked differences. The OBELIX collaboration at LEAR intend to make the comparison quantitative and, having determined the relative weights, to subtract spectra from one another to help isolate any new broad objects.

These techniques, exploiting the restricted angular momenta in annihilation at rest, can explore the mass range up to $2m_p$. To investigate the mass range above $2m_p$ production experiments with high incoming beam momenta will be necessary. In these measurements there will be a larger distribution of angular momenta in the initial state and so one will lose the possibility of direct comparison of data sets taken at the same energy with different angular momentum distributions.

In the detector aspect, these experiments hope to be able to pick up η, π^0 and neutral decay modes which have not been well studied in $\bar{p}p$ annihilations so far. Indeed, if the η, η' contain glue in their wave function, then there may be hope for isolating gluonic hadrons in $\bar{p}p \rightarrow$ "G" + η + ...

2.   Glueball Dynamics: Spectroscopy

In the absence of any good generally accepted datum for glueballs, any projections for glueball spectroscopy necessarily rely heavily on models. Different models differ in detail but there are some features on which they have greater or lesser agreement[8].

Potential models treat the gluon as a spin 1 effectively massive object by analogy with the way that quarks have been treated as if non-relativistic massive spin $\frac{1}{2}$ objects. They tend to predict that J^{PC} = $0^{++} \simeq 0^{-+}$ are the lightest states with 2^{-+}, 2^{++} somewhat heavier. In the bag model the 0^{++} is lightest, 0^{-+} near the 2^{++} with 2^{-+} rather heavier. Lattices have 0^{++} lightest, the 2^{++} may be nearby but there is controversy as to whether or not one can say more about other J^{PC} combinations. QCD sum rules and flux tube models also have 0^{++} lightest. The flux tube model has 1^{+-} relatively light: this is the only model where a C = – state appears among the lightest members.

The general concensus that we may abstract is that C = + states lie lowest with the 0^{++} the lightest of all. The 2^{++} is expected to be nearby and 0^{-+} should also be relatively light.

They all tend to predict that the lightest glueballs will occur in the 1-2 GeV mass region. We have gotten used to constructing models of quark-hadrons on a 1 GeV mass scale that there may be a Darwinian selection operating; or it may be a genuine and reliable result. A specific ilustration of what goes into predictions of glueball spectroscopy may help.

To give an example of gluonic mass spectroscopy and advertise the uncertainties I will choose the bag model as a guide. I do so because it has been used and quoted extensively and the uncertainties are often overlooked.

If you confine a spin $\frac{1}{2}$ massless quark in a spherical cavity you solve the Dirac equation subject to the boundary conditions that no quark current crosses the surface of the cavity. If the cavity is of the order of 1 fm radius, then the quark energy is some hundreds of MeV. Thus the "constituent mass", 350 MeV, of the naive quark model is motivated as the energy of a confined (nearly) massless quark.

Do the same for massless gluons by solving the Maxwell equation subject to analogous boundary conditions. You have to distinguish magnetic (1^+) and electric (1^-) modes and the energy turns out to be some 50% to 100% larger than for the lowest quark eigenmodes. Thus gluons act AS IF they have mass of some 500 MeV (1^+) or 800 MeV (1^-).

Thus one has, for orientation, the "ballpark" estimates

$$gg(1^+1^+ \to 0^+ + 2^+) \sim 1 \text{ GeV}$$
$$gg(1^+1^- \to 0^- + 2^-) \sim 1.3 \text{ GeV} \tag{2.1}$$
$$g q\bar{q}(0^{-+},1^{-+},2^{-+},1^{--}) \sim 1.5 \text{ GeV}$$

$$gqqq(\tfrac{1^+}{2}, \tfrac{3^+}{2}, \tfrac{5^+}{2}) \quad \sim 1.5 \text{ GeV}$$

hence in the accessible, and not well understood (for mesons) 1-2 GeV region. Hybrid baryons around 1.5 GeV would be accessible (and already eliminated by data!)

In quark hadron spectroscopy we know that spin dependent mass splittings (one gluon exchange) are very important and pull the low J states down while pushing high J up in mass (eg. $\tfrac{1^+}{2} N < \tfrac{3^+}{2}\Delta; 0^-\pi < 1^-\rho$). The mass shifts for spin 1 colour 8 gluons should be at least as large as the "smaller spin" ($\tfrac{1}{2}$), "less coloured" ($\underset{\sim}{3}$) quarks. If we take the ballpark masses seriously we would end up with a very light 0^+ gg (which one could perhaps live with by making excuses about vacuum quantum numbers causing subtleties) but also a low I=1 0^{-+} $q\bar{q}g$ and even a $\tfrac{1^+}{2}$ qqqg as light as the proton! Thus we see that hybrids can already constrain our intuition about confined gluons.

There is one thing we have overlooked (and which no one yet knows how to compute explicitly) – namely the self energies of quarks and gluons. In the case of quark hadrons this unknown is subsumed in the quark mass parameter m_q before one confines the quark in the bag; it is chosen to get the observed masses correct. In the case of gluonic hadrons we have no known states to guide us in the analogous effective parameter m_g. All that we can deduce, so far, is that it must be some hundreds of MeV if we are to avoid the "superlight" state paradoxes above. It is perhaps not unreasonable to suppose that these self energies are of $O(\alpha_s/R) \sim O(500 \text{ MeV})$ if $\alpha_s \simeq 2.2$ of the old MIT bag phenomenology is a guide and $R \simeq O(1 \text{ fm})^{(9-13)}$.

If one does this, one can expect gg states consistent with some well known states whose constitution is controversial.

$$0^+ f_o(975), \ 0^- \eta(1440) \ ; \ 2^+ f_2(1700) \ 2^-(A_3^!?)$$

There is also the possibility that $m_g^{effective}$ for magnetic and electric gluon modes differ, (a freedom exploited in ref 13).

The mass scale of gluonic hadrons relative to quark hadrons there depends rather critically on this unknown freedom! The question of which family of gluonic hadrons is the lowest lying depends on this. For example the qualitative guess might be that

$$M(q\bar{q}g) = M(q\bar{q}) + \frac{1}{2} M_G(gg) \qquad\qquad (2.2)$$

$$M(qqqg) \approx M(q\bar{q}g) + O(350 \text{ MeV})$$

in which case hybrids will be lighter or comparable to glueballs.

It is important to realise this. Many phenomenological discussions, such as $\psi \rightarrow \gamma X$, consider only the possibility of glueballs in X. This is economical because a glueball is a single meson; but a hybrid requires a whole nonet to exist – including I=1 states which are less subject to doubt than the I=0 sector.

Thus if we can convince ourselves that whole nonets of hybrids are absent, we can implicitly put a lower bound on the mass scale for gluonic hadrons and glueballs.

The relations above are rather qualitative but are shared by several models, (e.g. refs 10–13 and Isgur's lectures at this meeting).

The absence of I=1 0^{-+} states below 1.2 GeV puts a lower bound on the hybrid masses, and after hyperfine splittings were computed the isovector spectrum looked as follows

0^{-+}(1.1 GeV), 1^{-+}(1.4 GeV), 1^{--}(1.7 GeV), 2^{-+}(1.8 GeV) and the lightest hybrid baryons were $N^* \frac{1}{2}^+$ (1.5), $\frac{1}{2}^+$(1.7), $\frac{3}{2}^+$(1.75). These look tantalisingly like the Roper and other established states whose quark constitution is still debated.

However Barnes and Close (10c) have raised a question mark over this scenario. We discovered a selection rule, reminiscent of an old rule of Moorhouse, which forbids the photoexcitation of the lightest $\frac{1}{2}^+$

and $\frac{3^+}{2}$ (and $\frac{5^+}{2}$) hybrid N^* from __proton__ targets (though neutron excitation is allowed). The $\frac{1^+}{2}$ Roper $N^*(1470)$ is photoproduced from both neutron __and__ proton (with a ratio of $-\frac{2}{3}$ - equal to the magnetic moment ratio of static neutron and proton - which suggests that the Roper is indeed a radial excitation of the nucleon).

There is a possibility that the Roper is a mixture of hybrid and radial (in which case the orthogonal combination should be lurking somewhere - which makes this caveat seem somewhat contrived). Failing this, the lightest candidate is the $\frac{1^+}{2}(1710)$ and its $\frac{3^+}{2}$ partner the $N(1720)$ (Actually these masses could lie anywhere in the range 1680 to 1740 and 1690 to 1800 respectively). This makes for an interesting scenario of figs (1).

After flavour dependence is taken into account we could expect a pattern of fig (2). The splitting between I=1 and I=0 n$\bar{\text{n}}$g (labelled ρ and ω throughout) comes from a colour Coulomb repulsive force due to annihilation in the __colour octet__ q$\bar{\text{q}}$ I=0 channel, absent in the I=1 case.

I have assigned the mass scale such that 1.3 GeV coincides with the $\pi(1300)$. Notice the tantalising position of the ω state ($\eta(1440)$ - iota). The I=1 exotic $J^{PC} = 1^{-+}$ then arises at 1.5 GeV, which is interesting in view of the possible sighting of such an exotic state, reported here by Binon.

However, we anticipate that q$\bar{\text{q}}$q$\bar{\text{q}}$ states with $J^{PC} = 1^{-+}$ will occur here too. The general message is that a rich spectroscopy of states is expected in the 1.5-2 GeV region, over and above the q$\bar{\text{q}}$ states and as such offers prospects for experiment.

3.   Gluonic Hadrons: Phenomenology

The status of candidates produced in $\psi \rightarrow \gamma x$ has been discussed by Eigen here, so I will not duplicate that. There has been an interesting analysis of $\pi\pi$ phase shifts recently[14] which draws heavily on good data for central $\pi\pi$ production in pp $\rightarrow$ pp$\pi\pi$, and claims that a 0^{++} glueball may be hidden under the $f_o(975)$. Data at higher energy (Sp$\bar{\text{p}}$S collider) on p$\bar{\text{p}} \rightarrow$ pp could help confirm this.

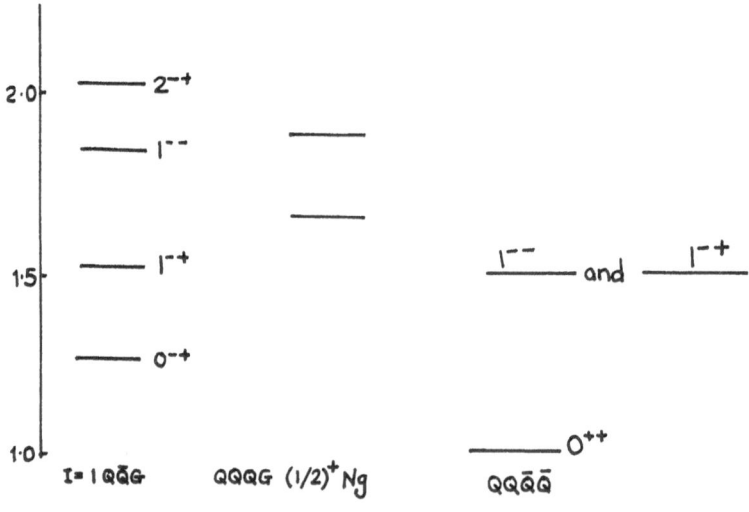

Fig. 1. $q\bar{q}B/q^3$ $B/q^2\,\bar{q}^2$ comparison

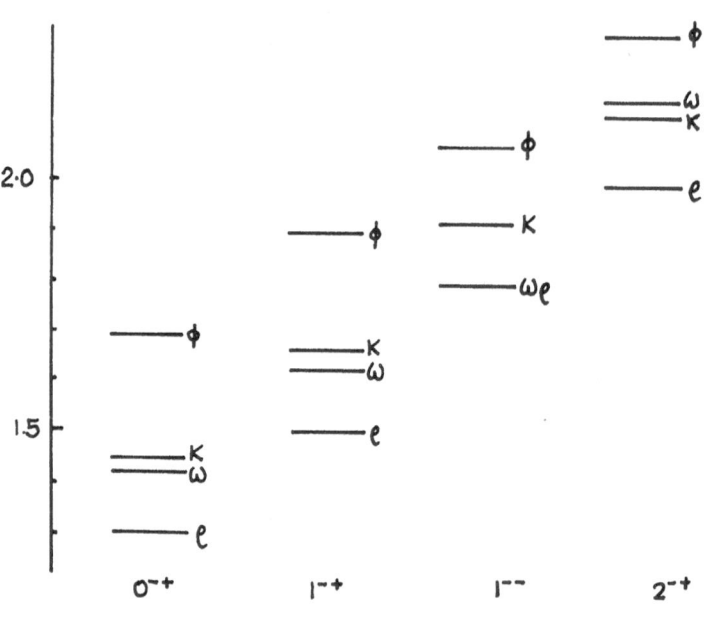

Fig. 2. $q\bar{q}$ flavour multiplets

There is a general problem in the 0^{++} sector that states around 980 MeV ($a_0(980)$ and $f_0(975)$) are not yet well understood. They may be $q\bar{q}$ with L=1, or $q^2\bar{q}^2$ or $K\bar{K}$ molecules. Until we are certain about the conventional $q\bar{q}$ spectroscopy we can have little hope of identifying gluonic states.

A direct probe of the flavour content of hadrons is to use photons. $\gamma\gamma$ and $\phi \to \gamma x$ can both help in the case of these 0^{++} states. The $\gamma\gamma^*$ (one photon off shell) couplings are now beginning to teach us about the 1^{++} states around 1300–1500 MeV. These can also be interesting in view of the possible GAMS 1^{-+} exotic in this region[15] and which, in principle, may couple to $\gamma\gamma^*$ too.

3.1   Messages from $\gamma\gamma$ Couplings: 2^+ 1^+ 0^+

For a given J^{PC}, the flavour dependence of the $\gamma\gamma$ widths is

$$(I=1) : n\bar{n} : s\bar{s} = 9 : 25 : 2 . \qquad (3.1)$$

A comparison of a_2 and $f_2(1270)$ is immediate because their masses are very similar and we may justifiably ignore any mass dependences. Data in KeV give

$$(I=1) : n\bar{n} : s\bar{s}=0.904 \pm 106 : 2.86 \pm 0.14 : (0.09 \pm 0.02)/(BR \to K\bar{K}) \qquad (3.2)$$

When comparing states with different masses the phase space behaves as m_H^3 but the coupling to the quarks gives a suppression m_q^{-2}. To the extent that $m_H(q\bar{q}) \sim m_q$ we expect

$$\Gamma(s\bar{s} \to \gamma\gamma) = \frac{2}{25} \frac{m_{s\bar{s}}}{m_{n\bar{n}}} \Gamma(n\bar{n} \to \gamma\gamma) \qquad (3.3)$$

which is well satisfied if $BR(f' \to K\bar{K}) \simeq 1$

In the $q\bar{q}$ L=1 states of the same flavour one predicts

$$\frac{\Gamma(\chi_2 \to \gamma\gamma)}{m_2^3} : \frac{\tilde{\Gamma}(\chi_1 \to \gamma\gamma^*)}{m_1^3} : \frac{\Gamma(\chi_0 \to \gamma\gamma)}{m_0^3} = 1: \frac{5}{3} : \frac{15}{4} \qquad (3.4)$$

53

where $\tilde{\Gamma}(\chi_1 \rightarrow \gamma\gamma^*) \equiv \frac{M^2}{Q^2} \Gamma(\chi_1 \rightarrow \gamma\gamma)$

So we expect

$$1^{++} : n\bar{n} \qquad \tilde{\Gamma}(\chi_1(n\bar{n}) \rightarrow \gamma\gamma) = (\frac{m_1}{m_2})^3_{n\bar{n}} \quad (4.77 \pm 0.23) \text{ KeV} \qquad (3.5)$$

$$s\bar{s} \qquad \tilde{\Gamma}(\chi_1(s\bar{s}) \rightarrow \gamma\gamma) = (\frac{m_1}{m_2})^3_{s\bar{s}} \quad \frac{(0.15 \pm 0.03)}{(BRf' \rightarrow K\bar{K})} \text{ KeV} \qquad (3.6)$$

$$0^{++} \qquad n\bar{n} \qquad \Gamma(\chi_0(n\bar{n}) \rightarrow \gamma\gamma) = (\frac{m_0}{m_2})^3_{n\bar{n}} \quad (1.07 \pm 0.45) \text{ KeV} \rightarrow 5.22 \pm 0.25 \text{ KeV}$$

$$\tag{3.7}$$

$$\Gamma(\chi_0(s\bar{s}) \rightarrow \gamma\gamma) = (\frac{m_0}{m_2})^3_{s\bar{s}} \quad \frac{(0.33 \pm 0.06)}{(BRf' \rightarrow K\bar{K})} \text{ KeV} \qquad (3.8)$$

$$\Gamma(\chi_0(L=1) \rightarrow \gamma\gamma) = (\frac{m_0}{m_2})^3 \quad (3.4 \pm 0.4) \text{ KeV} \rightarrow 1.55 \pm 0.19 \text{ KeV}$$

$$\tag{3.9}$$

where the widths quoted for the 0^{++} apply for $m_0 = 1$ GeV.

These results are interesting for the 0^+ sector where there has been much argument about the nature of the (S^*) $f_0(975)$ and (δ) $a_0(980)$ states. Are they $q^2\bar{q}^2$ [16], $K\bar{K}$ molecules [17] or simple L=1 $q\bar{q}$ states? In the latter case our above calculations apply.

Experimental data are rather far from these magnitudes.

The data are

$\Gamma(f_0(975) \rightarrow \gamma\gamma)$

$$\frac{<0.8 \text{ KeV}}{(BR(f_0 \rightarrow \pi\pi)} \qquad \text{Crystal Ball ref 18}$$

$$<0.8 \text{ KeV} \qquad \text{JADE ref 19}$$

$\Gamma(a_0(980) \rightarrow \gamma\gamma)$

$$\frac{0.19 \pm 0.07 \pm 0.10}{BR(a_0 \rightarrow \eta\pi)} \text{ KeV} \quad \text{Crystal Ball ref 20}$$

Barnes has discussed the implications of these data in some detail.

Both $q^2 \bar{q}^2$ and $K\bar{K}$ assignments lead to a ratio of unity for $\Gamma(f_0 \to \gamma\gamma)/\Gamma(a_0 \to \gamma\gamma)$, and the molecule picture predicts some 0.6 KeV for the individual widths. (The $q^2 \bar{q}^2$ widths are rather model dependent). A complimentary, and potentially clear cut, way of distinguishing among these three possibilities is to study $e^+e^- \to \phi \gamma_{a_0}$, γf_0 at a dedicated facility.

The ratios are predicted to be[21]

$$\Gamma(\phi \to \gamma_{a_0})/\Gamma(\phi \to \gamma f_0) = \begin{array}{cc} 0 & q\bar{q} \\ 1 & K\bar{K} \\ 9 & q^2 \bar{q}^2 \end{array} \qquad (3.10)$$

The absoloute branching ratio[21] is expected to be $O(10^{-5})$ for $q\bar{q}$ but could be as high as $\frac{1}{2}\%$ for $K\bar{K}$!

The prediction for the absolute widths of the χ_1 states is rather delicate in that $\chi_1 \not\to \gamma\gamma$ for a pair of real photons. Thus one obtains data for $\chi_1 \to \gamma\gamma^*(Q^2)$ and defines $\tilde{\Gamma} \equiv \frac{M^2}{Q^2}\Gamma$ in the low Q^2 limit and assumes that the residual Q^2 dependence is contained in a ρ-dominated form factor.

Gidal et al[22] report

$$\tilde{\Gamma}(f_1(1285) \to \gamma\gamma) = 8.2 \pm 2.2 \pm 1.5 \text{ KeV}$$

$$\tilde{\Gamma}(f_1(1423) \to \gamma\gamma) = 2.7 \pm 1.2 \pm 0.5 \text{ KeV}$$

$$\frac{\tilde{\Gamma}(f_1(1285) \to \gamma\gamma)}{\tilde{\Gamma}(f_1(1420) \to \gamma\gamma)} \equiv 3.0 \pm 1.6 \qquad (3.11)$$

(Aihara et al[5] quote $\tilde{\Gamma}(f_1(1423) \to \gamma\gamma) = 6 \pm 2 \pm 2 \text{ KeV})$

These absolute widths are rather larger than expected from the input 2^{++} widths. This may be the fault of assuming the naive ρ-dominance form factor. The ratio at eq (3.11) supports the $f_1(1285)$ as dominantly $n\bar{n}$

with $f_1(1420)$ s$\bar{\text{s}}$ but the result is far from the $\frac{25}{2}$ expected for ideal states. If the nonet deviates from ideal such that

$$f_1(1420) = \cos\lambda \, |s\bar{s}> \, - \, \sin\lambda \, |n\bar{n}>$$

then $\lambda = -14^0 + \begin{smallmatrix} 5^0 \\ - 10^0 \end{smallmatrix}$ can accommodate the observed ratio.

An exotic $J^{PC} = 1^{-+}$?

However these results do not seem to fit in with the signals[23] coming from $\psi \rightarrow \omega(K\bar{K}\pi)$, $\phi(K\bar{K}\pi)$ when $m(K\bar{K}\pi) \simeq 1440$. The angular distributions do not look like 0^{-+} and suggest $J^C = 1^+$. If it is the $f_1(1420)$ that is seen apparently recoiling against ω but not ϕ then this suggests that it is predominantly $n\bar{n}$ that is emerging in the $\psi \rightarrow \omega f_1$.

There is a growing feeling among many who have been following the emergence of these data that there are several states in this mass region (see e.g. Caldwell ref 24). However two $J^C = 1^+$ are hard to understand in a $q\bar{q}$ picture. This has led Chanowitz to point out that $q\bar{q}G$ has been predicted with $J^{PC} = 1^{-+}$ in this region and he proposes that this could be the state produced in $\gamma\gamma$.

Gidal et al[22] verify that $f_1(1285)$ is indeed 1^{++} but the $f_1(1423)$ is either 1^{-+} or 1^{++}. There may be a problem in claiming a $q\bar{q}G$ 1^{-+} in $\gamma\gamma$ for one would expect a significant 0^{-+} signal too, but none is seen.

Quite unrelated to this conundrum there have been claims for $J^{PC} = 1^{-+}$ in $\pi^- p \rightarrow \eta\pi n$. The GAMS collaboration[15] plot the cross section as a function of $m(\eta\pi)$ in the Gottfried Jackson frame and distinguish the two hemispheres where the π is produced forward or backward (relative to the direction of the incoming π beam). There is an excess of events around 1500 MeV when the η is produced forward; this asymmetry implying a P-wave signal, thus 1^{-+}. It is not yet established whether or not this is resonant.

If established as a resonance this would be the first clear signal for a state "beyond $q\bar{q}$". But this would in turn raise the question of whether it is a hybrid $q\bar{q}G$ or a multiquark state $q^2\bar{q}^2$ in P wave excitation. This latter possibility becomes interesting in view of a 1^{--} $\phi\pi^0$ state seen[25] in the $\pi^- p \rightarrow \phi\pi^0 n$ with mass 1480 ± 40 MeV. The decay into $\omega\pi$ is not seen and the authors suggest that this state is a $s\bar{s}(u\bar{u} - d\bar{d})$ state previously predicted[26].

The suggestive feature is the possible degeneracy of this 1^- and the GAMS 1^{-+} signal. This may suggest a common $q^2\bar{q}^2$ heritage[27].

This arises because the $q^2\bar{q}^2$ states are not eigenstates of G-parity, and go into one another under G. This can be shown by writing the four-particle states in terms of quark pairs in a definite configuration; e.g. $|(qq;^1S_0)\rangle$ and $|(qq;^1P_1)\rangle$ and similarly for the two antiquarks.

$$|(uu;^1P_1,I=1)(\bar{u}\bar{d};^1S_0,I=0)\rangle = G|(du;^1S_0,I=0)(\bar{d}\bar{d};^1P_1,I=1)\rangle \qquad (3.12)$$

where we have chosen the simplest angular momentum, color and isospin couplings which can give the desired quantum numbers for the four quark state; namely spin zero and antisymmetric color couplings for each pair, the two pairs in a relative s wave and one unit of internal orbital angular momentum in one pair. Both states in eq (3.12) are individually isospin eigenstates. They differ only in whether the quarks are in $^1S_0$ and the antiquarks in $^1P_1$ or vice versa. A similar relation holds for the two isovector states involving a strange quark pair

$$|(us;^1P_1)(\bar{s}\bar{d};^1S_0)\rangle = G|(us;^1S_0)(\bar{s}\bar{d};^1P_1)\rangle \qquad (3.13)$$

We can therefore construct the G eigenstates

$$|(un\bar{n}\bar{d};^1S_0,^1P_1,G=\pm)\rangle = |(uu;^1P_1,\bar{u}\bar{d};^1S_0)\rangle \pm |(du;^1S_0)(\bar{d}\bar{d};^1P_1)\rangle \qquad (3.14)$$

$$|(us\bar{s}\bar{d};^1S_0,^1P_1,G=\pm)\rangle = |(us;^1P_1,\bar{s}\bar{d};^1S_0)\rangle \pm |(us;^1S_0)(\bar{s}\bar{d};^1P_1)\rangle \qquad (3.15)$$

where n denotes nonstrange quark. The C which has $J^{PC} = 1^-$ for its neutral member and $J^{PG} = 1^{-+}$ for the multiplet is described by a linear combination of the states (3.14) and (3.15) with even G. But there is also the orthogonal linear combination with odd G which has the GP exotic quantum numbers $J^{PG} = 1^{--}$ for the multiplet and the CP exotic quantum numbers $J^{PC} = 1^{-+}$ for its neutral member. One might expect that the two states of the G doublet be nearly degenerate. It is therefore very interesting that a candidate for the G-partner of the C has now been found in an entirely different experiment which was not looking for a partner to the C. The exotic $\eta\pi$ resonance reported by GAMS at 1400 MeV has exactly the properties required for the G-doublet companion of the Serpukhov state.

4.   Gluonic Hadrons: Decays

4.1   Decays of hybrid $q\bar{q}G$ and multiquark $q^2\bar{q}^2$ states

Predictions for the relative branching ratios for $\eta\pi$ and $\eta'\pi$ decays of a $J^{PC} = 1^{-+}$ state are very different in $qq\bar{q}\bar{q}$ or $q\bar{q}G$ models. For the hybrid state one expects that $\eta\pi$ is suppressed relative to $\eta'\pi$. The dominant diagram is where the gluon in the hybrid creates $q\bar{q}$ in a colour octet state and no other gluons are exchanged.

$$|uG\vec{d}\rangle \rightarrow |u(q\bar{q})_8\vec{d}\rangle \rightarrow |(u\bar{q})_1(q\bar{d})_1\rangle \rightarrow |M_{u\bar{q}}(\vec{k}); M_{q\bar{d}}(-\vec{k})\rangle \qquad (4.1)$$

where q may be either a u or d quark, and $M_{u\bar{q}}(\vec{k})$ and $M_{q\bar{d}}(-\vec{k})$ denote mesons with quark constituents $u\bar{q}$ and $q\bar{d}$ with momenta $\vec{k}$ and $-\vec{k}$ respectively. Isospin symmetry requires the gluon to create the u and d pairs with equal amplitude. Thus

$$A\{(uG\bar{d}) \rightarrow M_{u\bar{d}}(k)M_{d\bar{d}}(-k)\} = A\{(uG\bar{d}) \rightarrow M_{u\bar{u}}(k)M_{u\bar{d}}(-k)\} \qquad (4.2)$$

The final pion has momenta $\vec{k}$, $-\vec{k}$ on the left and right hand sides respectively. This momentum reversal introduces a phase which depends upon the orbital angular momentum in decays of states with a definite L. Thus the decay amplitudes at a given value of L for the $u\bar{u}$ and $d\bar{d}$ components of the η and η' satisfy

$$A\{(uG\bar{d})_2 \rightarrow \pi^+\eta_{d\bar{d}}\} = PA\{(uG\bar{d})_2 \rightarrow \pi^+\eta_{u\bar{u}}\} \qquad (4.3)$$

where $P=(-)^L$ is the parity of the initial and final states. Thus far any isoscalar state η_i

$$A\{(uG\bar{d})_L \rightarrow \pi^+\eta_i\} = (1+p)\langle\eta_i|\eta_{u\bar{u}}\rangle A\{(uG\bar{d})_L \rightarrow \pi^+\eta_{u\bar{u}}\} \qquad (4.4)$$

which vanishes for all odd parity states.

Thus we appear to have eliminated both $\pi\eta$ and $\pi\eta'$. However, there is the possibility that the $G \rightarrow (q\bar{q})_8 \rightarrow (q\bar{q})_1$ by exchanging additional gluons but remaining a flavour singlet which then turns into η via the flavour singlet component.

$$|(uG\bar{d}> + |u(q\bar{q})_8\bar{d}> + |u(q\bar{q})_1\bar{d}> + |(q\bar{q})_1> |(q\bar{q})_1(u\bar{d})_1> |M_{q\bar{q}}(k)M_{u\bar{d}}(-k)>$$

$$(4.5)$$

This requires the hybrid → ηπ and η'π decays to go via the SU(3) singlet components of the η and η', thereby favouring the η':

$$A_8(q\bar{q}G \to \eta\pi) < A_8(q\bar{q}G \to \eta'\pi)$$

In the $q^2\bar{q}^2$ model the φπ and GAMS states are siblings. The production appears to be by π exchange: ππ → C → φπ and also B(C → φπ) > B(C → ωπ) suggest that the C is in $10-10^*$ flavour SU(3) representation – naturally leading to an additional state with the quantum numbers of the exotic ηπ GAMS resonance. The $10 \not\supset 8 \times 1$ and so this would feed the $\eta_8\pi$, in contrast to the q̄qG which feeds η, π. Here

$$A_{10}(qq\bar{q}\bar{q} \to \eta\pi) > A_{10}(qq\bar{q}\bar{q} \to \eta'\pi)$$

$$A_8(q\bar{q}G) \to \eta\pi) < A_8(q\bar{q}G \to \eta'\pi)$$

$$(4.6)$$

where $A_{8,10}$ denote the decay amplitudes without corrections for phase space. Hence we may be able to identify the nature of the GAMS state this way.

4.2   Decays of C = + glueballs: an anti-selection rule

A naive first orientation has been to argue that as gluons are flavourless they produce all flavours with equal facility, and hence have flavour symmetric decays. However we know from QCD inspired fits to hadron spectroscopy that single gluon exchange can exhibit dependence on the mass of the quarks to which it couples (the m^{-1} familiar in magnetic coupling) so one should expect that the coupling of timelike gluons to strange quarks, say, will differ from that to up or down. In the cavity perturbation theory inspired by the MIT bag such a flavour dependence occurs and has been taken by some to imply that, in certain circumstances, $s\bar{s}$ production may be **enhanced** relative to $u\bar{u}$ or $d\bar{d}$.

This may tell us more about the bag model than about nature! What

I intend to do here is allow g → s̄s to differ from g → d̄d without specifying whether it is enhanced or suppressed. Interesting selection rules still obtain and, insofar as they do not depend on the relative size or sign of the s̄s/d̄d coupling, may be hoped to have validity more general thay any particular model.

The analysis follows Lipkin[28] and extends it to obtain selection rules that may be used to discriminate against gluonic states.

Consider a c = + glueball - in particular a two gluon state where I label the gluons a and b to distinghish them. For simplicity I will restrict myself initially to the W diagram (fig 3) but the remarks generalise to the U topology (fig 4). As I will need to distinguish the quarks (e.g. s̄s versus ūu, d̄d) I have highlighted one pair - when necessary I will call these the strange quarks.

Depending on the G parity of the meson final state one adds (G = +) or subtracts (G = -) the two amplitudes in fig 3 . Thus

$$G \rightarrow \pi\pi, \ K\bar{K} \sim A + B$$
$$G \rightarrow \pi\rho, \ KK^* \sim A - B$$

$$(4.7)$$

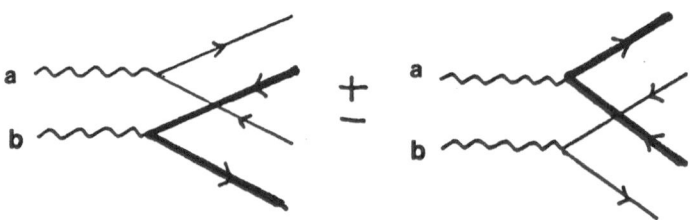

Fig. 3. W diagram for gg decay

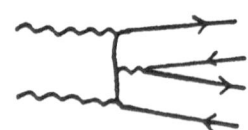

Fig. 4. U diagram for gg decay

If the amplitudes for

$$g_a \to q\bar{q} \equiv g_b \to q\bar{q} \qquad\qquad (4.8)$$

then $A \equiv B$ and the $\pi\rho$, KK^* decays will be forbidden. This is the standard G parity selection rule.

Now relax the constraint (4.8) in the case of strange quarks and write for the amplitudes

$$\frac{g_{a,b} \to s\bar{s}}{g_{a,b} \to d\bar{d} \text{ or } u\bar{u}} = s_{a,b}$$

Thus the branching ratios (apart from phase space effects) are

$$\frac{G \to K^+K^-}{G \to \pi^+\pi^-} = \left(\frac{s_a + s_b}{2}\right)^2$$

$$\frac{G \to KK^*}{G \to K\bar{K}} = \left(\frac{s_a - s_b}{s_a + s_b}\right)^2$$

So if $s_a \equiv s_b \equiv s$ we have Lipkin's result that

$$\frac{G \to K\bar{K}}{G \to \pi\pi} = s^2 \quad ; \quad G \to KK^* = 0$$

namely that KK^* is still forbidden even in presence of symmetry breaking.

Lipkin argued that $s_a = s_b$ since in the glueball rest frame the two gluons must balance momenta and, in some sense, be "equivalent". However in particular models this need not be the case. For example in cavity-bag models the 0^{-+} states involve one TE and one TM mode, whereas the 2^{++} for example involves gluons in the same mode, $(TE)^2$ or $(TM)^2$. Hence one's reaction to seeing a KK^* final state may depend upon the J^{PC} of the initial glueball candidate.

$$0^{-+}, \; 2^{-+} \to KK^* \quad (s_a \neq s_b)$$
$$0^{++}, \; 2^{++} \to KK^* \quad (s_a = s_b)$$

Thus $\iota \to KK^*$ could be compatible with ι being gluonic. However $f_2(1720)$ $\to KK^*$ would argue against a gluonic candidacy here. The ability to detect and identify K's (such as the Obelix experiment at LEAR) is therefore important.

Final states involving η and/or η' are not well studied and there is the possibility that our failure to find glueballs is because they have a propensity to decay into these channels. Indeed there are claims by the GAMS collaboration[29] that they see new states in $\eta\eta$ and $\eta\eta'$ channels.

I will work in the approximation that η and η' contain equal amounts of strange and nonstrange quarks ($n\bar{n} \equiv \frac{1}{\sqrt{2}}(u\bar{u} + d\bar{d})$

$$\eta = \frac{1}{\sqrt{2}}(n\bar{n} - s\bar{s})$$

$$\eta' = \frac{1}{\sqrt{2}}(n\bar{n} + s\bar{s})$$

In this approximation final states consisting of $K\bar{K}$, $\eta\eta$, $\eta'\eta'$ or $\eta\eta'$ each are 50:50 mixtures of strange and non-strange quarks, the flavours are merely distributed among the hadrons differently. So any symmetry breaking effects associated with the strange/non-strange quark production will be common to all and rather general conclusions may be anticipated.

Specialising to the case when $s_a \equiv s_b = s$ we have for the rates

$$\frac{G \to \eta\eta'}{G \to \eta\eta + \eta'\eta'} = \left(\frac{1-s^2}{1+s^2}\right)^2 \tag{4.9}$$

$$\frac{G \to \eta\eta}{G \to K^+K^-} = \frac{1}{8}\left(\frac{1}{s^2} + 2 + s^2\right) \tag{4.10}$$

which in the symmetry limit $s=1$ recover the familiar results

$$\eta\eta' \neq 0 \quad ; \quad \eta\eta = \frac{1}{2}K^+K^- = \frac{1}{4}K\bar{K}$$

The results, eqs (4.9,4.10) are plotted in fig 5. Notice that they are invariant under $s \leftrightarrow 1/s$ and so are true whether strange $q\bar{q}$ are suppressed or enhanced relative to non-strange. Notice also that

62

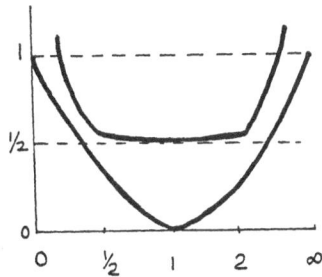

Fig. 5.   Ratio of $\eta\eta$ production (upper curve) and $2\eta'\eta/\eta\eta + \eta'\eta'$ (lower curve) plotted against s, the strange to nonstrange production ratio.

symmetry breaking always enhances $\eta\eta$ relative to $K\overline{K}$. This is because one of the $q\overline{q}$ productions is disfavoured relative to the other (suppose it is $s\overline{s}$). The $\eta\eta$ production can be fed by the "easy" route ($n\overline{n}$ in this case), whereas the $K\overline{K}$ necessarily involves both pairs to be present – hence paying a penalty for the disfavoured contribution.   Confronting eq (4,10) with data requires both good K identification (as in Obelix) and good neutral detection (as in the Crystal Barrel). So experiments at LEAR may be able to offer something when their results are compared.

The anaylsis has concentrated on the W diagram only. It holds more generally though. If there is intrinsic glue in η, η' this will enhance their production from glueballs, reinforcing the result that

$$G \to \eta\eta > \tfrac{1}{2} K^+ K^-$$

and bringing

$$G \to \eta\eta' < \tfrac{1}{2} (\eta\eta + \eta'\eta')$$

towards equality. In particular Gershtein et al.[30] have suggested that a large $\eta\eta'$ signal could be an indicator of glueball decay.

References

1.   P MacKenzie and G Lepage, Phys. Rev. Lett. 47, 1244 (1981). S. Brodsky, P. MacKenzie and G. Lepage, Phys. Rev. D28, 228 (1983).
2.   K. Konigsman, Physics Reports 139, 244 (1986).
3.   J. Perrier, SLAC-PUB-3436 (1984).

4. F.E. Close, An Introduction to Quarks and Partons, Academic Press 1978.

5. H. Aihara et al, Phys. Rev. Lett. 57, 51 (1986).

6. M. Krammer, Phys. Lett. 74B, 361 (1978).

7. C. Edwards et al, Phys. Rev. Lett. 49, 259 (1982)
    D. Scharre et al, Phys. Lett. 97B, 329 (1980); Phys. Rev. D23, 43 (1981).
    D. Coffman et al, SLAC-PUB-3720 (1986).

8. A detailed discussion may be found in "Gluonic Hadrons", F.E. Close to appear in Reports on Prog. in Physics.

9. T. de Grand et al, Phys. Rev. D12, 2060 (1975).

10. T. Barnes and F. Close, Phys. Lett. 116B, 365 (1982); ibid 123B, 89 (1983); 128B, 277 (1983).

11. T. Barnes, F. Close and F. de Viron, Nucl. Phys. B224, 241 (1983).

12. T. Barnes, F. Close and S. Monoghan, Phys. Lett. 110B, 159 (1982); Nucl. Phys. B198, 380 (1982).

13. M. Chanowitz and S. Sharpe, Nucl. Phys. B222, 211 (1983).

14. K. Au, D. Morgan and M. Pennington, Phys. Lett. 167B, 229 (1986).

15. M. Boutemeur, to be published in Hadrons, Quarks and Gluons, Proc. of XXII Rencontre de Moriond (1987);
    F. Binon, these proceedings.

16. R.L. Jaffe, Phys. Rev. D15, 281 (1977).

17. J. Weinstein and N. Isgur, Phys. Rev. Lett. 48, 659 (1982); Phys. Rev. D27, 588 (1983).

18. C. Edwards et al, Phys. Lett. 110B, 82 (1982).

19. J. Olsson, Proc. of Vth International Workshop on $\gamma\gamma$ Collisions, Aachen 1983 (Lecture Notes in Physics, 191, 45).

20. D. Antreasyan et al, Phys. Rev. D33, 1847 (1986).

21. F. Close and N. Isgur (in preparation).

22. G. Gidal et al SLAC-PUB-4274/4275 (1987).

23. L. Kopke (Mark III), Proc. 21st Rencontre de Moriond, Les Arcs (1986) Santa Cruz report SCIPP 86/61
    U. Mallik, SLAC-PUB-3946 (1986)
    S. Cooper, Proc. XXIII Internat. Conf. on High Energy Physics, Berkeley 1986; SLAC-PUB-4139.

24. D.O. Caldwell, "A possible solution to the E/i puzzle", Santa Barbara preprint UCSB-HEP-86-5.

25. S.I. Bityukov et al, Phys. Lett. B188, 383 (1987).

26. F.E. Close and H.J. Lipkin, Phys. Rev. Lett. 41, 1263 (1978).

27. F.E. Close and H.J. Lipkin WIS/87/8; RAL-87-046.

28. H.J. Lipkin, Phys. Lett. 106B, 114 (1981).

29. F. Binon et al, Nuovo Cimento 78A, 313 (1983).

30. S. Gershtein et al, Z. Phys. C24, 305 (1984).

SPIN-DEPENDENT FORCES BETWEEN QUARKS

Dieter Gromes

Institut für Theoretische Physik
Philosophenweg 16, D-6900 Heidelberg

ABSTRACT

We review our understanding of spin-dependent forces in quarkonia. The emphasis lies on an elementary presentation of the theoretical methods, involving a generalization of the Wilson loop formalism. Several open problems are pointed out.

1. A SHORT REVIEW OF HISTORY AND PHENOMENOLOGY

Spin-dependent forces between quarks show up everywhere in hadron spectroscopy - in old and new mesons as well as in baryons. The connection between theory and experiment is particularly close. Any theoretical ansatz can be directly inserted into the corresponding matrix element and the result immediately confronted with experiment. Let us begin with a general remark on the quark-antiquark potential. There are essentially **three regions** to be distinguished:

1) The region of **small distances:** This is the range of asymptotic freedom and perturbation theory. **One-gluon exchange** dominates, either with an effective fixed coupling constant or a **running coupling constant**. This region appears well understood, we will briefly discuss it in the following.

2) The region of large distances: This is the confinement regime where perturbation theory does not make sense any longer but where arguments originating from the strong coupling expansion of lattice gauge theory or string (flux tube) models lead to a consistent picture. The understanding of spin-dependent forces in this region will be the central part of this lecture.

3) The intermediate region. Here theory has very little to say; except some general theorems, quoted later, stating that the transition from small to large distances has to be smooth. It is embarassing that this intermediate region is the one relevant for spectroscopy. This is essentially clear from elementary quantum mechanics: The reduced wave function has to vanish for $r = 0$ and goes down exponentially for $r \to \infty$, therefore it tests essentially the intermediate region. The often quoted figure from Buchmüller and Tye[1] shows how various successful potentials are almost identical there while looking quite different outside.

We start by recalling how an effective potential can be extracted from a graph like one-gluon exchange between quark and antiquark in a meson. Besides the rather complicated nonrelativistic reduction of the Bethe-Salpeter equation there is a much simpler way leading to the same result: Start with a perturbative calculation or guess for the $q\bar{q}$ scattering amplitude and perform the non-relativistic limit (plus the relativistic corrections of order $1/c^2$). The effective $q\bar{q}$ potential (with relativistic corrections) is then obtained as spatial Fourier transform of the scattering amplitude with respect to momentum transfer. The last step is nothing but the inversion of the first Born approximation which, as is well known, gives the scattering amplitude as Fourier transform of the potential. In this way one photon exchange, considered up to order $1/m^2$, gives the well-known Breit-Fermi Hamiltonian. One gluon exchange between quark and antiquark in a color singlet state gives the same result up to the replacement $\alpha \to (4/3)\alpha_s$:

$$
\begin{aligned}
V = \frac{4}{3}\alpha_s \Big\{ & -\frac{1}{r} \\
& + \frac{1}{2r^3}\left(\frac{[\vec{r}\times\vec{p_1}]\cdot\vec{s_1}}{m_1^2} - \frac{[\vec{r}\times\vec{p_2}]\cdot\vec{s_2}}{m_2^2}\right) \\
& + \frac{1}{m_1 m_2 r^3}\left([\vec{r}\times\vec{p_1}]\cdot\vec{s_2} - [\vec{r}\times\vec{p_2}]\cdot\vec{s_1}\right) \\
& + \frac{8\pi}{3m_1 m_2}\delta^{(3)}(\vec{r})\vec{s_1}\cdot\vec{s_2} \\
& + \frac{3}{m_1 m_2 r^3}\left(\frac{(\vec{s_1}\cdot\vec{r})(\vec{s_2}\cdot\vec{r})}{r^2} - \frac{1}{3}\vec{s_1}\cdot\vec{s_2}\right) \\
& + \frac{\pi}{2}\left(\frac{1}{m_1^2} + \frac{1}{m_2^2}\right)\delta^{(3)}(\vec{r}) \\
& + \frac{1}{2m_1 m_2}\left(\vec{p_1}\frac{1}{r}\vec{p_2} + (\vec{p_1}\vec{r})\frac{1}{r^3}(\vec{r}\vec{p_2})\right) \Big\}.
\end{aligned}
\tag{1.1}
$$

Here $\vec{r} = \vec{r_1} - \vec{r_2}$. One recognizes the static Coulomb potential, then there are two different types of spin orbit terms, the first one proportional to the inverse squares of the masses and describing an interaction between the spin and the orbital angular momentum of a constituent, the second one proportional to the inverse product of the masses and describing the interaction of the spin of one constituent with the orbital angular momentum of the other. Finally there is the well-known contact spin-spin interaction and the tensor interaction, both of them proportional to $1/m_1 m_2$. They describe the interactions of the magnetic moments. There are also two spin-independent corrections, the Darwin term and the current current (or orbit orbit term) which will not be discussed in the following.

We next come to the long-range potential. From the strong coupling expansion of lattice gauge theory (Wilson's area law) one favors an asymptotically linear confining potential[2]. In the real world at large distances the creation of a quark-antiquark pair becomes possible thus breaking the color flux and screening the potential. The question arises how to calculate the relativistic spin-dependent corrections originating from the long-range potential. In the very early papers on the subject these corrections were simply ignored. The next simple assumption[3] was to replace the one gluon exchange by a general vector exchange, i.e. to use a dressed instead of a bare gluon propagator:

$$
\gamma^\mu \frac{1}{q^2}\gamma_\mu \Rightarrow \gamma^\mu D(q^2)\gamma_\mu.
\tag{1.2}
$$

The static potential $\varepsilon(r)$ then becomes the Fourier transform of $D(-\vec{q}^{\,2})$. The factors involving the momentum transfer $\vec{q}$ which originate from the expansion of the spinors up to order $1/m^2$ become differential operators in position space. In this way all spin dependent potentials can be given in terms of the static potential $\varepsilon(r)$:

$$
\begin{aligned}
V_{\text{vector}} = {}& \varepsilon(r) \\
&+ \frac{\varepsilon'(r)}{2r}\left(\frac{[\vec{r}\times\vec{p}_1]\cdot\vec{s}_1}{m_1^2} - \frac{[\vec{r}\times\vec{p}_2]\cdot\vec{s}_2}{m_2^2}\right) \\
&+ \frac{\varepsilon'(r)}{m_1 m_2 r}([\vec{r}\times\vec{p}_1]\cdot\vec{s}_2 - [\vec{r}\times\vec{p}_2]\cdot\vec{s}_1) \\
&+ \frac{2}{3m_1 m_2}\Delta\varepsilon(r)\vec{s}_1\cdot\vec{s}_2 \\
&+ \frac{1}{m_1 m_2}\left(\frac{\varepsilon'(r)}{r} - \varepsilon''(r)\right)\left(\frac{(\vec{s}_1\cdot\vec{r})(\vec{s}_2\cdot\vec{r})}{r^2} - \frac{1}{3}\vec{s}_1\cdot\vec{s}_2\right)
\end{aligned}
\tag{1.3}
$$

+spin independent corrections.

The general structure is identical to the Breit-Fermi Hamiltonian and contains the latter as a special case. There is, however, an elegant and convincing argument due to Schnitzer[4] showing that the splittings of the triplet P states of heavy quarkonia cannot be obtained with an effective vector exchange. Consider the ratios of the spittings of these states due to spin orbit and tensor (the spin-spin term gives a common contribution to the ^3P states and therefore does not split them). For a vector potential of the form

$$
\varepsilon(r) = -\frac{4}{3}\frac{\alpha_s}{r} + \lambda r + c
\tag{1.4}
$$

one obtains for the equal mass case (M_j is the mass of the ^3P_j state)

$$
r \equiv \frac{M_2 - M_1}{M_1 - M_0} = \frac{1}{5}\frac{8\alpha_s\langle r^{-3}\rangle + 7\lambda\langle r^{-1}\rangle}{2\alpha_s\langle r^{-3}\rangle + \lambda\langle r^{-1}\rangle}.
\tag{1.5}
$$

From this one has

$$
0.8 \le r \le 1.4
\tag{1.6}
$$

where the lower bound corresponds to a pure Coulomb potential and the upper bound to a purely linear potential. Note that the argument does not depend on the form of the wave function nor on the values of the parameters. Even if the potential is not exactly Coulomb plus linear the ratio r cannot lie far outside the region given above. The present data for charmonium and the two systems of P–states for bottomium are

$$
\begin{aligned}
r_c &= 0.48 \pm 0.01 \\
r_b &= 0.65 \pm 0.05 \\
r_{b'} &= 0.57 \pm 0.07.
\end{aligned}
\tag{1.7}
$$

They are in clear contradiction to the inequality (1.6) implying that the longe range potential cannot originate from a pure vector exchange.

For more than ten years an effective scalar[5] (where the γ^μ-matrices in (1.2) are replaced by 1) has been the most popular ansatz. The structure of the spin dependent corrections is now much simpler because there are no contributions $\sim 1/m_1 m_2$ which in the vector case came from the spatial components of the γ-matrices. There is neither a spin spin nor a tensor interaction but only one type of spin orbit term:

$$V_{\text{scalar}} = \varepsilon(r) - \frac{\varepsilon'(r)}{2r} \left(\frac{[\vec{r} \times \vec{p}_1] \cdot \vec{s}_1}{m_1^2} - \frac{[\vec{r} \times \vec{p}_2] \cdot \vec{s}_2}{m_2^2} \right). \tag{1.8}$$

For the ratio r, assuming a Coulomb vector and a linear scalar potential one now has

$$r = \frac{1}{5} \frac{8\alpha_s \langle r^{-3} \rangle - \frac{5}{2} \lambda \langle r^{-1} \rangle}{2\alpha_s \langle r^{-3} \rangle - \frac{1}{4} \lambda \langle r^{-1} \rangle} \tag{1.9}$$

leading to

$$
\begin{array}{lll}
r \leq 0.8 & \text{(if the Coulomb contribution dominates) or} & \\
r \geq 2 & \text{(if the linear contribution dominates).} &
\end{array} \tag{1.10}
$$

Most papers (and practically all serious papers) have used an effective scalar for the confining potential, resulting in a good fit to the experimantal spectrum.

One may ask what would happen for the other invariants[6]. For a pseudoscalar $\sim \gamma^5 \otimes \gamma^5$ the nonrelativistic limit vanishes and only a spin spin and tensor term survives. This is, of course, the well known reason why the deuteron is so loosely bound in spite of the (pseudoscalar!) pion exchange with its strong coupling and low mass. Axial vector or tensor exchange, finally, lead to a leading spin-spin term $\sim V(r)\vec{s}_1 \cdot \vec{s}_2$ in the nonrelativistic limit. Therefore hadron spectroscopy would look quite different if the dominant terms in the potential would be something else than vector or scalar. There are several consequences of scalar confinement which we sketch now:
1) The short range one gluon vector exchange and the long range confining scalar exchange contribute with opposite signs to the spin orbit term. This ensues a tendency of compensation of spin orbit terms allover in hadron spectroscopy. One exception is the antisymmetric or "threebody" effective spin orbit term arising in baryons where the signs of vector and scalar add up. We refer to the review[7] for a discussion and further references concerning this point.
2) There is no long range contribution to the spin-spin splitting. Therefore this splitting is due completely to the contact δ-function from one gluon exchange which does not contribute to P-states and higher angular momenta. The center of gravity of the triplet P-states, where the spin orbit and tensor terms cancel should therfore be degenerate with the singlet P-state:

$$M(^1P_1) = M(^3P_{COG}) \equiv \frac{5}{9}M_2 + \frac{3}{9}M_1 + \frac{1}{9}M_0. \tag{1.11}$$

Since the δ-function is a nonrelativistic approximation to a positive definite nonlocal operator, we would expect the lefthand side to be somewhat smaller. There are, however, other corrections like interactions with open or closed decay channels which may contribute to the splitting. A test of the relation (1.11) in ordinary mesons is difficult. Although we have the systems $a_2(1318), a_1(1275), a_0(983), b_1(1233)$ ($I = 1$) and $f(1274), f_1(1283), f_0(975), h_1(1190)$ ($I = 0$) with the quantum numbers of $^3P_2^{++}, {}^3P_1^{++}, {}^3P_0^{++}, {}^1P_1^{+-}$ which are in agreement with (1.11) the status of the scalars $a_0(983)$ (formerly called the δ) and $f_0(975)$ (formerly the S^*) is controversial because their widths appear too small for $q\bar{q}$ states. This may favour an interpretation as $K\bar{K}$ molecules[8]. In heavy quarkonia the $^1P_1^{+-}$ state cannot be seen in radiative decays from the ψ' or Υ' because of its negative charge conjugation quantum number. It is therefore a great experimental success that we now have candidates both for charmonium and for bottomonium. The R 704 experiment at the CERN ISR[9] has found 5 candidates for the $^1P_1$ charmonium state in $p\bar{p}$ annihilation. The mass value and, for comparison, the mass of the center of gravity are

$$M_c(^1P_1) = 3525.4\pm0.8 \pm 0.4 MeV$$
$$\bar{M}_c(^3P) = 3525.5\pm0.3 MeV. \tag{1.12}$$

At the CLEO detector at CESR[10] one has looked for the singlet P state of the Υ-system in the decay $\Upsilon(3s) \rightarrow \pi^+\pi^-\Upsilon((1^1P_1)$. The results are

$$M_b(^1P_1) = 9894.8 \pm 1.5 MeV$$
$$\bar{M}_b(^3P) = 9900.2 \pm 0.7 MeV. \tag{1.13}$$

The agreement with the theoretically predicted degeneracy is impressive and gives strong support to the picture of scalar confinement.

The degeneracy between $^1P_1$ and $^3P_{COG}$ has an analogy in baryon spectroscopy. It was already noticed in the pioneering work of De Rújula, Georgi and Glashow[11], for a review see ref. 7. Here the delta function of the spin spin interaction due to one gluon exchange gives a special pattern for the $SU(3)$–multiplets contained in the $SU(6)$ supermultiplet $[70,1]^-$ of negative parity P–states:

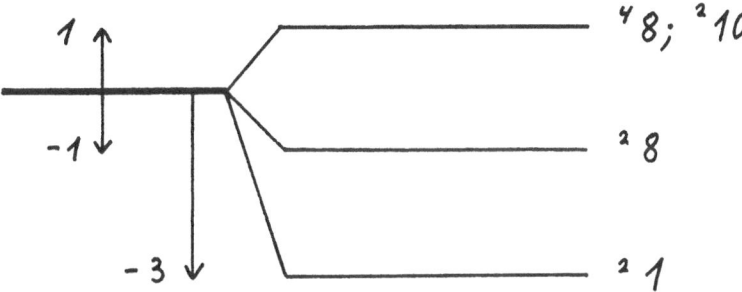

The ^{48} and ^{210} are degenerate and highest, in the middle comes the ^{28}, and the ^{21} (the two lowest lying Λ resonances are mostly $SU(3)$ singlets) is pushed down. Again this is in perfect agreement with observation.

For details we have to refer to the literature. For hadron spectroscopy in general as well as baryons see ref. 7, for mesons ref.12, baryons ref. 13, some recent reviews on heavy quarkonia are given in ref. 14.

2. THE WILSON LOOP APPROACH TO THE STATIC POTENTIAL

The powerful methods developed for a deeper theoretical understanding of the potential and, later on, of the spin dependent relativistic corrections rely heavily on one principle standing at the very beginning of the formulation of QCD:

Recall that in the more naive formulation we speak e.g. of a quark antiquark color singlet state with the quark at x_a and the antiquark at x_b, writing down an expression like $\psi_\alpha(x_b), \psi_\alpha(x_a)$. This expression is, however, not invariant under local gauge transformations, the reason being that the transformation may act quite differently at the different points x_a and x_b. A more physical argument shows a typical difficulty arising in non-abelian gauge theories in connection with static sources: A source is called static or external if its mass goes to infinity, thus it will not react to any forces and can be considered as fixed. This is, however, only true for the position in space but not for the color degree of freedom of a very heavy quark. Even in the limit of infinite mass the quark can emit or absorb gluons, thereby changing its color index. The latter will therefore always remain a dynamical degree of freedom.

The first task consists in writing down a manifestly gauge-invariant quark-antiquark state. This can be done in the standard way by inserting the famous path-ordered exponential

$$U(x_b, x_a; C) \equiv P \exp\{ig \oint_{x_a}^{x_b} A_\mu(z) dz^\mu\}. \tag{2.1}$$

Here we used the convenient matrix notation for the gauge fields

$$A_\mu \equiv \frac{\lambda^a}{2} A_\mu^a. \tag{2.2}$$

The operator P denotes the <u>path ordering prescription:</u> One has to decompose the path C connecting x_a with x_b into infinitesimal pieces, then take the exponential along the infinitesimal pieces which can be done in first order, and finally order the factors according to their appearance along the path:

$$U(x_b, x_a; C) = \lim[1 + ig\ A_\mu(x_b)(x_b - x_{n-1})^\mu]\cdots$$
$$\cdots[1 + ig\ A_\mu(x_1)(x_1 - x_a)^\mu]. \tag{2.3}$$

To obtain the transformation property of U, one applies an infinitesimal gauge transformation with gauge parameter θ. Consider the transformation of one of the factors in (2.3) in lowest order in θ and dx^μ: Using the transformation law

$$A_\mu \to A_\mu + i[\theta, A_\mu] + \frac{1}{g}\partial_\mu\theta$$

one has

$$1 + ig\ A_\mu dx^\mu \to 1 + ig\ A_\mu dx^\mu - g[\theta, A_\mu]dx^\mu + i(\partial_\mu\theta)dx^\mu\ .$$

With $(\partial_\mu\theta)dx^\mu = \theta(x_j) - \theta(x_{j-1})$ one gets in the order considered

$$1 + igA_\mu dx^\mu \to e^{i\theta(x_j)}[1 + igA_\mu dx^\mu]e^{-i\theta(x_{j-1})}.$$

In the product the intermediate exponentials all cancel due to the path ordering prescription, only the very left and right one survives, giving the transformation law

$$U'(x_b, x_a; C) = e^{i\theta(x_b)}U(x_b, x_a; C)e^{-i\theta(x_a)}. \tag{2.4}$$

Finally the expression

$$\overline{\psi(x_b)}\ U(x_b, x_a; C)\psi(x_a) \tag{2.5}$$

is gauge-invariant because the exponentials arising in the transformation law of U just cancel the exponentials coming from the transformation law of the quark fields.

We next discuss the general strategy for extracting an effective Hamiltonian in a manifestly gauge-invariant way. The starting point is the gauge-invariant four-point function

$$I = \langle 0|\bar\psi(\vec{x_1}, 0)U(x_1, x_2)\psi(\vec{x_2}, 0)\bar\psi(\vec{y_2}, T)U(y_2, y_1)\psi(\vec{y_1}, T)|0\rangle. \tag{2.6}$$

For simplicity the path-ordered exponentials are taken along the straight connection line, the times have been chosen equal to zero and T, respectively. The four-point function I can be interpreted as the time-evolution amplitude of the state $|\phi\rangle \equiv |\bar\psi U\psi|0\rangle$. We may therefore write

$$I = \langle\phi(\vec{x_1}\vec{x_2}; 0)|\phi(\vec{y_1}\vec{y_2}; T)\rangle = \langle\phi(\vec{x_1}, \vec{x_2}; 0)|e^{-iHT}|\phi(y_1, y_2; 0)\rangle. \tag{2.7}$$

Now one has to proceed in two steps:
  1) Find an approximation for the four-point function I in (2.6).
  2) Read off the Hamiltonian from the form (2.7).
To calculate I one first writes down the field theoretical path integral representation for this Green's function which reads

$$I = \int \bar\psi(\vec{x_1}, 0)U(x_1, x_2)\psi(\vec{x_2}, 0)\bar\psi(\vec{y_2}, T)U(y_2, y_1)\psi(\vec{y_1}, T)\ e^{i\int \mathcal{L}d^4x}\ \mathcal{D}[\psi]\mathcal{D}[\bar\psi]\mathcal{D}[A]. \tag{2.8}$$

Here

$$\mathcal{L} = \mathcal{L}_{YM} + \mathcal{L}_F = -\frac{1}{2}\operatorname{Tr} F_{\mu\nu}F^{\mu\nu} + \bar{\psi}(i\gamma^\mu D_\mu - m)\psi \qquad (2.9)$$

with

$$D_\mu \equiv \partial_\mu - igA_\mu \qquad (2.10)$$

the gauge covariant derivative. Because the action is quadratic in the quark fields, the fermion integration can be formally performed, resulting in

$$I = \int \{\operatorname{Tr} S(x_2, y_2; A)U(y_2, y_1)S(y_1, x_1; A)U(x_1, x_2)\det(A)$$

$$- \operatorname{Tr}(S(y_1, y_2; A)U(y_2, y_1))\operatorname{Tr}(S(x_2, x_1; A)U(x_1, x_2))\det(A)\}e^{i\int \mathcal{L}_{YM}d^4x}\mathcal{D}[A].$$
$$(2.11)$$

Here S denotes the quark propagator in the presence of the gluon field A_μ. It obeys the equation

$$(i\gamma_\mu D_x^\mu - m)S(x, y; A) = \delta^{(4)}(x - y). \qquad (2.12)$$

$Det(A)$ is the determinant of the operator $i\gamma^\mu D_\mu - m$ which in a perturbation expansion gives closed fermion loops. The two terms in eq. (2.11) can be graphically represented as

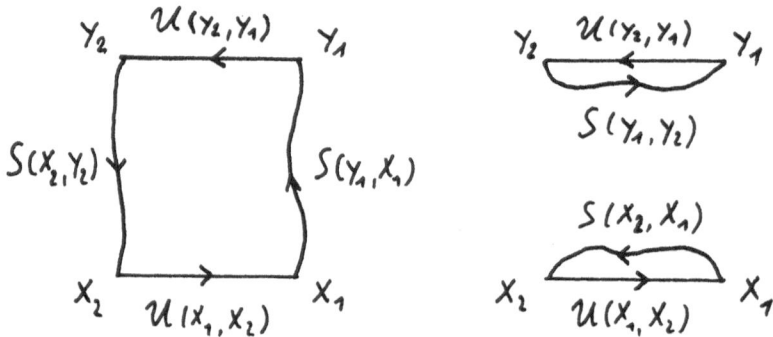

The second term appears for identical flavors only and describes quark-antiquark annihilation. This is a perturbative process dominated by two or three gluons, respectively, and will not be considered anymore in the following.

Equation (2.12) for the fermion propagator is a system of coupled partial differential equations which cannot be solved in closed form for arbitrary gauge field A_μ. The approximation studied by Brown and Weisberger[15] was to replace S by the static propagator S_0. The latter is obtained by dropping the spatial part of the gauge covariant derivative while keeping the time component. Obviously this approximation maintains manifest gauge invariance. The equation for the static propagator

$$(i\gamma_0 D_x^0 - m)S_0(x, y; A) = \delta^{(4)}(x - y) \qquad (2.13)$$

is an ordinary differential equation and can immediately be solved in closed form:

$$S_0(x, y; A) = -i\theta(x^0 - y^0)e^{-im(x^0-y^0)}\frac{1+\gamma^0}{2}U(x_0, y_0)\delta^{(3)}(\vec{x} - \vec{y})$$
$$(2.14)$$
$$-i\theta(y^0 - x^0)e^{-im(y^0-x^0)}\frac{1-\gamma^0}{2}U(x_0, y_0)\delta^{(3)}(\vec{x} - \vec{y}).$$

The fermion determinant becomes constant in the limit $m \to \infty$, therefore the four-point function I in the static limit becomes

$$I_0 \sim \delta^{(3)}(\vec{x_1} - \vec{y_1})\delta^{(3)}(\vec{x_2} - \vec{y_2})e^{-i(m_1+m_2)T}$$
$$\int \text{Tr} \, P \exp\{ig \oint A_\mu(z)dz^\mu\}e^{i\int \mathcal{L}_{YM}d^4x}\mathcal{D}[A]. \qquad (2.15)$$

The path–ordered exponentials along the horizontal lines, stemming from the gauge–invariant wave function, and those along the vertical lines, coming from the static propagators, have combined to the rectangular Wilson loop. If the integral in (2.15) behaves as $\exp(-i\epsilon(r)T)$ for large T one can identify $\epsilon(r)$ with the static quark–antiquark potential:

$$\epsilon(r) = -\frac{1}{i} \lim_{T \to \infty} \frac{1}{T} \ln \int \text{Tr} \, P \exp\{ig \oint A_\mu(z)dz^\mu\}e^{i\int \mathcal{L}_{YM}d^4x}\mathcal{D}[A]. \qquad (2.16)$$

Usually the whole procedure is performed in Euclidean space where T has to be replaced by $-iT$. The method of Brown and Weisberger has given a clear derivation of the manifestly gauge–invariant formula for the static potential.

In the abelian case there is a more intuitive argument for understanding the meaning of the Wilson loop: Consider a pair of static, point–like opposite charges located at $\vec{r_1}$ and $\vec{r_2}$. This corresponds to a current

$$j^\mu(x) = \delta^{\mu 0}\left(\delta^{(3)}(\vec{x} - \vec{r_1}) - \delta^{(3)}(\vec{x} - \vec{r_2})\right) \qquad (2.17)$$

and an interaction energy density

$$\begin{aligned} \mathcal{H}_I(x) &= gA_\mu(x)j^\mu(x) \\ &= gA_0(x)\left(\delta^{(3)}(\vec{x} - \vec{r_1}) - \delta^{(3)}(\vec{x} - \vec{r_2})\right). \end{aligned} \qquad (2.18)$$

Integration over space gives the interaction Hamiltonian

$$H_I(t) = \int \mathcal{H}_I(\vec{x},t)d^3x = g\left(A_0(\vec{r_1},t) - A_0(\vec{r_2},t)\right). \qquad (2.19)$$

Its expectation value is nothing but the change in energy which is caused by the sources, i.e. the static potential $\epsilon(r)$. Let the sources act for a time interval T and integrate over T. This results in

$$\langle \epsilon(r)T \rangle = \langle \int_0^T H_I(t)dt \rangle = \langle g \oint A_\mu(x)dx^\mu \rangle \quad \text{for} \quad T \to \infty, \vec{r} \text{ fixed}, \qquad (2.20)$$

where the last integration runs along the Wilson loop. The horizontal pieces in the loop have been added to make the integral gauge–invariant; they should not change the result for $\vec{r}$ fixed, $T \to \infty$. Physically they correspond to an instantaneous creation and annihilation of the charges. Finally, let us exponentiate within the expectation value (which again can be done for $T \to \infty$) to obtain

$$\langle \exp(ig \oint A_\mu(x)dx^\mu) \rangle \sim e^{-i\epsilon(r)T} \quad \text{for} \quad T \to \infty.$$

The representation (2.16) is particularly natural and useful on the lattice, where the dynamical variables are unitary matrices associated with the links. They may be interpreted as the path– ordered exponential of the gauge field along the link. The Wilson loop is then nothing but the trace of the product of the matrices along the rectangle. We mention that the idea of formulating a locally gauge–invariant theory on a lattice is originally due to F. Wegner[16] who performed this task for the Ising model. He also introduced the central idea of putting the dynamical variables on the links, the gauge transformations on the lattice sites, and using traces over loops as gauge–invariant objects.

One may finally ask what one has gained with the formulation (2.16). It is well known that a lowest order perturbative calculation (see e.g. ref. 17) gives the Coulomb potential, the strong coupling limit gives the long–range linear potential[2] while the whole potential can be computed by lattice Monte Carlo methods. We finally mention that from (2.16) one can derive exact relations by some simple manipulations and a use of Schwarz's inequality[18]. These are

$$V'(r) > 0, \quad V''(r) \leq 0 \qquad (2.21)$$

which means that the potential is monotonically increasing and convex. This is gratifying because it guarantees that nothing dramatic can happen in the intermediate region which dominates spectroscopy and where otherwise very little is known theoretically. Furthermore rigorous statements concerning the order of energy levels[19] can be derived from (2.21).

3. GENERALIZATION OF THE WILSON LOOP FORMALISM TO SPIN-DEPENDENT TERMS

A big step forward in the theoretical understanding of spin–dependent forces in QCD was done in the work of Eichten and Feinberg[20]. The underlying idea is quite simple. Remember that Brown and Weisberger[15] dropped the spatial part of the gauge covariant derivative in the equation for the quark propagator. Eichten and Feinberg (in the following EF) instead start a perturbative expansion with respect to this term. This expansion turns out to become an expansion with respect to the inverse quark masses, but it contains all orders of ordinary QCD perturbation theory with respect to the coupling constant. In this way one ends up with a formally exact representation of the spin–dependent potentials up to order $1/m^2$. Technically the procedure becomes rather involved and we only sketch the essential steps: First the differential equation for the propagator S is rewritten as an integral equation with the static propagator S_0 as inhomogeneous term. This integral equation is then solved by iteration, giving a series of terms involving integrals which contain pieces with the static propagator S_0 and spatial covariant derivatives $\vec{\gamma}\vec{D}$ acting on them. With the help of various manipulations these derivatives are then transformed into ordinary derivatives becoming momentum operators and into insertions of color electric and magnetic fields. The whole stuff is then inserted into the formula for the four–point function (compare eq. (2.11)) and the Hamiltonian extracted from the time evolution. The resulting formulae are complicated and demand for a compact notation. Define

$$\langle 0(x) \rangle = \int \text{Tr } P[\exp(ig \oint A_\mu(z)dz^\mu)0(x)]e^{i \int \mathcal{L}_{YM} d^4 x} \mathcal{D}[A]$$

as the expectation value of an operator insertion into a Wilson loop C. The loop C is identical to that before. The vertical pieces of the loop arise from the quark propagator, the horizontal pieces from the gauge–invariant definitions of the bound state wave functions. In particular one has

$$\langle 1 \rangle \sim e^{-i\epsilon(r)T} \quad \text{for} \quad T \to \infty. \tag{3.1}$$

The general structure for the Hamiltonian up to order $1/m^2$ becomes

$$
\begin{aligned}
V = &\epsilon(r) \\
&+ \left(\frac{[\vec{r} \times \vec{p_1}] \cdot \vec{s_1}}{2m_1^2} - \frac{[\vec{r} \times \vec{p_2}] \cdot \vec{s_2}}{2m_2^2} \right) \left(\frac{\epsilon'(r)}{r} + \frac{2V_1'(r)}{r} \right) \\
&+ \left(\frac{[\vec{r} \times \vec{p_1}] \cdot \vec{s_2}}{m_1 m_2} - \frac{[\vec{r} \times \vec{p_2}]\vec{s_1}}{m_1 m_2} \right) \frac{V_2'(r)}{r} \\
&+ \frac{\vec{s_1} \cdot \vec{s_2}}{3m_1 m_2} V_4(r) \\
&+ \frac{1}{m_1 m_2} \left(\frac{(\vec{r} \cdot \vec{s_1})(\vec{r} \cdot \vec{s_2})}{r^2} - \frac{\vec{s_1}\vec{s_2}}{3} \right) V_3(r) \\
&+ \quad \text{spin-independent corrections.}
\end{aligned}
\tag{3.2}
$$

The potentials V_1 to V_4 are given in terms of insertions of two color fields into the Wilson loop. (Integrals from $-T/2$ to $T/2$, limit $T \to \infty$ understood):

$$\frac{r^k}{r}V_1'(r) = -\frac{g^2 \epsilon ijk}{2T} \int\int (t'-t) \langle B^i(\vec{r_1},t)E^j(\vec{r_1},t')\rangle dt dt' / \langle 1 \rangle.$$

$$\frac{r^k}{r}V_2'(r) = -\frac{g^2 \epsilon^{ijk}}{2T} \int\int t' \langle B^i(\vec{r_2},t)E^j(\vec{r_1}t')\rangle dt dt' / \langle 1 \rangle. \tag{3.3}$$

$$\left(\frac{r^i r^j}{r^2} - \frac{\delta^{ij}}{3} \right)V_3(r) + \frac{\delta^{ij}}{3}V_4(r) = \frac{g^2}{T} \int\int \langle B^i(\vec{r_1},t)B^j(\vec{r_2},t')\rangle dt dt' / \langle 1 \rangle.$$

Only one spin–dependent term in (3.2) is directly connected to the static potential $\epsilon(r)$, namely the $\epsilon'(r)/r$ arising in the first spin orbit term. The other potentials are given by insertions of color fields in the following way:
V_1: Electric and magnetic field on the same line,
V_2: Electric and magnetic field on two lines,
V_3, V_4: Two magnetic fields on two lines

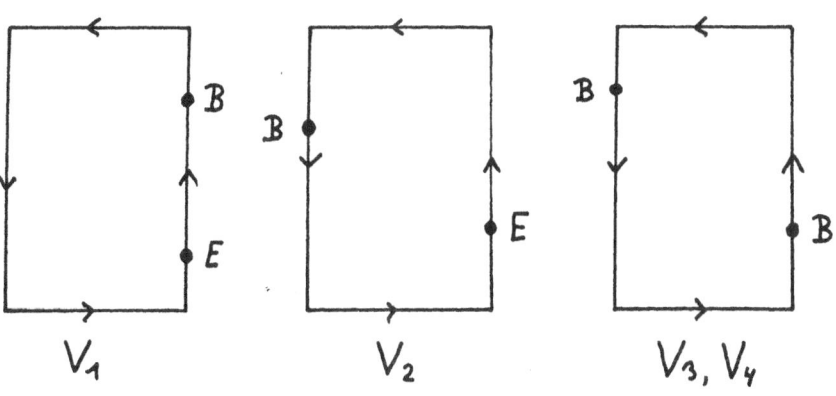

(EF claim the relation $\Delta V_2(r) = \frac{1}{2}V_4(r)$, which is, however, not generally true, since various terms were dropped in the derivation).

Before we go on in the discussion we briefly mention three other ways of arriving at these formulae in a less exact but physically more transparent way.

Method of Peskin[21]: In this method the gauge field A_μ in the Wilson loop is replaced by

$$A_\mu \Rightarrow A_\mu + \frac{i}{m}\tilde{F}_{\mu\nu}S^\nu \qquad (3.4)$$

with $\tilde{F}_{\mu\nu}$ the dual field tensor and S^ν the spin vector. (To our knowledge this modification has first been discussed by Wilczek and Zee[22] in connection with instantons). The replacement is the same which appears when the Dirac equation is cast into a form of the type of a Klein–Gordon equation. Then one has to expand with respect to the spin term, take into account quark motion etc. and arrives at (3.2), (3.3).

Method of Gromes[23]: Here the current corrsponding to static point sources is replaced by the current of Dirac spinors

$$j^\mu = \bar{u}\gamma^\mu u + \bar{v}\gamma^\mu v. \qquad (3.5)$$

To describe quarks of definite average position and momentum one can use appropriate Gaussians as wave functions. Expanding the currents up to order $1/m^2$ one then ends up with the following expectation value which is connected with the potential $V = \epsilon(r) + \delta V$:

$$\langle TrPe^{-igL}\rangle \sim e^{-i(\epsilon(r)+\delta V)T} \quad \text{for} \quad T \to \infty. \qquad (3.6)$$

Here

$$L = \left\{1 + \sum_{i=1}^{2}\frac{[\vec{\nabla}_i \times \vec{p}_i]\vec{s}_i}{2m_i^2}\right\}\oint_{c'}A_\mu dx^\mu$$

$$-\sum_{i=1}^{2}\frac{\vec{s}_i}{m_i}\int_{-T/2}^{T/2}\vec{B}(\vec{r}_i + \frac{\vec{p}_i}{m_i}t, t)dt, \qquad (3.7)$$

where C' is now a modified Wilson loop corresponding to moving quarks ($x_i(t) = r_i + \frac{p_i}{m_i}t$). The modification arising in (3.7) as compared to the static Wilson loop can be easily understood: The motion of the quarks is evident and due to the convective part of $\vec{j}$, the differential operator acting on the loop arises from the expansion of j^0, and the term involving the magnetic field from the spin-dependent part of $\vec{j}$. To show the equivalence with EF one has to rewrite the above expression by referring to the original static loop. This can be done by using the following lemma, expressing the path–ordered integral over a distorted path $\vec{x}_i(t) = \vec{r}_i + \vec{\rho}_i(t)$ by that over the static path, together with the insertion of an electric field:

$$\mathrm{Tr}\, P \exp ig \oint_{c'} A_\mu(z) dz^\mu$$

$$= \mathrm{Tr}\, P \exp ig \oint_c A_\mu(z) dz^\mu \{1 - ig \sum_{i=1}^{2} \int_{-T/2}^{T/2} \vec{\rho}_i(t)\vec{E}(\vec{r}_i, t)dt\} \tag{3.8}$$

(in first order in $\vec{\rho}_i(t)$).

The proof is trivial in the abelian case (consider the difference of the integrals along the distorted and the static path and use Stokes' theorem) and also straightforward in the nonabelian case. Manipulations of this kind also played an important role in the attempts to formulate QCD in terms of gauge invariant loop integrals only[24]. If we specialize to $\vec{\rho}_i(t) = \vec{r}_i + \frac{\vec{p}_i}{m_i}t$ we understand immediately the physical meaning of the electric field insertions together with the time factors in the representations for V_1 and V_2: They simply mean that the quarks are moving with velocities p_i/m_i.

<u>Method of Barchielli, Montaldi, and Prosperi[25]</u>: Here the starting point is a differential equation for the four–point function. It is solved in the form of a path integral representation up to order $1/m^2$. Some problems with respect to ordering of non–commuting operators arise in this approach. The results are again identical to the EF ones.

We now take up the discussion of the formulae (3.2), (3.3). After the appearance of the EF paper there was some confusion concering the sign of the long–range spin orbit term. To obtain an information about the long–range behaviour, where perturbation theory is no longer of any use, the original assumption of EF was that magnetic field correlations are short–ranged and should be calculable in perturbation theory. This would lead to

$$V_1^L = V_2^L = V_3^L = V_4^L = 0 \tag{3.9}$$

(The index L means long–range contribution). This would imply that only the spin orbit term $\sim \epsilon'(r)/r$ in (3.2) survives, while the other spin-orbit term as well as the spin-spin and tensor term vanish. Thus the result looks almost like that arising from an effective scalar exchange except for the wrong sign of the spin orbit term!

On the other hand W. Buchmüller[26] approached the problem in a very direct and intuitive way. His starting point is the picture of electric confinement assuming a purely electric field <u>in the momentary rest frame of the string</u> (note that for moving quarks the string is generally rotating). He then calculates the corresponding Thomas precession and ends up with the same result as above except that his sign differs from Eichten and Feinberg's and agrees with that of a scalar exchange! We generously conceal some further confusion caused by the present author[23] and come to the solution of the puzzle. The clue is again a fundamental principle of nature to be applied now[27].

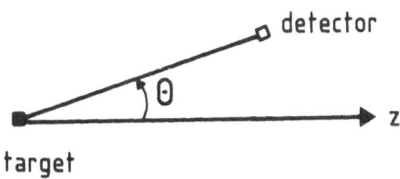

There is no privileged Lorentz frame in the game (except that the formalism is restricted to non–relativistic quark momenta). Therefore two observers whose coordinate frames differ by an infinitesimal boost should obtain the same expressions for the various potentials (Galilei invariance). Stated differently, one can also say that the vacuum expectation values in (3.3) or (3.6) can be evaluated in any reference system without changing the result. Applying an infinitesimal boost (we formulate in Minkowski space here, for convenience)

$$
\begin{aligned}
\vec{x}' &= \vec{x} - \vec{v}t; & \vec{B}' &= \vec{B} - \vec{v} \times \vec{E} \\
t' &= t - \vec{v}\vec{x}; & \vec{E}' &= \vec{E} + \vec{v} \times \vec{B}
\end{aligned}
\tag{3.10}
$$

to (3.6), (3.7) we find that the expression transforms into one of the same form, but with the replacements

$$
\begin{aligned}
\vec{p_i} &\Rightarrow \vec{p_i}' = \vec{p_i} - m_i \vec{v} \\
\vec{r} &\Rightarrow \vec{r}' = \vec{r} - [\vec{v} \times \vec{s_1}]/(2m_1) + [\vec{v} \times \vec{s_2}]/(2m_2).
\end{aligned}
\tag{3.11}
$$

The transformation of the momenta is obviously what one expects, the non-trivial transformation law for the quark-antiquark distance looks surprising at first sight. It is, however, completely in agreement with general investigations by Osborn, Close, and Copley[28] who showed that the relative coordinates for particles with spin have to be chosen carefully if they are to be correct up to order $1/c^2$. The expression derived by these authors transforms in the same way as (3.11). The total potential has to be invariant under (3.11). Introducing this into (3.2) and concentrating on the terms of order v/m, one finds that the change of momenta only involves the spin-orbit terms, while the change in the relative coordinate only affects the static potential. Invariance of V then leads to the exact relation

$$
V_2(r) - V_1(r) = \epsilon(r)
\tag{3.12}
$$

which connects the two spin–orbit potentials with the static potential $\epsilon(r)$.

The assumptions in ref. 20 lead to (3.9) for the long-range contributions, in contradiction with the exact relation (3.12). It is now quite clear what has gone wrong. The very assumption that magnetic fields can be treated perturbatively violates Lorentz invariance because in a different frame there also appears an electric field which cannot be treated perturbatively. Nevertheless the argument can be saved in an obvious way. The assumption appears unproblematic for V_3 and V_4 in (3.3) because the expectation values refer to quarks at rest, and there are no electric fields. In the formula for V_2 we can remove the electric field and refer to a loop where quark 1 is moving instead. Because the magnetic field is at $\vec{r_2}$, the assumption appears still reasonable. Doing the same for V_1, however, we have the moving quark at the same position $\vec{r_1}$ as the magnetic field. This means that in its rest frame there is also an electric field; the assumption of a perturbative treatment breaks down. If we accept the above arguments, we still have $V_2^L = V_3^L = V_4^L = 0$, which, using (3.12), then implies $V_1^L = -\epsilon$. Look what happens if this is introduced into (3.2).

The physical picture looks now quite transparent and consistent. In the long–range regime we have a purely electric field in the rest frame of the string (Buch-müller's ansatz[26], also supported by the area law applied to the formalism of ref. 23) which leads to a spin orbit term due to Thomas precession which is identical to that originating from a scalar exchange. The EF formalism, together with the exact relation $V_2 - V_1 = \epsilon$ originating from Lorentz invariance, leads to the same result under reasonable assumptions on magnetic field correlations. Agreement with the data is also in excellent shape now. The reader who wishes to stay with this nice consistent picture should stop here and not go on reading the concluding section.

4. OUTLOOK AND PROBLEMS

In this concluding part we briefly discuss five topics, two of them giving a nice support to the whole picture, one mentioning an open question, and two causing trouble.

1) Locality versus Nonlocality

People favouring the sum-rule approach always have emphasized that, due to the presence of condensates, there should be no such thing as a local potential[29]. How does this statement coexist with the potential $\epsilon(r)$ derived from the Wilson loop formalism and the local formulae for the spin–dependent potentials in (3.3)? This problematics, previously practically ignored by everybody, has been resolved by Marquard and Dosch[30]. The point is a careful discussion of the two time scales involved in the problem, namely the correlation time of the quark system and that of the gluon condensate. It was shown by these authors that for fixed distance and large quark mass the description by a local potential is correct.

2) Lattice Monte Carlo Calculations of $V_1 \cdots V_4$

In the same way as the static potential $\epsilon(r)$ and other quantities the potentials $V_1, \cdots, V_4$ can be computed by MC techniques, starting from the representations (3.3). Of course the whole procedure is rather subtle because the color fields have to be represented by loops surrounding a plaquette which can be done in different ways. Michael and Rakow[31] used a 16^4 lattice and $SU(2)$, Forcrand and Stack[32] a $6^3 \times 12$ lattice and $SU(3)$ and Campostrini, Moriarty and Rebbi[33] a $16^3 \times 32$ lattice and $SU(3)$. While the authors of ref. 32 find a vanishing V_1, the results of ref. 31 und 32 support the picture advocated for in the previous section: Short–range potentials V_2, V_3, V_4 and a long–range V_1. In particular C. Michael (second paper in ref. 31) has concentrated on V_1, V_2 and found fair agreement with eq. (3.12).

3) Baryons

The analysis à la EF has not yet been performed for baryons, multiquark states, gluonia, hybrids. Although there is no reason to expect big surprises, a general representation for the spin–dependent potentials and an information about the relations between them should certainly be performed. Work on baryons is in progress.

4) Spin–independent relativistic corrections

In principle these can be obtained by the same methods. Technically, however, the task becomes much harder. There are two reasons for this. The first point is that, while the spin–dependent terms start with order $1/m^2$ there are spin–independent terms of order $1/m$. The iteration of these gives a large number of terms of order $1/m^2$ which then have to be combined with other contributions. (The reader might wonder about the physical meaning of these $1/m$ terms which obviously never have shown up in the literatur. The answer is simple: They represent the kinetic energies which come out correctly albeit troublesome as spin–independent $1/m$ corrections in the formalism). The second point is even more ugly technically. Contrary to the spin–dependent corrections the spin–independent ones do not commute with the static potential $\epsilon(r)$ due to current–current–like terms involving p_1 and p_2. In expanding the exponentials $(\epsilon(r)+$ corrections) one has to use the Baker–Campbell–Hausdorff formula to a rather high order. We know of two calculations: In an unpublished diploma thesis[34] the resulting expressions when calculated in perturbation theory (even in lowest order in the abelian case) showed a wrong behavior for large T (higher powers, no finite limit in the formulae analogous to (3.3)). On the other hand, the authors of ref. 25 using their approach ended up with reasonable expressions leading back to the Breit–Fermi Hamiltonian in lowest order perturbation theory. They did, however, not perform a complete discussion of the long–range behavior, in particular it was not investigated whether these expressions give the corrections as expected from a scalar exchange. (Since there are various wrong expressions in the literatur for the latter, we emphasize that the correct one can be found in refs. 35). Whether the above discrepancy is due to the different methods or to some errors in the rather complicated calculations appears to be an open question.

5) One–loop perturbation theory

Pantalone, Tye, and Ng[36] have compiled their calculations, together with those of other authors, of the spin–dependent potentials in one–loop order of ordinary QCD perturbation theory. The results show a striking discrepancy when opposed to the general structure expected from the formulae of EF. The latter start with expressions proportional to $1/m^2$ and would proceed with higher powers of $1/m$. In contrast to this the formulae of PTN contain logarithmic mass dependences in $V_1, \cdots, V_4$ which have e. g. the form $\ln((m_1 m_2)^{1/2} r)$. Furthermore in the case of different masses there appears an additional term of the form

$$[(\vec{S}_1/m_1^2 - \vec{S}_2/m_2^2) \cdot \vec{L} + (\vec{S}_1 - \vec{S}_2)\vec{L}/m_1 m_2]V_5(m_1, m_2, r)$$

with $V_5(m_1, m_2, r) = \frac{c_F c_A \alpha_s^2}{4\pi r^3} \ln(m_1/m_2)$.

In spite of this discrepancy the PTN potentials fulfil relation (3.12). What has gone wrong? My personal feeling is that the formal EF expansion is correct as long as one does not perform the integration over the gluon fields, but that one would end up with divergent results when attempting to calculate the potentials $V_1 \cdots, V_4$ in (3.3) in one–loop order. Another point of trouble is that the one loop perturbative potential V_4 would lead to a divergent matrix element for S-states due to a singular behavior at $r \to 0$. Clearly there are essential points which have not yet been understood properly.

ACKNOWLEDGEMENT It is a pleasure to thank my friends T. Barnes, F. Close, N. Isgur, and J. M. Richard for stimulating discussions during this school. In particular I remind Frank Close that he owes me a barrel of Marsala.

REFERENCES

1. W. Buchmüller, S.-H. H. Tye, Phys. Rev. **D24**, 132 (1981)
2. K. G. Wilson, Phys. Rev. **D10**, 2445 (1974)
3. J. Pumplin, W. Repko, A. Sato, Phys. Rev. Lett., **35**, 1538 (1975); H. J. Schnitzer, Phys. Rev. **D13**, 74 (1976)
4. H. J. Schnitzer, Phys. Rev. Lett. **35**, 1540 (1975)
5. A. B. Henriques, B. H. Kellett, R. G. Moorhouse, Phys. Lett. **64B**, 85 (1976)
6. D. Gromes, Nucl. Phys. **B131**, 80 (1977)
7. D. Gromes, Ordinary Hadrons, in: Proceedings of the Yukon Advanced Study Institute, Whitehorse, Yukon, Canada: August 12–26, 1984 "The Quark Structure of Matter", Ed. N. Isgur, G. Karl, P.J. O'Donnell, World Scientific Publ. Comp., Singapore 1985
8. R.L. Jaffe, Phys. Rev. **D15**, 267, 281 (1977); R.L. Jaffe, K. Johnson, Phys. Lett. **60B**, 201 (1976); R.L. Jaffe, Phys. Rev. **D17**, 1444 (1978); J. Weinstein, N. Isgur, Phys. Rev. Lett. **48**, 659 (1982), Phys. Rev. **D27**, 588 (1983)
9. C. Baglin et al. Phys. Lett. **171B**, 135 (1986)
10. T. Bowcock et al. Phys. Rev. Lett. **58**, 307 (1987)
11. A. De Rújula, H. Georgi, S.L. Glashow, Phys. Rev. **D12**, 147 (1975)
12. S. Godfrey, N. Isgur, Phys. Rev. **D32**, 189 (1985)
13. A.J.G. Hey, R. Kelly, Phys. Rep. **96C**, No.2, 3; 71 (1983) S. Capstick, N. Isgur, Phys. Rev. **D34**, 2809 (1986)

14. S. Cooper SLAC–PUB–3819. (October 1985) and in: Proceedings of the Int. Europhys. Conf. on High Energy Physics, ed. L. Nitti, G. Preparata, Bari, Italy, 18/24 July 1985, SLAC–PUB–4139 (November 1986), Invited Talk at the XXIIIth Int. Conf. on High Energy Physics, Berkeley, California, July 16–23 1986 (also for light mesons)

    K. Königsmann, DESY 86–136 (October 1986), Invited Talk presented at the 6th International Conference on Physics in Collisions, Chicago, September 3–5 1986

    F.J. Gilman, SLAC–PUB–4253 (March 1987), presented at the SLAC Summer Institute on Particle Physics, Stanford, CA, July 28–August 8, 1986

    W. Buchmüller, S. Cooper, MIT preprint MIT–LNS–159 (March 1987)

15. L.S. Brown, W.I. Weisberger, Phys. Rev. **D20**, 3239 (1979)

16. F.J. Wegner, Journal of Math. Phys. **12**, 2259 (1971)

17. P. Becher, M. Böhm, H. Joos, "Gauge Theories", Wiley 1986

18. C. Bachas, Phys. Rev. D33, 2723 (1986)

19. B. Baumgartner, H. Grosse, A. Martin, Nucl. Phys. **B254**, 528 (1985)

20. E. Eichten, F. Feinberg, Phys. Rev. **D23**, 2724 (1981)

21. M.A. Peskin, SLAC–PUB–3273 (December 1983) 11th SLAC Summer Institute, July 18–29, 1983

22. F. Wilczek, A. Zee, Phys. Rev. Lett. **40**, 83 (1978)

23. D. Gromes, Zeitschrift f. Physik **C22**, 265 (1984)

24. A.A. Migdal, Phys. Rev. **102**, 199 (1983) and references therein

25. A. Barchielli, E. Montaldi, G.M. Prosperi, University of Milano preprint I.F.U.M. 322/F.T. (September 1986)

26. W. Buchmüller, Phys. Lett. **B112**, 479 (1982)

27. D. Gromes, Zeitschrift f. Physik **C26**, 401 (1984)

28. H. Osborn, Phys. Rev. **176**, 1523 (1968);
    F.E. Close, H. Osborn, Phys. Rev. **D2**, 2127 (1970);
    F.E. Close, L.A. Copley, Nucl. Phys. **B19**, 477 (1970)

29. M.B. Voloshin, Nucl. Phys. B154, 365 (1979);
    H. Leutwyler, Phys. Lett. **98B**, 447 (1981)

30. U. Marquard, H.G. Dosch, Phys. Rev. **D35**, 2238 (1987)

31. C. Michael, P.E.L. Rakow, Nucl. Phys. **B256**, 640 (1985);
    C. Michael, Phys. Lett. **56**, 1219 (1986)

32. P. de Forcrand, J.D. Stack, Phys. Rev. Lett. **55**, 1254 (1985)

33. M. Campostrini, K. Moriarty, C. Rebbi, Phys. Rev. Lett. **57**, 44 (1986)

34. R. Tafelmeyer, Diplomarbeit, Heidelberg 1986 (unpublished)

35. T. Barnes, G.I. Ghandour, Phys. Lett. **118B**, 411 (1982)
    R. McClary, N. Byers, Phys. Rev. **D28**, 1692 (1983)
    F. Gesztesy, H. Grosse, B. Thaller, Phys. Rev. **D30**, 2189 (1984)

36. J. Pantalone, S.H.M. Tye, Y. J. Ng, Phys. Rev. **D33**, 777 (1986)

RESONANCES

Lucien Montanet

CERN, European Organization for Nuclear Research, Geneva
Switzerland

1. INTRODUCTION

These lectures are an introduction to various techniques used to
analyze experimental data on resonances.

We first summarize briefly the main theoretical ideas relative to this
field of physics to underline its interest and complexity.

We limit the discussion to meson resonances since this school is
organized around the physics which can be done at LEAR.

The fundamental idea in meson spectroscopy is to assume that mesons are
made of one pair of constituent quarks. $q\bar{q}$. This is the famous "naive
relativistic quark model" [1] which is so successful in meson
classification. N. Isgur [2] has shown that this model is not so naive and
can be funded on rather safe theoretical grounds (QCD).

For a set of fermion–antifermion pair like $q\bar{q}$, the orbital angular
momentum L, spin S, parity P and charge conjugation C must satisfy the
following relations:

$$P = (-1)^{L+1} \qquad C = (-1)^{L+S}$$

Introducing the usual nuclear spectroscopy notation,

$$(2S+1)L_J$$

this gives for the lowest states:

$^1S_0$ (L = 0) $J^{PC} = 0^{-+}$

$^3S_1$ 1^{--}

$^1P_1$ (L = 1) 1^{+-}

$^3P_0$ 0^{++}

$^3P_1$ 1^{++}

$^3P_2$ 2^{++}

$^1D_2$ (L = 2) 2^{-+}

$^3D_1$ 1^{--}

$^3D_2$ 2^{--}

$^3D_3$ 3^{--} etc.

If the quarks were appearing in one variety (one flavour), f, one would expect to observe one meson $f\bar{f}$ for each set of allowed quantum numbers.

With two varieties of quarks, u and d, one should observe quartets of mesons of a given J^{PC}:

$u\bar{u}$ $u\bar{d}$ $d\bar{u}$ $d\bar{d}$

Five varieties of quarks have been discovered, and very likely a sixth one exists. This leads, in principle, to 36 states for a given J^{PC}.

Limiting ourself to the three lightest quarks (u, d, s), the $q\bar{q}$ model requests the existence of 9 mesons for each J^{PC}:

$u\bar{d}$ $u\bar{s}$ $d\bar{u}$ $d\bar{s}$ $s\bar{u}$ $s\bar{d}$ $u\bar{u}$ $d\bar{d}$ $s\bar{s}$.

The last three combinations, having not net flavour, can mix to form states which are linear combinations of $u\bar{u}$, $d\bar{d}$, $s\bar{s}$.

Are there meson states which do not fall into the fundamental classication provided by the $q\bar{q}$ model?

Numerous theoretical arguments suggest that mesons with no constituent quarks (glueballs), or with several pairs of constituent quarks: $(q\bar{q})^2$, or also with a mixture of quarks and gluons (hybrids) [3] may be formed.

How to identify these "exotic" states?

The most convincing evidence for the existence of, say a $(q\bar{q})^2$ state, would be the observation of (additive) quantum numbers (such as the electric charge Q, the hypercharge Y, etc.) which are not accessible to a $q\bar{q}$ pair (i.e. Q = 2, S = 2, I = 3/2 ...). These mesons are called exotic of the <u>first kind</u>. None have been observed until now. Indeed, their absence has been taken as a strong evidence for the validity of the $q\bar{q}$ model (see however sect. 8, the I = 1/2 sector).

The <u>second kind</u> of exotism is defined by a set of quantum numbers, J^{PC}, which is forbidden for a $q\bar{q}$ pair. These are the pseudoscalar, $J^{PC} = 0^{--}$ and the series of natural parity mesons with "wrong" C parity:

$$P = (-1)^J \quad \text{and} \quad CP = -1,$$

i.e. $J^{PC} = 0^{+-}, 1^{-+}, 2^{+-}$, etc.

These quantum numbers are open to glueballs with three "constituent" gluons or to hybrids such as $(q\bar{q}g)$. A recent experiment [4] has presented some evidence in support of the observation of a 1^{-+} state.

A <u>third kind</u> of exotism could be defined by the observation of states with quantum numbers open to a $q\bar{q}$ system but for which the nonet is already filled. It may be the case of the $f_2(1720)$. However, an additional possibility exists within the $q\bar{q}$ model. Like the familiar structure observed for the positronium, the quarkonium $(q\bar{q})$ can exibit radial excitations. One therefore expects not only one, but several nonets for a given set of J^{PC}. The energy spacing of these nonets depend upon the potential which governs the system. The equal spacing of the harmonic oscillator potential seems to apply to the $q\bar{q}$ system.

Before taking for granted that the observation of a tenth meson with given J^{PC} is exotic, one must be sure that it does not fit with the energy level expected for a radial excitation. A detailed description of the energy level expected for the $q\bar{q}$ meson spectrum has been given by S. Godfrey and N. Isgur [5]. It will be used as a reference frame all along these lectures.

We have already mentioned the possibility that four quark states: $(q\bar{q})^2$ are present in meson spectroscopy. With the bag model, R. Jaffe [6] has found that the $a_o(980)$ and $f_o(990)$ may be good candidates for the 4-quark states $(s\bar{s}\ u\bar{d})$ and $(s\bar{s}\ u\bar{u})$. However, if the mass of the $(q\bar{q})^2$ state is above the appropriate threshold, one may expect that the $(q\bar{q})^2$ state will not form a bound state but will fall apart into two singlets of colour which, in turn, will generate ordinary mesons:

$$(q\bar{q})^2 \rightarrow (q\bar{q})_1 + (q\bar{q})_2 \rightarrow \pi + \pi,$$

$$\pi + K$$

$$K + \bar{K} \quad \text{etc.}$$

In some cases, however, the conditions can be met so that $(q\bar{q})_1$ and $(q\bar{q})_2$ form an S-wave bound state, called a molecule by T. Barnes [7]. We have therefore to analyze carefully the phenomena which occur near the $\pi\pi$, $K\pi$, $K\bar{K}$, K^*K, etc. thresholds in view of identifying these "molecules".

A complementary approach to the search for exotic states is to study their coupling to specific initial or final states. For instance, it is generally assumed that glueballs, made of gluons, have no coupling (at least in first order) to photons. Therefore a relatively large coupling to $\gamma\gamma$ disfavours the glueball hypothesis. Vice-versa, it is expected that states produced in radiative decay of the J/ψ,

$$J/\psi \rightarrow \gamma + X$$

may have a dominant coupling to two gluons and are therefore favoured glueball candidates.

One sees that meson spectroscopy is rich of possibilities and that the search for exotic states is a real challenge. One still needs to accumulate clean and complete information on the fundamental $q\bar{q}$ spectrum, which is far from being complete in the 1 to 3 GeV mass range. One must perform rather detailed spin-parity analysis, involving often several channels. One has to measure the couplings of the candidates to several initial or final states, analyzing simultaneously the results of complementary experiments.

All along these lectures, we shall adopt the new nomenclature which has been proposed by the Particle Data Group [8].

The name of a meson depends basically upon its quantum numbers: J, P, C, I.

The first step is to distinguish the states with CP = +1 from those with CP = -1. Within each of these two classes, and using the $q\bar{q}$ model as a guideline, one separates states with L even from states with L odd.

These propositions lead to the following table:

table 1

		$I = 1$	$I = 1/2$	$I = 0$
CP = -1	L even	0^{-+}, 2^{-+}, ... π	K	η
	L odd	1^{+-}, 3^{+-}, ... b		h
CP = +1	L even	1^{--}, 2^{--}, ... ρ	K^{*}	$\omega(\phi)$
	L odd	0^{++}, 1^{++}, 2^{++} a		f

The spin J is indicated as a subscript (except for the "ground states": 0^{-+} and 1^{--}):

ex.: $a_0(980)$, $a_1(1270)$, $a_2(1320)$ ...

     $f_0(990)$, $f_2(1270)$ ...

The exotic states (wrong C parity) are named as follows:

$\pi \rightarrow 0^{-+} \rightarrow 0^{--} \rightarrow \hat{\pi}$
(η) $\qquad\qquad\qquad\quad (\hat{\eta})$

$\rho \rightarrow 1^{--} \rightarrow 0^{-+} \rightarrow \hat{\rho}$
(ω) $\qquad\qquad\qquad\quad (\hat{\omega})$

$\alpha_2 \rightarrow 2^{++} \rightarrow 2^{+-} \rightarrow \hat{a}_2$
(f_2) $\qquad\qquad\qquad\quad (\hat{f}_2)$

etc.

Everything else is named: X.

2. SIMPLE PHENOMENOLOGICAL APPROACH - MASS - WIDTH - ANGULAR DISTRIBUTIONS

What is the most reasonable resonant shape to employ in order to define the mass and the width of a resonance?

Following D. Jackson [9], we consider first 2-body resonances: $R \rightarrow 1 + 2$.

Standard invariant phase space for n particles:

$$dF_n = \int \frac{d^3p_1 \, d^3p_2 \, \ldots \, d^3p_n}{(2n)^{3n} \, (2E_1)(2E_2) \, \ldots \, (2E_n)} \, \delta^4(P_1 + P_2 + \ldots \, P_n - P) \qquad (2.1)$$

Single out the two-body phase space particles 1 and 2 in the rest frame of (2.1,2.2)

$$dF_n = f_{12} \times \int \frac{d^3Q \, d^3p_3 \, \ldots \, d^3p_n}{(2\pi)^{3(n-1)}(2Q_o)(2E_3) \, \ldots \, (2E_n)} \, \delta^4(Q + P_2 + \ldots \, P_n - P) \qquad (2.2)$$

where

$$f_{12} = \frac{1}{4(2\pi)^3} \frac{9}{\omega} \, d\Omega_{12} \, d\omega^2 \qquad (2.3)$$

$$\vec{Q} = \vec{P}_1 + \vec{P}_2$$

$$\omega^2 = -(P_1 + P_2)^2 = \text{inv. mass squared}$$

$$Q_o^2 = \vec{Q}^2 + \omega^2$$

q = 3-momentum of each member of the pair 1, 2 in system (2.1,2.2).

One goes from the production cross section of a <u>stable</u> (S) particle of mass ω: $d\sigma_S(\omega)$ to the production cross section of a <u>resonance</u> (R) of mass ω: $d\sigma_R(\omega)$ by introducing a propagator and a decay vertex amplitude:

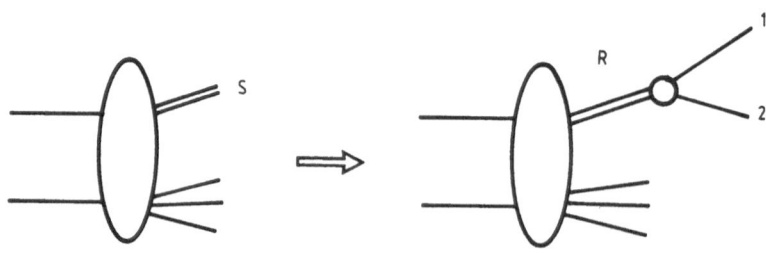

$$d\sigma_R(\omega) = d\sigma_s(\omega) \left[\frac{\pi^{-1}\omega_o \Gamma(\omega)}{(\omega_o^2 - \omega^2)^2 + \omega_o^2\Gamma^2(\omega)} \right] d\omega^2 \qquad (2.4)$$

Definition:

ω_o : mass of the resonance

$\Gamma(\omega)$: width of the resonance:

$$\Gamma(\omega) = \frac{1}{32\ \pi^3}\ \frac{1}{2J + 1}\ \Sigma \int |V|^2\ \frac{q}{\omega}\ d\Omega_{12} \qquad (2.5)$$

J : angular momentum of R

the sum is extended over the spin states of the outgoing particles, the integration is over the decay angles of the two-body phase space. (2.4) is valid only if the resonance is produced clearly, with no interference effects in the final state.

If the resonance has several decay modes and only one mode is considered to evaluate $\sigma_R(\omega)$, then the width in the numerator of (2.4) is the partial width Γ_{12} for this mode, while the width in the denominator is the total width.

The elastic scattering of the pair (1, 2) via the resonant state gives rise to a scattering cross section

$$\sigma_{scat}(\omega) = \frac{4\pi}{q^2}\ \frac{2J + 1}{(2j_1 + 1)(2j_2 + 1)}\ \frac{\omega_o^2\ \Gamma^2(\omega)}{(\omega_o^2 - \omega^2)^2 + \omega_o^2\Gamma^2(\omega)} \qquad (2.6)$$

j_1, j_2 are the angular momenta of (1, 2).

The elastic scattering (2.6) and production cross section (2.4) can be expressed in terms of a resonant phase shift: $\delta(\omega)$:

$$tg\ \delta = \frac{\omega_o\Gamma(\omega)}{\omega_o^2 - \omega^2} \qquad (2.7)$$

Then (2.4) can be written

$$d\sigma_R = d\sigma_s\ \frac{sin^2\delta}{\pi\ \omega_o\ \Gamma(\omega)} \qquad (2.8)$$

At the resonance mass, $\omega = \omega_o$, $\delta = \pi/2$.

<u>Remarks on $\Gamma(\omega)$:</u>

When the natural width of R becomes comparable to the experimental resolution, one needs to distinguish:

- the full width at half maximum, or "observed width": Γ_o,
- the full width at half maximum or "resolution function": Γ_R,
- the full width at half maximum of the resonance or "natural width": Γ_N,

if the errors are gaussian, as they should in principle:

$$\Gamma_R = 2 \, \sigma_R \sqrt{2\log 2} = 2.35 \, \sigma_R,$$

and if the natural width is also gaussian (it is not!)

$$\Gamma_N^2 = \Gamma_o^2 - \Gamma_R^2. \tag{2.9}$$

If the errors are assumed to follow a Breit–Wigner behaviour, like the natural width of the resonance,

$$\Gamma_N = \Gamma_o - \Gamma_R. \tag{2.10}$$

The truth is between the two limits (2.9) and (2.10).

The energy dependence of the natural width depends on the dynamics of the decay vertex. For a two-body decay with an angular momentum ℓ,

$$\Gamma_N(\omega) \simeq \Gamma \left(\frac{q}{q_o} \right)^{2\ell+1} \tag{2.11}$$

where q, q_o are the 3-momenta of the daughters (in the CM frame of the resonance) at ω and ω_o, respectively. (2.11) is a crude approximation which needs refinements for low lying broad resonances (like the $\rho(770)$). A more accurate expression of $\Gamma_N(\omega)$ can be written:

$$\Gamma_N(\omega) = \Gamma \left(\frac{q}{q_o} \right)^{2\ell+1} \frac{\rho(1)}{\rho(\omega_o)} \tag{2.12}$$

where $\rho(\omega)$ is given, in lowest order perturbation theory:

$$1^- \rightarrow 0^- 0^- \quad : \quad \rho(\omega) = \omega^{-1}$$

$$1^- \rightarrow 0^- 1^- \quad : \quad \rho(\omega) = \omega \quad , \text{ etc.}$$

Final state interaction - Watson's formula

One goes from n body phase space (2.1) to the resonance production (2.4) by inserting the factor:

$$\phi_R(\omega) = C \frac{\omega}{q} \frac{\Gamma(\omega)}{(\omega_o^2 - \omega^2)^2 + \omega_o^2 \Gamma^2(\omega)} = C' \frac{\omega \sin^2 \delta}{q\Gamma(\omega)} \qquad (2.13)$$

where C, C' are normalization constants.

ϕ_R is essentially the enhancement factor of Watson which states that $\phi_R(\omega)$ is proportional to the absolute square of

$$\frac{e^{i\delta} \sin\delta}{q^{\ell+1}} \quad .$$

Angular distribution

For a two-body decay, define two angles:

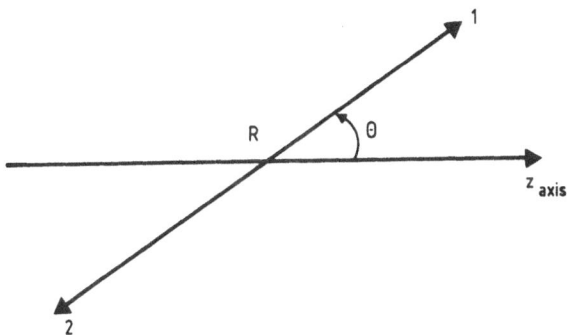

θ,   defined with respect to z axis; if the z axis is given by the direction of the "incident particle" (assuming that the resonance is produced in a two-body collision), then θ is the Gottfried-Jackson angle θ_{GJ}.

φ,   is defined by choosing as y axis the normal to the production plane. With this definition, φ is equivalent to the Treiman-Yang angle which is the angle between the decay and the production planes as seen in the incident particle rest-frame.

For a three-body decay (like $\omega(783)$) one replaces the direction of 1 by the normal to the decay plane.

The angular distribution $W(\theta, \phi)$ will depend on the spin of the resonance and on its production properties: For a resonance with spin $J = 1$ decaying into two spinless particles:

$$W(\theta, \phi) = \frac{3}{4\pi} [\rho_{00} \cos^2\theta + \rho_{11} \sin^2\theta - \rho_{1-1} \sin^2\theta$$

$$- \rho_{1-1} \sin^2\theta \cos2\phi - \sqrt{2} \, Re\rho_{10} \sin^2\theta \cos\phi] \quad (2.14)$$

$\rho_{m,m'}$ are the spin density matrix elements of R.

Gottfried and Jackson [10] have proved that, in the case of a $J = 1^-$ resonance produced by a 0^- incident particle with a 0^- exchange, only ρ_{00} is different from zero. Then

$$W(\theta) \sim \cos^2\theta.$$

For a natural parity exchange (1^-, 2^+ ...), only ρ_{11} and ρ_{1-1} are different from zero. For unnatural parity exchange (1^+, 2^- ...) all elements of $\rho_{mm'}$ are different from zero in general.

3. THE S MATRIX

In the previous chapter, we have discussed the mass, the width and the decay angular distributions of a resonance in a phenomenological way. In particular, we have introduced the Breit-Wigner formula (2.4) without much justification.

We shall now try to define a resonance in a more rigourous way. This will lead us to associate resonance phenomena to poles of the S matrix.

In the Schrödinger picture, the time evolution of a state vector ψ_t will satisfy:

$$i \frac{\partial}{\partial t} |\psi_t> = H |\psi_t> \tag{3.1}$$

with H (hamiltonian) $= H_{free} + H_{int} = H_o + H_{int}$. Solutions of (3.1)

$$|\psi_t> = e^{-iH(t-t_o)} |\psi_o>$$

$$|\phi_t> = e^{-iH_o(t-t_o)} |\phi_o>$$

If H is hermitian, $e^{-iH(t-t_o)}$ is unitary, i.e. $|\psi_t>$ keeps its module but rotates in the abstract vector state space as t evolves.

One can rewrite $|\psi_t>$:

$$|\psi_t> = e^{-iH(t-t_o)} [e^{iH_o(t-t_o)} \cdot e^{-iH_o(t-t_o)}] |\psi_o>$$

and, if much before the interaction:

$$|\psi_o> \equiv |\phi_o>,$$

$$|\psi_t> = e^{-iH(t-t_o)} \cdot e^{iH_o(t-t_o)} \cdot e^{-iH_o(t-t_o)}] |\phi_o>$$

Write:

$$e^{-iH_o(t-t_o)} |\phi_o> = |\phi_t>$$

$$e^{-iH(t-t_o)} \cdot e^{iH_o(t-t_o)} = S(t,t_o)$$

then

$$|\psi_t> = S(t, t_o) |\phi_t> \tag{3.2}$$

Take $t_o \to -\infty$ and $t \to +\infty$, the S matrix allows to go from "in" states to "out" states:

$$|\beta, out> = S|\alpha, in> \tag{3.3}$$

or

$$<\beta, out|\alpha, in> = <\beta, in|\alpha, out> = S_{\alpha\beta} \tag{3.3'}$$

Probability conservation - Time reversal invariance

The conservation of probability gives:

$$\langle \alpha, in | \alpha, in \rangle = 1, \qquad i.e.: \quad \sum_{\beta} |S_{\alpha\beta}|^2 = 1$$

$$\langle \beta, in | \alpha, in \rangle = 0, \qquad i.e.: \quad \sum_{\gamma} S^{*}_{\alpha\gamma} S_{\beta\gamma} = 0$$

or

$$\sum_{\gamma} S^{*}_{\alpha\gamma} S_{\beta\gamma} = \delta_{\alpha\beta}$$

or

$$SS^{+} = I$$

If the time direction does not matter,

$$S_{\alpha\beta} = S_{\beta\alpha}$$

or

$$SS^{+} = S^{+}S = I \qquad \text{(unitarity of S)} \qquad (3.4)$$

These constraints reduce the number of independent parameters of a general (complex) $N \times N$ matrix: $2N^2$ to $1/2\ N(N + 1)$. This is the number of parameters found in a real $N \times N$ symmetric matrix, like the K matrix:

$$K = i\ \frac{1 - S}{1 + S} \qquad (3.5)$$

Resonance: a resonance can be defined as a metastable state of energy ω_o such that the time delay undergone by the interacting wave with respect to the free wave is maximum:

$$\tau = \frac{\langle \alpha, in | r | \alpha, in \rangle - \langle \beta, out | r | \beta, out \rangle}{\langle \alpha, in | v | \alpha, in \rangle} \qquad (3.6)$$

r: distance operator

v: radial speed of the wave from the diffusion center.

With (3.3), write

$$\langle \beta, out | r | \beta, out \rangle = \langle \alpha, in | S^{+}rS | \alpha, in \rangle$$

then numerator of (3.6) becomes:

$$\langle \alpha, in | S^{+}Sr - S^{+}rS | a, in \rangle = \langle \alpha, in | S^{+}[S, r] | \alpha, in \rangle$$

and (3.6) becomes

$$\tau = \langle \alpha, \ in| \ -i \ \frac{S^+ \partial S}{\partial \omega} | \ \alpha, \ in \rangle$$

where we have used the quantum mechanics relation:

$$\frac{\partial S}{\partial P} = -i[r, \ S]$$

elastic scattering case: The S matrix reduces to a 1 × 1 matrix which can be written:

$$S = e^{2i\delta}$$

then

$$\tau = 2 \ \frac{\partial \delta}{\partial \omega} = \frac{2}{v} \ \frac{\partial \delta}{\partial p} \tag{3.7}$$

In a potential of range R, a free particle would spend the time:

$$\tau_o \sim \frac{2R}{v}$$

Causality (the particle cannot escape before entering) leads to the Wigner's inequality:

$$\tau + \frac{2R}{v} > 0$$

or

$$\frac{\partial \delta}{\partial p} > -R \tag{3.8}$$

To represent a resonance, we have to introduce a maximum of $\partial \delta / \partial p$.
Write:

$$\frac{1}{t} = a + bx^2$$

with $x = \omega - \omega_o$.

$$\tau = 2 \ \frac{\partial \delta}{\partial \omega} = \frac{1}{a + bx^2}$$

$$S = e^{2i\delta} \qquad \ell nS = 2i\delta$$

$$\ell nS = i \int \frac{dx}{a + bx^2} = \frac{i}{\sqrt{ab}} \ artg \ \frac{\sqrt{a/b}}{\omega_o - \omega_o}$$

or

$$\delta = \gamma \ artg \ \frac{\Gamma/2}{\omega_o - \omega}$$

with

$$\gamma = \frac{1}{2\sqrt{ab}}$$

$$\Gamma = 2\sqrt{a/b}$$

At $\omega = \omega_0$, $\delta = 90°$ and, near the resonance energy ω_0, δ moves fast and in a positive way with ω.

What are the consequence for S?

Introducing complex values for the energy ω:

$$\text{Re } \omega = \omega_0 - \omega \qquad\qquad \text{Im } \omega = \frac{\Gamma}{2}$$

$$\ln(\omega_0 - \omega + i\,\frac{\Gamma}{2}) = \ln(\omega_0 + \omega) + i\,\text{artg}\,\frac{\Gamma/2}{\omega_0 - \omega}$$

$$\ln(\omega_0 - \omega + i\,\frac{\Gamma}{2}) = \ln(\omega_0 - \omega) + i\,\text{artg}\,\frac{\Gamma/2}{\omega_0 - \omega}$$

$$S = \left(\frac{\omega - \omega_0 - i\Gamma/2}{\omega - \omega_0 + i\Gamma/2}\right)^{\gamma} \qquad\qquad (3.9)$$

Analycity of S imposes $\gamma = 1, 2...$ For $\gamma = 1$, a resonance corresponds to a pole of S at

$$\omega = \omega_0 - i\Gamma/2$$

(For $\gamma = 2$, we would have a dipole).

From the general expression of σ_{el}:

$$\sigma_{el} = \frac{4\pi}{q^2}\,(2\ell + 1)\,\sin^2\delta$$

one obtains, for a resonance as defined above, with

$$\delta = \text{artg}\,\frac{\Gamma/2}{\omega_0 - \omega}\,,$$

$$\sigma_{el} = \frac{4\pi}{q^2}\,(2\ell + 1)\,\frac{\Gamma^2/4}{(\omega_0 - \omega)^2 + \Gamma^2/4} \qquad\qquad (3.10)$$

This is the non-relativistic Breit–Wigner formula for elastic scattering.

4. SCATTERING (S), TRANSITION (T) AND REACTION (K) MATRICES

In the previous chapter, we have introduced the scattering matrix S and have shown that a resonance corresponds to a pole of the S matrix in the complex energy plane.

We have also introduced the K matrix (eq. (3.5)) and have underlined that the number of independent parameters needed to describe the K matrix corresponds to the number of parameters expected to describe a N coupled channel scattering once unitarity and time reversal are satisfied.

We shall now elaborate in more details the properties of these matrices, as well as of the transition (T) matrix which is very useful, in particular, to visualize the presence of a resonance on the "Argand plots".

A standard scattering experiment is sketched below:

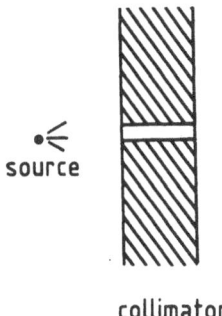

source

collimator

The collimator is not essential but it shields the detector from the source.

Assume a monochromatic perfectly collimated steady beam (no time dependence). Schrödinger:

$$[\nabla^2 + k^2 - U(r)]\psi(r) = 0 \tag{4.1}$$

$$k = \frac{p}{\hbar} = \frac{1}{\lambda} \qquad\qquad U(r) = \frac{2m}{\hbar^2} V(r)$$

p: c.m. momentum

m: reduced mass $\left(= \dfrac{m_1 m_2}{m_1 + m_2} \right)$

Solution without the target:

$$\psi(r) = e^{ikz} \tag{4.1'}$$

With the target, a scattered wave appears:

$$\psi(r) \underset{r\to\infty}{\to} e^{ikz} + \frac{e^{ikr}}{r} f(\theta, \phi)$$

(4.1")

$f(\theta, \phi)$: scattering amplitude

and the cross section at θ, ϕ is

$$\sigma(\theta, \phi) = |f(\theta, \phi)|^2$$

(4.2)

Flux conservation (no source, no absorption):

$$\sigma_{tot} = \frac{4\pi}{k} \text{ Im } f(0) \qquad \text{(optical theorem)}$$

(the target casts a shadow in the forward direction: the forward current
is reduced by the amount which appears in the scattering process).

Partial wave decomposition

For the time being, we still limit the discussion to elastic scattering
in a central symmetric potential.

The Schrödinger wave function can be decomposed into a radial and
angular part

$$\psi(r) = \frac{1}{kr} \sum_{\ell=0}^{\infty} U_\ell(r) Y_\ell(r)$$

(4.3)

$Y_\ell(\theta)$ are the spherical harmonics, eigenfunctions of ℓ^2. $U_\ell(r)$ is the
solution of the radial equation

$$\left[\frac{d^2}{dr^2} + k^2 - U(r) - \frac{\ell(\ell + 1)}{r^2} \right] U_\ell(r) = 0$$

(4.4)

$U(r)$ being a potential of limited range, the asymptotic solutions of
(4.4) for $r > R$ (R = range of $U(r)$) are:

$$F_\ell(kr) = kr \, j_\ell(kr)$$

called the "regular" solution and

$$G_\ell(kr) = -kr n_\ell(kr)$$

called the "irregular" solution.

100

If $kr \gg \ell$, $F_\ell(kr)$ and $G_\ell(kr)$ take the form

$$F_\ell(kr) \to \sin(kr - \frac{1}{2}\ell\pi)$$

$$G_\ell(kr) \to \cos(kr - \frac{1}{2}\ell\pi)$$

(4.5)

if there is no scattering

$$U_\ell(r) = F_\ell(kr)$$

if there is scattering, $U_\ell(r)$ (at $r > R$) is a linear combination of $F_\ell(kr)$ and $G_\ell(kr)$

$$U_\ell(r) = a \, F_\ell(kr) + b \, G_\ell(kr)$$

(4.6)

(a and b are complex but b/a is real if $V(r)$ is real).

With help of (4.5), (4.6) can be written $(r > R)$

$$U_\ell(r) = a \sin(kr - \frac{1}{2}\ell\pi) + b \cos(kr - \frac{1}{2}\ell\pi)$$

(4.7)

$$= (a^2 + b^2)^{1/2} \sin(kr - \frac{1}{2}\ell\pi + \delta_\ell)$$

δ_ℓ is a __real__ phase shift:

$$\delta_\ell = \text{artg}\,(b/a)$$

(4.8)

The effect of the scattering process is to shift the phase of the radial wave function. In absence of scattering, $U_\ell(r)$ take the form of a regular solution.

Comparing (4.2) to (4.7) one gets the norm

$$(a^2 + b^2)^{1/2} = (e^{i\delta_\ell}).$$

Finally:

$$U_\ell(r) = e^{i\delta_\ell} \sin(kr - \frac{1}{2}\ell\pi + \delta_\ell)$$

(4.9)

Alternative ways to write the radial wave function: first approach to S, T, K. (4.9) can also be written:

$$U_\ell(r) = e^{i\delta_\ell}[\cos\delta_\ell \, F_\ell(kr) + \sin\delta_\ell \, G_\ell(kr)],$$

(4.10)

i.e., the scattering process introduces into the radial wave function (beyond $r > R$) a contribution from the irregular solution $G_\ell(kr)$. The relative amplitude of this contribution is

$$K_\ell = (tg \ \delta_\ell) \tag{4.11}$$

One can also write (4.9)

$$U_\ell(r) = F_\ell(kr) + e^{i\delta_\ell} \sin\delta_\ell [G_\ell(kr) + i \ F_\ell(kr)]$$

$$= \sin(kr - \frac{1}{2} \ell\pi) + e^{i\delta_\ell} \sin\delta_\ell \ \exp \ i(kr - \frac{1}{2}\pi\ell) \tag{4.12}$$

i.e. the scattering process adds to the free particle regular solution an outgoing wave which amplitude is

$$T_\ell = e^{i\delta_\ell} \sin\delta_\ell = \frac{e^{2i\delta_\ell} - 1}{2i} = \frac{1}{\cot g \ \delta_\ell - i} \tag{4.13}$$

Finally, (4.9) can be written:

$$U_\ell(r) = \frac{1}{2} \ i \left[\exp - i(kr - \frac{1}{2}\pi\ell) - e^{2i\delta_\ell} \exp + i(kr - \frac{1}{2}\pi\ell) \right] \tag{4.14}$$

Comparing the ingoing to the outgoing wave, the scattering process introduces a phase:

$$S_\ell = e^{2i\delta_\ell} \tag{4.15}$$

(4.11), (4.13) and (4.15) give the definition of the K, T, S matrix element in the case of elastic scattering in the ℓ partial wave.

Using (4.1") for the wave function, its partial wave expansion (4.3) and (4.9), the scattering amplitude can be written

$$f(\theta) = -\frac{1}{k} \sum_\ell (2\ell + 1) \ e^{i\delta_\ell} \sin\delta_\ell \ P_\ell(\cos\theta)$$

$$= -\frac{1}{k} \sum_\ell (2\ell + 1) \ T_\ell \ \ell_\ell(\cos\theta) \tag{4.16}$$

and the wave function is written:

$$\psi(r) = \frac{1}{kr} \sum_\ell (2\ell + 1)i^\ell \ U_\ell(r) \ P_\ell(\cos\theta) \tag{4.17}$$

The elastic cross section is

$$s_{e\ell} = \frac{4\pi}{k^2} \sum_\ell (2\ell + 1) \left| \frac{e^{2i\delta_\ell} - 1}{2i} \right|^2 \tag{4.18}$$

Scattering length - Effective range

We have shown above that the phase shift δ_ℓ plays a central role in particle scattering. Quantum theory of scattering should provide guidelines on the behaviour of δ_ℓ. Experiments should measure δ_ℓ.

Lets first comment on the expected behaviour of δ_ℓ.

From the radial wave function (4.6)

$$U_\ell(r) = a\, F_\ell(kr) + b\, G_\ell(kr).$$

δ_ℓ can be computed, imposing continuity of the wave function at $r = R$:

$$L(R) = \frac{F'_\ell(kR) + \dfrac{b}{a} G'_\ell(kR)}{F_\ell(kR) + \dfrac{b}{a} G_\ell(kR)} = \frac{F'_\ell(KR)}{F_\ell(KR)}$$

where

$$K^2 = k^2 - U.$$

One gets:

$$tg\delta_\ell = \frac{b}{a} = \frac{-F'_\ell(kR) + L(R)F_\ell(kR)}{G'_\ell(kR) - L(R)G_\ell(kR)} \qquad (4.19)$$

Scattering by a hard sphere:

$$K^2 \to -\infty \qquad\qquad L(R) \to \infty$$

$$tg\delta_o = \frac{-F_o(kR)}{G_o(kR)} = -tg(kR)$$

$$\delta_o = -kR$$

$$f(\theta) = -\frac{1}{k} e^{-ikR} \sin kR \qquad (\text{S-wave})$$

$$\sigma_{tot} = \frac{4\pi}{k^2} \sin^2 kR \sim 4\pi R^2$$

The S-wave scattering by a hard sphere of radius R leads to a total cross section four times larger than the classical one.

Low energy limit: Using (4.19) and the limits for $kr \ll \ell$

$$F_\ell(kr) = \frac{(kr)^{\ell+1}}{(2\ell + 1)!!}$$

$$G_\ell(kr) = \frac{(2\ell - 1)!!}{(kr)^\ell}$$

One gets

$$tg\delta_\ell \sim (k^{2\ell+1}) \qquad (4.20)$$

The tangent of δ_ℓ approaches zero as $k^{2\ell+1}$. For the S-wave

$$tg\delta_o = \frac{1 - RL_o(R)}{L_o(R)} \qquad k = -k\alpha$$

α is called the "zero energy scattering length" (or "scattering length").

The radial wave function can be written, in this approximation:

$$U_o(r) \rightarrow e^{i\delta_o} \sin(kr + \delta_o)$$

$$\rightarrow k(r - \alpha)$$

α can be interpreted as the "effective radius" of the target at zero energy $(k \rightarrow 0)$.

In the case that $\delta_o \sim \pi/2$: $tg\delta_o \rightarrow \infty$, $\rightarrow |\alpha|$ is large.

A large scattering length implies a large admixture of the irregular function into the external wave function.

If $k > 0$, α large and negative, we are in presence of a resonance near threshold.

If $k < 0$, α large and positive, we have a bound state near zero.

To extend this S-wave formalism beyond $k \sim 0$, use a development in k.

If $|\alpha|$ is large;

$$k \cot g\delta_o = -\frac{1}{\alpha} + \frac{1}{2}\rho k^2 - P\rho^3 k^4 \quad \ldots \quad (|\rho/\alpha| \ll 1) \qquad (4.21)$$

ρ is the "effective range"
P the "shape parameter".

if $|\alpha|$ is not large, use instead:

$$\frac{1}{k} tg\delta_o = -\alpha - \frac{1}{2}\rho \alpha^2 k^2 + \ldots \quad (|\rho/\alpha| \gg 1) \qquad (4.21')$$

Absorption

We shall now introduce the possibility that another "reaction" channel compete with "elastic" scattering.

The outgoing wave will not only be shifted (by $2\delta_\ell$), but also attenuated (attenuation factor η_ℓ, real between 1, no attenuation, and 0: full absorption): (4.14) becomes

$$U_\ell(r) = \frac{1}{2} i[exp - i(kr - \frac{1}{2} \pi\ell) - \eta_\ell \, e^{2i\delta_\ell} \, exp + i(kr - \frac{1}{2} \pi\ell)] \qquad (4.22)$$

S_ℓ, which was $e^{2i\delta_\ell}$, becomes $\eta_\ell e^{2i\delta_\ell}$

T_ℓ, which was $e^{i\delta_\ell} sin\delta = \dfrac{e^{2i\delta_\ell} - 1}{2i}$ becomes $\dfrac{\eta_\ell e^{2i\delta_\ell} - 1}{2i}$

δ_ℓ and η_ℓ are real. One can introduce, equivalently, a complex parameter, Δ_ℓ:

$$\Delta_\ell = \delta_\ell + i\gamma_\ell \quad \text{with} \quad \eta_\ell = e^{-2\gamma_\ell}$$

so that, with absorption, S_ℓ, T_ℓ, K_ℓ can still be written

$K_\ell = tg\Delta_\ell$

$$T_\ell = e^{i\Delta_\ell} sin\Delta_\ell = \frac{1}{cotg\Delta_\ell - i} = \frac{K_\ell}{1 - iK_\ell} \qquad (4.23)$$

$$S_\ell = e^{2i\Delta_\ell} = \frac{1 + iK_\ell}{1 - iK_\ell}$$

The elastic, reaction and total cross sections, with absorption, are written:

$$\sigma_{ela} = \frac{4\pi}{k^2} \sum_\ell (2\ell + 1) \left| \frac{\eta_\ell \, e^{2i\delta_\ell} - 1}{2i} \right|^2 \qquad \text{(compare to (4.18))} \qquad (4.24)$$

$$\sigma_{rea} = \frac{\pi}{k^2} \sum_\ell (2\ell + 1)(1 - \eta_\ell^2) \qquad (4.25)$$

$$\sigma_{tot} = \sigma_{ela} + \sigma_{rea} = \frac{\pi}{k^2} \sum_\ell (2\ell + 1) \, 2(1 - \eta_\ell \, cos2\delta_\ell) \qquad (4.26)$$

σ_{rea} is maximum, of course, for $\eta_\ell = 0$ (full absorption)

$$\sigma_{reamax}^\ell = \frac{\pi}{k^2} (2\ell + 1)$$

then $\sigma_{ela}^\ell = \sigma_{rea}^\ell$ and

$$\sigma_{reamax}^\ell = \frac{\pi}{k^2} (2\ell + 1)$$

σ_{ela} is maximum for $\eta_\ell = 1$ (no absorption) and $\delta_\ell = \pi/2$:

$$\sigma_{elamax}^\ell = \frac{4\pi}{k^2} (2\ell + 1)$$

Of course, the optical theorem is still satisfied

$$\text{Im } f(0) = \frac{1}{k} \sum_{\ell} (2\ell + 1) \frac{1}{2} (1 - \eta_{\ell} \cos 2\delta_{\ell})$$

$$= \frac{k}{4\pi} \sigma_{tot}$$

Argand plots

We have seen that the transition scattering amplitude for a given partial wave can be written in terms of two real parameters: δ, η:

$$T_{ela} = \frac{\eta_{\ell} e^{2i\delta} - 1}{2i}$$

η and δ must be measured experimentally as function of the energy ω of the system. The evolution of T_{ela} can be followed on an "Argand plot" which displays:

$$\text{Re } T_{ela} = \frac{\eta}{2} \sin 2\delta$$

$$\text{Im } T_{ela} = \frac{1}{2} (1 - \eta \cos 2\delta)$$

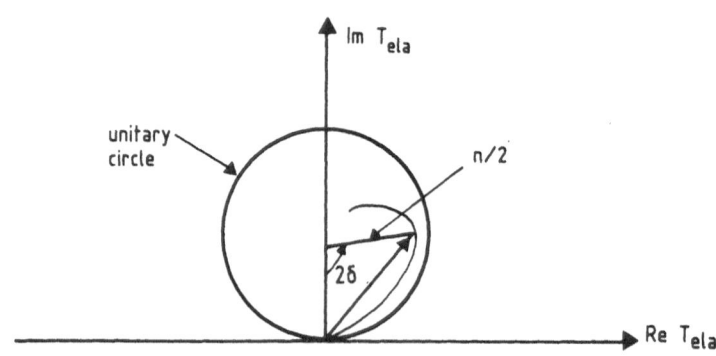

If no absorption, T_{ela} will stay on unitary circle: a resonance will appear as a loop going anticlockwise (Wigner's inequality) with a maximum speed at the resonance. The mass of the resonance will be defined by the energy at $\delta = \pi/2$.

In practice, a background term and absorption will introduce distorsions of the loop. But the resonances will still be identified by the general features given above (look at p. 295 and following of [8]).

106

The K and M matrix formalism

We shall now describe the K (and M) matrix formalism which is used in the data analysis to establish the existence of new resonances.

First lets introduce the reduced transition amplitude $\mathcal{G}$ and the corresponding $\mathcal{H}$ amplitude

$$T = k^{\ell+1/2}\, \mathcal{G}\, k^{\ell+1/2} \qquad\qquad (4.27)$$

$$K = k^{\ell+1/2}\, \mathcal{H}\, k^{\ell+1/2}$$

This transformation has the advantage of underlying the kinematical singularities which occur at threshold of new channels.

In a multi-channel analysis, $\mathcal{G}$ and $\mathcal{H}$ are matrices related by:

$$\mathcal{G} = \frac{\mathcal{H}}{1 - i\, k^{\ell_a+1/2}\, \mathcal{H}\, k^{\ell+1/2}} \qquad\qquad (4.28)$$

$k^{\ell+1/2}$ is a diagonal kinetic matrix

$$k^{\ell+1/2} = \begin{vmatrix} k_1^{\ell_1+1/2} & & \\ & k_2^{\ell_2+1/2} & \\ & & \ddots \\ & & & \ddots \end{vmatrix} \qquad\qquad (4.29)$$

with

$$k_i = \sqrt{\frac{s - m^2}{s}} \qquad\qquad m^2 = (m_a + m_b)^2$$

below threshold $k_i = i|k_i|$.

The $\mathcal{M}$ matrix is the inverse of $\mathcal{H}$:

$$\mathcal{M} = \mathcal{H}^{-1} \qquad\qquad (4.30)$$

and

$$\mathcal{G} = \frac{1}{\mathcal{M} - ik^{2\ell+1}} \qquad\qquad (4.31)$$

Consider an S-wave two channels situation (like $\pi\pi \to \pi\pi$, $\pi\pi \to K\bar{K}$).

The elastic channel (channel 1) will be written (remember (4.3), (4.10)):

$$r\psi_1' = \frac{\sin k_1 r}{k_1} - \mathcal{H}_{11} \cos k_1 r$$

and the inelastic channel

$$r\psi_2^1 = - \mathcal{H}_{21} \cos k_2 r$$

is a real symmetric 2 × 2 matrix

$$\mathcal{H} = \begin{vmatrix} \alpha & \beta \\ \beta & \gamma \end{vmatrix}$$

then:

$$r\psi_1^1 = \frac{\sin k_1 r}{k_1} - \alpha \cos k_1 r = \frac{\sin k_1 r}{k_1} + \mathcal{G}_{11} e^{ik_1 r}$$

(4.32)

$$r\psi_2^1 = -\beta \cos k_2 r = \mathcal{G}_{21} e^{ik_2 r}$$

The wave function must satisfy boundary conditions at $r = 0$ such that

$$(r\psi_1^1)_{r=0} = A(r\psi_1^1)'_{r=0}$$

with $A = a + ib$. Using (4.32)

$$a = -\alpha + \frac{\beta^2 \gamma\, k_2^2}{1 + \gamma^2 k_2^2}$$

(4.33)

$$b = \frac{\beta^2\, k_2}{1 + \gamma^2\, k_2^2}$$

$$\mathcal{G}_{11} = \frac{A}{1 - ik_1 A}$$

(4.34)

which gives:

$$k \cot g \Delta = \frac{1}{A}$$

(this is the zero effective range approximation). Then:

$$\sigma_{ela} = \frac{4\pi}{k^2} |T_{ela}|^2 = 4\pi |\mathcal{G}_{11}|^2$$

$$= 4\pi \frac{a^2 + b^2}{1 + 2 k_1 b + k_1^2 (a^2 + b^2)}$$

and:

$$\mathcal{G}_{21} = \frac{\beta^2}{1 + \gamma^2 k_2^2} \frac{1}{|1 - ik_1 A|^2}$$

$$= \frac{b/k_2}{1 + 2 k_1 b + k_1^2(a^2 + b^2)}$$

giving:

$$\sigma_{rea} = \frac{4\pi}{k_1^2} |T_{21}|^2 = \frac{4\pi}{k_1} \left| \frac{T_{21}}{k_1^{1/2} k_2^{1/2}} \right|^2 k_2$$

$$= \frac{4\pi}{k_1} \frac{b}{1 + 2 k_1 b + k_1^2(a^2 + b^2)}$$

A resonance will occur when T has a pole, i.e:

$$1 - ikA = 0$$

(with $A = a + ib$):

$$1 + k_R b - ik_R a = 0$$

introducing the complex energy:

$$\omega_R = m_R - i\Gamma/2$$

$$\omega_R = m_a + m_b + \frac{k_R^2}{2\mu} \qquad (\mu = \frac{m_a m_b}{m_a + m_b})$$

$$m_R = (m_a + m_b) \left[1 - \frac{a^2 - b^2}{2 m_a m_b (a^2 + b^2)^2} \right]$$

$$\Gamma = - \frac{(m_a + m_b)}{m_a m_b} \frac{2ab}{(a^2 + b^2)^2} \qquad\qquad (4.35)$$

The resonance parameters (m_R, Γ) are given in terms of a, b, themselves function of the $\mathcal{H}$ matrix elements (4.33).

Using the $\mathcal{MG}$ matrix with elements

$$\mathcal{M} = \begin{pmatrix} m_{11} & m_{12} \\ m_{12} & m_{22} \end{pmatrix}$$

$$T = \frac{k_1^{1/2} k_2^{1/2}}{D} \begin{pmatrix} m_{22} - ik_2 & -m_{12} \\ -m_{12} & m_{11} - ik_1 \end{pmatrix}$$

$$D = \det(\mathcal{M} - ik) = (m_{11} - ik_1)(m_{22} - ik_2) - m_{12}^2$$

the poles of T will be the zeroes of D. The transition amplitudes in terms of $\mathcal{M}$ are:

$$T_{11} = \frac{k_1}{D} (m_{22} - ik_2)$$

$$T_{12} = T_{21} = \frac{-k_1^{1/2} k_2^{1/2}}{D} m_{12}$$

$$T_{22} = \frac{k_2}{D} (m_{11} - ik_1)$$

and the cross section are defined, as usual, by

$$\sigma(i \to j) = \frac{4\pi}{k^2} |T_{ij}|^2$$

so that σ_{11}, σ_{12}, σ_{22} can be related to the elements of $\mathcal{M}$.

5. THE I = 0 S-WAVE ($J^P = 0^+$) AROUND 1 GeV

To determine the phase shift $\delta_o = \delta$ and $\eta_o = \eta$ of the I = 0 $\pi\pi$ S-wave ($J^{PC} = 0^{++}$), the ideal experiment would be to measure with high precision the cross section and angular distribution of the processes

$$\pi\pi \to \pi\pi$$

$$\pi\pi \to K\bar{K}$$

$$K\bar{K} \to K\bar{K} \qquad\qquad (5.1)$$

$$\pi\pi \to \eta\eta \qquad \ldots\ldots$$

In practice, we do not have direct access to these reactions and one must extract the information from more complex processes like

$$\pi N \rightarrow \pi \pi N, \qquad \pi \pi \Delta$$

$$\pi N \rightarrow K \bar{N}, \qquad K \bar{K} \Delta \qquad\qquad (5.2)$$

$$\bar{p} p \rightarrow \pi \pi X, \qquad K \bar{K} X.$$

More attractive reactions are perhaps those in which the initial state is controlled by reasonably well understood mechanisms. This is the case of the "two-photon production"

$$e^+ e^- \rightarrow (e^+ e^-) \gamma \gamma$$

and

$$\gamma \gamma \rightarrow \pi \pi, \; K \bar{K} \; \text{etc.} \qquad\qquad (5.3)$$

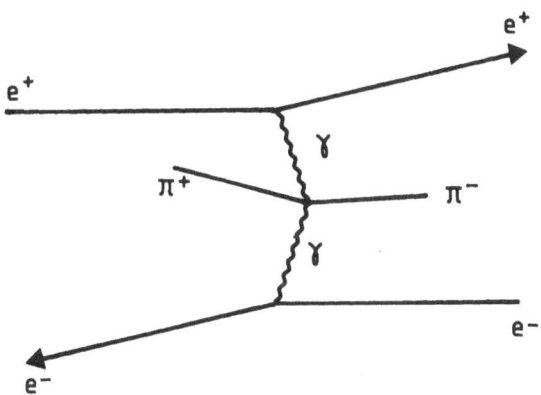

This is also the case of the "two pomeron exchange":

$$pp \rightarrow (pp) \; PP$$

$$PP \rightarrow \pi \pi, \; K \bar{K}, \; \text{etc.} \qquad\qquad (5.4)$$

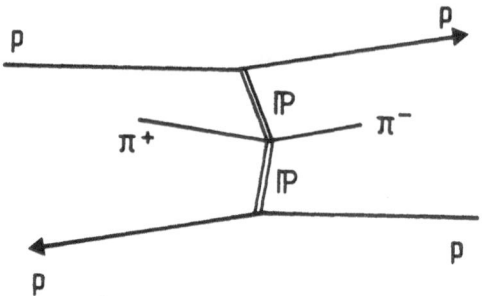

Before studying in more details the results of (5.4), let us comment on reactions (5.2), which have provided in the 70's most of the experimental information on the ππ reaction. These reactions are assumed to go via the "one pion exchange" mechanism provided that the ππ final state is produced at low momentum transfer with respect to the incident π:

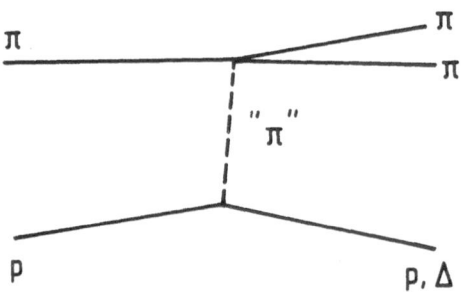

The method relies on the extrapolation of the data to the π-pole $(t = \mu^2)$. But, for the reactions of the type

πp → ππp,

the amplitude has a zero between $t = 0$ and $t = \mu^2$ (Adler's zero) which complicates the extrapolation.

In the case of

πp → ππΔ,

there are no such zeroes, but the extrapolation to the pion pole is further away from the physical region.

Assuming these difficulties overcome, and having extracted from the data the total $\sigma_{\pi\pi}$ cross section and the spherical harmonics at $t = \mu^2$, one can in turn express these quantities in terms of the S, P, D, F waves:

$$S_{\pi\pi} = \frac{4\pi}{k^2} \left(|S|^2 + 3|P|^2 + 5|D^2| + \ldots \right)$$

$$\langle Y_1^0 \rangle = \frac{4\pi}{k^2 \sigma_{\pi\pi}} \left(\sqrt{\frac{3}{\pi}} \, Re(S^*P) + 2\sqrt{\frac{3}{\pi}} \, Re(P^*D) \right)$$

For S, D ... waves, we have in addition to take into account the isospin decomposition

$$S = \frac{2}{3} T_0^0 + \frac{1}{3} T_0^2 \qquad\qquad (\pi^+\pi^- \text{ final state})$$

$$P = T_1^1$$

The transition matrix elements are written (see (4.13))

$$T_\ell^I = \frac{1}{2i} \left(\eta_\ell^I \, e^{2i\delta_\ell^I} - 1 \right)$$

There are in general too many unknowns to determine at each energy all the parameters η_e^I, δ_ℓ^I. One therefore performs energy dependent analysis, imposing smooth energy dependence of η_ℓ^I and δ_ℓ^I and introducing, for the waves already known (like the P-wave), the known energy dependence.

Recently, K.L. Au, D. Morgan and M. Pennington [11] have analyzed the data obtained at the ISR colliders at CERN [12] on reaction (5.4), using the K matrix formalism discussed in sect. 4.

Consider the I = 0 S-wave with two channels, 1 and 2:

$\pi\pi \rightarrow \pi\pi$: reduced transition amplitude: $\mathcal{G}_{11}$.

$\pi\pi \rightarrow K\bar{K}$: reduced transition amplitude: $\mathcal{G}_{12}$ $\mathcal{G}_{12} = \mathcal{G}_{21}$ by CPT.

$K\bar{K} \rightarrow K\bar{K}$: reduced transition amplitude: $\mathcal{G}_{22}$.

The reduced transition amplitudes are related to the transition amplitudes T and the S matrix through the relations (4.13, 4.15, 4.27):

$$S = I + 2i \, k^{1/2} \, \mathcal{G} \, k^{1/2}, \tag{5.5}$$

with the unitarity relation

$$Im \; \mathcal{G}_{ij} = \sum_r k_r \, \mathcal{G}_{ir}^* \, \mathcal{G}_{rj} \tag{5.6}$$

This quadratic unitarity relation can be described by the diagrams:

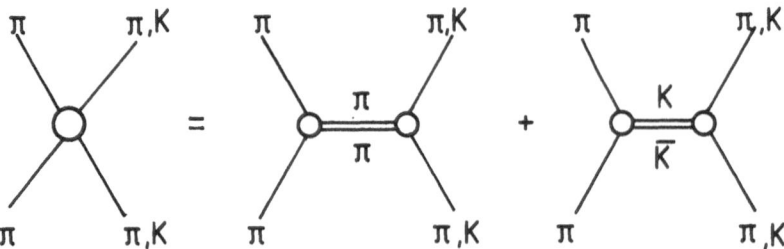

giving explicitly

$\pi\pi \to \pi\pi$ $\text{Im } \mathcal{G}_{11} = k_1 \mid \mathcal{G}_{11}\mid^2 + k_2\mid \mathcal{G}_{12}\mid^2$

$\pi\pi \to K\bar{K}$ $\text{Im } \mathcal{G}_{12} = k_1 \mathcal{G}_{11}^* \mathcal{G}_{12} + k_2 \mathcal{G}_{12}^* \mathcal{G}_{22}$

with

$$k_1 = \sqrt{1 - \frac{4m_\pi^2}{\omega^2}}$$

$$k_2 = \sqrt{1 - \frac{4m_K^2}{\omega^2}}$$

Using the $\mathcal{M}$ matrix [(4.36), (4.37), (4.38)]:

$$\mathcal{G}_{11} = \frac{m_{22} - ik_2}{D_m} \tag{5.7}$$

$$\mathcal{G}_{22} = \frac{m_{11} - ik_1}{D_m} \tag{5.8}$$

$$\mathcal{G}_{12} = \frac{-m_{12}}{D_m} \tag{5.9}$$

$$D_m = (m_{11} - ik_1)(m_{22} - ik_2) - m_{12}^2 \tag{5.10}$$

One can also use the $\mathcal{H}$ matrix elements, related to the $\mathcal{M}$ matrix through:

$$\frac{k_{22}}{D_K} = m_{11} \qquad \frac{k_{11}}{D_K} = m_{22} \qquad \frac{k_{12}}{D_K} = -m_{12} \tag{5.11}$$

$$D_K = k_{11}k_{22} - k_{12}^2 . \tag{5.12}$$

$$\mathcal{G} = \frac{\mathcal{H}}{1 - ik^{1/2}\mathcal{H}k^{1/2}} \tag{5.13}$$

with (5.5)

$$S = \frac{1 + i\, k^{1/2}\, \mathcal{H}\, k^{1/2}}{1 - i\, k^{1/2}\, \mathcal{H}\, k^{1/2}} \; . \tag{5.14}$$

As we have seen in sect. 3, the resonances correspond to poles of the S-matrix, i.e. poles of the $\mathcal{G}$ matrix, i.e.:

$$1 - ik^{1/2}\, \mathcal{H}\, k^{1/2} = 0 \tag{5.15}$$

or, in terms of the $\mathcal{M}\mathcal{G}$ matrix:

$$D_m = 0 \tag{5.16}$$

To find these poles, Au, Morgan and Pennington parametrize the $\mathcal{M}\mathcal{G}$ matrix (or the $\mathcal{G}$ matrix) function of the energy squared $\omega^2 = s$, with a polynomial and first order poles:

$$\mathcal{H}_{ij} = \sum_{\text{poles}} \frac{f_i^p\, f_k^p}{s_p - s} + \sum_n c_{ij}^n (s)^n \tag{5.17}$$

The number of poles introduced in this parametrization is arbitrary: indeed, several acceptable solutions are obtained with different number of poles. The coefficients of the pole terms are written in a factorizable way to help, as we shall see later, in the determination of the couplings to the two final states.

We have now to introduce the production amplitudes for the process:

$$\mathbb{P}\mathbb{P} \to \pi\pi,\ K\bar{K}$$

($\mathbb{P}$ is for pomeron exchange).

The basic hypothesis here is that the pomerons do not participate as intermediate states in the unitarity diagrams:

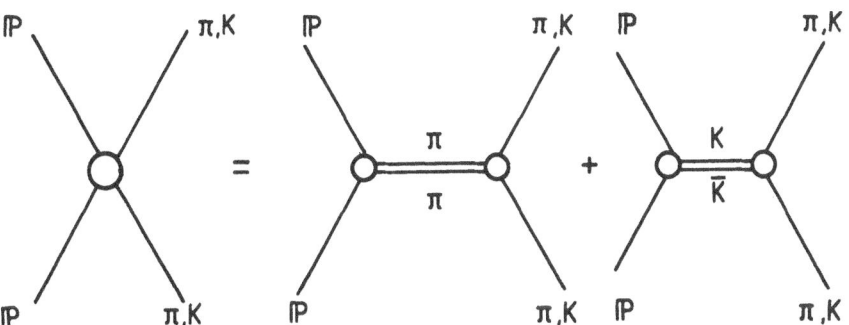

The production amplitudes A_1 (for $\pi\pi$) and A_2 (for $K\bar{K}$) satisfy then the unitarity relations:

$$\mathrm{Im}\, A_1 = k_1 A_1^* \, \mathcal{G}_{11} + k_2 A_2^* \, \mathcal{G}_{21}$$

$$\mathrm{Im}\, A_2 = k_1 A_1^* \, \mathcal{G}_{12} + k_2 A_2^* \, \mathcal{G}_{22}$$

(5.18)

Note here the linearity of these relations, in contrast to the quadratic form of (5.6).

To satisfy (5.18) and (5.6), we write the production amplitudes:

$$A_i = \sum_j \alpha_j \, \mathcal{G}_{ji}$$

where the α_j's are real functions of s. For $\mathbb{PP} \to \pi^+\pi^-$:

$$A_1 = \sqrt{\tfrac{2}{3}} \, (\alpha_1 \, \mathcal{G}_{11} + \alpha_2 \, \mathcal{G}_{21})$$

and for $PP \to K^+K^-$:

$$A_2 = \tfrac{1}{\sqrt{2}} \, (\alpha_1 \, \mathcal{G}_{12} + \alpha_2 \, \mathcal{G}_{22})$$

For the S-wave production one gets the production amplitudes from the Mueller-Regge analysis of the double pomeron production:

$$\sigma_{\mathbb{PP}}(\omega) = \frac{1}{2\,\alpha^{12}\,\beta^4_{\mathbb{PPP}}(t)} \cdot \frac{(m_\rho^2 + \omega^2)^2}{\omega^3} \cdot \frac{d^3\sigma}{dt_1\,dt_2\,d\omega}$$

$$|A(\omega)|^2 = \frac{1}{16\pi} \frac{\omega^3}{\sqrt{\omega^2 - 4\mu^2}}\,\sigma_{\mathbb{PP}}(\omega) \qquad (\mu = m_\pi,\, m_K)$$

In principle if $\mathcal{G}_{11}$ and $\mathcal{G}_{12}$ were accurately known, the double pomeron measurements would just serve as a measurement of the couplings α_1 and α_2. In practice, the data obtained by [12], which are strongly dominated by S-wave at the small momentum transfers which are observed, help to contraint the amplitudes $\mathcal{G}_{11}$ and $\mathcal{G}_{12}$ around the $K\bar{K}$ threshold region where the $\pi\pi$ S-wave is not dominant in the more traditional reactions like (5.2).

All the experimental data available on $\mathcal{G}_{11}$ and $\mathcal{G}_{12}$ will therefore be used to find the best parameterization of the $\mathcal{H}$ or $\mathcal{M}$ matrix. Basically, the experimental observations provide detailed information on the phase shift $\delta_{11}(\omega)$, the elasticity $\eta_{11}(\omega)$ and the phase $\phi_{12}(\omega)$ which enter in this formulation of $\mathcal{G}_{11}$ and $\mathcal{G}_{12}$:

$$\mathcal{C}_{11} = \frac{\eta_{11} e^{2i\delta_{11}} - 1}{2ik_1}$$

$$\mathcal{C}_{12} = \frac{(1 - \eta_{11}^2)^{1/2} e^{i\phi_{12}}}{2 k_1^{1/2} k_2^{1/2}}$$

(5.19)

To justify the form of (5.19), let us remember that (5.6)

$$\text{Im } \mathcal{C}_{11} = k_1 |\mathcal{C}_{11}|^2 + k_2 |\mathcal{C}_{12}|^2$$

giving

$$k_2 |\mathcal{C}_{12}|^2 = \text{Im } \mathcal{C}_{11} - k_1 |\mathcal{C}_{11}|^2$$

but

$$k_1 |\mathcal{C}_{11}|^2 = \frac{1 + \eta_{11}^2 - 2\eta_{11} \cos\delta_{11}}{4k_1}$$

and

$$\text{Im } \mathcal{C}_{11} = \frac{1 - \eta_{11} \cos 2\delta_{11}}{2k_1}$$

therefore:

$$k_2 |\mathcal{C}_{12}|^2 = \frac{1 - \eta_{11}^2}{4k_1}$$

$$\mathcal{C}_{12} = \frac{(1 - \eta_{11}^2)^{1/2} e^{i\phi_{12}}}{2k_1^{1/2} k_2^{1/2}}$$

$|\mathcal{C}_{12}|$ is relatively well constrained by the measurements of the CERN-Munich $\pi^+\pi^-$ production [13], as well as by the measurements of Cohen et al. [14] and Etkin et al. [15] of the reactions:

$$\pi^- p \rightarrow K_s^0 K_s^0 n$$

and

$$\pi^- p \rightarrow K^+ K^- n$$

The phase ϕ_{12} is more difficult to measure unambiguously. Usually, it is measured, relative to a well-known wave, like the D-wave which is dominant in the 1.0 to 1.4 GeV mass region (the $f_2(1250)$):

$$\phi_{SD} = |\phi_{12} - \phi_D|$$

ϕ_D being given by the resonant behaviour of $f_2(1250)$:

$$\text{tg } \phi_D = \frac{\Gamma/2}{\omega_R - \omega}$$

Au, Morgan and Pennington find several good fits to all the 258 data in the mass range from $\pi\pi$ threshold up to 1.7 GeV. In terms of the matrix parameterization, they find good fits if at least one pole is introduced.

But the number of poles introduced in the $\mathcal{H}$ matrix depend on the complexity of the polynomial associated to the parameterization (5.17). Equally good fits are obtained with 3 poles in the $\mathcal{H}$ matrix and relatively simpler polynomials. This situation underlines that the poles of the $\mathcal{H}$ matrix are not necessarily related to S-matrix poles and should not be taken as evidence for resonances. Using the $\mathcal{MG}$ matrix parameterization, they find a good fit with one pole which is strongly related to the pole found in the $\mathcal{H}$ matrix. The small differences observed between these three good solutions are highlighted by looking at the Argand plot of the $\pi\pi$ S-wave elastic amplitude: $k_1 \mathcal{G}_{11}$:

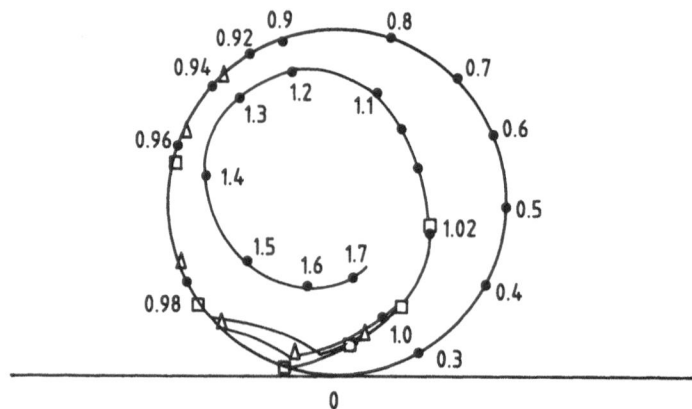

Having obtained a parameterization of the $\mathcal{H}$ matrix which gives a good interpretation of the set of data available on the S-wave around the $K\bar{K}$ threshold, one must deduce the poles of the S-matrix, in the complex energy plane, which will be taken as genuine resonances if they occur not too far from the physical region, and with a negative imaginary part (eq. (3.9)).

Given a complex energy ω, we get 2 determinations of k_2:

$$2k_2 = \pm\sqrt{\omega^2 - 4m_k^2} \qquad (5.20)$$

Therefore, a pole in ω will have two images in the k_2 plane, such that

$$k_2(1) = -k_2(2),$$

unless a pole in ω occurs on the real axis of ω (as a virtual bound state).

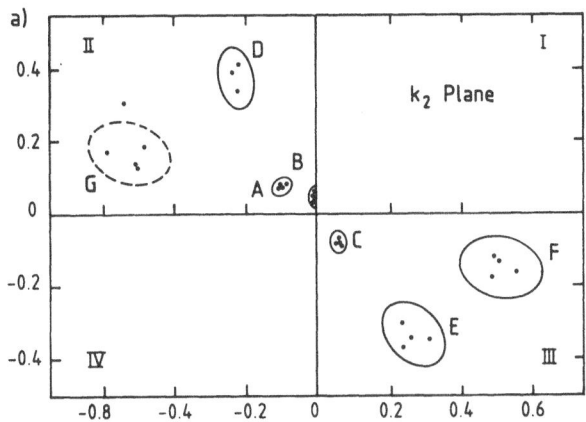

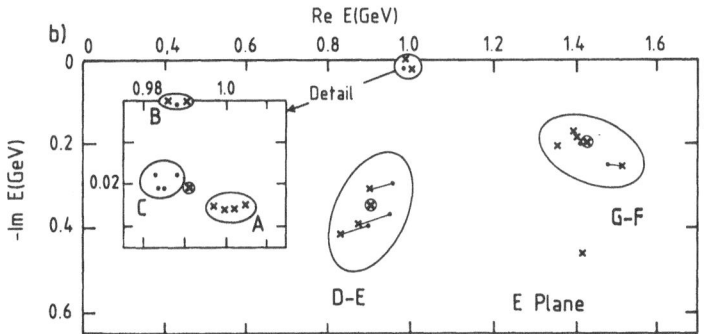

Au, Martin and Pennington find that their solutions correspond to 7 poles in the k_2 plane; 6 of them can be associated 2 by 2, according to (5.20), the seventh one occurring on the real axis, just below the $K\bar{K}$ mass threshold.

Following David Morgan, one can define the resonance position as the complex energy which satisfies:

$$m_{11} - ik_1 = 0 \tag{5.21}$$

Four resonances are then given with the following mass and width:

f_o (991)	M =	991 MeV	$\Gamma =$ 42 MeV
f_o (910)		910 MeV	350 MeV
f_o (1430)		1430 MeV	200 MeV
f_o (988)		988 MeV	0 MeV

The couplings of these resonances to the $\pi\pi$ and $K\bar{K}$ channels are given, according to the definition:

$$\mathcal{C}_{ij} = \frac{\gamma_i \gamma_j}{s_R - s} \, , \qquad\qquad s = \omega^2 \tag{5.22}$$

as the limit:

$$\gamma_i \gamma_j = \lim \, (s_R - s) \, \mathcal{C}_{ij} \qquad s \rightarrow s_R \tag{5.23}$$

Applied for instance to the $\mathcal{MG}$ matrix formalism, these definitions lead to:

$$\gamma_1^2 = \frac{-m_{22} + ik_2}{D'_m}$$

$$\gamma_1 \gamma_2 = \frac{m_{12}}{D'_m}$$

$$\gamma_2^2 = \frac{-m_{11} + ik_1}{D'_m}$$

with

$$D'_m = \frac{\partial}{\partial s} \, D_m$$

D_m being given by (5.10).

The physical couplings, g_i, will be taken as the modulii of the γ_i:

$$g_i = |\gamma_i| \tag{5.24}$$

and are found to be equal to:

f_o (991)	g_π = 0.22 GeV	g_K = 0.28 GeV
f_o (910)	0.52	0.27

f_o (1430)	0.58	0.16
f_o (988)	0.02	0.35

Accepting for the time being, that the four resonances claimed by Au, Martin and Pennington are well established, their couplings to the $\pi\pi$ and $K\bar{K}$ channels may give precious indications on their "nature".

Consider the couplings expected from SU(3).

For a singlet:

$$1 \to 8 \times 8$$

$$\varepsilon_1 = \frac{1}{\sqrt{2}} K\bar{K}, \qquad \sqrt{\frac{3}{8}} \pi\pi, \qquad -\sqrt{\frac{1}{8}} \eta_8 \eta_8$$

For an octet member:

$$\varepsilon_\gamma = \frac{1}{\sqrt{2}} K\bar{K}, \qquad -\sqrt{\frac{3}{5}} \pi\pi, \qquad -\sqrt{\frac{1}{5}} \eta_8 \eta_8$$

But an isoscalar singlet can mix with an isoscalar octet and form, by ideal mixing, a pure $s\bar{s}$ object:

$$\varepsilon_s = s\bar{s} = \sqrt{\frac{1}{3}} \varepsilon_1 - \sqrt{\frac{2}{3}} \varepsilon_8$$

i.e.

$$\varepsilon_s = 0 \ \pi\pi, \quad -\sqrt{\frac{6}{5}} K\bar{K}, \qquad \sqrt{\frac{8}{15}} \eta_8 \eta_8$$

Then the non-strange partner has the couplings:

$$\varepsilon_{ns} = \frac{u\bar{u} + d\bar{d}}{\sqrt{2}} = \sqrt{\frac{2}{3}} \varepsilon_1 + \sqrt{\frac{1}{3}} \varepsilon_8$$

i.e.

$$\varepsilon_{ns} = -\frac{3}{\sqrt{5}} \pi\pi, \quad -\frac{3}{\sqrt{15}} K\bar{K}, \quad \frac{1}{\sqrt{15}} \eta_8 \eta_8$$

To summarize, the relative branching amplitudes are:

	$\pi\pi$	$K\bar{K}$	$\eta_8 \eta_8$
ε_1	1	$2/\sqrt{3}$	$-1/\sqrt{3}$
ε_8	1	$-1/\sqrt{3}$	$1/\sqrt{3}$
ε_{ns}	1	$\sqrt{1/3}$	$-\sqrt{1/27}$
ε_s	0	1	-0.67

Note the relative importance of $\eta_8 \eta_8$ for ε_8 and ε_s.

A priori, the obvious candidate for a ε_s object is $f_o(988)$, since it does not couple to $\pi\pi$.

But if $f_o(988)$ were the $s\bar{s}$ member of an ideally mixed nonet, on should expect to observe the non-strange partner, ε_{ns}, at $\sim$ 730 MeV, according to the empirical rule that the mass splitting is essentially due to the mass difference between the strange and non-strange quarks and is given by:

$$m_\phi^2 - m_\omega^2 = 0.43 \text{ GeV}^2.$$

No such isoscalar, scalar resonance is observed at $\sim$ 730 MeV: the identification of the $f_o(988)$ to a $s\bar{s}$ isoscalar, $J^P = 0^+$ resonance, is therefore doubtful.

What is it? According to T. Barnes, it could be a virtual $K\bar{K}$ bound state [7] and then not enter into the $q\bar{q}$ classification.

The next attempt is to associate the $f_o(991)$, for which $g_\pi/g_k \sim 0.8$, to an ε_1 non-mixed singlet ($g_\pi/g_k = 0.87$ for ε_1), and the $f_o(910)$, to the isoscalar octet ($g_\pi/g_K = 1.7$ compared to 1.73 as expected for ε_s). This reduction of $f_o(991)$ and $f_o(910)$ to the two isoscalar members of the $J^P = 0^+$ nonet is attractive, in particular if we accept the assumption that the isovector scalar partner is the $a_o(980)$. According to the Gell-Mann-Okubo mass formula for a $q\bar{q}$ octet;

$$m_f^2 = \frac{1}{3} (4m_k^2 - m_a^2)$$

we would need a (not yet observed) scalar K resonance in the 900–950 MeV mass range.

The $f_o(1430)$ could then be taken as a radial excitation of the $f_o(910)$.

These assignments do not fit well with the $q\bar{q}$ classification of Godfrey and Isgur [5] which would expect the isoscalar members of the fundamental ^3P_o nonet at $\sim$ 1.09 and $\sim$ 1.38 GeV, with a radial excitation not appearing below $\sim$ 1.80 GeV.

This contradiction may indicate that the scalar mesons are subject to additional complications (like mixing with glueball states). But before drawing too far fetch conclusions, let us underline that this analysis, although being the most comprehensive and most detailed one, still suffers from limitations which may affect its conclusions: it is the result of the analysis of a mixture of data where bias, systematic experimental errors and resolution vary from experiment to experiment: it is difficult to estimate the consequences of these effects. The analysis depends on the parameterization of the production amplitudes. It would be highly desirable to accumulate large statistics and accurate information on other reactions, like i.e.

$$\bar{p}p \rightarrow \pi^+\pi^- X, \quad K^+K^- X, \quad \eta\eta X$$

to repeat the analysis with completely different production mechanisms.

This coupled channel analysis is limited to 2 channels: $\pi\pi$ and $K\bar{K}$. What is the role of the $\eta\eta$ and 4π channels?

The 3 resonance solution in the 1 GeV region is giving better interpretation of the data than a 2 resonance solution for one reaction,

$$PP \rightarrow K^+K^-$$

on which the experimental information is still rather limited. It is essential to improve the experimental information on the K^+K^- final state to consolidate (or disprove) the need of a third resonance.

6. INTERNAL STRUCTURE - SYMMETRIES - MIXING

In the previous chapter, we have analyzed the $I = 0$ S-wave, following Au, Morgan and Pennington, and have concluded that four resonances may coexist with the quantum numbers $I = 0$, $J^{PC} = 0^{++}$. It is unreasonable to propose that these four resonances are the two members of the fundamental $^3P_0$ nonet and of its first radially excited one. One expects completely different mass relations and couplings for these four expected resonances. Mixing of isoscalars may help to accommodate the expected properties to the observations (not enough, however, in this case).

In this chapter, we shall briefly review the main ideas which justify the importance given to symmetries and the role of mixing in meson spectroscopy. A systematic application of these basic ideas, supplemented by QCD inspired potentials, have led S. Godfrey and N. Isgur to propose [5] a relativized quarkonium model which describes in a unified framework all mesons, from pion to the heaviest upsilon.

The hadrons, as already mentioned in the introduction, are not elementary particles like leptons. The hadrons are extended objects of bound constituents. The constituents are spin 1/2 particles, the "quarks", which obey Dirac's equation and the Pauli exclusion principle.

The mesons (hadrons with integer spins) are assumed to be $q\bar{q}$ states. The intrinsic parity of a $q\bar{q}$ pair is negative (Dirac equation) and if an orbital angular momentum L is attached to this $q\bar{q}$ pair, the parity of the system will be given by:

$$P = (-1)^{L+1} \tag{6.1}$$

Introducing the operator C which changes a particle into its antiparticle

C|particle> = |antiparticle>,

one can define for particles which are "neutral" (charge, strangeness, etc.), and therefore eigenstates of C, the eigen values of C:

$C = \pm 1$.

For a $q\bar{q}$ system with total spin S and angular momentum L,

$$C = (-1)^{L+S} \tag{6.2}$$

For example, with L = 0,

if: S = 0   J = L + S = 0 : $^{1}S_{0}$

            P = -1   C = +1 (π°)

if: S = 1   J = L + S = 1 : $^{3}S_{1}$

            P = -1   C = -1 (ρ°)

124

To extend the C parity to charged states (still with no strangeness, no charm, etc.), introduce the G parity:

$$G = C (-1)^I \qquad (6.3)$$

G is a multiplicative quantum number:

$$\pi^{\pm}: \quad I = 1 \qquad G = + (-1)^1 = -1$$

and $\rho \to 2\pi \to G_\rho = +1$.

Consequences of C and I conservation. Clearly, $u\bar{u}$ is an eigenstate of C and has

$$C(u\bar{u}) = (-1)^{L+S}$$

We shall keep track of the eigenvalues of C by writing

$$u\bar{u} \pm \bar{u}u \qquad \text{for } C = \pm$$

We shall keep track of the isospin of the neutral members of the $q\bar{q}$ system by writing:

$$\frac{1}{\sqrt{2}} (d\bar{d} \pm u\bar{u}) \quad \text{for } I = 0 \text{ or } 1,$$

respectively.

One can therefore define four neutral states:

$$\pi^o \text{ like: } C = +1, I = 1 : \frac{1}{2} [(d\bar{d} - u\bar{u}) + (\bar{d}d - \bar{u}u)] = n\bar{n}_1^{-1} \qquad (6.4)$$

$$\rho^o \text{ like: } C = -1, I = 1 : \frac{1}{2} [(d\bar{d} - u\bar{u}) - (\bar{d}d - \bar{u}u)] = n\bar{n}_1^{-1} \qquad (6.5)$$

$$\eta^o \text{ like: } C = -1, I = 0 : \frac{1}{2} [(d\bar{d} - u\bar{u}) + (\bar{d}d + \bar{u}u)] = n\bar{n}_o^{-1} \qquad (6.6)$$

$$\omega^o \text{ like: } C = -1, I = 0 : \frac{1}{2} [(d\bar{d} - u\bar{u}) - (\bar{d}d + \bar{u}u)] = n\bar{n}_o^{-1} \qquad (6.7)$$

For the charged states,

$$G|\pi^+> = -|\pi^->, \qquad G|\rho^+> = |\rho^->$$

$$\pi^+ \text{ like:} \quad G = -1, \ I = 1 : \ \tfrac{1}{\sqrt{2}}(u\bar{d} + \bar{d}u) \tag{6.8}$$

$$\pi^- \text{ like:} \quad G = -1, \ I = 1 : \ -\tfrac{1}{\sqrt{2}}(\bar{u}d + d\bar{u}) \tag{6.9}$$

$$\rho^+ \text{ like:} \quad G = +1, \ I = 1 : \ \tfrac{1}{\sqrt{2}}(u\bar{d} - \bar{d}u) \tag{6.10}$$

$$\rho^- \text{ like:} \quad G = +1, \ I = 1 : \ -\tfrac{1}{\sqrt{2}}(\bar{u}d - d\bar{u}) \tag{6.11}$$

We get for the two-body decay of $q\bar{q}$ mesons, assuming two flavours only, u and d, and no mixing:

$$u\bar{u} \to (u\bar{u})(u\bar{u}) + (u\bar{d})(d\bar{u})$$

$$\bar{u}u \to (\bar{u}u)(\bar{u}u) + (\bar{u}d)(\bar{d}u)$$

$$d\bar{d} \to (d\bar{u})(u\bar{d}) + (d\bar{d})(d\bar{d})$$

$$\bar{d}d \to (\bar{d}u)(\bar{u}d) + (\bar{d}d)(\bar{d}d)$$

(we have just introduced a $q\bar{q}$ pair with I = 0 and C = +1 to form a two-body decay). Conservation of parity is not (yet) introduced.

Applying these decompositions to the four neutral states defined above, we arrive at the following result (notation: $I^C_{I_3}$).

For the C = +1, I = 1 states (π^0-like):

$$1^+_0 \to 0^+1^+_0 + 0^-1^-_0 + 1^+_+1^-_- + 1^-_+1^+_- \tag{6.12}$$

thus

$$a_2(1320) \to \eta\pi^0, \ \omega\rho^0, \ \pi^+\rho^-, \ \pi^-\rho^+$$

but

$$a_2(1320) \not\to \pi^0\pi^0, \ \eta\eta, \ \pi^0\rho^0.$$

For the C = −1, I = 1 states (ρ^0 like)

$$1^-_0 \to 0^+1^-_0 + 0^-1^+_0 + 1^+_+1^+_- + 1^-_+1^-_1 \tag{6.13}$$

thus:

$$\rho^0(1690) \not\to \eta\rho^0, \ \omega\pi^0, \ \pi^+\pi^-, \ \rho^+\rho^-$$

but

$$\rho^0(1600) \not\to \eta\eta, \ \omega\omega, \ \pi^0\pi^0.$$

For the C = +1, I = 0 states (η like):

$$0^+ \to 0^+0^+ + 0^-0^- + 1^+_01^+_0 + 1^-_01^-_0 + 1^+_+1^+_- + 1^-_+1^-_- \tag{6.14}$$

thus:

$$\eta(1440) \to \eta\eta, \ \omega\omega, \ \pi^\circ\pi^\circ, \ \rho^\circ\rho^\circ, \ \pi^+\pi^-, \ \rho^+\rho^-$$

but

$$\eta(1440) \not\to \rho^+\pi^-, \ \rho^\circ\pi^\circ.$$

For the C = −1, I = 0 states (ω like):

$$0^- \to 0^+0^- + 1^+_\circ 1^-_\circ + 1^+_\circ 1^-_\circ + 1^+_+ 1^+_- + 1^-_+ 1^+_- \tag{6.15}$$

thus:

$$\omega(1670) \to \eta\omega, \ \pi^\circ\rho^\circ, \ \pi^+\rho^-, \ \pi^-\rho^+$$

but

$$\omega(1670) \not\to \eta\eta, \ \pi^\circ\pi^\circ, \ \rho^\circ\rho^\circ, \ \pi^+\pi^-.$$

In addition to the constraints imposed by the conservation of isospin and C parity, we must consider the consequences of the conservation of P, the spatial parity.

For instance, by I and C conservation, $\eta^\circ \to \pi^\circ\pi^\circ$ is allowed (eq. (6.14)) but $J^P(\eta) = 0^-$ and $P(\pi^\circ\pi^\circ) = +1$ therefore, by P conservation

$$\eta^\circ \not\to \pi^\circ\pi^\circ.$$

The $\eta(550)$ decay

We have just seen that $\eta \not\to \pi^\circ\pi^\circ$. Since the mass of $\eta(550)$ is smaller than $4m_\pi$, $\eta(550)$ has no strong decay mode ($\eta \to 3\pi$ would violate G parity).

We have therefore to consider the electromagnetic decays, represented by the annihilation diagram:

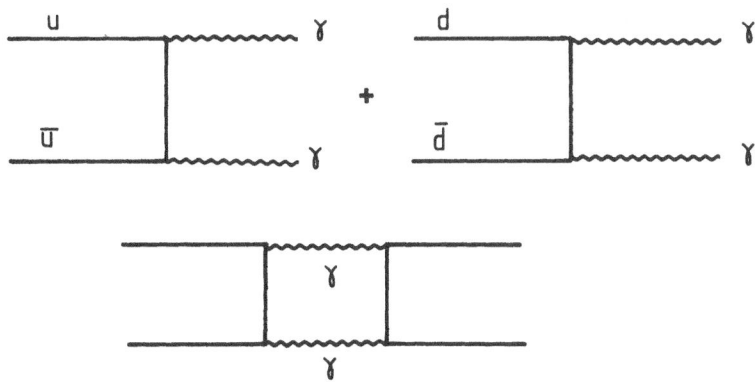

$$u\bar{u} \rightarrow \left(\frac{2}{3}\right)^4 u\bar{u} + \left(\frac{2}{3}\right)^2 \left(\frac{1}{3}\right)^2 d\bar{d}$$

$$d\bar{d} \rightarrow \left(\frac{2}{3}\right)^2 \frac{1}{3}^2 u\bar{u} + \left(\frac{1}{3}\right)^4 d\bar{d} \; .$$

The output of the annihilation is a mixture of $I = 0$ ($u\bar{u} + d\bar{d}$) and $I = 1$ ($u\bar{u} - d\bar{d}$) components: by electromagnetic decay, an $I = 1$ term is generated, which can decay into 3π. In fact, the width of $\eta(550)$ is typical of electromagnetic decays:

$$\Gamma_\eta = 1.05 \pm 0.15 \text{ keV}$$

and its branching ratios underline the importance of the electromagnetic interaction:

$\eta \rightarrow \gamma\gamma$	: 39%
$\eta \rightarrow 3\pi^0$	: 32%
$\eta \rightarrow \pi^+\pi^-\pi^0$	: 24%
$\eta \rightarrow \pi^+\pi^-\gamma$	: 5%

The $\omega(783)$ decay

The $\omega(783)$ is also relatively narrow for a meson

$$\Gamma = 9.8 \pm 0.3 \text{ MeV},$$

although, in this case, we have manifestly a strong decay. The $\rho(770)$, with a similar mass, has a width of 153 MeV.

The $\omega(783)$ is narrow because its allowed decay mode is

$$\omega \rightarrow \rho\pi \qquad \text{(eq. (6.15))}$$

but this decay is phase space limited.

$$\omega \rightarrow "\rho"\pi \rightarrow \pi^+\pi^-\pi^0 : 89.6\%$$

The next important decay mode is the electromagnetic one:

$$\omega \rightarrow \pi^0\gamma \qquad : 8.7\%$$

and some 2π decay has been observed, via isospin mixing:

$$\omega \rightarrow \pi^+\pi^- \qquad : 1.7\% \; .$$

If the $\rho(770)$ and $\omega(783)$ are produced coherently, one may expect $\omega-\rho$ interference effects due to the electromagnetic interaction:

$$A = \frac{2}{3}\left[\frac{1}{\sqrt{2}}(\omega - \rho)\right] - \frac{1}{3}\left[\frac{1}{\sqrt{2}}(\omega + \rho)\right] = \frac{1}{\sqrt{2}}\left[\frac{1}{3}\omega - \rho\right].$$

128

Introduction of a third flavour - Mixing

We introduce now a third quark flavour: the strange quark s. The flavour composition of the $3 \times \bar{3} = 9$ mesons with a given J^{PC} is the following:

$$
\begin{array}{ccc}
u\bar{s} & & d\bar{s} \\
& u\bar{u},\ d\bar{d} & \\
u\bar{d} & & d\bar{u} \\
& s\bar{s} & \\
s\bar{d} & & s\bar{u}
\end{array}
$$

All these mesons should have the same mass, picking up the same energy from gluon exchange: m_o. The mesons containing s quark will have an additional contribution due to the mass difference which is generally assumed between the s quark and the u, d quarks: m_s. Finally, the flavour neutral members ($u\bar{u}$, $d\bar{d}$, $s\bar{s}$) may pick up some energy from the virtual gluon intermediate state possible in these cases: m_a

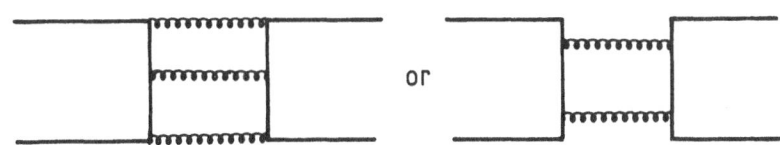

For instance, for $u\bar{u}$, one may have:

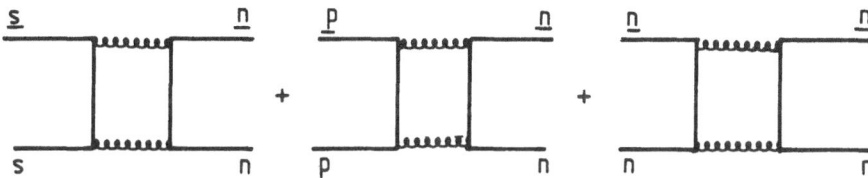

giving:

$$M|u\bar{u}> = (M_o + S + A)|u\bar{u}> + A|d\bar{d}> + A|s\bar{s}> \tag{6.16}$$

and similar expressions for $|d\bar{d}>$ and $|s\bar{s}>$.

One sees that the state

$$|u\bar{u} + d\bar{d} + s\bar{s}> \tag{6.17}$$

has a mass given by: $m_o + m_s + 3m_a$, the states

$$|d\bar{d} - u\bar{u}> \quad \text{and} \quad |u\bar{u} + d\bar{d} - 2s\bar{s}> \tag{6.18}$$

have a mass given by $m_o + m_s$, like the charged states (6.8)–(6.11) as well as the strange states $u\bar{s}$, $d\bar{s}$, $s\bar{d}$, $s\bar{u}$. The nonet splits into an octet of mesons with mass $m_o + m_s$ and a singlet with mass $m_o + m_s + 3m_a$.

Of the three flavour neutral states defined by (6.17) and (6.18), $|d\bar{d} - u\bar{u}>$ has obviously the structure expected for the neutral member of the isovector which associates $|d\bar{d} - u\bar{u}>$, $|u\bar{d} - \bar{d}u>$, $|\bar{u}d - d\bar{u}>$. The two other flavour neutral states must therefore contain the combination $|d\bar{d} + u\bar{u}>$ but the adjonction of $s\bar{s}$ can be introduced with a free mixing parameter: α. The two isoscalar states of the nonet can therefore be written:

$$|I = 0> = |d\bar{d} + u\bar{u} + \alpha\, s\bar{s}> \qquad (6.19)$$

where α takes two values. From (6.16) and similar expressions for $|d\bar{d}>$ and $|s\bar{s}>$, α can be expressed as a function of the ratio of the quark mass difference m_s to the annihilation terms m_a:

$$p = \frac{m_s}{m_a}$$

$$\alpha = \frac{1}{2}\,(p - 1 \pm \sqrt{(p - 1)^2 + 8})$$

if $p \ll 1$, i.e. if $m_s \ll m_a$ (SU3 limit where all quark mesons are equal):

$$\alpha = \frac{1}{2}\,(-1 \pm 3) \to 1, -2.$$

One recovers the Su3 isoscalar states (6.17) and (6.18)

$$\omega_1 = \frac{1}{\sqrt{3}}\,|u\bar{u} + d\bar{d} + s\bar{s}>$$

$$\omega_8 = \frac{1}{\sqrt{6}}\,|u\bar{u} + d\bar{d} - 2s\bar{s}>$$

if $p \gg 1$, i.e. the annihilation term is small, compared to the splitting between the u, d and the s quarks, we get

$$\omega = \frac{1}{\sqrt{2}}\,|u\bar{u} + d\bar{d}> \qquad (6.20)$$

$$\phi = |s\bar{s}>. \qquad (6.21)$$

This is known as the "ideal mixing" of the isoscalars SU3 eigenstates ω_1 and ω_8. In general, the physical states ω, ϕ can be any orthogonal mixture of the unitarity members ω_1, ω_8:

$$\omega = \omega_8 \cos\theta + \omega_1 \sin\theta$$

$$\phi = -\omega_8 \sin\theta + \omega_1 \cos\theta$$

(6.22)

We recover the ideal mixed states (6.21) if

$$tg\theta = \frac{1}{\sqrt{2}}, \quad \theta = 35.03°$$

(6.23)

In terms of θ, the masses of the nonet members obey the Gell-Mann Okubo mass relation:

$$4\ m_k^2 = m_\rho^2 + 3(m_\omega^2 \cos^2\theta + m_\phi^2 \sin^2\theta).$$

From the measured masses of the 0^{-+}, 1^{--}, 2^{++} and 3^{--} nonets, one gets:

$$
\begin{array}{ll}
0^{-+} & \theta = -10° \\
1^{--} & \theta = 39° \\
2^{++} & \theta = 28° \\
0^{-+} & \theta = 29°
\end{array}
$$

(6.24)

The 1^{--}, 2^{++} and 3^{--} nonets show nearly ideal mixing.

This is not the case of the 0^{-+} nonet. We shall come back on this point in the next chapter when the two photon couplings will be discussed.

7. TWO-PHOTON PRODUCTION AND DECAY OF MESON RESONANCES

The analysis of two-photon production or decay of meson resonances brings interesting constraints on the quantum numbers of the coupling states and allows to test some of the symmetry properties of the $q\bar{q}$ nonets.

A state coupling to 2 photons must have

$$C = +1.$$

(7.1)

Following C.N. Yang [16], let us consider three transformations:

$$R_\phi, \quad R_x, \quad P$$

which are applied to the photon state:

$$|k, \lambda> \qquad (7.2)$$

k: momentum along $\vec{z}$

λ: helicity along $\vec{z}$

R_ϕ is a rotation around $\vec{z}$ by an angle ϕ

R_x is a rotation around $\vec{x}$ by an angle 180°

p is the spatial parity operator.

$$R_\phi|k, \lambda> = e^{i\lambda\phi} |k, \lambda>$$

$$R_x|k, \lambda> = \quad |-k, \lambda> \qquad (7.3)$$

$$P |k, \lambda> = \quad |-k, -\lambda>$$

For two photons (in the rest frame of the resonance):

$$|k, \lambda_1>, \quad |-k, \lambda_2>$$

Yang defines four possible states:

$$\psi_1 = \frac{1}{\sqrt{2}} \, [|k, +>| - k, +> + |k, ->|-k, ->]$$

$$\psi_2 = \frac{1}{\sqrt{2}} \, [|k, +>| - k, +> - |k, ->|-k, ->]$$

$$\qquad (7.4)$$

$$\psi_3 = |k, +>|-k, ->$$

$$\psi_4 = |k, ->|-k, +>$$

Under R_ϕ, R_x, P one finds the following eigenvalues for ψ_1, ψ_2, ψ_3, ψ_4:

	ψ_1	ψ_2	ψ_3	ψ_4
R_ϕ	1	1	$e^{2i\phi}$	$e^{-2i\phi}$
P	1	-1	1	1

$$\qquad (7.5)$$

The total helicity λ are 0 for ψ_1 and ψ_2, and 2 for ψ_3 and ψ_4.
Accordingly, $J^{PC} = 2^{++}$ coupled to ψ_3, ψ_4 (helicity 2). For helicity 0
states (ψ_1, ψ_2) $J^{PC} = 0^{++}$ and 0^{-+} as well as 2^{++} and 2^{-+} are coupled, but
the states with $J^{PC} = 1^{++}$, 1^{-+} are forbidden since $R_x = 1$ for ψ_1 and ψ_2
whereas Y_1^0 is odd under R_x.

In summary, we have the following possibilities:

$$J^{PC} = 0^{-+} \qquad \lambda = 0$$

$$= 0^{++} \qquad = 0$$

$$= 2^{-+} \qquad = 0$$

$$= 2^{++} \qquad = 0 \text{ and } 2.$$

For $J > 2$, all combinations of P and λ are allowed.

Therefore, two photon couplings serve as a discriminator, not only against states with $C = -1$, like the ${}^3S_1$ and ${}^1P_1$ $q\bar{q}$ nonets, but also against the ${}^3P_1$ nonet.

Note also that the isoscalars of the ${}^1D_2$ $q\bar{q}$ nonet ($J^{PC} = 2^{-+}$), not yet discovered, couple to the helicity 0 of two photon states. The $J^{PC} = 2^{++}$ states present the difficulty of having a priori two possible couplings ($\lambda = 0$, $\lambda = 2$), making the analysis of the results more difficult.

The Yang analysis applies only to real photons. For off-shell photons, the selection rules described above do not apply.

Analysis of the pseudoscalar mesons

The two-photon coupling of the π^o has been taken as a corner stone for colour gauge theory since, in the limit $m_{\pi^o} = 0$, the coupling $g_{\pi\gamma\gamma}$ depends lineary on the number of colours, N_c:

$$g_{\pi\gamma\gamma} = \frac{\sqrt{2\alpha}}{\pi f_\pi} N_c <e_q^2>_{\pi^o} \tag{7.6}$$

f_π: pion decay constant ($\sim$ 95 MeV).

With:

$$\pi^o = \frac{1}{\sqrt{2}} (u\bar{u} - d\bar{d})$$

$$<e_q^2>_{\pi^o} = \frac{1}{\sqrt{2}}$$

and one finds:

$$\Gamma(\pi^\circ \to \gamma\gamma) = 7.63 \text{ eV if } N_c = 3.$$

$$\Gamma(\pi^\circ \to \gamma\gamma) = 0.85 \text{ eV if } N_c = 1.$$

Experimentally,

$$\Gamma(\pi^\circ \to \gamma\gamma) = 7.33 \pm 0.20 \text{ eV}, \tag{7.7}$$

a nice confirmation that $N_c = 3$.

The partial widths of the pseudoscalars will be given, in terms of the couplings $g^2_{P\gamma\gamma}$ and of the masses m_p by the relation:

$$\Gamma_{P\gamma\gamma} = \frac{m_p^3}{64\pi} g^2_{P\gamma\gamma} \tag{7.8}$$

The partial width of $\eta(550)$ has been accurately measured in e^+e^- two-photon experiments [17]

$$e^+e^- \to e^+e^-\gamma\gamma \to e^+e^-\eta \to e^+e^-\gamma\gamma \qquad (\text{or } e^+e^-\pi^+\pi^-\pi^\circ) \tag{7.9}$$
$$\Gamma(\eta \to \gamma\gamma) = 0.56 \pm 0.04 \text{ keV}.$$

This result is quite different from the earlier value obtained from photoproduction experiments, using the Primakoff effect:

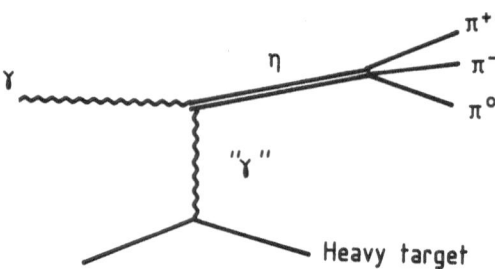

The partial width of $\eta'(958)$ is also measured in e^+e^- two-photon experiments [17]:

$$e^+e^- \to e^+e^-\gamma\gamma \to e^+e^-\eta' \to e^+e^-\rho\gamma \to e^+e^-\pi^+\pi^-\gamma \tag{7.10}$$
$$\Gamma(\eta' \to \gamma\gamma) = 4.3 \pm 0.27 \text{ keV}.$$

In this experiment, the decay mode which is observed:

$$\eta' \to \rho\gamma \to \pi^+\pi^-\gamma$$

needs to be treated carefully to take into account the detector acceptance for a three-body final state where the photon spectrum goes like k_γ^3 and the $\rho \to \pi^+\pi^-$ angular distribution like $\sin^2\theta$.

What do we expect for these partial widths? In the exact SU3 limit (6.19):

$$\frac{\Gamma(\pi^0 \to \gamma\gamma)}{m_\pi^3} \quad : \quad \frac{\Gamma(\eta_8 \to \gamma\gamma)}{m_{\eta_8}^3} \quad : \quad \frac{\Gamma(\eta_1 \to \gamma\gamma)}{m_{\eta_1}^3} \tag{7.11}$$

$$= \quad 3 \quad : \quad 1 \quad : \quad 8$$

But we know that SU3 is broken ($m_s > m_u,\ m_d$), and that a mixing must be introduced (6.21). Depending on the mixing angle θ, we expect:

$$\frac{\Gamma(\eta \to \gamma\gamma)}{\Gamma(\pi^0 \to \gamma\gamma)} = \frac{1}{3}\,(\cos\theta - 2\sqrt{2}\sin\theta)^2 \left(\frac{m_\eta}{m_{\pi^0}}\right)^3 \tag{7.12}$$

Using the experimental measurements, one gets

$$\theta = -18.4° \pm 1.8, \tag{7.13}$$

not too far from the mixing angle ($\theta = -10°$) obtained from the mass relation.

One can also compare π^0 to η':

$$\frac{\Gamma(\eta' \to \gamma\gamma)}{\Gamma(\pi^0 \to \gamma\gamma)} = \frac{1}{3}(\sin\theta + 2\sqrt{2}\,\cos\theta)^2 \left(\frac{m_{\eta'}}{m_{\pi^0}}\right)^3 \tag{7.14}$$

giving

$$\theta = -22.8° \pm 2.2, \tag{7.15}$$

in fair agreement with the mixing angle (7.13) obtained from the comparison of η to π^0.

One can also use the knowledge of $\Gamma(\pi^0 \to \gamma\gamma)$ and $\Gamma(\eta \to \gamma\gamma)$ to get $\Gamma(\eta' \to \gamma\gamma)$ independently of θ:

$$\frac{\gamma(\eta')}{m_{\eta'}^3} = 3\frac{\Gamma(\pi^0)}{m_{\pi^0}^3} - \frac{\Gamma(\eta)}{m_\eta^3} \tag{7.16}$$

giving $\Gamma(\eta' \to \gamma\gamma) = 4.88 \pm 0.3$ keV, to be compared to the measured width of 4.3 ± 0.27 keV. This agreement is remarkable, considering the large normalization factors introduced by the m^3 terms.

135

Analysis of the tensor mesons ($J^{PC} = 2^{++}$)

As shown in the introduction, two couplings must be considered in this case: $\lambda = 0$ and $\lambda = 2$. To study their relative contributions, one can analyze the decay angular distributions. For example, considering the $f_2(1270)$ meson:

$$\gamma\gamma \rightarrow f_2(1270) \rightarrow \pi^+\pi^- \tag{7.17}$$

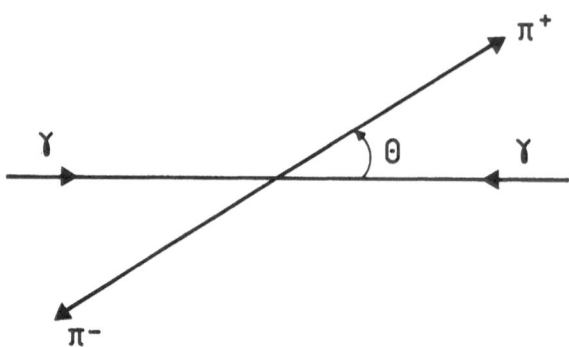

one expects, for $\lambda = 2$, $W(\theta) \sim \sin^2\theta$ and for $\lambda = 0$, $W(\theta) \sim 3\cos^2\theta - 1$.

These measurements show that $\lambda = 2$ dominates, and one can assume, for the comparison of the couplings, that one amplitude dominates ($\lambda = 2$).

Reaction (7.17) has been measured. The $\pi^+\pi^-$ mass spectrum shows a clear enhancement, but displaced in mass with respect to the PDG mass value (~ 1220 MeV instead of 1274 MeV). To explain this mass shift, G. Menessier [18] has considered that three kinds of diagrams contribute to the $\pi^+\pi^-$ spectrum:

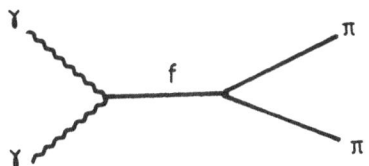

The $f_2(1270)$ partial width is:

$\Gamma(f_2 \to \gamma\gamma) = 2.78 \pm 0.14$ keV

The $a_2(1320)$ partial width is measured via two final states:

$$\gamma\gamma \to a_2(1320) \to \eta\pi \quad \to \gamma\gamma\pi \tag{7.18}$$

$$\to \rho^0\pi^0 \to \pi^+\pi^-\pi^0,$$

yielding (assuming $\lambda = 2$ dominance for angular distributions):

$\Gamma(a_2 \to \gamma\gamma) = 0.95 \pm 0.14$ keV.

The measurement of the $f_2'(1525)$ partial width raises two problems – first its branching ratio into the observed final states ($K_s^0 K_s^0$ or K^+K^-) is not well known. Moreover, the three 2^{++} resonances; $f_2(1270)$, $a_2(1320)$ and $f'(1525)$ have a non-zero branching ratio into $K_s^0 K_s^0$ (K^+K^-) and interfer in such a way that the $K_s^0 K_s^0$ and the K^+K^- mass spectra look different:

$$\sigma_{\gamma\gamma \to K\bar{K}} \sim |A_{f_2} \pm A_{a_2} + A_{f_2'}|^2$$

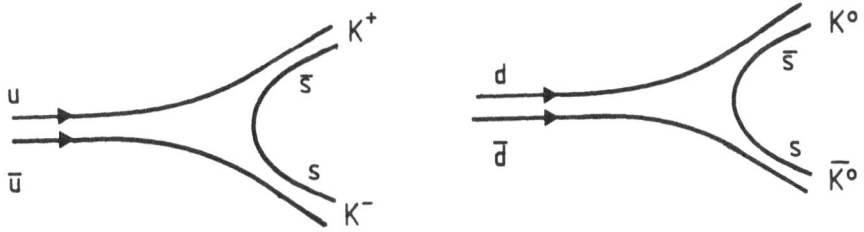

Since $|f_2\rangle = 1/\sqrt{2} \ (u\bar{u} + d\bar{d})$ and $|a_2\rangle = 1/\sqrt{2} \ (u\bar{u} - d\bar{d})$.

$f_2 - a_2$ will interfer positively in the K^+K^- final state and negatively in the $K_s^0 K_s^0$ system.

One gets, for $f'_2(1525)$:

$$\Gamma(f'_2 \to \gamma\gamma) \cdot B(f'_2 \to K\bar{K}) = 0.11 \pm 0.02 \text{ keV}.$$

The 2^{++} nonet being nearly ideally mixed (6.24), one expect instead of (7.11):

$$\frac{\Gamma(a_2 \to \gamma\gamma)}{m_{a_2}^3} \quad : \quad \frac{\Gamma(f_2 \to \gamma\gamma)}{m_{f_2}^3} \quad : \quad \frac{\Gamma(f'_2 \to \gamma\gamma)}{m_{f'_2}^3} \tag{7.19}$$

$$= 9 \qquad : \qquad 25 \qquad : \qquad 2.$$

In particular, $\Gamma(f_2 \to \gamma\gamma)/\Gamma(a_2 \to \gamma\gamma) = 2.48$, to be compared to 2.93 ± 0.45 as observed experimentally. Note, however, that this ratio is not very sensitive to the mixing angle. If, instead of (7.19), one uses (7.14):

$$\frac{\Gamma(f_2 \to \gamma\gamma)}{\Gamma(a_2 \to \gamma\gamma)} = \frac{1}{3}(\sin\theta + 2\sqrt{2}\,\cos\theta)\left(\frac{m_{f_2}}{m_{a_2}}\right)^3 \tag{7.20}$$

one finds that this ratio varies only from 2.4 to 2.75 when θ goes from 0 to 40°. The ratio $\Gamma(f'_2 \to \gamma\gamma)/\Gamma(a_2 \to \gamma\gamma)$ is more sensitive to the mixing angle. Assuming

$$B(f'_2 \to K\bar{K}) > 50\%,$$

θ is restricted to $24.3° < \theta < 35°$ at 90% CL by the measured partial widths.

These constaints imply that the non-strange quark content of the $f'_2(1525)$ is very small:

$$\left|\frac{u\bar{u} + d\bar{d}}{s\bar{s}}\right|^2_{f'_2} = \left(\frac{1/\sqrt{2}\,\cos\theta - \sin\theta}{\sqrt{2}\,\cos\theta + \sin\theta}\right)^2$$

with $\theta > 24.3°$, the $f'_2(1525)$ must consist of at least of 98% of $s\bar{s}$ quarks.

Two-photon partial width of the scalar mesons

As discussed in sect. 5, the status of the 0^{++} mesons is rather confusing. One may hope that the analysis of their coupling to two photons will shed some light on their nature.

138

The Crystall Ball experiment [19] has reported on the measurement of the isovector scalar meson $a_o(980)$ into two photon production:

$$\gamma\gamma \to a_o(980) \to \eta\pi.$$

Since the branching ratio of $a_2(980) \to \eta\pi$ is not well known, one quotes the results as the product:

$$\Gamma(a_o \to \gamma\gamma) \cdot B(a_o \to \eta\pi) = 0.19 \pm 0.10 \text{ keV}.$$

With respect to the isoscalars f_o in the 1 GeV mass region and coupling to $\pi\pi$ (sect. 5), the Crystall Ball experiment quote an upper limit:

$$\Gamma(f_o \to \gamma\gamma) \cdot B(f_o \to \pi\pi) < 0.8 \text{ keV} \qquad (95\% \text{ CL})$$

These measurements do not help us to clarify the composition of the ^3P_o nonet. The absolute values or the limits observed for the two-photon couplings of the scalar mesons seems to be rather small, compared to the couplings of the tensors. In the positronium theory, one expects huge widths for the 0^{++}:

$$\Gamma(^3P_o \to \gamma\gamma) = \frac{15}{4} \Gamma(^3P_2 \to \gamma\gamma). \qquad (7.21)$$

This fact has been taken as a serious argument to postulate that the $a_o(980)$ and associated f_o resonances are not the members of the ^3P_o $q\bar{q}$ nonet but rather candidates for $2q2\bar{q}$ states since, in this case, one may expect [20]:

$$\Gamma(2q2\bar{q} \to \gamma\gamma) \sim \alpha_s^2 \, \Gamma_{\gamma\gamma}(q\bar{q})$$

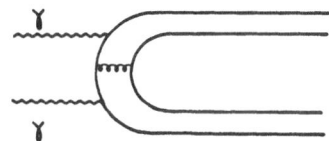

yielding $\Gamma(2q2\bar{q} \to \gamma\gamma) \sim 0.27$ keV, not in contradiction with the measurements.

Two-photon coupling of the $\eta(1440)$ and $f_1(1420)$

It is well known that two resonances with similar mass and decay mode have been observed in the 1430 MeV mass region, the $\eta(1440)$ and the $f_1(1420)$ [8]. The $\eta(1440)$, a $J^{PC} = 0^{-+}$ resonance first observed in $p\bar{p}$ annihilations at rest [21]

$$\eta(1440) \to K^o K^{\pm} \pi^{\mp}$$

with ~ 50% $K^*\bar{K}$ and ~ 50% $a_o(980)\pi$ in the final state. This resonance is also observed in the radiative decay of the J/ψ:

$$J/\psi \rightarrow \gamma\eta(1440) \rightarrow \gamma K\bar{K}\pi$$

and, due to this observation, raises the question of its relation with glueballs since resonances produced in the radiative decay of the J/ψ are expected to be rich in gluons:

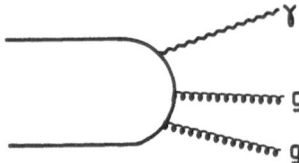

In other hadronic processes, a 1^{++} resonance, the $f_1(1420)$ has been observed:

$$f_1(1420) \rightarrow K^o K^{\pm} \pi^{\mp},$$

with the $K^*\bar{K}$ final state dominant. However, no phase variation has yet been measured for the $f_1(1420)$, whereas the resonant nature of the $\eta(1440)$ has been established [22]. We shall come back on this question in sect. 8.

Recently, the PEP-TPC-two-photon experiment [23] has observed a peak in the $K^o K^{\pm} \pi^{\mp}$ mass spectrum with

$$M = 1417 \pm 13 \text{ MeV}$$

$$\Gamma = 35 \pm ^{47}_{20} \text{ MeV},$$

in one tag photon data ($\gamma\gamma^*$) whereas no peak is observed in no tag photon data ($\gamma\gamma$). The implication of these results is that the resonance observed has spin 1, since, according to the Yang theorem discussed above, two (quasi) real photons cannot couple to spin 1, whereas off-shell photon can. Indeed, preliminary spin-parity analysis of this resonance prefers $J^{PC} = 1^{++}$.

The absence of $\eta(1440)$ signal gives the upper limit:

$$\Gamma(\eta(1440) \rightarrow \gamma\gamma) \cdot B(\eta \rightarrow K^o K^{\pm} \pi^{\mp}) < 1.5 \text{ keV},$$

which does not help us to elucidate the nature of $\eta(1440)$, since two-photon couplings of glueballs could be expected to be in the 0.1 to 1 keV partial width range.

8. PRESENT STATUS OF THE $q\bar{q}$ SPECTROSCOPY

In this last chapter, we shall review briefly the present status of meson spectroscopy. Established meson states will be compared to the predictions of the $q\bar{q}$ model of S. Godfrey and N. Isgur [5]. States which are not predicted by the $q\bar{q}$ model could be taken as evidence for glueballs, hybrids or multi-quark states as indicated in the introduction (sect. 1).

The I = 1 sector

According to [5], ~ 17 $q\bar{q}$ isovector states are expected below 2 GeV. These are:

		J^{PC}			
n = 1	$^1S_0$	0^{-+} (π)	m = 0.14 GeV	E	
	$^3S_1$	1^{--} (ρ)	0.76	E	
n = 1	$^1P_1$	1^{+-} (b_1)	1.22	E	
	$^3P_0$	0^{++} (a_0)	1.09	S	
	$^3P_1$	1^{++} (a_1)	1.25	E	
	$^3P_2$	2^{++} (a_2)	1.31	E	
n = 2	$^1S_0$	0^{-+} (π)	m = 1.35	E	
	$^3S_1$	1^{--} (ρ)	1.42	V	
n = 1	$^1D_2$	2^{-+} (π_2)	1.70	E	
	$^3D_1$	1^{--} (ρ)	1.67	V	
	$^3D_2$	2^{--} (ρ_2)	1.71	A	
	$^3D_3$	3^{--} (ρ_3)	1.72	E	
n = 2	$^1P_1$	1^{+-} (b_1)	1.80	A	
	$^3P_0$	0^{++} (a_0)	1.79	S	
	$^3P_1$	1^{++} (a_1)	1.84	A	
	$^3P_2$	2^{++} (a_2)	1.82	A	
n = 3	$^1S_0$	0^{-+} (π)	1.90	E?	

The last column indicates the experimental status = E is for "established".
Including the n = 3 $^1S_0$ $\pi(1770)$ in the list of "established" mesons, one
sees that 9 of the 17 isovectors with m < 2 GeV are known. Only four of
them are not yet observed (A). Among those unobserved states, three are
members of the second radial excitation of the P states and only one, the
n = 1 $^3D_2$ isovector, is missing among the n = 1 nonets.

Searches for this J^{PC} = 2^{--} state need the investigation of relatively
complex final states since, with G = +1,

$$\rho_2 \not\to 3\pi$$

and, by parity conservation,

$$\rho_2 \not\to 0^-0^-$$

The ρ_2 should be looked for in $\omega\pi^+$, $K^*\bar{K}$ final states.

Two isovectors are marked "V". At least one resonance, $\rho(1600)$, is
well established but its mass and widths are given with large errors. This
leaves its assignment, n = 2 $^3S_1$ versus n = 1 $^3D_1$ still an open question.

Concerning the 0^{++} states, we have already underlined the difficulties
met with these scalar particles (sects 5, 7). For the I = 1 J^{PC} = 0^{++}
n = 1 $^3P_0$ state, one obvious candidate could be the $a_0(980)$ which couples
to the $\eta\pi$ and $K\bar{K}$ channels. However, this resonance occurs near the $K\bar{K}$
threshold and accurate data are missing to allow a detailed coupled channel
analysis so that the presently observed couplings of $a_0(980)$ to $\eta\pi$ and $K\bar{K}$
cannot be usefully compared to the SU3 $q\bar{q}$ couplings. We have also already
indicated that the small coupling of $a_0(980)$ to two photons (sect. 7)
favours a 2q2$\bar{q}$ interpretation. Most likely, what is called the $a_0(980)$ is
a mixture of $q\bar{q}$ and 2q2$\bar{q}$ states. One needs more accurate data to clarify
this complex situation.

All in all, with the exception of the missing n = 1 $^3D_2$ an isovector
which is predicted with a mass of ~ 1710 MeV, the experimental status of
the isovector meson resonances agrees remarkably well with the $q\bar{q}$ model of
Godfrey and Isgur.

Are there I = 1 candidates which do not fit the $q\bar{q}$ model? Recently,
Bityukov et al. [24] have observed a $\phi\pi$ state in the reaction:

$$\pi^-p \to \phi\pi^0n$$

142

at 32.5 GeV/c. The mass and width are (1480 ± 40) and (130 ± 60)MeV. The authors note that the slope of the t' distribution is consistent with the one-pion exchange mechanism; this would imply $J^{PC} = 1^{--}$, 3^{--} for this new state. The decay properties (Gottfried-Jackson angles) for the ϕ agree with $J^P = 1^-$ and rule out $J = 0$ and $J = 3$. If it were not for its OZI forbidden decay mode, this state could be a candidate for the $n = 2\ ^3S_1$, $J^{PC} = 1^{--}$ resonance expected at 1.42 GeV in the $q\bar{q}$ model. But, for such a $q\bar{q}$ state, one would expect a large $\omega\pi$ decay, compared to $\phi\pi$. The $\omega\pi$ decay is not seen:

$$B(X(1.480) \rightarrow \omega\pi)/B(X(1480) \rightarrow \phi\pi) < 2 \qquad (8.1)$$

at 95% CL.

Therefore X(1480) is likely to be a $2q2\bar{q}$ state (with one pair of strange quarks).

An alternative interpretation is to assume that X(1480) is a hybrid state (with $J^{PC} = 1^{--}$ or 1^{-+}). F. Close, however, has shown that it is unlikely that hybrids are formed with masses much lower than 1700 MeV [3].

Another $I = 1$ exotic resonance has been proposed by the GAMS experiment [4,25]. It is observed in the $\pi\eta$ system produced in the reaction:

$$\pi^- p \rightarrow \pi\eta n$$

at 100 GeV/c. The $\pi\eta$ spectrum is dominated by the $a_2(1320)$ but the forward-backward asymmetry, evaluated in the Gottfried-Jackson frame, shows a significant structure in the a_2 mass region, which may be interpreted as an interference effect of a P-wave with D-wave a_2. This P-wave would present a maximum at ~ 1350 MeV and its width would be ~ 280 MeV.

An $I = 1$ $\pi\eta$ P-wave has $J^{PC} = 1^{-+}$, i.e. a set of quantum numbers outside the $q\bar{q}$ model. This $\hat{\rho}(1350)$ would be the first evidence for a genuine exotic meson resonance.

One may note that hybrids [3] are expected to occur in pairs, like 1^{-+} and 1^{--}. The X(1480) and $\hat{\rho}(1350)$ could be the first example of such a pair. However, before reaching such important conclusions, the X(1480) and $\hat{\rho}(1350)$ must be confirmed. Their couplings to other final states must be measured. N. Isgur [26] has pointed out that hybrids are not likely to decay into two ground state nonets ($X \rightarrow \pi\eta$) and H. Lipkin [27] has argued that 1^{-+} hybrid could not couple to $\pi\eta$ due to OZI and parity conservation.

The I = 1/2 sector

According to [5], 16 $q\bar{q}$ isodoublets (strange) states are expected below 2 GeV. These are:

		J^{PC}			
n = 1	$^1S_0$	0^-	m = 0.52 GeV	E (0.50)
	$^3S_1$	1^-	0.89	E (0.89)
n = 1.	$^1P_1$	1^+	1.38	E (1.28)
	$^3P_1$	1^+	1.36	E (1.40)
	$^3P_2$	2^+	1.43	E (1.43)
	$^3P_0$	0^+	1.25	S (1.35)
n = 2	$^1S_0$	0^-	1.50	E (1.46)
	$^3S_1$	1^-	1.56	E (1.41)
n = 1	$^1D_2$	2^-	1.80	E (1.58)
	$^3D_2$	2^-	1.83	E (1.77)
	$^3D_1$	1^-	1.79	E (1.79)
	$^3D_3$	3^-	1.81	E (1.78)
n = 2	$^3P_1$	1^+	1.92	A
	$^1P_1$	1^+	1.95	A
	$^3P_0$	0^+	1.91	A
	$^3P_2$	2^+	1.94	A

As for the isovectors, the experimental status of the $q\bar{q}$ isodoublets is in very good shape. Except for the n = 2 P states, all the predicted $q\bar{q}$ states are "established", if we include in this category the 0^- K(1460) and the 1^- K^*(1790).

Note that C parity does not separate the singlet from the triplet state with the same spin-parity (i.e. $^1P_1$ vs. $^3P_1$, $^1D_2$ vs. $^3D_2$) and that the observed states can be a mixture of the pure $q\bar{q}$ SU3 states.

Beyond 2 GeV, a few K, K^* resonances have also been established, like the $K^*_4(2060)$ with $J^P = 4^+$, expected at 2.14 GeV ($n = 1$, $^3F_4$), the $K_2(2250)$ with $J^P = 2^-$ (which could be one or both of the $n = 2$ $^3D_2$, $^1D_2$ expected at 2.26–2.28 GeV), the $K_3(2320)$ with $J^P = 3^+$ (which could be one or both of the $n = 1$, $^3F_3$, $^1F_3$ expected at 2.15–2.17 GeV) and even a $K_4(2500)$ with $J^P = 4^-$, which could be one or both of the $n = 1$ $^3G_4$, $^1G_4$ expected at 2.45–2.47 GeV.

The unique evidence for an isodoublet which is not expected in the $q\bar{q}$ model is a narrow meson at 3.1 GeV observed by the WA62 experiment at CERN with a Σ^- beam [28], decaying into $\Lambda\bar{p}$ + pions with $Q = +1$, 0 or -1. The width is compatible with the experimental resolution (24 MeV). Having strangeness $S = -1$ and being observed in the charged state $Q = +1$, this meson cannot be a $q\bar{q}$: it is compatible with the $2q2\bar{q}$ hypothesis, one of the quarks being a strange one. A narrow state in this mass range which may be a multiquark state is reminiscent of the baryonium spectroscopy and of the T and M quarkonia states advocated by Chan and Högasen [29]. After a long period of controversial results, it seems that none of these narrow states have been firmly established. Experimental confirmation is needed before drawing firm conclusions.

The I = 0 sector

The isoscalar sector of the $q\bar{q}$ model is more difficult to analyze than the $I = 1$ and $I = 1/2$ ones. The model predicts of course the existence of two singlets per nonet. The mass predictions given by the $q\bar{q}$ potential may be affected by the annihilation terms, via gluons (sect. 6):

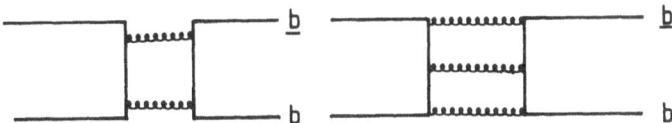

The $q\bar{q}$ objects may also mix with glueballs: gg, ggg, so that the masses, widths and couplings of the observed resonances may be quite different from the predicted properties.

The $q\bar{q}$ model of Godfrey and Isgur provides room for ~ 30 isoscalar states below 2 GeV! Instead of giving the list of the predicted and observed states, we shall review the experimental situation according to the J^{PC} classification of these states.

The $I = 0$ $J^{PC} = 0^{++}$ mesons

The $q\bar{q}$ model expects four states:

$n = 1$ $^3P_0$ $f_0 = 1090$ MeV

$f'_0 = 1380$

$n = 2$ $^3P_0$ $f_0 = 1790$

$f'_0 = 2020$

We have already discussed the results of the latest $\pi\pi$, $K\bar{K}$ coupled channel analysis [11] which conclude at the existence of four $I = 0$ $J^{PC} = 0^{++}$ resonances:

$f_0(991)$	$m = 991$ MeV	$\Gamma = 42$ GeV	$g_\pi = 220$ MeV	$g_K = 280$ MeV
$f_0(910)$	910	350	520	270
$f_0(1430)$	1430	200	580	160
$f_0(988)$	988	0	20	350

We have pointed out that $f_0(988)$ may be interpreted as a $K\bar{K}$ molecule (or $2q2\bar{q}$ state).

One may try to associate $f_0(991)$ and $f_0(910)$ to the $n = 1$ $^3P_0$ nonet, although the masses are not in agreement with the predictions. It is however possible that threshold effects ($\eta\eta$, $\eta'\eta$), as pointed out by Tornqvist (30) shift the masses of the resonances as much as by 300 MeV.

$f_0(1430)$ could be the first radial excitation ($n = 2$ $^3P_0$). It would have to be shifted by ~ 350 MeV to agree with the $q\bar{q}$ expectations.

Another source of mass shifting may take place if scalar glueballs are formed in this mass range. The need for glueballs becomes more apparent if one adds, to the four scalar identified by Au, Martin and Pennington, the evidence for another 0^{++} state, the $f_0(1590)$, observed by GAMS in the $\eta\eta$, the $\eta'\eta$ as well as in the $4\pi^0$ final state [4]:

$f_0(1590)$ $m = 1586 \pm 16$ MeV $\Gamma = 286 \pm 50$ MeV

$\eta\eta'/\eta\eta = 2.7 \pm 0.8$

$\pi^0\pi^0/\eta\eta < 0.3$

$K\bar{K}/\eta\eta < 0.6$.

These abnormal decay branching ratios suggest that the $f_0(1590)$ is a glueball candidate or, at least, a mix state ($q\bar{q}$ and glue) particularly rich in glue.

Another $\eta\eta$ state has also been obseved by GAMS [4], with
m = 1755 ± 8 MeV and a width compatible with the experimental resolution
(Γ < 50 MeV). The spin-parity of this state is not uniquely determined (it
could be $J^P = 0^+$ or 2^+). The $J^P = 2^+$ assignment would make it comparable
to the $f_2(1720)$ observed in the radiative decay of the J/ψ (but the width
of the $f_2(1720)$ is 134 ± 19 MeV). As for the $f_0(1590)$, the X(1755) seems
to prefer $\eta\eta$ to $\pi^0\pi^0$ decay:

$$B(X \rightarrow \pi^0\pi^0)/B(X \rightarrow \eta\eta) < 0.3. \qquad (8.2)$$

To complete the description of the 0^{++} resonances, one must add that,
in a partial wave analysis of the $K_s^0 K_s^0$ system, Etkin et al. (31) have
observed two S-wave resonances:

$$f_0(1240) \rightarrow K_s^0 K_s^0$$
$$m = 1240 \pm 22 \text{ MeV}$$
$$\Gamma = 140 \pm 22 \text{ MeV}$$

and

$$f_0(1730) \rightarrow K_s^0 K_s^0$$
$$m = 1730 \pm 20 \text{ MeV}$$
$$\Gamma = 80 \pm 50 \text{ MeV}$$

These observations need confirmation but they contribute to a total of
7, maybe 8 scalar mesons observed below 2 GeV whereas the $q\bar{q}$ model predicts
the existence of 4 states. The unavoidable conclusion is that non $q\bar{q}$
resonances (glueballs, multiquark states, $K\bar{K}$ molecules) are probably present
(possibly mixed with $q\bar{q}$ states) in the available experimental results.

The I = 0 $J^{PC} = 2^{++}$ mesons

The experimental status is nearly as complex as for the scalar mesons.

The $q\bar{q}$ model predicts 6 states between 1.25 and 2.25 GeV.

$$
\begin{array}{llll}
n = 1 & {}^3P_2 & f_2 = 1280 \text{ MeV} \\
 & & f_2' = 1510 \\
n = 2 & {}^3P_2 & f_2 = 1820 \\
 & & f_2' = 2040 \\
n = 1 & {}^3F_2 & f_2 = 2070 \\
 & & f_2' = 2270.
\end{array}
$$

Two resonances are perfectly well established and fit nicely with the $q\bar{q}$ predictions for $n = 1\ ^3P_2$:

the $f_2(1270)$ and the $f'_2(1515)$

Their properties fit with a mixing angle which is nearly "ideal":
$\theta_{mix} = 28°$.

Another well established resonance is the $f_2(1720)$ which has very stricking production properties since it is only seen in the radiative decay of the J/ψ:

$$J/\psi \to \gamma\ f_2(1720), \text{ with } M = 1720 \pm 7 \text{ MeV and } \Gamma = 132 \pm 15 \text{ MeV.} \qquad (8.3)$$

but not in photon-photon reactions (whereas $f_2(1270)$ and $f_2(1515)$ are observed):

$\Gamma_{\gamma\gamma}$ (keV)	:	$f_2(1270)$	:	$f_2(1515)$	:	$f_2(1720)$
		2.6	:	~ 0.1	:	< 0.2

Moreover, the $f_2(1720)$ is not observed in the $s\bar{s}$ favoured coupling channel (LASS experiment):

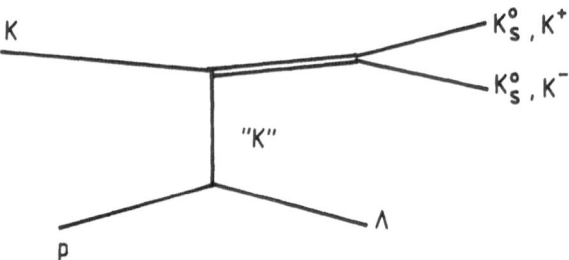

Its branching ratios are not well established, but it has been observed in the decay modes:

$K\bar{K}$, ηη, ππ,

showing that it is not a pure $s\bar{s}$ state.

Since the radiative decay mode of the J/ψ is favouring glue coupling:

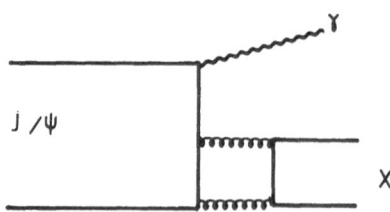

and that the coupling of two photons to glue is assumed to be small (it needs a quark loop):

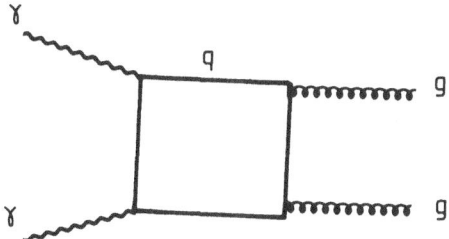

the ratio:

$$S_X = \frac{J/\psi \to \gamma X}{X \to \gamma\gamma}$$

discriminates between states rich in glue. The comparison of $f_2(1720)$ to $f_2(1270)$ and $f_2'(1515)$ points out the special nature of $f_2(1720)$

	f(1720)	:	$f_2(1270)$	:	$f_2'(1515)$
S_X = > 20		:	1	:	3

Another $I = 0$ $J^{PC} = 2^{++}$ resonance, not yet firmly established, has been observed at ~ 1810 MeV with Γ ~ 200 MeV. From the analysis of Cason et al. [32], this resonance could have a branching ratio

$$B(f_2(1810) \to \pi\pi) = 0.44 \pm 0.03.$$

With these characteristics, the $f_2(1810)$ (which needs confirmation) could be the $n = 2$ $^3P_2$ $q\bar{q}$ state predicted to occur at 1820 MeV.

This resonance may have received some confirmation by the GAMS experiment [4], which observes, in the $4\pi^0$ system, a 2^{++} state with

$$M = 1810 \pm 15 \quad \text{and} \quad \Gamma = 170 \pm 30 \text{ MeV.}$$

However, the GAMS Collaboration finds for this object branching ratios which do not agree well with the expectations for a $q\bar{q}$ state:

$B(X \to 2\pi^0)$	:	$B(X \to 4\pi^0)$	:	$B(X \to \eta\eta)$
1	:	5	:	~ 5.

This resonance does not seem to be observed by the LASS group, indicating that the coupling of the $f_2(1810)$ to $s\bar{s}$ is small.

All these results converge to the same conclusion: the $f_2(1720)$ is probably not a good $q\bar{q}$ candidate. The $f_2(1810)$ is a much better one, for the non strange partner of the $n = 2$ $^3P_2$ state. Therefore, what is the $f_2(1720)$?

In addition, a 2^{++} resonance has been observed at 2230 MeV as a narrow $K\bar{K}$ spike in the radiative decay of the J/ψ:

$$J/\psi \rightarrow \gamma \; K^+K^-$$
and
$$J/\psi \rightarrow \gamma \; K^0_s K^0_s \qquad\qquad (8.4)$$

but also by GAMS, in

$$\pi^- p \rightarrow \eta\eta' n \qquad\qquad (8.5)$$

by the MIS experiment at IHEP (Serpukhov), in

$$\pi^- p \rightarrow K^0_s K^0_s n \qquad\qquad (8.6)$$

and by LASS in

$$K^- p \rightarrow K^0_s K^0_s \Lambda$$
and
$$K^- p \rightarrow K^+ K^- \Lambda \; . \qquad\qquad (8.7)$$

There is an embarassing experimental difficulty with this resonance. The Mark III group quote a very narrow width for their observations:

$$\Gamma = 26^{+20}_{-16} \pm 17 \text{ MeV in } K^+K^-$$
and
$$\Gamma = 18^{+23}_{-15} \pm 10 \text{ MeV in } K^0_s K^0_s$$

with rates which are relatively high:

$$B = (4.2 \pm 2) \times 10^{-5} \text{ for } K^+K^-$$
and
$$B = (3.1^{+1.6}_{-1.3} \pm 1.7) \times 10^{-5} \text{ for } K^0_s K^0_s$$

The DM2 group, studying the same reaction with comparable statistics, do not see these narrow spikes and quote

$$\Gamma > 30 \text{ MeV}, \qquad B < 4 \times 10^{-5} .$$

Gams give only an upper limit for the width.

$$\Gamma < 60 \text{ MeV}$$

and MIS-IHEP quote $\Gamma \sim 70$ MeV.

150

If we accept for this state:

M ~ 2230 MeV Γ ~ 60 MeV,

with a large branching ratio into $K\bar{K}$, one cannot exclude that it is the n = 1 ${}^3F_2$ $q\bar{q}$ predicted state.

To complete this short review on the I = 0 J^{PC} = 2^{++} objects, one must
also mention two more results which are not easy to interpret in terms of the $q\bar{q}$ model.

At BNL, an MPS experiment [33] has studied the exclusive reaction:

$$\pi^-p \rightarrow \phi\phi n \tag{8.8}$$

at 22 GeV/c and found an accumulation of events near the $\phi\phi$ threshold. A partial wave analysis of the $\phi\phi$ system shows that three waves contribute to this effect: the $\phi\phi$ S-wave, with total spin 2, and two $\phi\phi$ D-waves, one with total spin 2, the other with total spin 0. The two D-waves show a moving phase with respect to the S-wave. The authors need to introduce 3 poles in the K-matrix to get a good interpretation of the data and conclude at the observation of 3 resonances, all with I = 0 J^{PC} = 2^{++} distributed in the mass range 2000 to 2350 MeV. All 3 resonances have widths of order 200 MeV. As discussed in sect. 5, a resonance is defined by a pole in the S-matrix, and there is not necessarily a one to one correspondance between the S-matrix and the K-matrix poles. But the essential point made by the authors is to underline that the production of these $\phi\phi$ resonances violate the OZI rule. This is a sign that these resonances may have a large coupling to glue.

One should however make the remark that a I = 0 J^{PC} = 2^{++} resonance has been observed in proton-antiproton annihilations into $2\pi^o$ [34]

$$\bar{p}p \rightarrow \pi^o\pi^o \tag{8.9}$$

with M ~ 2150 MeV, Γ ~ 250, therefore not completely excluding that one of the $\phi\phi$ resonances is a $q\bar{q}$ object (or a 4-quark state as it could be the case in processes like (8.9)).

Finally, one must mention that the double pomeron reaction already studied in sect. 5:

$$pp \rightarrow p\pi^+\pi^-p \tag{8.10}$$

at ISR [12] cannot give a good interpretation of the $\pi^+\pi^-$ mass spectrum if an I = 0 $J^{PC} = 2^{++}$ resonance is not introduced, with a mass and width:

M ≃ 1480 ± 50 MeV.

Γ ≃ 150 ± 40 MeV.

However, the f_2(1270) is not observed, and this is surprising for a reaction like (8.10). The f_2(1480) needs serious confirmation but, if real, it appears again at a mass where the $q\bar{q}$ model does not need any new state. In fact, the partial wave analysis of the $K_s^o K_s^o$ system seems to demand a resonance D-wave at 1410 MeV.

In summary, the $q\bar{q}$ model predicts the existence of 6 states between 1250 and 2270 MeV and 7 to 8 states have been observed, with, indeed, not the same degree of confidence: the f_2(1270), f_2(1515), f_2(1720) are firmly established — the f_2(1810), f_2(2150) and f_2(2230) have a rather safe status. The f_2(1410), f_2(2010) and f_2(2340) need confirmation. It seems unlikely that the $q\bar{q}$ model alone will be sufficient to understand the I = 0, $J^{PC} = 2^{++}$ spectroscopy. The f_2(2010), which is the lowest $\phi\phi$ resonance of Etkin et al. [33], is also the one which correspond to an orbital angular momentum L = 0 between the two ϕ particles: one can therefore speculate that it is a $\phi\phi$ molecule, like the f_o(988) was tentatively proposed to be a K$\bar{K}$ S-wave molecule [7].

The I = 0 $J^{PC} = 1^{++}$ and 1^{+-} mesons

We choose to study simultaneously the C = +1 and C = −1, $J^P = 1^+$ resonance since, in the K$\bar{K}\pi$ final state, which provides most of the information, both C = +1 and C = −1 may be present.

According to the $q\bar{q}$ model, we expect:

n = 1	$^1P_1$ ($J^{PC} = 1^{+-}$)	h_1(1220)	h_1(1480)
n = 2	$^1P_1$	h_1(1800)	h_1(2020)
n = 1	$^3P_1$ ($J^{PC} = 1^{++}$)	f_1(1250)	f_1(1500)
n = 2	$^3P_1$	f_1(1840)	f_1(2050)

One state is well established: this is the f_1(1285), seen in πp interactions, in $\bar{p}$p annihilations (except $\bar{p}$p annihilations at rest), also observed, but with a much smaller coupling, in Kp interactions, and not seen in J/ψ decays and in photon-photon interactions. All these results are consistent with the f_1(1285) being $J^{PC} = 1^{++}$.

152

The $f_1(1285)$ was first observed in the final state $K^0K^{\pm}\pi^{\mp}$, but, its mass being below the threshold for KK^*, it is not straightforward to determine its branching ratios (a KK^* on an S-wave would lead to $J^{PC} = 1^{++}$). Its $\eta\pi^+\pi^-$ decay mode has now been firmly established, and it has been shown that it goes via the quasi-two-body process:

$$f_1(1285) \rightarrow a_0(980)) \rightarrow \eta\pi\pi$$

The decay mode $a_0(980)\pi \rightarrow K\bar{K}\pi$ is therefore expected, leading to a $K\bar{K}$ threshold enhancement, as observed.

Another important decay mode of the $f_1(1285)$ is 4π (experimentally, it is not easy to separate the $f_1(1285)$ from the $f_2(1270)$, but one may play with the large difference observed for the width of these resonances).

Thus the branching ratios of the $f_1(1285)$ strongly suggest that this is the non-strange quark member of the $n = 1$ $I = 0$ $^3P_1$ nonet. Furthermore, its mass nearly coincides with the mass of the $I = 1$ $^3P_1$ member, the $a_1(1270)$, as it should if the mixing is ideal.

Where is, then, the $s\bar{s}$ member of this nonet?

The $q\bar{q}$ model expect it at $M \sim 1500$ MeV.

For many years, a $I = 0$ $J^{PC} = 1^{++}$ resonance has been observed in $\bar{p}p$ annihilations (but not in $\bar{p}p$ annihilations at rest), in $\pi^{\pm}p$ interactions and, to some extend, in $K^{\pm}p$ interactions. The first detailed spin-parity analysis was made by Dionisi et al. [35], studying the $K^0K^{\pm}\pi^{\mp}$ Dalitz plot density. With the limited statistics available in 1978, one could not undertake a complete partial wave analysis, with phase determination. More recently, with much larger statistics, and studying the $K^0K^{\pm}\pi^{\mp}$ system produced centrally, in the reactions:

$$\pi^+p \rightarrow \pi^+ (K_s^0K^{\pm}\pi^{\mp})p$$

and   (8.11)

$$\pi^+p \rightarrow \pi^+ (K_s^0K^{\pm}\pi^{\mp})p$$

Armstrong et al. [36], have also analyzed the Dalitz plot density and conclude that a 1^{++} resonance is observed, with

$$M = 1425 \pm 2 \text{ MeV}, \qquad \Gamma = 62 \pm 5 \text{ MeV}$$

and is consistent with being 100% K^*K (on an S-wave).

Being a pure K^*K object, it would be natural to conclude that the $f_1(1425)$ is the $s\bar{s}$ member of the $^3P_1$ nonet. However, one would then expect to observe the f_1 in $K^\pm p$ reactions more easily than the $f_1(1285)$. This is not the case. Indeed, in the reaction

$$K^-p \rightarrow K^+K^o\pi^-\Lambda, \tag{8.12}$$

Gavillet et al. [36] have pointed out that they observe a 1^{++} K^*K resonance, not at 1425 MeV, but at 1530 MeV, with $\Gamma = 107 \pm 15$ MeV. This result is now confirmed by the LASS group.

We may have therefore observed already three $I = 0$ $J^{PC} = 1^{++}$ states in the 1270–1550 MeV mass range, whereas the $q\bar{q}$ model predicts the existence of two states, which could be the $f_1(1285)$ and the $f_1(1530)$, which have the masses and branching ratios expected. It would have the $f_1(1420)$ as an extra state. Again, here, it is tempting to assume that the $f_1(1420)$ is an S-wave K^*K molecule (or a 4 quark state).

In the same experiment as mentioned above (8–12), the LASS group, performing a partial wave analysis of the $K\bar{K}\pi$ system, not only confirms the existence of the 1^{++} $f_1(1530)$, but, in addition, gets evidence for a 1^{+-} $h_1(1400)$ at m ~ 1400 MeV with Γ ~ 100 MeV.

This new resonance, being observed in reaction (8.12), is a natural $s\bar{s}$ candidate for the $n = 1$ $^1P_1$ nonet for which the other (non-strange) isoscalar partner is already known as being the $h_1(1190)$ (seen in 3π final state).

Nothing is reliably known on $I = 0$ $J^{PC} = 1^{++}$ and 1^{+-} states above ~ 1550 MeV.

In summary, here again the $q\bar{q}$ model seems to be in good shape, with the two isoscalars of the two $^1P_1$ and $^3P_1$ ground state identified. We have however to find another interpretation for the $f_1(1420)$, which may be a K^*K molecule.

The $I = 0$, $J^{PC} = 0^{-+}$ mesons

This sector contains the well-known ground states, the η and η', members of the $\eta = 1$ $^1S_0$ nonet. We have already discussed their mixing properties in sects 6 and 7.

Experimentally, two 0^{-+} states have been established above the η and η'. These are the $\eta(1275)$ and the $\eta(1420)$.

The $\eta(1275)$ is observed in the partial wave analysis of the $\eta\pi\pi$ system.

Ando et al. [37] have performed a full partial wave analysis of the $\eta\pi^+\pi^-$ system:

$$\pi^-p \rightarrow \eta\pi^+\pi^-n \qquad\qquad (8.13)$$

They not only confirm the $\eta(1275)$, with m = 1279 ± 5 MeV and Γ = 32 ± 10 MeV, but also find that the phase of the $a_0(980)\pi$ amplitude moves again in a motion characteristic of a resonance in the 1420 MeV mass region. This is to be compared to the results of a full partial wave analysis of the $K\bar{K}\pi$ system produced in π^-p, K^-p and $\bar{p}p$ interactions at 6–8 GeV/c, performed by S.U. Chung et al. [38]: the 0^{-+} $a_0(980)\pi$ wave exhibits a phase motion characteristic of a resonance at m ~ 1420 MeV (a Breit–Wigner fit yields m = 1421 ± 1 MeV, Γ = 73 ± 6 MeV). Furthermore, a significant $\eta(1275)$ is also observed.

None of these two isoscalar, 0^{-+} resonances, $\eta(1275)$ and $\eta(1420)$ have an important $s\bar{s}$ coupling. They are not observed in the LASS experiment. In fact, the earliest results on the $\eta(1420)$, which were obtained in 1965 [39], analyzing the reactions

$$\bar{p}p \rightarrow K^oK^{\pm}\pi^{\mp}\pi^+\pi^-$$

and $\qquad\qquad (8.14)$

$$\bar{p}p \rightarrow K^oK^{\pm}\pi^{\mp}\pi^o\pi^o$$

with annihilations at rest, had already concluded that

$$m(0^{-+}) = 1425 \pm 7 \text{ MeV}$$
$$\Gamma \quad = \quad 80 \pm 10 \text{ MeV}$$

and
$$K^*K/\text{total} < 50\%$$

It seems therefore unlikely that the $\eta(1420)$ is the $s\bar{s}$ partner of the $\eta(1275)$. This hypothesis becomes even more evident if we remember that an $\eta(1440)$ seems to be copiously produced in the radiative decay of the J/ψ, pointing out that the $\eta(1440)$ (which we assume to be the same object than the $\eta(1420)$) may have a large coupling to gluons.

Exotic mesons

Some of the results already discussed are of course candidates for being exotic mesons, i.e. mesons which do not fall into the $q\bar{q}$ classification.

In this respect, three candidates for hadron-hadron molecules (or four quark states) have been proposed: the $f_o(988)$ (a $K\bar{K}$ molecule), the $f_1(1420)$ (a $\bar{K}K^*$ molecule), may be the $f_2(2010)$ (a $\phi\phi$ molecule).

The $f_2(1720)$ has properties which are difficult to reconcile with the $q\bar{q}$ model, as well as the $f_o(1590)$.

In the isovector sector, we have already mentioned the surprising result observed by Bityukov et al. [24], a $\rho_1(1480)$ which could be accommodated by the $q\bar{q}$ model if it were not for its OZI forbidden decay mode

$$\rho_1 \rightarrow \phi\pi,$$

which make it a possible candidate for a 4-quark object (with a pair of strange quarks).

Another $I = 1$ exotic resonance has been observed by Gams [4,25]. They find, in a partial wave analysis of the $\eta\pi$ system, a significant contribution of the $J^{PC} = 1^{-+}$ amplitude: if confirmed this $\hat{\rho}(1350)$ would be the first $I = 1$ exotic meson (of the second kind) observed in meson spectroscopy.

The other candidate for a genuine exotic meson is found in the $I = 1/2$ sector [28], a 3.1 GeV narrow meson being observed to decay into $\bar{\Lambda}p$ + pions with $Q = +1$, which is incompatible with the $q\bar{q}$ model. This could be an example of exotic meson of the first kind.

9. CONCLUSIONS

After several decades of research on meson spectroscopy we arrive at the following situation: by far and large, the relativistic QCD inspired $q\bar{q}$ model [2,5] give a natural and simple interpretation of the observations made on hundreds of levels, couplings, symmetries, etc.

It would be good, of course, that all the levels predicted by the $q\bar{q}$ model are discovered. In the I = 1 sector, the $^3D_2$ state is still missing. The status of the I = 0 sector is much more difficult to establish (due to the mixing of $q\bar{q}$ states themselves and possible mixing with mesons of other nature like glueballs). In general, one observes more states than predicted by the $q\bar{q}$ model. However, before concluding that we observe glueballs, we have to investigate the detailed properties of these states. Some of them, occurring near thresholds, present the properties expected for what has been called "molecules" by T. Barnes and N. Isgur.

Some hints, however, are now available which may point to the existence of exotic mesons. Several detailed, high statistics measurements, with various initial and final states are still needed to reach a satisfactory interpretation of meson spectroscopy.

REFERENCES

[1] G. Morpurgo, Physics 2 (1965) 95

[2] R. Dalitz, High energy physics, Les Houches, 1965, etd. C. de Witt, M. Jacob (Gordon-Breach).

[3] N. Isgur, these proceedings.

[4] F. Close, these proceedings.

[5] F. Binon, these proceedings.

[6] S. Godfrey and N. Isgur, Phys. Rev. D32 (1985) 189.

[7] R. Jaffé, Phys. Rev. D15 (1977) 267.

[8] T. Barnes, these proceedings.

[9] Review of Particle Properties (PDG), Phys. Lett. 170B (1986) 1.

[10] D. Jackson, Nuovo Cimento 34 (1964) 1644.

[11] K. Gottfried and J.D. Jackson, Phys. Lett. 8 (1964) 144.

[12] K.L. Au, D. Morgan and M. Pennington, RAL 86-076.

[13] T. Akesson et al., Nucl. Phys. B264 (1986) 154.

[14] G. Grayer et al., Nucl. Phys. B75 (1974) 189.

[15] D. Cohen et al., Phys. Rev. D22 (1980) 2595.

[16] A. Etkin et al., Phys. Rev. D28 (1982) 1786.

[17]   C.N. Yang, Phys. Rev. 77 (1950) 242.

[18]   For a review on two photon physics, see Ch. Berger and W. Wagner,
       Phys. Rep. 146 (1987) 1.

[19]   C. Menessier, Z. Phys. C16 (1983) 241.

[20]   D. Antreasyan et al., Phys. Rev. D33 (1986) 847.

[21]   N.N. Achasov et al., Phys. Lett. B108 (1982) 134.

[22]   P. Baillon et al., Nuovo Cimento 50A (1967) 393.

[23]   S. Protoposcu, Second hadron spectroscopy Conference, KEK (1987;

[24]   A. Ando et al., Second hadron spectroscopy Conference, KEK (1987.

[25]   H. Aihara et al., Phys. Rev. Lett. 57 (1986) 51, 2500.

[26]   S.I. Bityukov et al., Phys. Lett. B188 (1987) 383.

[27]   M. Boutemeur, Proc. 22nd Rencontre de Moriond (1987)
       (Ed. Frontieres, Tran Than Van).

[28]   N. Isgur and J. Patou, Phys. Lett. B124 (1983) 247.

[29]   H.J. Lipkin, 22nd Rencontre de Moriond (1987).

[30]   M. Bourquin et al., Phys. Lett. B172 (1986) 113.

[31]   H.M. Chan and H.A. Hogaasen, Nucl. Phys. B136 (1978) 401.

[32]   N. Tornqvist, Phys. Rev. Lett. 49 (1982) 624.

[33]   A. Etkin et al., Phys. Rev. D25 (1982) 2446.

[34]   N.M. Cason et al., Phys. Rev. D28 (1983) 1586.

[35]   A. Etkin et al., Phys. Lett. B165 (1985) 217.

[36]   R.S. Dulude et al., Phys. Lett. B79 (1978) 335.

[37]   C. Dionisi et al., Nucl. Phys. B169 (1980) 1.

[38]   T.A. Armstrong et al., Phys. Lett. B146 (1984) 273.

[39]   A. Ando et al., Phys. Rev. Lett. 57 (1986) 1296.

[40]   S.U. Chung, Proc. of 20th Rencontre de Moriond (1985) 489,
       (Ed. Frontieres, Tran Than Van).

[41]   P. Baillon et al., Nuovo Cimento 50A (1967) 393.

EXPERIMENTAL SEARCH FOR GLUONIC MESONS

F. Couchot

Laboratoire de l'Accélérateur Linéaire
Orsay, France

Introduction

The gluonic mesons were first mentionned in 1972, at the very beginning of Quantum Chromo Dynamics (Q.C.D.), when H. Fritzsch and M. Gell-mann[1] suggested that "meson states would appear that act as if they were made of gluons rather than qq̄ pairs". These states would build "a sequence of extra SU(3) singlet meson states" ; flavour SU(3) nonets could no longer describe all the low mass mesons.

Glue, gluonia or glueballs are the present names of gluonic mesons, and they would be the only example of matter built only with bosons. Their prediction initiated many experimental and theoretical investigations, and the first candidates have been discovered in the beginning of the eighties [2].

The present experimental situation is very rich and rather complex[3], so this paper is not a full revue of the state of the art, but an approach of several methods that have been used to reach the world of glue.

I. THEORETICAL APPROACH

In the Lagrangian of Q.C.D., the self-energy term of the gluon field generates terms of orders 3 and 4 in the field, which manifest themselves as pointlike couplings of 3 and 4 gluons. These couplings lead to a non trivial gluon theory, whereas an electronless theory of electromagnetism would have only free photons.

Up to 1987, we had no direct experimental evidence for the existence of the three gluons vertex. UA1[4] and UA2[5] experiments on the pp̄ collider at CERN have given one through the study of the production of two low invariant mass jets. If that result is not a proof of the existence of gluons bound states, it strenghtens the motivation for their search.

1. Evidence for the 3 gluons vertex in the pp̄ minijets

The events p+p̄ → jet+jet come from the diffusion of a constituent of the proton on a constituent of the anti-proton. These constituents are the valence and sea quarks and the gluons, each type carrying about half of the total momentum of the proton. The distributions of their momenta are the proton structure fonctions.

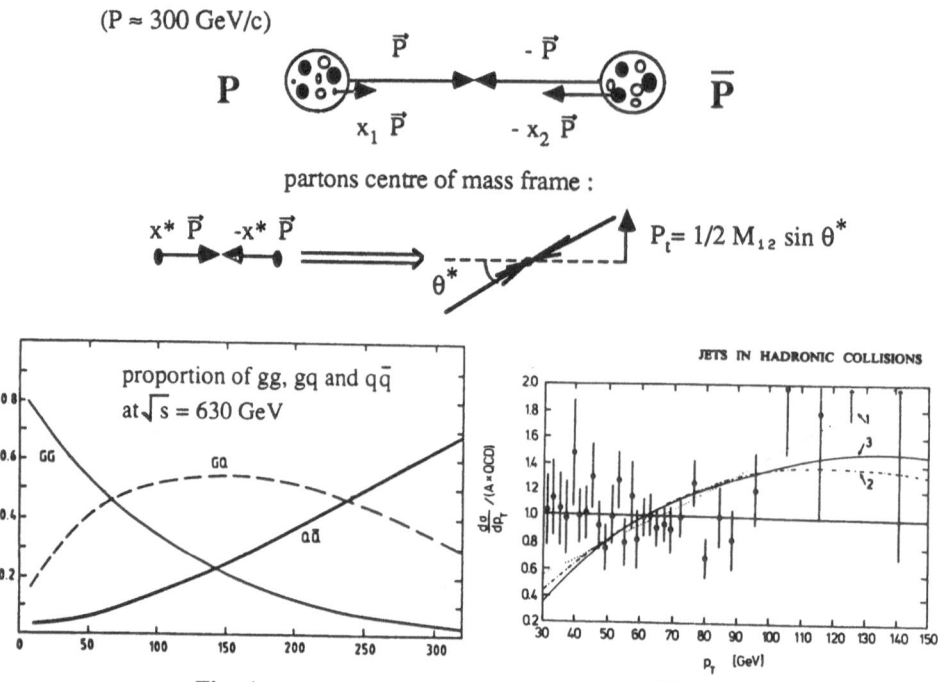

Fig. 1.   Fig. 2.

With the above defined kinematical variables, neglecting the masses with respect to the momenta, the invariant mass M_{12} of the two jets is equal to $2 P \sqrt{x_1 x_2} = 2 P x^*$. That gives, for example, at $M_{12} = 25$ GeV, $x^* \approx 25/600 \approx 0.04$.

The gluons dominate the structure function of the proton at low x (x < 0.1). Then, most of the events at low M_{12} mass are of the type g+g → jet+jet, as shown on figure 1[4].

Figure 2[5] shows the measured cross section for p+p̄ → jet+jet , as a function of P_t, normalized to the Q.C.D. prediction. Below 40 GeV, it is dominated by the process g+g → jet+jet. It is well described by Q.C.D., whereas several models without gluon interactions lead to curves 1 to 3, in disagreement with the data.

The description of the process g+g→jet+jet in QCD is given by the following diagrams :

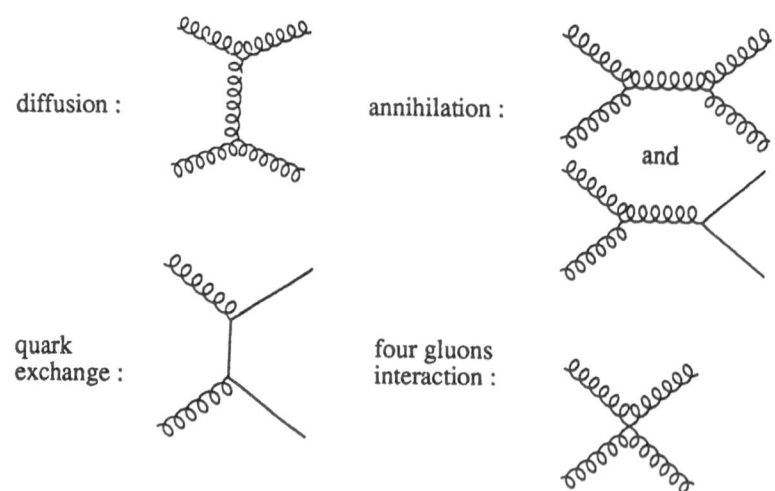

diffusion :

annihilation :   and

quark exchange :

four gluons interaction :

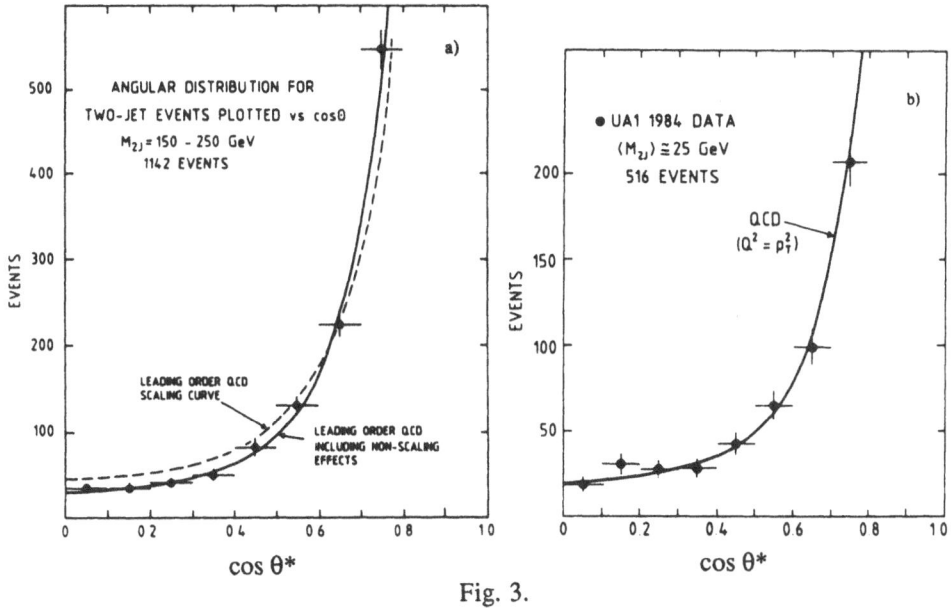

Fig. 3.

Each of these diagrams has its own angular dependence. The diffusion term is proportional to $1/(1-\cos\theta^*)^2$ and dominates at $\cos\theta^* > .5$. This angular dependence is characteristic of the exchange of a vector through the t channel. Small angle Bhabha scattering, Rutherford scattering which are dominated by one photon exchange, and the process $q+\bar{q}\to$jet+jet at high values of $\cos\theta^*$, also dominated by the exchange of a gluon, have the same behaviour.

UA1 has measured the angular distribution of jet pairs at values of $\cos\theta^*$ high enough to show the presence of the diffusion term. Figure 3[4] shows the $\cos\theta^*$ dependence of the jets for two sets of jet-jet invariant masses : the lower one selects g+g→jet+jet, and the higher one selects $q+\bar{q}\to$jet+jet events.

These distributions are well fitted with the QCD diffusion term, which is a strong argument for the presence of the exchange of a gluon in the gluon-gluon diffusion, due to the three gluons coupling.

2. Origin of the gluon couplings

In the Lagrangians of Q.E.D., Q.C.D. and of the electroweak theory , the kinetic energy of the gauge field A_μ is given by

$$\mathcal{L}_g = -1/4 \, \text{Tr} \, F_{\mu\nu}^2, \text{ with } F_{\mu\nu} = \delta_\mu A_\nu - \delta_\nu A_\mu + g[A_\mu,A_\nu] \text{ [6]}.$$

In Q.E.D., the gauge group is U(1), A_μ is the quadrivector of the electro-magnetic potential, and $[A_\mu,A_\nu]=0$. $\mathcal{L}_g = 1/2 \, (E^2+B^2)$ is the free photon energy.

In the two other cases, the gauge group is non abelian, and the A_μ are tensor objects. The commutator of the fields is proportional to their product and introduces couplings between the gauge fields.

In the electroweak theory, the gauge group is SU(2)×U(1) and there are 4 gauge fields to describe the $W^\pm$, the Z° and the photon.

For Q.C.D., the gauge group is color SU(3) and $A_\mu = \sum_{i=1}^{8} \lambda_i A_\mu^i$, where the λ_i are 3×3 matrices generators of SU(3), and the A_μ^i are 8 quadri-potentials that describe the 8 gluon fields. The gluons energy term is given by

$$\mathcal{L}_g = -1/4 \sum_{i=1}^{8} \mathrm{Tr} \, (\delta_\mu A_\nu^i - \delta_\nu A_\mu^i + g \, f_{jk}^i \, A_\mu^j \, A_\nu^k)^2 \,, \text{ where the } f_{jk}^i \text{ are the SU(3) structure}$$

constants. In the development of $\mathcal{L}_g$, the terms $g \, f_{jk}^i \, A_\mu^j \, A_\nu^k \, \delta^\mu A^{\nu i}$ are the source of the 3 gluon coupling and the terms $g^2 \, f_{jk}^i A_\mu^j \, A_\nu^k \, f_{lm}^i A^{\mu l} A^{\nu m}$ the source of the 4 gluon coupling.

3. The glueball hypothesis

These couplings imply that the gluon physics is not a free field one. This is not sufficient to prove that gluon bound states exist. Actually, the $Z^\circ W^+ W^-$ vertex, in the electroweak theory, does not lead to the existence of bound states of intermediate bosons. The size of such states should have the order of magnitude of the weak interaction range, which is given by the Z° mass ($d \approx \hbar c / M_{z^0} \approx 10^{-18}$m). On the other side, the Heisenberg uncertainty principle would force each constituent to have a momentum of the order of $p \approx M_{z^0} c$, which means a kinetic energy of the order of 100 GeV. Then, the existence of bound states needs binding energies greater than 100 GeV, out of the reach of the weak interactions.

The same argument is true for the other types of interactions. It shows that the existence of atoms is partly due to the infinite range of the electromagnetic interactions (i.e. zero mass of the photon). In the case of low mass states bound by strong interactions, the forces have a large range and the binding energies are of the same order of magnitude as the effective masses of the constituents. These two features favour the existence of bound states of gluons.

To go further and predict masses, widths, quantum numbers, decays and production processes for the gluonic states, five models have been used : perturbative Q.C.D.(1), potential models (2), bag models (3), Q.C.D. sum rules (4) and lattice gauge calculations (5). The last four models are different ways of taking into account the non perturbative aspect of strong interactions. They all predict the existence of gluonic mesons in the same mass region as ordinary light quark mesons (.5 to 2. GeV/c^2)[7], with J^{PC} equal to 0^{++}, 0^{-+} and 2^{++}.

4. Glueball properties

The first consequence of the existence of glue, as announced in the introduction, is the fact that flavour SU(3) cannot describe the whole set of mesons. Unfortunately, this does not imply the existence of pure glue states, because the additionnal degree of freedom offered by glue is not orthogonal to others, all these states being expected in the same mass region. Then, the physical states are likely to be mixings of glue and $q\bar{q}$ states. Therefore, even if the glue has very peculiar properties, it is possible that the observed mesons don't exhibit them.

So, the search for glue is strongly related to the knowledge of ordinary mesons.

Quantum numbers

The glueball candidates must have no electric charge and a zero isospin.

Some of the excited states of the 2 and 3 gluons systems have exotic J^{PC}, unaccessible to the $q\bar{q}$ systems[8]. 1^{-+}, 0^{--} 0^{+-} and 2^{+-} are some of these exotic J^{PC} [9]. The discovery of such a particle could not be explained by the classical $q\bar{q}$ model. However, it would not be a proof of the existence of glue, since other systems, like $qq\bar{q}\bar{q}$ or $q\bar{q}g$, can have the same exotic J^{PC}.

Anyway, no candidates of that kind have yet been observed and the search is mainly turned toward the lowest mass states, with J^{PC} equal to 0^{++}, 0^{-+} and 2^{++}.

Glueball production

The safest theoretical prediction on glue concerns the way to produce it. The arguments are qualitative and come from perturbative Q.C.D. They favor processes where the expected gluonic state cannot share quarks with the other particles involved. There are different ways to get that situation :

- by construction :

a) in decays like $c\bar{c}$ or $b\bar{b} \to g+X$ (1), where X has a mass too low to contain b or c quarks

b) in reactions of the type $a+b \to a+b+X$ (2), where X is produced by two pomerons exchange and is nearly at rest in the centre of mass of a and b

- by selection of special decay channels :

c) in the reaction $\pi^-+p \to n+X$ (3) followed by the decay $X \to \phi\phi$, the $\phi\phi$ system contains only s and $\bar{s}$ quarks, absent in the π^-, the proton and the neutron. This reaction is said to violate the O.Z.I. rule[10].

In the following, we will mainly concentrate on the process (1) applied to the 1^{--} ground state of the charmonium, the $J/\Psi(3097)$, where a lot of experimental results are available.

Widths and decays

The widths of the lower lying states depend on the coupling of glue to the world of quarks. It could be less than the width of the ordinary mesons with the same quantum numbers, which are allowed to decay directly[11]. For instance, (α) contains two $q\bar{q}g$ coupling absent in (β) .

In fact, the diverging character of this coupling at low momenta forbids, for the time being, safe predictions, even for pure glue states.

Excited glue states can decay into other glue states and are likely to be wider[11].

The partial width into two photons would be zero in a world without quarks, since gluons are not sensitive to the electro-magnetic interaction. It is given by the box diagram (γ) where the gluons to photons coupling is obtained through a quark loop. At first sight, this process leads to $\gamma\gamma$ widths one order of magnitude below the $\gamma\gamma$ widths of ordinary mesons [12], but, as for the total widths, the non perturbative aspect of the strong interaction weakens the prediction[13].

The fact that glue is flavourless should give the clearest criteria to recognize it in hadronic decays. The coupling gluon-quark is indeed the same for all flavours, and if one neglects the mass differences between the u, d and s quarks, all the decays of glue are left unchanged by any permutation of them. On the contrary, the decays of $q\bar{q}'$ states keep memory of the initial flavours. For instance, after correction of phase space effects, the three branching ratios $(gg \to \rho\rho, \omega\omega$ or $\phi\phi)$ are equal, whereas $(s\bar{s} \to \rho\rho$ or $\omega\omega)$ and $(u\bar{u}\pm d\bar{d} \to \phi\phi)$ are forbidden by the O.Z.I. rule[10]. Similarly, for 0^{++} and 2^{++} gluonic mesons, one expects comparable rates into pseudoscalar pairs $\pi\pi$, $K\bar{K}$ and $\eta\eta$.

For the same reason, if they are accessible to the experiment, the radiative modes (δ) have rates respectively proportional to $(q_u-q_d)^2/2$, $(q_u+q_d)^2/2$ and q_s^2, that is to say 9, 1 and 2, since they only differ by the the mean electric charge of the quarks in the vector meson.

(δ)

Otherwise, all the decay modes allowed by phase space and conservation laws exist a priori[14]. It can be noticed that the decays of the η_c(2980), pseudoscalar ground state of the $c\bar{c}$ system, should look much like the decays of a two gluons state, since it is a flavour SU(3) singlet that decays through two gluons[15]. Today, about only 20% of its decays are known, among which $\rho\rho$ and $\phi\phi$, which are, as expected, comparable[16].

The theoretical approach of the mass region from 1 to 2 GeV/c^2 is still lacking in precision, which is due to the non perturbative aspect of the strong interaction and to the lack of experimental data to test the models. The experimental approach must then be as wide as possible. Gathering information is needed to evidence this new state of matter.

II. EXPERIMENTAL APPROACH

1. The J/Ψ, a source of gluonic mesons

The J/Ψ(3097), discovered in 1974, is the lower lying $c\bar{c}$ vector meson. Its mass is below the threshold of open charm production, which gives it remarkable properties. An ordinary 3 GeV meson would have a width of several hundred MeV, whereas the J/Ψ width is only 63 keV. Electromagnetic interactions play an important part in its decays which are described by the following five diagrams :

where the percentages are the approximate rates of each type of decay, (2) and (4) being supposed not to interfere. This description neglects graphs with more than three gluons.

Radiative decays, as described by diagram (5), produce a two gluons system with an invariant mass spectrum extending from 0 to 3 GeV/c^2. This system has a positive charge conjugation, because both the J/Ψ and the photon have negative charge conjugations. For on shell gluons, the spin-parity of this system splits up in fonction of its mass as shown on figure 4, computed from first order Q.C.D.[29]

Spin-parity 2^+ dominates, the 0^+ and 0^- components are equal and there is a small contribution of 4^+ at low masses.

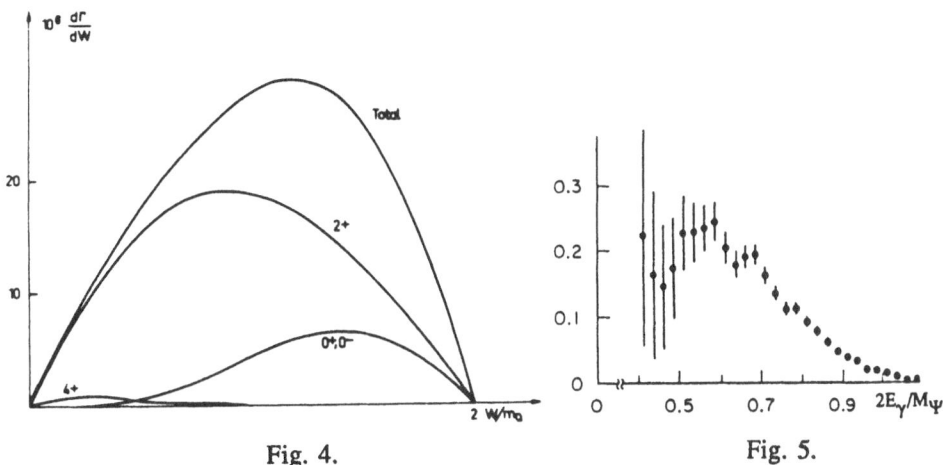

Fig. 4.   Fig. 5.

With massive gluons, there are additional 1^{++} [30] and 1^{-+} [31] components.

This two gluon system is expected to be strongly coupled to the two gluon bound states of same quantum numbers and less coupled to ordinary mesons with positive charge conjugation and zero isospin[28]. The radiative decays of the J/Ψ are therefore a potential source of scalar, pseudoscalar and tensor glueballs. Moreover, as it will be shown in the next paragraph, the important rate of these decay modes favors their experimental study.

2. Calculation of the radiative decay rate

The J/Ψ decays description by processes (1) to (5) allows an evaluation of the inclusive branching ratio (J/$\Psi \rightarrow \gamma X$). This description uses the fact that processes (1) to (3) are known from experiment, and that graphs (4) and (5) are both related to (1) by the theory.

- The processes (1), J/$\Psi \rightarrow$ e⁺e⁻ and J/$\Psi \rightarrow \mu^+\mu^-$, have been measured[17]. Both branching ratios amount to 6.9 ± 0.9 %. As they must be equal, up to terms of the order of $(M_\mu/M_\Psi)^2$ [18], the error on their value becomes $0.9 / \sqrt{2} = 0.64$ %.

- (2) is related to (1) through the ratio R = σ(e⁺e⁻ → hadrons)/σ(e⁺e⁻ → $\mu^+\mu^-$) whose value is 2.59 ± 0.15 ± 0.08 [19] at 3 GeV center of mass energy.

- The decay (3) towards the η_c has a rate of 1.27 ± 0.36 %[20].

- The processes (4) and (5) have been calculated at the time of the discovery of the J/Ψ in first order Q.C.D. They are both proportional to (1). The corresponding relations are :

$$\Gamma_{\Psi \rightarrow 3g} = \frac{5 \, (\pi^2 - 9)}{18\pi} \frac{\alpha_s^3}{\alpha^2} \Gamma_{\Psi \rightarrow \mu\mu} \qquad \text{(a) [21]},$$

and
$$\Gamma_{\Psi \rightarrow \gamma gg} = \frac{16}{5} \frac{\alpha}{\alpha_s} \Gamma_{\Psi \rightarrow 3g} \qquad \text{(b) [22]}.$$

When neglecting the interference between (2) and (4) and the effect of higher order processes, we get :

$$(\Psi \rightarrow \gamma\eta_c) + (\Psi \rightarrow e^+e^-) + (\Psi \rightarrow \mu^+\mu^-) + (\Psi \rightarrow q \, \bar{q}) + (\Psi \rightarrow 3g) + (\Psi \rightarrow \gamma gg) = 100 \ \%,$$

where the parentheses stand for branching ratios.

165

Using the preceeding relations, one gets

$$(1.27 \pm .36)\ 10^{-2}+(6.9 \pm .6)\ 10^{-2}.[2+(2.59 \pm .17)+\frac{5}{18}\ \frac{\pi^2-9}{\pi}\ \frac{\alpha_s^3}{\alpha^2}\ (1+\frac{16}{5}\ \frac{\alpha}{\alpha_s})] = 1 \quad (A)$$

which gives the following equation in α_s: $\quad \alpha_s^3+ 2.34\ 10^{-2}\ \alpha_s^2 = (6.74 \pm 0.90)\ 10^{-3}$ (B)

hence $\quad \underline{\alpha_s(M\Psi) = 0.181 \pm 0.008 \text{ and } (J/\Psi\rightarrow\gamma gg) = 7.7 \pm 0.4\ \%.}$

which, if true, is very precise. Actually, we have used two different kinds of approximations which could considerably alter the result :

- the effect of higher order terms in α_s in the perturbative development of Q.C.D. must be examined,

- relations (a) and (b) are obtained in a non relativistic approximation. The relativistic effects[25] involve corrections of the order of $<v^2/c^2>$ (≈ 0.2 at the J/Ψ) which have not yet been precisely computed[24] but which are probably important for relation (a) and small for relation (b).

The next perturbative order in α_s gives additionnal terms[25] :

$$(10.2 \pm 0.5)\ \alpha_s(M_\Psi)\ /\ \pi \text{ to the } \Gamma_{\Psi\rightarrow 3g}/\Gamma_{\Psi\rightarrow\mu\mu} \text{ ratio}$$

and $\quad (4.4 \pm 0.4)\ \alpha_s(M_\Psi)\ /\ \pi \text{ to the } \Gamma_{\Psi\rightarrow\gamma gg}/\Gamma_{\Psi\rightarrow 3g} \text{ ratio.}$

Equation (B) is now :

$$\alpha_s^3(1+(10.2 \pm .5)\alpha_s/\pi) + 2.34\ 10^{-2}\ \alpha_s^2(1+(4.4 \pm .4)\alpha_s/\pi) = (6.74 \pm .90)\ 10^{-3}$$

and its solution is $\alpha_s = 0.159 \pm 0.006$, only 12% lower than the value obtained at first order. The radiative branching ratio is even less affected by this correction, since the lowering of α_s is balanced by the additional positive term in the $\Gamma_{\Psi\rightarrow\gamma gg}/\Gamma_{\Psi\rightarrow 3g}$ ratio. It amounts now to (7.0±0.4)%. The result is therefore stable against the Q.C.D. radiative corrections. A few remarks can be made on this subject.

$\alpha_s(Q^2)$ is measured here in the time-like region and its Q^2 dependence is probably not the same as in the space-like region, where diffusion processes take place[26]. Its low value is compatible with the other existing measurements[70].

As the sum of all perturbative orders is directly related to physical quantities, the second order calculation depends on the choice of a renormalisation scheme. Several ways have been investigated to find a scheme that minimizes the contribution of the non computed higher order terms[24]. The result presented uses the standard MS scheme. Another convention, the so called "effective charges" or FAC scheme[27] chooses the renormalisation point which cancels the second order correction. It leads to $\alpha_s(M_\Psi) = 0.15 \pm 0.01$ and to the same branching ratio as in the first order calculation.

To avoid the relativistic corrections, the relation (b) where they are supposed to cancel can be associated with the measurement $(\Psi\rightarrow 3g)+(\Psi\rightarrow\gamma gg) = 6\pm3\ \%$. $\alpha_s(M_\Psi)$ has now to be known.

With $\Lambda_{QCD}= 200\pm100$ MeV, $\alpha_s(M_\Psi)=.21\pm.04$, (b) gives $(\Psi\rightarrow 3g)/(\Psi\rightarrow\gamma gg) = .11\pm.02$ and the radiative branching ratio amounts to $6.6 \pm 1.1\%$, which is very close to the preceeding values.

If the Q.C.D. second order correction is now included in (b), with the same value for $\alpha_s(M_\Psi)$, $(J/\Psi \to \gamma gg) = 5.2 \pm 1.0\%$. Again, the result is quite stable.

The conclusion of this discussion is that neither the Q.C.D. higher orders, nor the relativistic corrections have a large effect on the naive first order prediction. Taking these corrections into account, one gets :

$$\boxed{(J/\Psi \to \gamma gg) = 6 \pm 2\%.}$$

This calculus can be also applied to the $\Upsilon(9460)$ which is the analog of the J/Ψ for the $b\bar{b}$ system. But, in this case, the ratio of the partial widths into γgg and $3g$ is 4 times lower than for the J/Ψ because the charge of the b quark is half the charge of the c quark. In spite of this lower rate, the radiative decays of the Υ potentially produce gluonia and high statistics analyses would extend this study to masses higher than 3 GeV/c².

3. Inclusive measurements

e⁺e⁻ allows high statistic studies of J/Ψ decays, one month of data taking corresponding to about one million J/Ψ (1MΨ) produced on the existing machines (about 3MΨ on DCI). The J/Ψ has been studied in Hamburg on DORIS by PLUTO and DASP, in Stanford on SPEAR by MARK I, MARK II, CRYSTAL BALL (2.2 MΨ) and MARK III (5.7 MΨ) and in Orsay on DCI by DM2 (8.6 MΨ).

MARK II[32] has given an inclusive measurement of the radiative mode. Such an analysis suffers from a large background which comes from hadronic decays with π°'s or η's. Due to detection inefficiencies, the decays of π° and η to two photons can simulate isolated photons. This background has been separately measured and subtracted from the energy spectrum of the whole set of photons. The spectrum is shown on figure 5.

For $E_\gamma < .6\ M_\Psi/2$, the result is not precise. For $E_\gamma > .6\ M_\Psi/2$, or $M_X < 2$ GeV/c², the branching ratio of the process J/$\Psi \to \gamma X$ amounts to 4.1±.8%, near to the expected rate for the whole spectrum.

CRYSTAL BALL has analyzed the radiative modes with more statistics and a better photon energy resolution. The lower granularity of this detector makes it more difficult to extract the π° and η backgrounds. On the inclusive spectrum shown on figure 6, only several classes of non radiative events have been removed, and the masses lower than about 500 MeV/c² have been cut out[2].

This spectrum shows structures that have been studied in exclusive analyses.

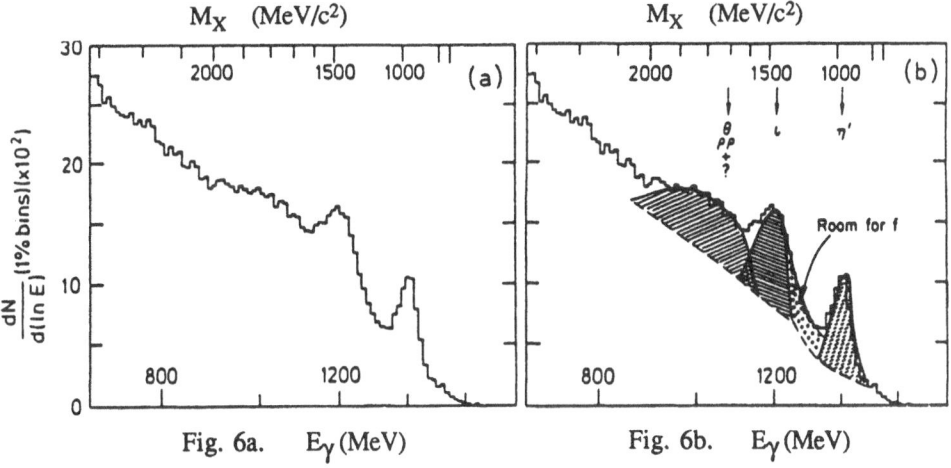

Fig. 6a. E_γ (MeV)   Fig. 6b. E_γ (MeV)

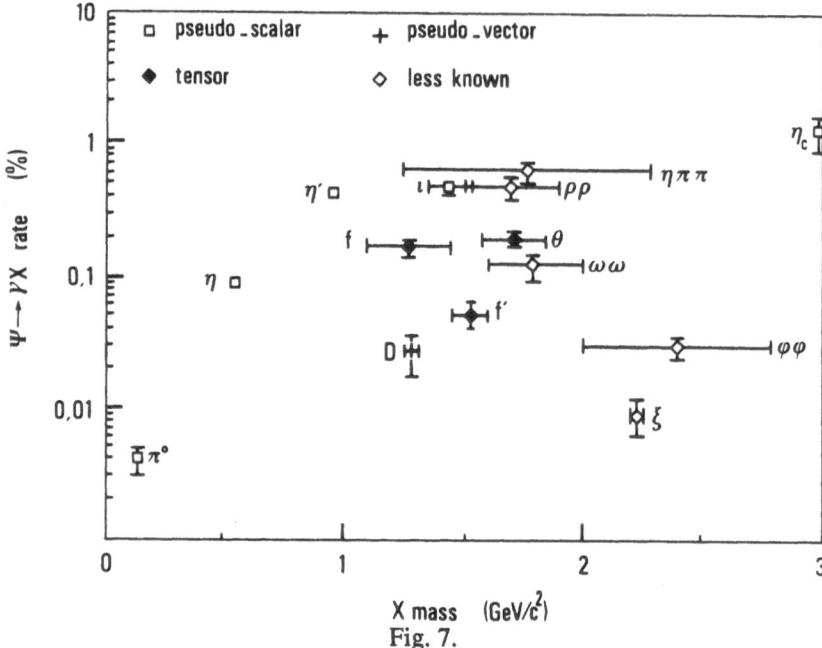

X mass (GeV/c²)

Fig. 7.

4. Exclusive measurements

Figure 7 shows the present status of exclusive studies on J/Ψ radiative decays.

Most of them will be discussed in the following. The old names of particles have been used to simplify the figure. ι stands for η(1440), D for f_1(1285), f for f_2(1270), θ for f_2(1720) and f' for f'$_2$(1525).

On figure 6b, the η(1440) and η' are seen with large and comparable rates, which is confirmed on figure 7. The f_2(1720) and the wide states seen in ρρ and ηππ can explain the excess of events between 1.6 and 2. GeV/c².

The most copiously produced states are a priori the best glueball candidates.

There is yet no measurements on scalar mesons.

Except for the η_c, the sum of all measured radiative branching ratios amounts to 2.6±.4%. There is some room left for yet unmeasured processes, if the theoretical prediction or the MARK II inclusive measurement are to be taken at face value.

5. Glue component in the η'

The η' mass is much higher than that of the other members of the pseudoscalar nonet (π, K and η). It is very abundant in the radiative decays of the J/Ψ. These two properties could be related to the presence of a glue component in the η'.

There are several ways to determine its wave function[33]. One of them is based upon the coupled analyses of all the J/Ψ decays into one vector (V) and one pseudoscalar (PS). These decays can be described by the following diagrams :

(a) hadronic   and   (b) electromagnetic

in which the vector and the pseudoscalar must have the same flavour content.

The vectors are the ρ, ω, ϕ and K^* and the pseudoscalars are the π, η and η'. The vector nonet is supposed to be ideally mixed (the ϕ is a pure $s\bar{s}$) and the nonet symmetry is supposed to be conserved. So, only the $s\bar{s}$ component of the pseudoscalars couples to the ϕ, which forbids the $\phi\pi^\circ$ decay. Similarly, the ρ and ω select the u and d quarks in the pseudoscalars.

The ten possible channels are $\phi(\eta$ or $\eta')$, $\omega(\pi^\circ, \eta$ or $\eta')$, $\rho(\pi, \eta$ or $\eta')$, $K^{*\pm} K^\mp$ and $K^{*\circ} K^\circ$.

The calculation in the SU(3) framework depends on the 8 following parameters :

- one amplitude for the electromagnetic term,

- two amplitudes for the hadronic term (one conserves flavour SU(3), the other breaks it through a mass term),

- one phase between the electromagnetic and the hadronic terms,

- the four (X,Y) components of the η and η' on the states $|N\rangle = 1/\sqrt{2}|u\bar{u}+d\bar{d}\rangle$ and $|S\rangle = |s\bar{s}\rangle$[35].

MARK III measured 9 of these modes and set limits on $\phi\pi^\circ$ and $\rho\eta'$. The 8 parameters have been optimized according to the 9 branching ratio values. The χ^2 probability is 14% and the result is :

$$X_\eta^2 + Y_\eta^2 = 1.1 \pm 0.2 \quad \text{and} \quad X_{\eta'}^2 + Y_{\eta'}^2 = 0.65 \pm 0.18 .$$

So, the η wave function is saturated by a mixing of two states $|N\rangle$ and $|S\rangle$, whereas the 2σ effect on the η' implies with a 97% probability the presence in it of an extra state which will be called $|G\rangle$ in the following discussion.

Besides, this analysis finds out that the electromagnetic term contributes for about 10%, the same as for the general decays of the J/Ψ.

The decay into $\rho\eta'$ has been seen by DM2[68] and its branching ratio is in agreement with what is expected from this analysis.

Discussion

- This result assumes that all measurements are independent. It would not be the case if systematics errors on different channels were correlated. The values could then be slightly wrong and the errors underestimated.

- This description does not take into account the "simply" and "doubly disconnected" diagrams (c) and (d) which couple the singlet components of V and PS without flavour correlation :

The absence of the $\phi\pi^\circ$ mode is only due to the nonet symetry and the absence of a mixing between ϕ and π° that would give them a common flavour component[36]. It is not an argument against the presence of disconnected diagrams since (c) and (d) can produce only zero isospin particles.

Recent analyses of the same kind[67], including diagram (c) and assuming that there are no extra components in the η and η', fit well the data and lead to a η - η' mixing angle in good agreement with current values[71]. The contribution of diagram (c) is found to be as large as the electromagnetic decay (b) in the case of the η' which has an important isosinglet component. Some theoretical predictions go in the same direction[37,38]. So, if there is any extra component in the η', it is certainly smaller than in the above analysis.

- The extra $|G\rangle$ that saturates the η' wave function could be of gluonic or $q\bar{q}$ nature. The third physical state involved in the $|N\rangle, |S\rangle, |G\rangle$ mixing could be the $\eta(1440)$[36].

If $|G\rangle$ is an excited $q\bar{q}$ state, then the interference between the ground and excited states in the production diagrams (a) and (b) has to be destructive. For instance, with $|\eta'\rangle = X_{\eta'}|N\rangle + Y_{\eta'}|S\rangle + Z_{\eta'}|G\rangle$, $|G\rangle$ being of the $|N\rangle$ type, $|X_{\eta'}+Z_{\eta'}|^2 + Y_{\eta'}^2 = 0.65 \pm 0.18$ and with $X_{\eta'}^2 + Y_{\eta'}^2 + Z_{\eta'}^2 = 1$ it follows that $|X_{\eta'}+Z_{\eta'}|^2 < |X_{\eta'}|^2 + |Z_{\eta'}|^2$.

If $|G\rangle$ is of gluonic nature, this analysis assumes that it does not couple to the vectors. If this is true for diagrams (a) and (b), this is less sure for (c) and (d). Indeed, if the coupling to a $q\bar{q}$ pair on the PS side is removed, the V can be produced together with a two gluon bound state :

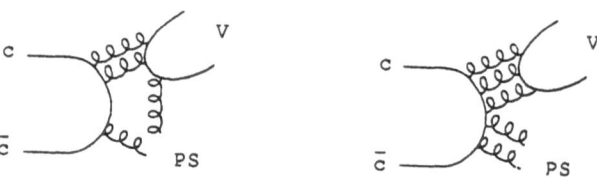

This mechanism, which would be responsible for the production of glue associated with an ω or a ϕ in the decays of the J/Ψ, has not been calculated yet.

- This approach can be seen as an exercise on the η and η' system. It is recent and can be improved on both theoretical and experimental sides. A better measurement of all V+PS channels would help to go further.

6. Coupled study of $J/\Psi \to (\gamma,\omega,\phi) + X$

The preceeding study can be applied to other particles than the π°, η or η'. If a particle X is observed in the radiative mode, it can be searched for in association with an ω or a ϕ. If X is a pure gluonic meson, $J/\Psi \to \omega X$ and $J/\Psi \to \phi X$ can only proceed through disconnected diagrams. Their branching ratios are then expected to be smaller than for $J/\Psi \to \gamma X$ in which X is strongly coupled to the gluon pair. If X is a mixing of components (X_x, Y_x, Z_x) in the basis $(|N\rangle, |S\rangle, |G\rangle)$, where $|G\rangle$ is a glue state, then, the disconnected diagrams being neglected, $\Gamma(J/\Psi \to \omega X)$ is proportional to X_x^2 (and independent of Y_x and Z_x), $\Gamma(J/\Psi \to \phi X)$ is proportional to Y_x^2, and $\Gamma(J/\Psi \to \gamma X)$ is an unknown function of (X_x, Y_x, Z_x).

If, in the picture of §5, only the main contribution (i.e. the SU(3) conserving hadronic term) is taken into account, then $\Gamma(J/\Psi \to \phi X)/\Gamma(J/\Psi \to \omega X) = Y_x^2/X_x^2$ up to kinematical terms that probably do not differ too much between the two modes.

This naive first order picture relates the Y_x/X_x ratio very directly to experimental measurements. It can be applied to the data on J/Ψ decays from MARK III and DM2. Up to now, nine particules have been looked for in association with a photon, an ω or a ϕ. Figure 8 shows the results.

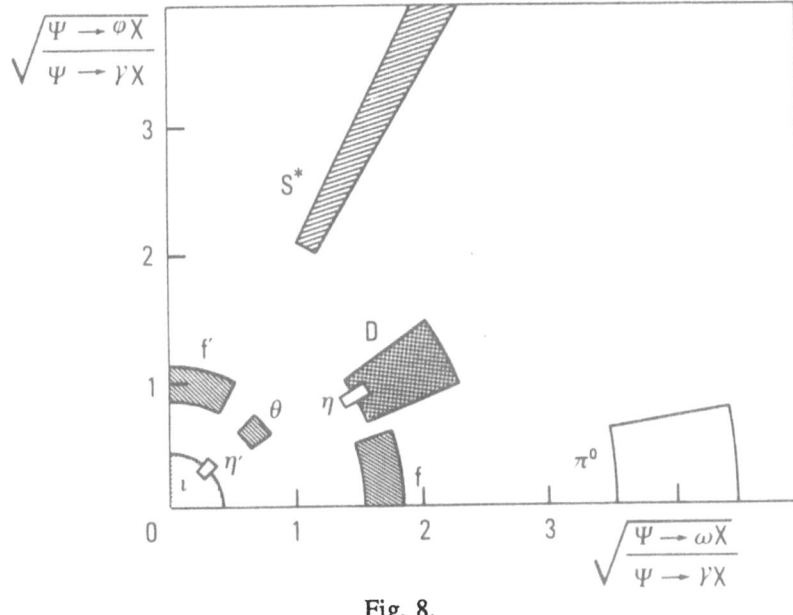

Fig. 8.

In this figure, the mean polar angle θ of each particle is such as tg $\theta = Y_x/X_x$ and the square of the mean radius is the ratio

$$\frac{\Gamma(J/\Psi \rightarrow \phi X) + \Gamma(J/\Psi \rightarrow \omega X)}{\Gamma(J/\Psi \rightarrow \gamma X)}$$

which measures the ratio of the X couplings to q$\bar{\text{q}}$ pairs and to gluon pairs. This ratio is expected to be small for gluonic mesons.

As in figure seven, the old names of particles have been used. S^* stands for $f_0(975)$, D for $f_1(1285)$, f for $f_2(1270)$, θ for $f_2(1720)$, f' for $f'_2(1525)$ and ι for $\eta(1440)$.

If the disconnected diagrams were important, all particles would be aligned on the first diagonal ($\theta = 45°$).

This simple picture agrees with our present knowledge on these particles :

- X_f and Y_f are compatible with zero as expected, since the f' is nearly a pure $|S\rangle$ and the f nearly a pure $|N\rangle$.

- The η and η' positions are near the first diagonal, which is coherent with the pseudo-scalar mixing angle that gives them comparable X and $Y^{[37]}$.

This analysis brings also new informations on the $f_0(975)$ and the $f_1(1285)$ which can no more be considered as pure $|N\rangle$ or $|S\rangle$ states.

The comparison of the radial positions contains also information. For instance, the $\eta(1440)$ and η' are near the origin, where glue is suspected to be.

This naive description which seems to work well can be improved. The angular analysis would give the contribution of S, P and D waves in each decay and allow to calculate the kinematical terms involved in the branching ratios. This would give better values for the Y_x/X_x ratios. The apparently weak contribution of disconnected diagrams has to be theoretically explained. Theoretical predictions of $\Gamma(J/\Psi \rightarrow \gamma, \omega, \phi + X)$ for known q$\bar{\text{q}}$ mesons could allow a deeper understanding of the meaning of figure 8.

In the next paragraphs, the experimental results leading to this coupled study are shown in more details. Most of these results come from the MARK III and DM2 experiments, which generally agree remarkably. The data of these two experiments were taken from 1982 to 1985. The radiative channels have been first analysed. The two other studies are more recent (1985 for the ϕ and 1986 for the ω) because they are associated to more complex final states ($\phi \rightarrow K\bar{K}$ and $\omega \rightarrow \pi^+\pi^-\pi^\circ$).

A coupled analysis of the decays $J/\Psi \rightarrow (\rho, K^*)+X$ would give the same kind of informations on non zero isospin states.

7. $X \rightarrow K\bar{K}\pi$

Figure 9[39)] shows the mass spectra of $K\bar{K}\pi$ systems associated with the photon, the ω and the ϕ as measured by MARK III. The $\eta(1440)$, originally called ι and discovered by MARK II[40)], is seen in the radiative spectrum. It is the most abundantly produced particle in the J/Ψ radiative decays and this is the reason why it was the first glueball candidate. It is a pseudoscalar[41)] with the following mass and width parameters :

$$M=1458\pm5 \text{ MeV/c}^2 \text{ and } \Gamma=100\pm10 \text{ MeV/c}^2.$$

The comparison between the three channels is more complex than expected.

The absence of signal associated with the ϕ proves that the potential $s\bar{s}$ component in the $\eta(1440)$ is small. Actually, at the 90% confidence level, one gets :

$$\frac{\Gamma(J/\Psi \rightarrow \phi\, \eta(1440))}{\Gamma(J/\Psi \rightarrow \gamma\, \eta(1440))} < 4\%.$$

On the contrary, a state X is seen against the ω ; its parameters are incompatible with those of the $\eta(1440)$. Its mass and width are smaller, its $K\bar{K}$ mass spectrum is not as peaked at low values and the angular analysis favours a non zero spin[42)]. Restricted to the $K\bar{K}\pi$ decay channel, $\Gamma(J/\Psi \rightarrow \omega X)$ is about 15% of $\Gamma(J/\Psi \rightarrow \gamma\, \eta(1440))$.

According to the above defined criteria and notations, X is likely to be a $q\bar{q}$ state of the $|N\rangle$ type. It could be part of what is called $\eta(1440)$ in the radiative mode.

Then, at least some fraction of the $\eta(1440)$ fits the glueball selection criteria which have been defined above. If it is really a gluonic meson, it must be observed into other decay channels, for instance $\eta\pi\pi$, $K\bar{K}\pi\pi$, 4π. The $\eta\pi\pi$ mode is especially expected because the $\eta(1440)$ decay into $K\bar{K}\pi$ seems to be dominated by the intermediate state $a_0(980)\pi$ which gives $\eta\pi\pi$ when the $a_0(980)$ decays into $\eta\pi$.

8. $X \rightarrow \eta\pi\pi$

The first study of the $J/\Psi \rightarrow \gamma\eta\pi\pi$ channel has been done by the CRYSTAL BALL. Instead of the expected $\eta(1440)$, it showed a large production of a broad resonance (cf. fig.7) without dominance of the $a_0(980)\pi$[43)] decay. MARK III and DM2 confirm this result with more statistics and they show that this strong $\eta\pi\pi$ production is not due to only one resonance.

The present limit on the $\eta(1440)$ branching ratio into $\eta\pi\pi$ has been given by part of the MARK III statistics. The limit is $\Gamma(\eta(1440) \rightarrow \eta\pi\pi) / \Gamma(\eta(1440) \rightarrow K\bar{K}\pi) < .2$, at the 90% confidence level. The angular analysis of the $\eta\pi\pi$ system would probably lower this limit[44)].

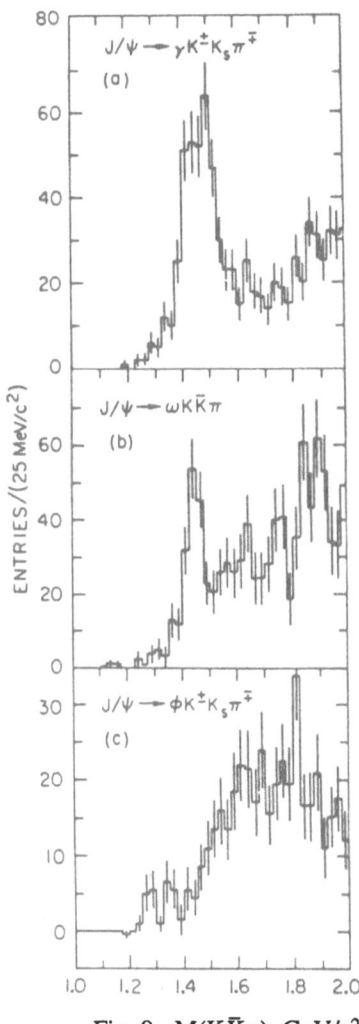

Fig. 9. M(K$\bar{\text{K}}\pi$) GeV/c^2

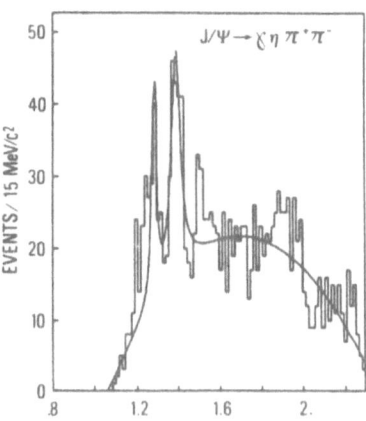

Fig. 10. M($\eta\pi\pi$) GeV/c^2

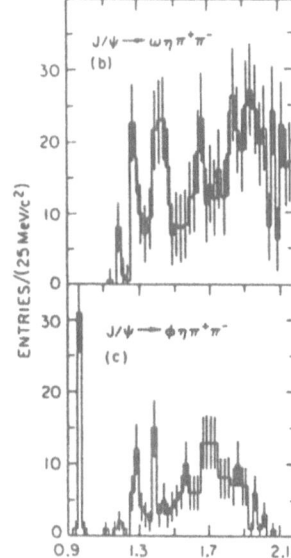

Fig. 11. M($\eta\pi\pi$) GeV/c^2

The $\eta\pi\pi$ suppression is not well understood at present. A possible explanation could be the destructive interference in this mode between $\eta\epsilon$ and $a_0(980)\pi$, the only contribution to $K\bar{K}\pi$ coming from $a_0(980)\pi$. Actually, even the $a_0(980)\pi$ dynamics of the $K\bar{K}\pi$ channel is not conclusive.

Figure 10 shows the DM2 $\eta\pi\pi$ mass spectrum with one $a_0(980)$ required in the $\eta\pi$ masses. The $f_1(1285)$, which is also seen in the channel $J/\Psi\to \gamma 4\pi$, is probably seen. The two measurements of $J/\Psi\to \gamma f_1(1285)$ from DM2 are consistent[46] and in good agreement with the perturbative Q.C.D. predictions for pseudovectors[30].

Another peak is found with a width of about 50 MeV/c^2 and a mass of 1391±3 MeV/c^2. It corresponds to a yet unknown particle which may have been recently observed in a π^-p experiment at KEK[69]. In the $\eta\pi\pi$ decay mode, the branching ratio for $J/\Psi\to \gamma X(1390)$ is an order of magnitude below the branching ratio of $J/\Psi\to \gamma\eta(1440)$ in the $K\bar{K}\pi$ mode. This X state could be looked for in the $K\bar{K}\pi$ mode with an $a_0(980)$ cut on the $K\bar{K}$ mass, however, it is probably small compared to the $\eta(1440)$ signal.

Several other structures can be guessed at higher masses, due to the very similar spectra from MARK III and DM2. To learn more about them, a full angular analysis would be necessary. Such an analysis has been attempted by MARK III on part of the statistics[44]; it is very difficult, especially due to the large background. Nevertheless, if it is done on the whole statistics, it might bring new results. For instance, the $\eta(1275)$, which has to be confirmed, might be hidden under the $f_1(1285)$ peak.

Figure 11, from MARK III, shows the production associated with an ω and a ϕ. The $f_1(1285)$ is seen again, with rates comparable with the radiative one (cf.fig.8). The X(1390) is absent. In its place, coupled to the ω, appears as in the $K\bar{K}\pi$ mode a state around 1420 MeV/c^2 with a width of about 50 MeV/c^2. This state could be the $f_1(1420)$ known as a $q\bar{q}$ pseudovector[39].

9. $X \rightarrow \eta\eta$

This channel has been studied by only one experiment[72], CRYSTAL BALL, which discovered there the $\theta(1650)$[47]. The θ was later seen in the $K\bar{K}$ mode on the right side of the $f'_2(1525)$[48], with a higher mass than in the $\eta\eta$ mode. After that, CRYSTAL BALL improved its analyses and explained its $\eta\eta$ signal by the presence of both the $f'_2(1525)$ and the θ, called now $f_2(1720)$. The corresponding mass spectrum is shown on figure 12.

CRYSTAL BALL also studied the $\eta\eta'$ mode and set the limit

$$\Gamma(f_2(1720) \rightarrow \eta\eta') / \Gamma(f_2(1720) \rightarrow \eta\eta) < .63$$

that favours an isosinglet $f_2(1720)$ since the $\eta\eta'$ combination is pure octet[11]. Anyway, the obervation of $f_2(1720) \rightarrow \eta\eta'$ with an important rate would have favoured its interpretation as a gluonic meson[49].

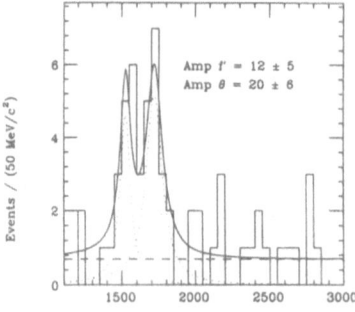

Fig. 12. M($\eta\eta$) MeV/c^2

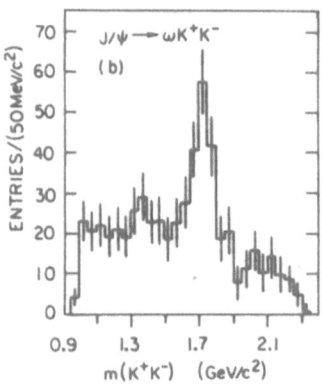

Fig. 14. M(K+K−) GeV/c^2

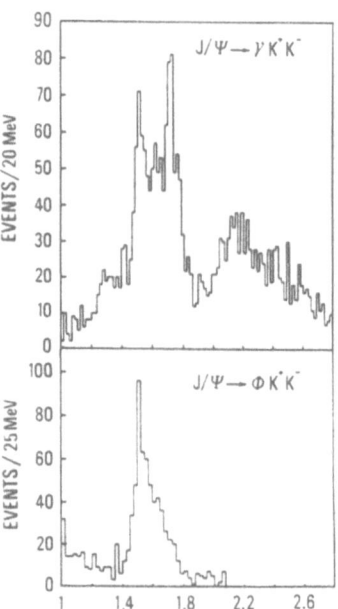

Fig. 13. M(K+K−) GeV/c^2

10. $X \rightarrow K\bar{K}$

This mode is at present time the dominant decay mode of the $f_2(1720)$. Its full angular analysis has been made by MARK III and DM2. The $f_2(1720)$ is clean and it counts a few hundreds of events. However, the conclusions of the angular analysis are not clear : the 2^{++} hypothesis is less favoured against the 0^{++} hypothesis than expected from the simulation of a 2^{++} state produced with the helicity parameters found for the $f_2(1720)$.

This fact might mean that, as the $\theta(1650)$ was the superposition of the $f'_2(1525)$ and the $f_2(1720)$, the $f_2(1720)$ could be a superposition of two or more states with different angular distributions. The present statistics does not allow to separate them.

If, as it will be the case in the following, the $f_2(1720)$ is considered as one particle, its helicity parameters are very different from those of the $f_2(1270)$ and the $f'_2(1525)$. This gives a place apart to the $f_2(1720)$.

Figure 13 from DM2[46] compares the radiative and the $\phi K\bar{K}$ channels. In the last one, the right shoulder of the $f'_2(1525)$ can be interpreted as the $f_2(1270)$ if both particles are allowed to interfere.

On figure 14 from MARK III[39], the $f_2(1720)$ is produced alone against the ω. The $f'_2(1525)$ is absent because it is nearly a pure $s\bar{s}$ state (cf.fig.8).

These measurements lead to suspect the presence of u, d and s quarks in the $f_2(1720)$. However, several arguments plead for a glue component in this meson :

- the comparison of the three branching ratios $J/\Psi \rightarrow (\gamma,\omega,\phi)+f_2(1720)$ shows that the $f_2(1720)$ is produced more strongly in association with a photon than with an ω or a ϕ (cf.fig.8),

- its decays show the flavour symmetry expected for a gluonic meson : its decay into $\pi\pi$[51] may have been observed and its decay rates into $\pi\pi$, $K\bar{K}$ and $\eta\eta$ would be respectively proportional to 1, 4 and 1,

- the present limit on the photon-photon width of the $f_2(1720)$ is small ; it is due to TASSO which gives $\Gamma(f_2(1720) \rightarrow K\bar{K}) \Gamma_{\gamma\gamma}(f_2(1720)) < 0.28$ keV[52].

The $f_2(1720)$ has been unsuccessfully looked for in pion and kaon exchange experiments ($\pi^- p \rightarrow K\bar{K}n$ and $K^- p \rightarrow K\bar{K}\Lambda$) where a large signal was expected due to its important partial width into $\pi\pi$ and $K\bar{K}$. The absence of a signal could be explained if a large part of the $f_2(1720)$ decay modes had up to now escaped to analysis[53], which is unlikely.

On the contrary, signals have been seen in the $f_2(1720)$ mass region in the $\pi\pi$[54] and $K\bar{K}$[55] channels in central production experiments, and in $\eta\eta$ in the reaction $\pi^- p \rightarrow \eta\eta n$[56].

11. $X \rightarrow \pi\pi$

The scalar gluonic mesons masses are the lowest predicted ones[7]. They could be around 1 GeV/c^2. The dominant decay mode should be $\pi\pi$. This is where they have been searched for.

The charged mode study is impossible below 1 GeV/c^2 as shown on the DM2[50] figure 15, due to a very large background around the $\rho^0(770)$ mass. This background comes from the very abundant decay $J/\Psi \rightarrow \rho^0\pi^0$ where the π^0 can decay into an asymmetrical photon pair. Then the photon that carries almost all the π^0 momentum mimics perfectly the radiative photon of the decay $J/\Psi \rightarrow \gamma\pi^+\pi^-$ and the other one has a too low energy to be detected.

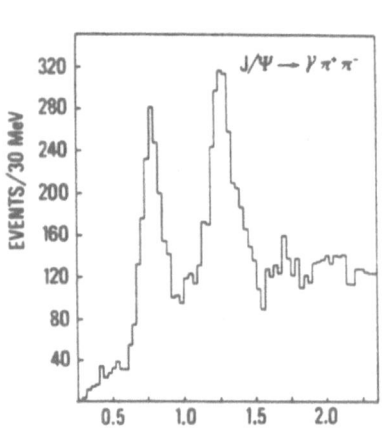

Fig. 15. M($\pi^+\pi^-$) GeV/c²

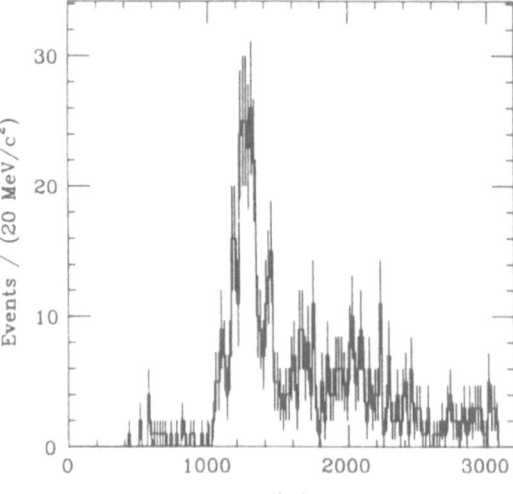

Fig. 16. M($\pi^0\pi^0$) MeV/c²

The channel J/Ψ→γπ⁰π⁰ does not suffer from this background because ρ°→π°π° is forbidden by C-conservation. The best limit has been given by CRYSTAL BALL in this all neutral decay[72]. Figure 16 shows the π⁰π⁰ mass spectrum. The dominant signal is as in the charged decay mode due to the $f_2(1270)$. This detector cannot study the π⁰π⁰ masses below 500 MeV/c² for which, in most cases, the four photon showers are merged. The limit, for a state between 500 and 1000 MeV/c², with a width smaller than 100 MeV/c², is :

$$(J/\Psi \to \gamma X \to \gamma \pi^0 \pi^0) < 1.3 \; 10^{-5}.$$

This limit is very strong compared to the measured radiative branching ratios (cf.fig.7). It seems to exclude totally the existence of a gluonic meson between 500 and 1000 MeV/c². However, it is worth recalling that the two gluon system produced in the radiative decay of the J/Ψ is weakly coupled to the low mass scalars (cf.fig.4), which can explain the low $f_0(975)$ radiative production.

Besides, a narrow glueball candidate has been claimed in an interpretation of data on ππ and K$\bar{\text{K}}$ production by two pomeron exchange[59]. This positive result might not be in contradiction with the CRYSTAL BALL limit.

There are no limits for masses below 500 MeV/c² where anyway no gluonic mesons are expected, and for masses over 1000 MeV/c² where much more statistics would be necessary to extract a scalar particle from the other present signals. A full angular analysis of the whole mass spectrum would be useful, particularly to get sure that the signal around 1700 MeV/c² is really due to the $f_2(1720)$.

The shape of the right shoulder of the $f_2(1270)$ is not well fitted by a Breit-Wigner. The full angular analysis of the π⁺π⁻ spectrum confirms the presence around 1440 MeV/c² of a narrow object with helicity parameters very different from those of the $f_2(1270)$[73]. This fact cannot be explained by the presence of the $f'_2(1525)$ interfering with the $f_2(1270)$ because both particles have similar angular distributions.

The π⁺π⁻ spectra associated to an ω and a φ, coming from DM2, are shown on figures 17 and 18[46]. In the φπ⁺π⁻ mode, the $f_0(975)$ is clearly seen. Its shape is affected by the opening of the K$\bar{\text{K}}$ channel. It does not fit with a Flatté distribution but it is well described by a final state interactions model optimized on ππ phase shift data, which relates these apparently disconnected fields[74].

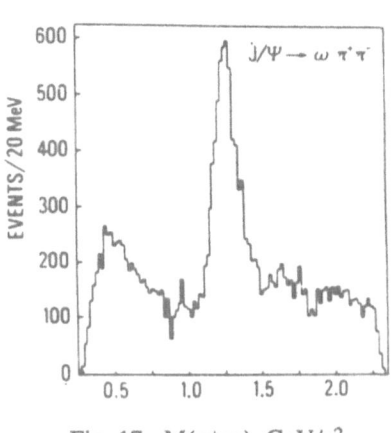

Fig. 17. M($\pi^+\pi^-$) GeV/c^2

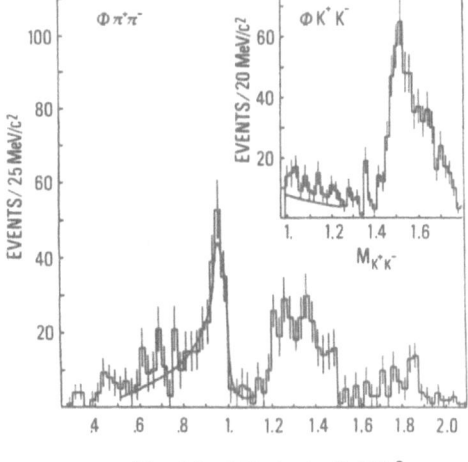

Fig. 18. M($\pi^+\pi^-$) GeV/c^2

The enhancement at low $\pi\pi$ masses in the $\omega\pi\pi$ channel can be explained in the Q.C.D. framework[75].

The f$_2$(1720), seen in the channels $\omega K\bar{K}$ and $\gamma\pi\pi$[51] should appear in the $\pi\pi$ spectrum produced against an ω. Its observation would give a cross check of these analyses. For the time being, the background level is too high to draw conclusions.

12. $X \to \phi\phi$

Two pion diffusion experiments on protons[60] (fig.19) and on Beryllium[61] (fig.20) have observed wide resonances decaying into $\phi\phi$ pairs with masses between 2. and 2.4 GeV/c^2.

The argument for their gluonic nature is the fact that the O.Z.I. rule is violated in their production(cf.I.4). In fact, all their characteristics can be explained if they are interpreted as mixings of two 2P excited states of the $|N\rangle$ and $|S\rangle$ type, $|S\rangle$ being dominant[62].

A signal has been observed in the same decay channel in the radiative decays of the J/Ψ. Figure 21 shows the DM2 spectrum[63]. A part of this production is compatible with 2^{++} and the branching ratio of the process J/$\Psi \to \gamma\phi\phi$ for masses below 2.9 GeV/c^2 is about 3. 10^{-4}.

It can be attempted to relate the two observation, but the present statistics on the J/Ψ decays is two small to draw conclusions. Nevertheless, these states must be looked for in other decay modes, for example $\rho\rho$ and $\omega\omega$ which have been studied and $\eta'\eta'$ and $K^*\bar{K}^*$ which have not.

13. $X \to \rho\rho$ and $X \to \omega\omega$

The study of these two channels in the radiative mode has been made by MARK III[64,65] and DM2[50,46]. Figure 22 shows the DM2 multi-channel analysis of the process J/$\Psi \to \gamma 4\pi^\pm$. Figure 23 shows the MARK III $\omega\omega$ signal.

Both analyses show the presence of structures between 1.6 and 2. GeV/c^2. The low mass peak at about 1.6 GeV/c^2 can be attributed to the η(1440) distorted by the presence of the $\rho^0\rho^0$ mass threshold[66]. If this interpretation is correct, it gives one more argument for the gluonic nature of the η(1440).

On part of its data, MARK III set limits of a few 10^{-4} on the branching ratios to the pp and ωω 2^{++} projections between 2.1 and 2.4 GeV/c. The φφ branching ratio being of the same order of magnitude, these limits don't tell more about the φφ states.

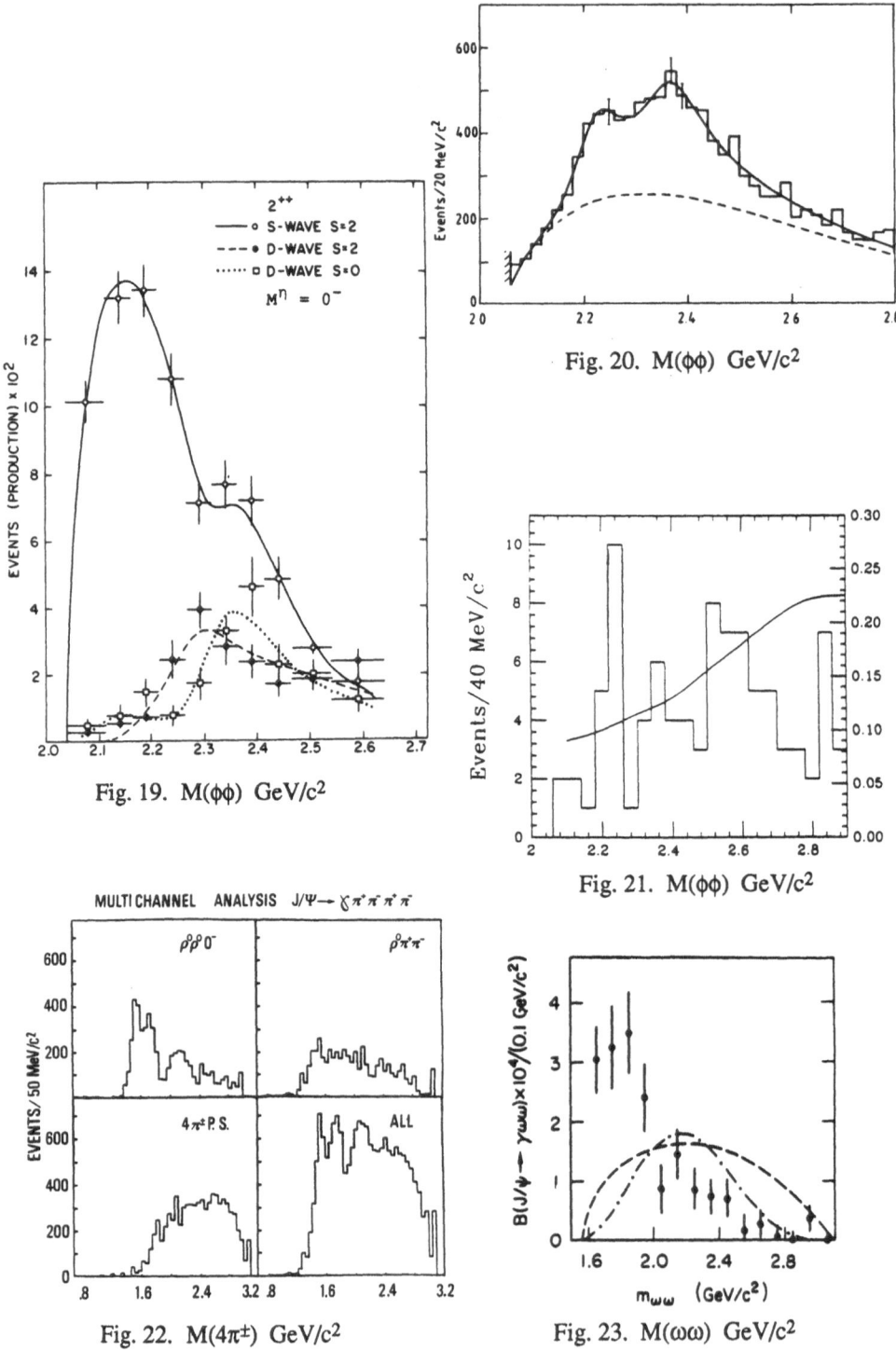

Fig. 19. M(φφ) GeV/c²

Fig. 20. M(φφ) GeV/c²

Fig. 21. M(φφ) GeV/c²

Fig. 22. M(4π±) GeV/c²

Fig. 23. M(ωω) GeV/c²

Conclusions

If the gluonic mesons are firmly predicted by theory, their theoretical and experimental search is not an easy task. It is strongly related to the $q\bar{q}$ spectroscopy in the 1 GeV/c^2 mass region, which has not yet been fully explored. It needs gathering information on the candidates by :

- the study of all the accessible decay channels,

- the diversification of experimental approaches,

- the clarification of glueball signature criteria.

The J/Ψ decays play in this search an important part. They give to the same experiment access to all the possible decay channels. Many new states are seen, particularly the η(1440) which fills all the conditions required to be a gluon bound state. The present analyses of MARK III and DM2 leave many questions unsolved. A new generation of experiments will have to answer them, with higher statistics and better angular acceptance to allow the separation of the different partial waves in every channel.

Acknowledgements

This course was originally prepared for the 1986 Gif-sur-Yvette school in France. I thank the organizers of this workshop for the opportunity they gave me to present it in Erice. I thank for their help and advice during the preparation and the writing G. Mennessier, F.M. Renard, M. Fontannaz and A. Le Yaouanc for the theory, J. Lefrançois and L. Fayard for the p$\bar{p}$ minijets part, the DM2 collaboration members and especially B. Jean-Marie and G. Szklarz for the experimental aspect, with a special thank to J.E. Augustin who directed this work.

I thank Ms. H. Blattmann, A. Pottier and N. Mathieu for the practical realisation and G. Szklarz and B. Le Treut for the reading of the english translation.

References

1. H. Fritzsch, M. Gell-Mann, Proceedings of the XVI International Conference on High Energy Physics, Chicago-Batavia, Illinois, Vol.2, 135 (1972).

2. E.D. Bloom, comptes rendus de la 21ème Conférence Internationale de Physique des Hautes Energies, Paris, France, JOURNAL DE PHYSIQUE, C3-407 (1982).

3. B. Dieckman, CERN EP/86-112 ;
    J.E. Augustin, Proceedings of The Quark Structure of Matter, Strasbourg-Karlsruhe, 323 (1985) ;
    J.E. Augustin, Proceedings of Physics in Collisions 5, Autun, France, 243 (1985) ;
    N. Wermes, Proceedings of Physics in Collisions 5, Autun, France, 221 (1985) ;
    S. Cooper, Proceedings of the International Europhysics Conference on High Energy Physics, Bari, Italy, 945 (1985)

4. F. Cerapibi, Proceedings of the 23rd International Conference on High Energy Physics, Berkeley, California (1986).

5. L. Di Lella, Ann. Rev. Nucl. Part. Sci., 35, 107 (1985).

6. R.P. Feynman, Les Houches, 121 (1976) ;
    J. Iliopoulos, CERN 76-11 ;
    E. De Rafael, Ecole de Gif, tome 2 (1978).

7. S.R. Sharpe, A.I.P., Conf. Proc. n°121, 1 (1984).

8. M.S. Chanowitz, Proceedings of the Summer Institute on Particle Physics, SLAC-245, 41 (1981).

9. S. Meshkov, to be published in Proceedings of the Aspen Winter Physics Conference, (1986).

10. S. Okubo, Phys. Lett. $\underline{5}$, 165 (1963) ;
    G. Zweig, CERN TH-401 et TH-402 (1964) ;
    J. Iizuka, Prog. Theor. Phys. Supp. $\underline{37\text{-}38}$, 21 (1966).

11. F. Renard, unpublished paper refering to :
    H.J. Lipkin, Phys. Lett. $\underline{109B}$ 326 (1982) ;
    J.M. Cornwall, A. Soni, Phys. Lett $\underline{120B}$ 431 (1982) ;
    J.J. Coyne, P.M. Fishbane and S. Meshkov, Phys. Lett. $\underline{91B}$, 259 (1980).

12. H. Kolanoski, Proceedings of the 1985 International Symposium on Lepton and Photon Interactions at High Energies, Kyoto, Japan, 90.

13. J.F. Donoghue, Phys. Rev. $\underline{D30}$, 114 (1984) ;
    M.S. Chanowitz, Proceedings of the VI International Workshop on Photon-Photon Collisions, Lake Tahoe, California, 95 (1984).

14. J.D. Bjorken, Proceedings of the International Conference on High Energy Physics (E.P.S.), Geneva, vol 1, 245 (1979)
    J.F. Donoghue, A.I.P. Conf. Proc. n°81, 97 (1981).

15. E.D. Bloom, Proceedings of the Summer Institute on Particle Physics, SLAC-245, 1 (1981) ;
    E.D. Bloom, C.W. Peck, Ann. Rev. Nucl. Part. Sci., $\underline{33}$, 143 (1983)
    R. Barbieri et al., Phys. Lett $\underline{106B}$, 497 (1981) ;
    T. Appelquist et al., Phys. Rev. Lett. 34, 365 (1975).

16. B. Jean-Marie, LAL 86/29, Proceedings of the 23rd International Conference on High Energy Physics, Berkeley, California (1986).

17. S.M. Boyarski et al., Phys. Rev. Lett. $\underline{34}$, 1357 (1975).

18. T.Appelquist, R.M.Barnett and K.Lane, Ann.Rev.Nucl.Part.Sci.,414 (1978).

19. J.L. Siegrist et al., Phys. Rev. $\underline{D26}$, 969 (1982).

20. J.E. Gaiser, Thèse, SLAC-255 (1982).

21. T. Appelquist, H.D. Politzer, Phys. Rev. Lett. $\underline{34}$, 43 (1975).

22. M.S. Chanowitz, Phys. Rev. $\underline{D12}$, 918 (1975).

23. W. Buchmuller, S.H.H. Tye, Phys. Rev. $\underline{D24}$, 132 (1981).

24. D.W. Duke, R.G. Roberts, Phys. Rep. $\underline{120}$, 275 (1985).

25. P.B. Mackenzie, G.P. Lepage, A.I.P. Conf. Proc. n°74, 176 (1981).

26. M.R. Pennington, R.G. Roberts and G.G.Ross Nucl. Phys. $\underline{B242}$, 69 (1984).

27. A. Grunberg, Phys. Rev. $\underline{D29}$, 2315 (1984).

28. S.J. Brodsky et al., Phys. Lett. $\underline{73B}$, 203 (1978) ;
    K. Koller, and T. Walsh, Nucl. Phys. $\underline{B140}$, 449 (1978).

29. A. Billoire, R. Lacaze, A. Morel, H. Navelet, Phys. Lett. $\underline{80B}$, 381 (1979).

30. J.C. Körner et al., Nucl. Phys. $\underline{B229}$, 115 (1983) ;
    J.C. Körner et al., Zeit. f. Phys. $\underline{C16}$, 279 (1983).

31. R. Lacaze, H. Navelet, Nucl. Phys. $\underline{B186}$, 247 (1981).

32. D.L. Sharre et al., Phys. Rev. $\underline{D23}$, 43 (1981).

33. J.L. Rosner, Proceedings of the 1985 International Symposium on Lepton and Photon
    Interactions at High Energies, Kyoto, Japan, 448 ;
    J. Haïssinski, to be published in Proceedings of the VIITH Workshop on Photon-Photon
    Collisions, PARIS (1986).

34. R.M. Baltrusaitis et al. Phys. Rev. $\underline{D32}$, 2883 (1985).

35. J.L. Rosner Phys. Rev. $\underline{D27}$, 1101 (1983).

36. H.E. Haber and J. Perrier, Phys. Rev. $\underline{D32}$, 2961 (1985).

37. H. Fritzsch and J.D. Jackson Phys. Lett. $\underline{66B}$, 365 (1977).

38. S.S. Pinsky, Phys. Rev. $\underline{D31}$, 1753 (1985).

39. L. Köpke, Proceedings of the 23rd International Conference on High Energy Physics,
    Berkeley, California (1986).

40. D.L. Sharre et al., Phys. Lett. $\underline{97B}$, 329 (1980).

41. C. Edwards et al., Phys. Rev. Lett. $\underline{49}$, 259 (1982) ;
    J.D. Richman, Ph.D., CALT-68-1231 (1985).

42. L. Köpke, XXIème Rencontre de MORIOND $\underline{2}$, 437 (1986).

43. C. Edwards et al., Phys. Rev. Lett. $\underline{51}$, 859 (1983).

44. J.J. Becker, Ph.D., University of Illinois at Urbana-Champaign (1984).

45. W.F. Palmer et al., Phys. Rev. $\underline{D30}$, 1002 (1984) .

46. B. Jean-Marie, LAL 86/33, Proceedings of the 23rd International Conference on High
    Energy Physics, Berkeley, California (1986).

47. C. Edwards et al., Phys. Rev. Lett. $\underline{48}$, 458 (1982).

48. M.E.B. Franklin, Ph.D., SLAC-254 (1982).

49. J.Schechter, Phys. Rev. $\underline{D27}$, 1109 (1983) ;
    S.S. Gerstein et al., Zeit. f. Phys. $\underline{C24}$, 305 (1984).

50. J.E. Augustin et al., LAL/85-27.

51. K.F. Einsweiler, Ph.D., SLAC-Report-272 (1984) ;
    D.M. Coffman et al., SLAC-PUB-3720.

52. M. Althoff et al., Zeit. f. Phys. $\underline{C29}$, 189 (1985).

53. S. Cooper , Proceedings of the 23rd International Conference on High Energy Physics, Berkeley, California (1986).

54. T. Akesson et al., Nucl. Phys. B264, 154 (1986).

55. T.A. Armstrong et al., Phys. Lett. 167B, 133 (1986).

56. D. Alde et al., Nucl. Phys. B269, 485 (1986) ;
    D. Alde et al., CERN EP/86-140.

57. R.A. Lee, Ph.D., SLAC-Report-282 (1985).

58. K. Wacker, XVIIIème Rencontre de MORIOND 2, 91 (1983).

59. K.L. Au et al., XXIème Rencontre de MORIOND 2, 455 (1986).

60. A.Etkin et al., Phys. Lett. 165B, 217 (1985).

61. P.S.L. Booth et al., CERN-EP 85-138 (1985).

62. S.Ono , O. Pene and F. Schöberl, Prog. Theo. Phys. 74, 545 (1985).

63. D. Bisello et al. Phys. Lett. 179B, 294 (1986).

64. R.M. Baltrusaitis et al., Phys. Rev. D33, 1222 (1986) ;

65. R.M. Baltrusaitis et al., Phys. Rev. Lett. D33, 1723 (1986).

66. N. Wermes, Proceedings of Physics in Collisions 5, Autun, France, 221 (1985).

67. A. Seiden, H.F.-W. Sadrozinsky, H.E. Haber, SCIPP 86/73 ;
    H.F.-W. Sadrozinsky in Hadrons, to be published in Quarks and Gluons,
    Rencontres de Moriond 1987 ;
    A. Falvard, to be published in Int. Eur. Conf. on High Energy Physics, Uppsala,
    Sweden (1987).

68. G.Szklarz, to be published in the 2nd Int. Conf. on Hadron Spectroscopy,
    Tsukuba,Japan (1987).

69. S. Fukui et al., contributed paper to the 2nd Int. Conf. on Hadron Spectroscopy,
    Tsukuba,Japan (1987).

70. see for instance F. Sciulli, Proceedings of the 1985 International Symposium on Lepton
    and Photon Interactions at High Energies, Kyoto, Japan, 7.

71.F.J. Gilman, R. Kauffman, Slac-Pub-4301 (1987).

72. This channel is being analysed by DM2 and results will appear in F. Couchot,
    Thèse d'état, to be published as a LAL report.

73. V. Lepeltier et al., LAL 87/10, submitted to Zeit. f. Phys.

74. G. Mennessier, Zeit. f. Phys. C16, 241 (1983).

75. H.G. Dosch and D. Gromes, HD-THEP-87-1, Heidelberg (1987).

EXPERIMENTAL STATUS OF J/ψ DECAYS

Gerald Eigen

High Energy Physics
California Institute of Technology
Pasadena, California 91125

1. INTRODUCTION

The J/ψ was discovered simultaneously at Brookhaven and at SLAC in November 1974. At BNL, the MIT-BNL collaboration studied the process $p + Be \rightarrow e^+e^- + X$ using a high intensity beam of $\sim 2 \times 10^{12}$ protons/pulse from the BNL 30 GeV synchrotron.[1] The e^+e^- pairs were measured with a high resolution double arm spectrometer. The observed e^+e^- invariant mass spectrum displayed in Fig. 1 shows a narrow resonance at 3.1 GeV which was called the J particle. At the e^+e^- storage ring, SPEAR, the SLAC-LBL group scanned the energy region around 3.1 GeV with the SPEAR Magnetic Detector, the first 4π multipurpose detector.[2] A huge increase in the cross section for multihadronic events was observed at a mass $m = 3105 \pm 3$ MeV, as shown in Fig. 2. The error resulted from a 0.1% uncertainty in the absolute energy calibration of the ring. This resonance, named the ψ, also appeared in the mass spectra of e^+e^- and $\mu^+\mu^-$ events. Its width was determined to be $\Gamma < 1.3$ MeV, less than the machine resolution. These results were immediately confirmed by groups at the ADONE and DORIS storage rings.[3] Shortly thereafter, another narrow state, the $\psi(3685)$, was found at SLAC.[4] The discovery of narrow heavy states with a huge production cross section in e^+e^- collisions indicated new physics.

These resonances are bound states of a charm quark and its antiquark, forming the charmonium system.[5] Figure 3 shows the level diagram of the $c\bar{c}$ family. Except the $^1P_1$ state, all other states are well established. The J/ψ is the $^3S_1$ ground state with the quantum numbers of the photon. Its mass and width are now precisely measured to be $m_\psi = 3096.93 \pm 0.09$ MeV and $\Gamma_{tot} = 63 \pm 8.6$ keV.[6] A comparison of quasi-hadronic two-body decays shows that the J/ψ is an isoscalar and an SU(3) singlet.[7]

The study of J/ψ decays thirteen years after its discovery still provides interesting physics, which is motivated in Chapter 2. After a summary of the main properties of the Mark III detector in Chapter 3, I discuss the results of two-pseudoscalar meson systems produced in radiative and hadronic J/ψ decays in Chapter 4. In Chapter 5, I report on the status of the $\xi(2230)$. Chapter 6 is devoted to the $J/\psi \rightarrow \gamma 1^{--}1^{--}$ decays. In Chapter 7, I present the measurements of 3-pseudoscalar meson systems produced in radiative and hadronic J/ψ decays. The data on hadronic two-body decays are summarized in Chapter 8, and a conclusion is given in Chapter 9. Since DM2 results were presented separately by F. Couchot, I primarily discuss Mark III results using data from other experiments for comparison.[8]

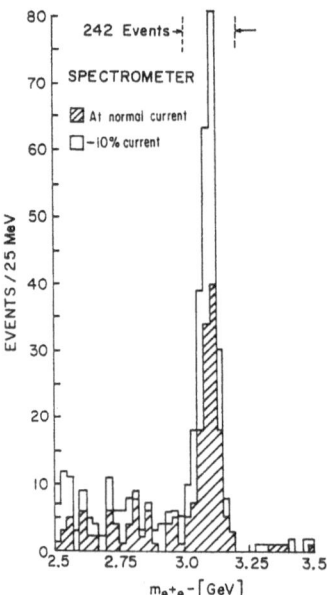

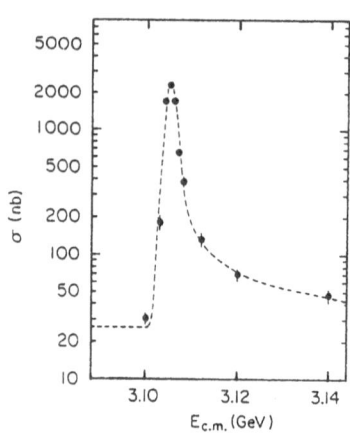

Fig.2. The cross section for $e^+e^- \rightarrow$ *hadrons* showing the ψ particle.

Fig.1. The e^+e^- mass spectrum from $pBe \rightarrow e^+e^-X$ showing the J particle.

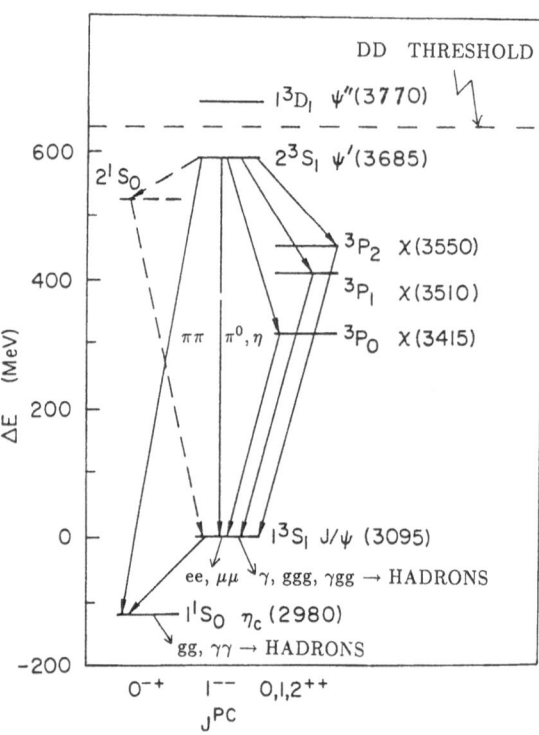

Fig.3. Level diagram for the charmonium system.

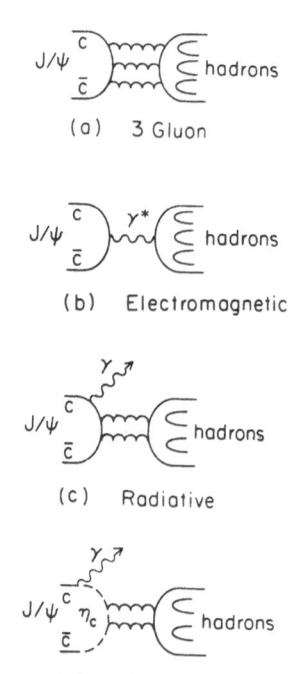

Fig.4. Lowest order diagrams for J/ψ decays.

184

2. PHENOMENOLOGY OF J/ψ DECAYS

2.1 General Remarks

The narrowness of the J/ψ is an immediate consequence of the Okubo, Zweig, Iizuka (OZI) rule,[9] which states that decays with disconnected quark lines are suppressed relative to decays with continuous quark lines. The $c\bar{c}$ system annihilates through a photon or gluons into lighter quarks. The lowest order diagrams are depicted in Fig. 4, in order of strength. The largest decay mode proceeds through a three-gluon exchange (Fig. 4a), as the one-gluon exchange is forbidden by color conservation, and the two-gluon annihilation violates charge-conjugation invariance. The three-gluon decay rate with first order QCD corrections is given by:[10]

$$\Gamma_{ggg} = \frac{40}{81\pi}(\pi^2 - 9)\alpha_s^3(q^2)\frac{|\psi(0)|^2}{m_V^2}[1 + \frac{3.8\alpha_s}{\pi}(q^2)], \qquad (1)$$

where $\alpha_s(q^2)$ is the running strong coupling constant, m_V is the mass of the initial vector meson and $\psi(0)$ is the wave function at the origin. At the J/ψ, where $\alpha_s(m_\psi^2) \approx 0.21$, the 3 gluon rate accounts for $\sim 62\%$ of all decays.

The electromagnetic (EM) decay proceeding through a virtual photon (Fig. 4b) is the next largest mode. Including first order QCD corrections, the EM decay rate is:[11]

$$\Gamma_\gamma = 16\pi^2\alpha^2 e_c^2(|\psi(0)|^2/m_V^2) \times (2 + R) \times [1 - (16/3\pi)\alpha_s(q^2) + \Delta], \qquad (2)$$

where α is the fine structure constant, e_c is the charge of the c quark and Δ estimates relativistic and unknown higher order QCD corrections. The factor $(2 + R)$ accounts for final state lepton and quark pairs, where R is defined in terms of cross sections by $R = \sigma(e^+e^- \rightarrow \text{hadrons})/\sigma(e^+e^- \rightarrow \mu^+\mu^-)$. In J/ψ decays, where $R = 2$, the electromagnetic rate contributes at the 30% level.

The radiative process shown in Fig. 4c proceeds at a rate:[12]

$$\Gamma_{\gamma gg} = \frac{32}{9\pi}(\pi^2 - 9)\alpha_s^2(q^2)\alpha e_c^2\frac{|\psi(0)|^2}{m_V^2}[1 + \frac{6\alpha_s}{\pi}(q^2)]. \qquad (3)$$

The second term in square brackets is the first order QCD correction. The radiative diagram represents $\sim 7\%$ of all J/ψ decays. Finally, $\sim 1\%$ of the J/ψ decays through an M1 photon transition into the η_c as shown in Fig. 4d.

Since the J/ψ decays dominantly through gluons into hadrons, one expects to be able to learn about the quark-gluon coupling and other gluon properties. In addition, various states predicted below 3 GeV may be observable in J/ψ decays:

- Ordinary $q\bar{q}$ states should be copiously produced, thus providing a new approach to the light quark spectroscopy and allowing testing of SU(3) relations.
- $q\bar{q}q\bar{q}$ states and $q\bar{q} - q\bar{q}$ molecules have been predicted.[13] The questions are whether such states exist and whether they are produced in J/ψ decays. Ted Barnes will give a detailed discussion on this subject.[14]
- QCD predicts gg and ggg bound states called glueballs. The observation of a glueball would provide strong support for QCD.
- QCD also predicts $q\bar{q}g$ and $qqqg$ bound states, which are called hybrids, hermaphrodites or meiktons. So far no hybrid has been identified, though candidates exist.

These prospects make the study of J/ψ decays very interesting.

The J/ψ is an excellent laboratory to perform these studies for several reasons:

- The J/ψ is a narrow resonance with a huge production cross section in e^+e^-.

- The nonresonant background is very small ($< 1\%$ at SPEAR).
- The J/ψ is produced at rest. The energy and momentum of the initial state are well defined, which is important for kinematic fits.
- The J/ψ is a flavor singlet carrying the quantum numbers of the photon. It is produced with an incoherent mixture of both $J = \pm 1$ states.
- The initial state is devoid of light quarks.
- High statistics can be accumulated easily ($\sim 5 \times 10^4$ J/ψ's per day at SPEAR), yielding high sensitivity to processes with small branching ratios. Currently, Mark III can detect branching ratios of $\gtrsim 10^{-5}$.

2.2 The $q\bar{q}$ Spectrum

The study of the $q\bar{q}$ system is by itself an important and interesting task. In order to locate and identify glueballs, hybrids and other exotic states it is necessary to understand the "ordinary" mesons in detail. This means that each state in the $q\bar{q}$ nonets must be identified unambiguously and that all its properties have to be well measured. Once the nonets are established, one can probe the predictions of relativistic QCD-inspired potential models which presently provide an excellent description of both light and heavy mesons.[15] Other interesting issues include the study of SU(3) breaking patterns, relativistic and threshold effects, contributions from higher order QCD corrections, non-perturbative QCD effects and dynamics effects.

After ~ 30 years of study, the $q\bar{q}$ system is still not perfectly understood. Figure 5 summarizes the current status of the $q\bar{q}$ nonets. The individual boxes represent the $I = 1$, $I = 1/2$ and the two $I = 0$ ($u\bar{u} + d\bar{d}$ and $s\bar{s}$) members of each nonet. Solid lines indicate established mesons, dashed lines denote particles which need confirmation and wavy lines represent ambiguous assignments. Currently, only 4 nonets are completely filled: the 0^{-+}, 1^{--}, 2^{++} and 3^{--} nonets. If the $h_1'(1380)$[16] is confirmed, the 1^{+-} nonet will also be complete. Four other nonets have all positions filled with possible candidates. In the 0^{++} nonet, questions remain about the $a_0(980)$ and the $f_0(975)$. In the 1^{++} nonet, the interpretation of the $f_1(1420)$ is unclear. In the first radial excitation of the 0^{-+} nonet, the $K(1830)$ and the radially excited η' need to be established. In the 4^{++} nonet, the $a_4(2040)$ and the $f_4'(2300)$ have to be confirmed. The remaining nonets are in worse shape, since many candidates are missing and most of those shown need confirmation. Hopefully, the status of the $q\bar{q}$ nonets will become much clearer in the next few years.

2.3 Glueballs and Hybrids

Since Close[17] will give a detailed discussion on glueballs and hybrids, I will discuss only a few important facts, starting with glueballs. Glueballs are bound states of 2 or 3 gluons in a color singlet. The gg states have even C-parity, while for ggg states both even and odd C parities are allowed. Assuming that gluons inside a glueball are massive, the states shown in Fig. 6 can be formed.[18] In the gg sector, the lowest states are 0^{++}, and 2^{++}, where the scalar glueball is the ground state. In the ggg sector, the lowest allowed states are 0^{-+}, 1^{--} and 3^{--}. In the figure, the scale between gg and ggg states is arbitrary. However, the gg states are expected to be at lower masses than the ggg states. In the gg sector, the first oddball, a 1^{-+} state, occurs with $L = 1$. Oddballs are glueballs with J^{PC} quantum numbers like 0^{--}, 1^{-+}, 3^{-+},...,0^{+-}, 2^{+-},... which do not exist in the $q\bar{q}$ sector. However, if gluons inside glueballs are in fact massless, $J = odd$ states are forbidden in the gg sector by Yang's theorem,[19] implying that oddballs are absent in the gg sector, although they may exist in the ggg sector.

The mass predictions for glueballs are essentially based on 3 methods: Lattice Gauge Theories,[20,21] Bag Models,[22] and Potential Models.[23] Figure 7 shows a representative sample of glueball-mass predictions in the gg and ggg sectors. In lattice

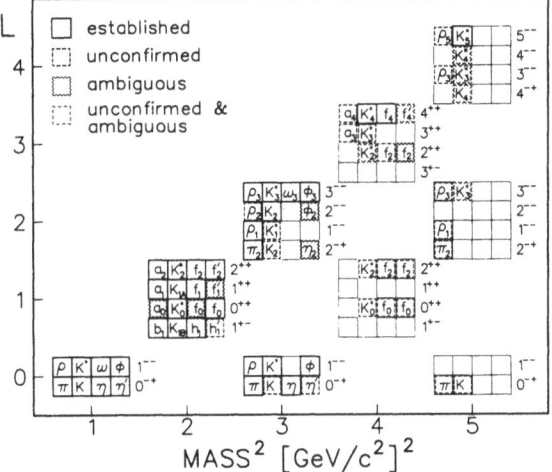

Fig.5. Current status of the light $q\bar{q}$ nonets as a function of mass squared for different orbital angular momenta L and different principal quantum numbers N.

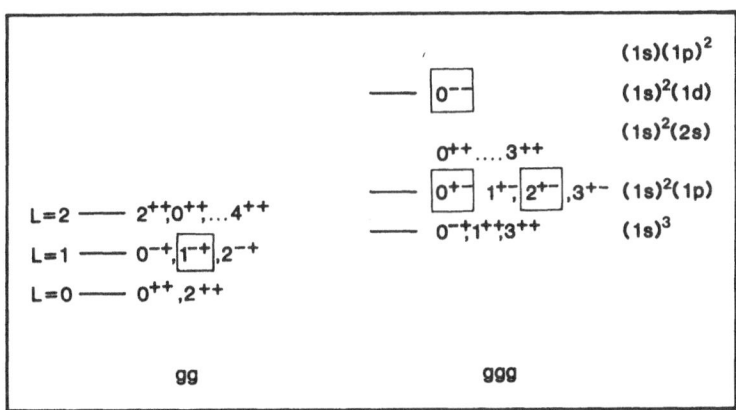

Fig.6. Levels for gg and ggg glueballs.

calculations the mass scale of the glueball spectrum depends crucially on the location of the scalar glueball, which is the ground state in all models. While earlier lattice calculations [20] predicted the 0^{++} glueball mass between 0.6 and 1.0 GeV, newer lattice calculations constrain $m(0^{++})$ to the 1.15-1.4 GeV mass region.[21] The predictions depend on the value chosen for the string tension, which is typically $K \simeq 0.18$ (GeV)^2. In bag models, mass predictions depend strongly on the ratio $r_C = C_{TE}/C_{TM}$ of the gluon self energy C in the transverse electric (TE) and transverse magnetic (TM) cavity modes. Mass calculations exist for $r_C = 1/2$, 1 and 2. But as pointed out by Carlson et al.,[22] only for $r_C = 1/2$ is the TE kinetic energy larger than the TE gluon self energy. The predictions of Chanowitz and Sharpe and Carlson et al., shown in Fig. 7 are, therefore, those for $r_C = 1/2$. In the two models this choice yields masses of the 0^{++} state of 0.65 and 0.78 GeV, respectively. For larger values of r_C the glueball masses increase, yielding e.g. $m(0^{++}) = 1.21$ GeV for $r_C = 1$ in the Chanowitz-Sharpe model. The potential model of Cornwall and Soni predicts the mass of the scalar glueball at 1.15 GeV.

The mass predictions for the tensor glueball vary between 1.4 and 1.9 GeV, where the new lattice calculations yield masses of 1.4 GeV (Berg et al.), 1.68 GeV (Parisi et al.) and 1.84 GeV (de Forcrand et al.). For the pseudoscalar states, the mass predictions range from 1.15 to 1.45 GeV. Currently, no newer lattice predictions exist for the 0^{-+} state. In the Chanowitz and Sharpe model, the 0^{-+} state is constrained to 1.44 GeV. The predictions for $m(0^{-+})$ and $m(2^{++})$ do not improve even if the uncertainty of the 0^{++} mass is removed by defining the ratio $r(J^{PC}) = m(J^{PC})/m(0^{++})$. The mass ratios obtained are $1.2 \leq r(2^{++}) \leq 2.6$ and $1.3 \leq r(0^{++}) \leq 2.1$. However, most models predict the level ordering of the $0^{++}, 0^{-+}$ and 2^{++} states to be: $m(2^{++}) > m(0^{-+}) > m(0^{++})$. Note that Fig. 7 also shows three predictions for a 1^{-+} oddball. In the Cornwall-Soni[23] model a mass of 1.4 GeV is predicted, which is close to a new lattice calculation by de Forcrand et al., yielding 1.28 GeV. For a different set of parameters in the lattice calculation the mass moves up to 2.38 GeV. An earlier prediction by the same group gave a mass at ~ 1.8 GeV. The predictions for the ggg states are even more vague. In conclusion, the predictive power of these models is currently very limited. All one can say is that the low lying glueballs are expected in the $600 - 2500$ MeV mass range.

The situation with respect to width predictions is even more controversial. Robson[24] suggests that the glueball width is given in terms of the OZI-allowed rate Γ_{ordinary} and the OZI-suppressed rate Γ_{OZI} by: $\Gamma = \sqrt{\Gamma_{\text{ordinary}} \cdot \Gamma_{\text{OZI}}}$. This is based on the assumption that the glueball couples through the gluonic constituents to pure quark states and that, in perturbative QCD, the OZI-violating processes are mediated by intermediate gluons. In this scheme, the typical widths are of the order of $10-30$ MeV. However, many theorists disagree with this model. Cornwall and Soni,[25] for example, estimate the width of the scalar glueball in the $250-1000$ MeV range. For the pseudoscalar state, they obtain a width of $\Gamma(0^{-+}) \sim 50 - 200$ MeV. In addition, mixing with $q\bar{q}$ states leads to broader widths. In summary, the question of glueball widths remains uncertain.

Since the mass and width predictions are so unreliable, one has to look for other properties to tag glueballs. Unfortunately, no "smoking gun" tests exist except for oddballs. These characteristics are:

1) For a glueball candidate there is no room in the $q\bar{q}$ nonet.
2) Glueballs are copiously produced in gluon-enhanced channels, such as radiative J/ψ decays, OZI-forbidden hadronic processes (e.g.: $\pi^- p \to \phi\phi n$) or pomeron-exchange reactions (e.g.: $pp \to ppX$).
3) Glueball production is suppressed in ordinary hadronic processes, such as $\pi^- p \to K\bar{K}\pi n$, since glueballs are devoid of quarks.
4) Glueballs are not produced in $\gamma\gamma$ collisions since they have no charge. A useful

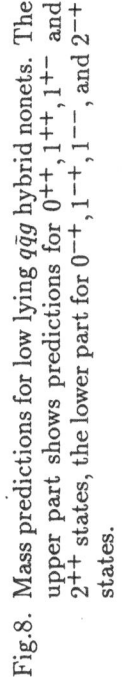

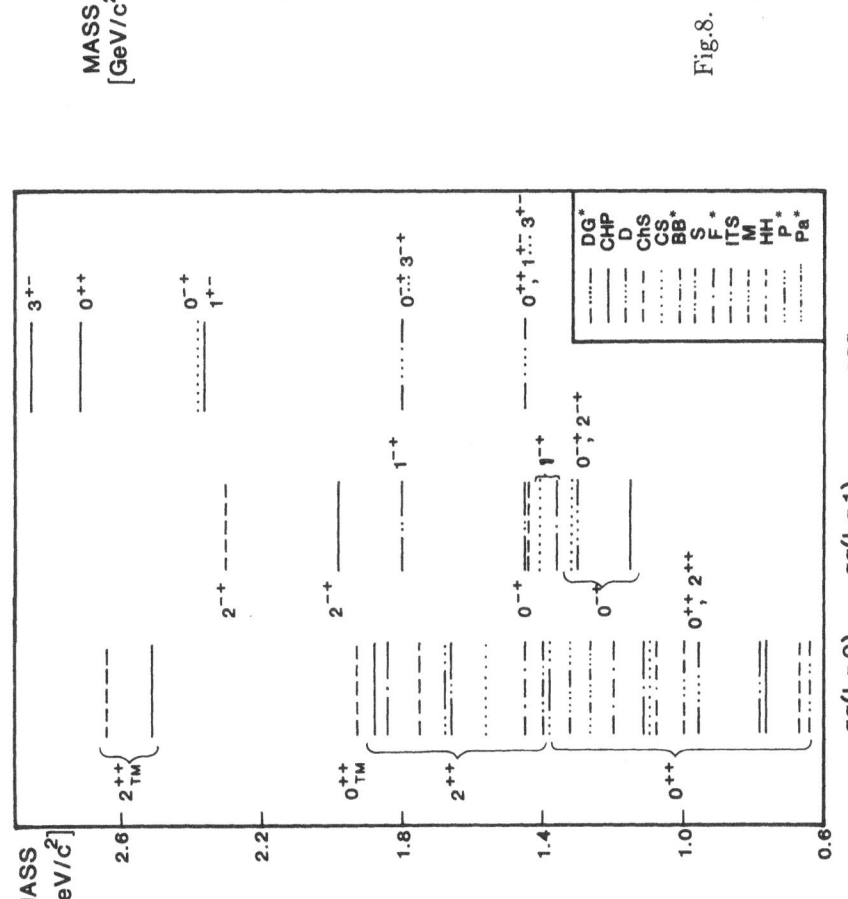

Fig.8. Mass predictions for low lying $q\bar{q}g$ hybrid nonets. The upper part shows predictions for 0^{++}, 1^{++}, 1^{+-} and 2^{++} states, the lower part for 0^{-+}, 1^{-+}, 1^{--}, and 2^{-+} states.

Fig.7. Mass predictions for low lying glueballs: CHP: Carlson et al., CHS: Chanowitz et al., D: Donoghue (Bagmodels); CS: Cornwall et al. (Potential model); BB: Berg et al., DG: DeGrand, F: de Forcrand et al., HH: Hamber et al., ITS: Ishikawa et al., M: Michael et al., P: Parisi et al., Pa: Patel et al., S: Seo (Lattice Gauge Theories). An * marks a recent prediction.

189

quantity is the stickiness introduced by Chanowitz:[26]

$$S = \frac{\Gamma(J/\psi \to \gamma X) \cdot P(X \to \gamma\gamma)}{\Gamma(X \to \gamma\gamma) \cdot P(J/\psi \to \gamma X)}, \tag{4}$$

where P is the Lorentz invariant phase space. The stickiness is a rough measure of the ratio of the color to electric charges with phase space effects factored out. Note that, in the measured quantity, unknown branching ratios for decays of the state X cancel. In addition, pseudoscalar and tensor meson production in $\gamma\gamma$ collisions is well understood.

5) Glueballs are SU(3) singlets in color and flavor.

6) For $J \neq 0$ glueballs, decays should be flavor symmetric, since the coupling to u, d, s quarks is expected to be equal. For $J = 0$ glueballs, final states with strange quarks should be preferentially produced, since in perturbative QCD the amplitude is a function of the ratio of quark mass and glueball mass: [27]

$$A(gg \to q\bar{q})_0 \propto (m_q/M_G)\{a + b\ln(M_G/m_q) + \ln^2(M_G/m_q)\}, \tag{5}$$

where $a = \pi^2/12 \, (2\ln 2 + \pi^2/12)$ for 0^- (0^+) states and $b = -(i\pi + 2)$ for both $J=0$ states.

If an observed resonance satisfies all these requirements, it is very likely a glueball. However, the actual experimental situation may be more complex, since glueballs may mix with nearby $q\bar{q}$ states. In this case, criteria 3), 4) and 6) may not apply stringently.

I turn now to hybrid states, where I restrict myself only to $q\bar{q}g$ states. Currently, two techniques are used to determine masses: Bag Models[28,29] and QCD sum rules.[30] In the bag model, the lowest hybrid states are constructed from a $q\bar{q}$ pair, which is either in a spin singlet with $J^{PC} = 0^{-+}$ or in a spin triplet with $J^{PC} = 1^{--}$, combined with a TE gluon having $J^{PC} = 1^{+-}$. This yields four $q\bar{q}g_{TE}$ nonets with $J^{PC} = 1^{--}$ and $(0,1,2)^{-+}$. Note that several exotic 1^{-+} states occur in the spectrum. The other possibility of combining the $q\bar{q}$ pair with a TM gluon with $J^{PC} = 1^{--}$ gives four nonets having $J^{PC} = 1^{+-}$ and $(0,1,2)^{++}$. Like all observable particles, hybrids are color singlets.

Figure 8 shows two mass predictions for the $q\bar{q}g_{TE}$ states by Chanowitz and Sharpe (ChS)[28] and by Barnes and Close (BC).[29] Also shown are the $q\bar{q}g_{TM}$ mass spectra predicted by Chanowitz and Sharpe. The hybrid states are labeled in analogy to vector mesons to indicate their $q\bar{q}$ contents. The $q\bar{q}g_{TE}$ hybrids lie in the 1.2-2.0 GeV mass range, while the mass range for the $q\bar{q}g_{TM}$ states is ~ 500 MeV higher. The predictions again depend on the ratio r_C. As before, a value of $r_C = 1/2$ is chosen in Fig. 8. For $r_C = 1$, the $q\bar{q}g_{TE}$ masses increase by ~ 200 MeV, while the $q\bar{q}g_{TE}$ values decrease by ~ 200 MeV. In the $q\bar{q}g_{TE}$ nonets, the variation between both predictions ranges from 0 up to ~ 100 MeV. In the Chanowitz and Sharpe model, the level ordering follows the vector meson pattern, while in the BC model for the $(0,1,2)^{-+}$ nonets, the ωg and K^*g states are interchanged. The reason is that the color Coulomb effect shifts the mass of the ωg and the ϕg states in the $(0,1,2)^{-+}$ nonets upward. This effect is included in both models, but the explicit values for α_s and the strange quark mass determine the relative position of the K^*g to the ωg states. The QCD sum rules only determine the center of gravity of the nonets, yielding 1.3 GeV for the 1^{-+} and 1^{+-} states. The question of the hybrid widths is currently still open.

It is interesting to compare the production rates of a hybrid $H = |q\bar{q}g\rangle$, a glueball $G = |gg\rangle$ and a meson $M = |q\bar{q}\rangle$ in J/ψ decays. The individual diagrams are shown in Fig. 9. For the radiative decay, one obtains the hierarchy of rates: [31]

$$\Gamma(J/\psi \to \gamma G) > \Gamma(J/\psi \to \gamma H) > \Gamma(J/\psi \to \gamma M), \tag{6}$$

where each rate is reduced by a factor proportional to $\alpha_s(q^2)$. In hadronic decays, the hierarchy of production rates is:

$$\Gamma(J/\psi \to HM) > \Gamma(J/\psi \to GM) \sim \Gamma(J/\psi \to MM'), \qquad (7)$$

provided that the HM final state is produced via the diagram in Fig. 9f. If the hybrid is produced through the doubly-disconnected diagram in Fig. 9g, the rate $J/\psi \to HM$ is reduced by $O(\alpha_s)$ compared to the two other rates in Eq. (7).

The $q\bar{q}g$ states are likely to decay by $g \to q\bar{q}$ into two $q\bar{q}$ mesons. Since the TE gluon cannot decay into an S-wave $q\bar{q}$ pair because of its parity, one either gets two $L = 0$ mesons in a relative P-wave, or an $L = 1$ and an $L = 0$ meson in a relative S-wave. Examples are $\rho g(0^{-+}) \to \eta a_0(980)$ or $\rho g(0^{-+}) \to \rho\omega$, respectively. A long list of decay modes can be found in Ref. 28. It is worth noting that hybrids may have multi-kaon decays, since the TE gluon s-channel coupling is approximately flavor symmetric. For $q\bar{q}g_{TM}$ hybrids, the s-channel coupling to $s\bar{s}$ states is enhanced according to the bag model,[32] yielding preferentially kaons in the final state. Of special interest are the doubly OZI-violating decays $H \to \omega\phi$, $\omega\rho$... which are expected in both $q\bar{q}g_{TE}$ and $q\bar{q}g_{TM}$ decays. These decays result from a soft gluon exchange between the two color octet $q\bar{q}$ pairs transforming them into singlets.

The identification of hybrids in experiments is different than glueball tagging. Since hybrids contain quarks, their properties are closer to those of $q\bar{q}$ states. They appear in nonets like $q\bar{q}$ states and can be produced in ordinary hadronic processes and in $\gamma\gamma$ collisions. Though criteria 1-2 still remain true, the main effort has to be put into a detailed study of individual decay modes. One must search especially for the doubly OZI-violating decays. From a comparison of decay properties of individual modes, one may be able to find these states.

2.4 Radiative and Hadronic J/ψ Decays

Radiative J/ψ decays may be an excellent hunting ground to look for glueballs, hybrids and other new states, as indicated by Eq. (6). The gg system is in a color singlet and has even C-parity. In order to find the allowed decay modes, the generalized C-parity is a useful quantity.[33] It is the SU(3) generalization of the ordinary charge conjugation for a singlet by assigning C of the third or eight component to the entire octet. Therefore, states with $J^{PC} = 0^{++}, 0^{-+}, 1^{++}, 1^{-+}, 2^{++}, ...$ are accessible, including the low-lying glueballs and hybrids, while all negative C-parity states are forbidden. Experimentally, one performs a systematic study of entire decay families, like $J/\psi \to \gamma PP$, $J/\psi \to \gamma VV$, $J/\psi \to \gamma 3P$..., where P denotes a pseudoscalar meson and V a vector meson. In these modes, thresholds are low and the phase space is large, thus probing a large mass region in 2 and 3 meson systems. In fact, many new structures have been observed in radiative J/ψ decays, such as the $\iota/\eta(1440)$, the $\theta/f_2(1720)$, the $\xi(2230)$ and several pseudoscalar states which will be discussed in the following chapters. Another issue is whether radiative decays are flavor independent.

Hadronic J/ψ decays are well suited to perform a systematic analysis of the low-lying mesons and baryons. For example, by examining 2-body hadronic decays, the following issues are addressed:

i) SU(3) violations can be studied in decays like $J/\psi \to PP$, $J/\psi \to VV$ or $J/\psi \to B_8\bar{B}_{10} + $ c.c., where $B_8(B_{10})$ is an octet (decuplet) baryon. These modes are forbidden by SU(3) and can proceed only via the EM decay in Fig. 4c.

ii) The quark contents of the η and η' and the pseudoscalar mixing angle can be determined from a comparison of $J/\psi \to VP$ decays.

iii) The properties of tensor mesons (T) can be measured in $J/\psi \to VT$ decays.

iv) The properties of scalar mesons (S) can be obtained from a study of $J/\psi \to VS$ decays. The interesting questions are whether the $f_0(975)$ and $a_0(980)$ are scalar mesons, $q\bar{q}q\bar{q}$ states or $K\bar{K}$ molecules.

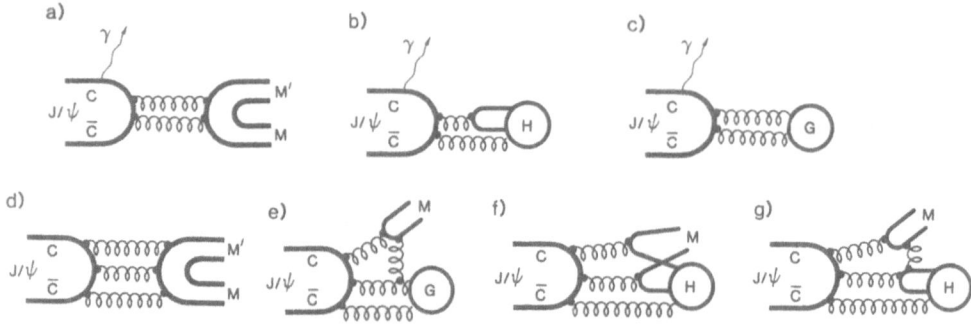

Fig.9. Lowest order radiative diagrams for producing a) mesons, b) a $q\bar{q}g$ hybrid, c) a glueball. Lowest order hadronic annihilation diagrams for producing d) 2 mesons, e) a meson plus a glueball and f) a meson plus a hybrid. Diagram g denotes hybrid production via a doubly disconnected diagram. Dots indicate a coupling α_s.

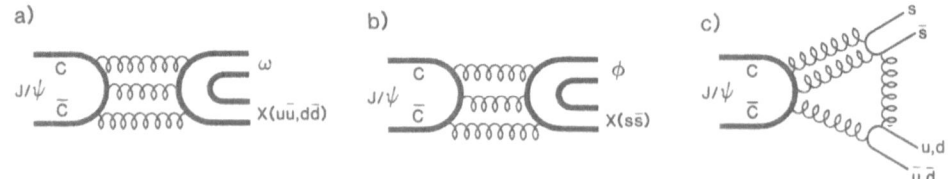

Fig.10. Feynman diagrams for hadronic quasi 2 body decays; a) recoil against an ω meson, b) recoil against a ϕ meson, c) doubly disconnected diagram.

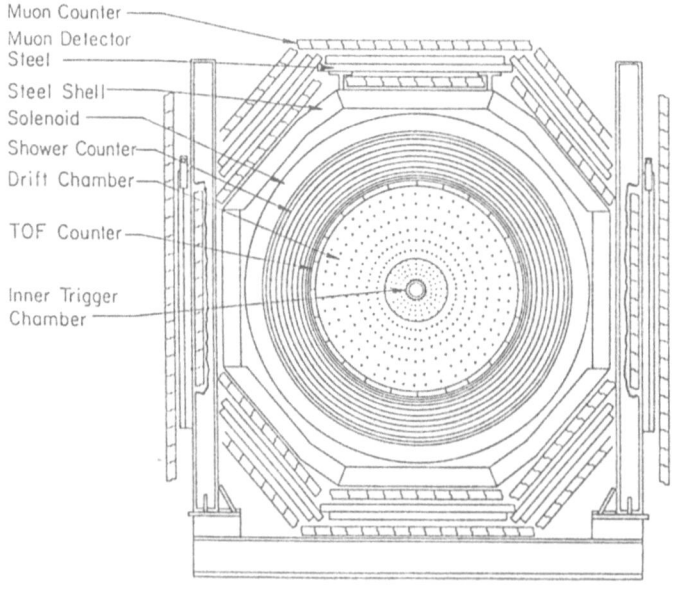

Fig.11. An endview of the Mark III detector.

v) The properties of the axial vector states can be studied in $J/\psi \to V1^{++}$ decays. The most interesting issues concern the $f_1(1420)$ and $f_1(1285)$ mesons.

vi) Finally, a very important issue is whether glueball candidates, such as the $\iota/\eta(1440)$ and $\theta/f_2(1720)$, are produced in hadronic decays.

Another interesting subject is the study of quark correlations in the hadronic decays $J/\psi \to \omega X$ and $J/\psi \to \phi X$. Figure 10 shows the lowest order diagrams for these processes. The ω and ϕ mesons are ideally mixed, i.e., $|\omega\rangle = |u\bar{u} + d\bar{d}\rangle/\sqrt{2}$ and $|\phi\rangle = |s\bar{s}\rangle$. Hence, a state X recoiling against an ω meson contains $u\bar{u}$ and $d\bar{d}$ quarks, whereas a state recoiling against a ϕ meson contains $s\bar{s}$ quarks. In the doubly-disconnected graph in Fig. 10c, no correlation between the vector meson and the recoiling state exists. Contributions from such graphs are expected to be small, as confirmed by the observation of $B(J/\psi \to \omega f_2(1270)) > 30 \cdot B(J/\psi \to \omega f_2'(1525))$. Therefore, the diagrams in Figs. 10a,b are assumed to dominate, implying that the quark contents of X and the vector mesons are correlated. But in some cases, the presence of doubly-disconnected diagrams may be noticeable because of interference effects.

The ω and ϕ mesons have other virtues. Both particles have relatively small widths, which is experimentally advantageous for background reduction. They are isoscalars and have the same quantum numbers. Therefore, a comparison of these processes yield further clues about the state X. For this matter it is useful to introduce the reduced rate $\tilde{\Gamma}$ (and equivalently the reduced branching ratio $\tilde{B}$) as the ordinary rate (branching ratio) divided by the phase space factor. If X is a pure SU(3) singlet, the ratio of reduced rates $R_V = \tilde{\Gamma}(J/\psi \to \omega X)/\tilde{\Gamma}(J/\psi \to \phi X)$ is predicted to be $R_V = 2$, while for a pure octet state one expects $R_V = 1/2$. Due to nonet mixing and SU(3) breaking effects, which are larger for ωX than for ϕX, the measured ratio for an SU(3) singlet may deviate from $R_V = 2$. If X is a singlet, a comparison with the radiative decay may provide some information about the glueball contents. For a $q\bar{q}$ or a hybrid state, one expects $R_\gamma = \tilde{\Gamma}(J/\psi \to VX)/\tilde{\Gamma}(J/\psi \to \gamma X) \gtrsim O(1)$, while for glueball candidates, $R_\gamma << 1$ is expected.

In conclusion, the hadronic decays $J/\psi \to \omega X$ and $J/\psi \to \phi X$ are well suited to measure the quark content of mesons and hybrids, and to check whether glueball candidates contain quarks.

3. THE MARK III DETECTOR

The Mark III detector,[34] shown in an end view in Fig. 11, is a general purpose spectrometer at SPEAR designed to operate in the low energy region of the charmed particles. It operates at a magnetic field of 0.4 Tesla. The beryllium beam pipe is surrounded by a small trigger chamber containing 4 axial layers. The resolution is $350 \mu m$. Next follows the main drift chamber which has 24 axial sense wire planes and 6 stereo sense wire planes. It covers 84% of the total solid angle. The momentum resolution is $\sigma_p/p = 1.5\%\sqrt{(1 + p^2)}$, where the first term is due to multiple scattering. The first 12 axial wire planes are used for dE/dx measurements. The dE/dx system is useful for particle identification for momenta up to ~ 600 MeV/c, where K/π separation is about 1 standard deviation (s.d.). Behind the drift chamber lies the time-of-flight (TOF) system which consists of an array of 48 five-cm-thick scintillators covering 80% of 4π sr. The TOF resolution was originally 191 ps degrading to 220 ps in 1985 because of counter degradation and an increase in the bunch length. These resolutions provide a $2\sigma K/\pi$ separation up to 1.2 GeV/c and 1.1 GeV, respectively. Behind the TOF system, but inside the coil, lies the barrel electromagnetic (EM) calorimeter which consists of 24 layers of Pb proportional tube shower counters with charge division readout. Each side is closed by an endcap EM calorimeter increasing the solid angle to 94% of 4π sr. The calorimeter is 12 radiation lengths deep and has an energy resolution of $\sigma_E/E = 17\%/\sqrt{E(\text{GeV})}$. The photon detection efficiency is

> 99% for energies > 100 MeV, falling off rapidly with decreasing photon energies. Outside the coil lies the flux return steel, with the muon detection system providing a coverage of 62% of 4π sr.

The detector is therefore well suited to perform full kinematic reconstructions of exclusive J/ψ decay modes with high reconstruction efficiency and good resolution. Mark III has collected 5.8×10^6 events at the J/ψ in three run periods.

4. SYSTEMATIC STUDY OF $J/\psi \to 1^{--}PP$

This chapter reviews recent results from the radiative channels $\gamma K\bar{K}$, $\gamma\pi\pi$ and $\gamma\eta\eta$ and from the hadronic modes $\omega K\bar{K}$, $\phi K\bar{K}$, $\omega\pi\pi$ and $\phi\pi\pi$. In the $\gamma K\bar{K}$ channel, I will focus only on the mass region below 2 GeV, since the $\xi(2230)$ is reviewed in the next chapter. Points of interest are the $\theta/f_2(1720)$ and $f_0(975)$. The $\theta/f_2(1720)$ was first observed by the Crystal Ball (CB)[35] in $J/\psi \to \gamma\eta\eta$. The resonance parameters were measured to be: $M_\theta = 1640 \pm 50$ MeV, $\Gamma_\theta = 220 \, ^{+100}_{-70}$ MeV and $B(J/\psi \to \gamma\theta) \cdot B(\theta \to \eta\eta) = (4.9 \pm 1.4 \pm 1.0) \times 10^{-4}$, where the $f_2'(1525)$ contribution was not resolved. A spin-parity analysis yielded $J^{PC} = 2^{++}$. The $\theta/f_2(1720)$ has been considered as a candidate for the tensor glueball.

4.1 $J/\psi \to \gamma\{K\bar{K}, \pi\pi, \eta\eta\}$

Figure 12a shows the Mark III K^+K^- invariant mass plot.[36] The data have been selected by four constraint (4C) kinematic fits to the hypothesis $J/\psi \to \gamma K^+K^-$, using an initial sample of 2.7×10^6 J/ψ events. Below 2 GeV, the $f_2'(1525)$ and the $f_2(1720)$ are the only apparent resonances. The solid curve represents a fit of 2 incoherent relativistic Breit-Wigner (BW) amplitudes and a parametrization of the 3 body phase space. The $\gamma K_s^0 K_s^0$ decay mode is observed in the $\gamma\pi^+\pi^-\pi^+\pi^-$ final state using 4C kinematic fits. The resulting $K_s^0 K_s^0$ invariant mass spectrum is similar to the mass plot in Fig. 12a. The resonance parameters measured in the two modes are summarized together with the DM2 results[37] in Table 1. The branching ratios for γK^+K^- and $\gamma K_s^0 K_s^0$ indicate that the $\theta/f_2(1720)$ is an isoscalar. An update of the $\gamma K\bar{K}$ channels using the entire J/ψ data sample is in progress.

Figure 12b displays the most recent Mark III $\pi^+\pi^-$ invariant mass spectrum obtained from a 4C kinematic fit to the hypothesis $J/\psi \to \gamma\pi^+\pi^-$.[38] Several significant structures are observed. The largest peak corresponds to the $f_2(1270)$. On the high mass side, there is a shoulder which is identified as the $f_2'(1525)$. The $\theta/f_2(1720)$ and a structure at 2.1 GeV are also visible. The ρ peak is due to background feeding in from the hadronic decay $J/\psi \to \rho^0\pi^0$. The other background source is $\rho^\pm\pi^\mp$ events, which produce a broader distribution centered at higher $\pi^+\pi^-$ masses. The solid curve shows a maximum likelihood fit of 3 coherent relativistic Breit-Wigner line shapes for the $f_2(1270)$, $f_2'(1525)$ and $\theta/f_2(1720)$, a non-interfering Breit-Wigner curve for the $X(2100)$ and a $\rho\pi$ background parametrization determined from a Monte Carlo simulation. The masses and widths of the 3 interfering Breit-Wigner amplitudes were fixed to the values determined by the Particle Data Group.[6] The individual branching ratios from this fit and the corresponding DM2 results[37] are given in Table 1. The relative phase between the $f_2(1270)$ and $f_2'(1525)$ amplitudes comes out to be $\sim \pi$, as expected for two isoscalars in an ideally-mixed nonet. A comparison with the $\gamma K\bar{K}$ channel after phase space correction yields $\tilde{B}(f' \to \pi\pi)/\tilde{B}(f' \to K\bar{K}) = (2.4 \pm 1.3)\%$.

Mark III has also searched for a narrow state in the $\gamma\pi^+\pi^-$ channel.[38] No structure is observed in the $\pi^+\pi^-$ invariant mass spectrum. For any object with $M_X < 0.6$ GeV and $\Gamma_X < 2$ MeV, an upper limit at 90% C.L. is

$$B(J/\psi \to \gamma X) \cdot B(X \to \pi^+\pi^-) < 2.1 \times 10^{-5}.$$

The Crystal Ball has analyzed the radiative decay $J/\psi \to \gamma\pi^0\pi^0$. The resulting $\pi^0\pi^0$ invariant mass spectrum looks very similar to the $\pi^+\pi^-$ mass distribution in Fig. 12b. The results for the $f_2(1270)$, $\theta/f_2(1720)$ and $X(2100)$ are consistent with those from the $\gamma\pi^+\pi^-$ analysis, as shown in Table 1.

Figure 12c plots the $\eta\eta$ mass distribution obtained from a re-analysis of $J/\psi \to \gamma\eta\eta$ in the 5-photon final state by the Crystal Ball.[39] The $f_2'(1525)$ and the $\theta/f_2(1720)$ are well separated. The solid curve represents a fit to 2 incoherent Breit-Wigner line shapes with masses and widths fixed to their nominal values. The resulting branching ratios are given in Table 1. An analysis of this mode by Mark III is in progress.

For the $f_2(1270)$, the $f_2'(1525)$ and the $\theta/f_2(1720)$, the spin-parity is determined using a maximum likelihood fit.[36] Due to Bose statistics and generalized C-parity, only the quantum numbers $J^{PC} = (\text{even})^{++}$ are allowed for the PP system. In this analysis, contributions of spins up to $J = 4$ are included. The angular distributions in the helicity formalism depend on two parameters: $x \equiv A_1/A_0$ and $y \equiv A_2/A_0$, where A_i is the amplitude for producing the state X with helicity i. For spin 0, x and y are identically zero. For $J \geq 2$, x and y are a priori unknown and are determined from the maximum likelihood fit. The results of the fit are:

$$\theta : \quad x_\theta = -1.07 \pm 0.16 \quad y_\theta = -1.09 \pm 0.15 \quad \phi_x = 0.6 \pm 0.6 \quad \phi_y = -0.1 \pm 0.5$$

$$f : \quad x_f = 0.96 \pm 0.07 \quad y_f = 0.06 \pm 0.13 \quad \phi_x = -0.5 \pm 0.7 \quad \phi_y = -0.4 \pm 1.9$$

$$f' : \quad x_{f'} = 0.63 \pm 0.09 \quad y_{f'} = 0.17 \pm 0.15 \quad \phi_x = 1.3 \pm 0.6 \quad \phi_y = 2.6 \pm 0.9$$

All three particles have $J^P = 2^+$. For the $f_2(1270)$ and the $f_2'(1525)$, the helicity-2 amplitudes are absent, while for the $\theta/f_2(1720)$, all 3 amplitudes are equal. All phases are consistent with zero.

4.2 $J/\psi \to \omega K\bar{K}$ and $J/\psi \to \phi K\bar{K}$

Mark III has analyzed the hadronic mode $\omega K\bar{K}$ in both the K^+K^- and $K_s^0 K_s^0$ recoil systems.[40] A 5C kinematic fit is applied to the $\pi^+\pi^-\pi^0 K^+K^-$ final state. Figure 13a shows the resulting K^+K^- invariant mass spectrum. The efficiency is 20%, independent of mass. The background indicated by the shaded region is estimated from sidebands above and below the ω mass. The only significant structure is a bump at 1730 MeV which is consistent with the $\theta/f_2(1720)$. No evidence for the $f_0(975)$, the $f_2'(1525)$ or the $X(2100)$ is found. The results for the $\theta/f_2(1720)$ are listed in Table 2. In addition, an upper limit on $f_2'(1525)$ production is given. The $\omega K_s^0 K_s^0$ channel, which is analyzed by performing 5C kinematic fits to the hypothesis $J/\psi \to \gamma 3(\pi^+\pi^-)\pi^0$, yields similar results, as shown in Fig. 13b and Table 2. Though the background level is low, the statistics are rather small. A comparison with the ωK^+K^- mode shows that the $\theta/f_2(1720)$ is consistent with the isoscalar hypothesis. The total branching ratio for $\omega K\bar{K}$ production is:

$$B(J/\psi \to \omega K\bar{K}) = (1.72 \pm 0.08 \pm 0.34) \times 10^{-3}.$$

Mark III has performed a new analysis of the $\phi K\bar{K}$ mode in the final states $K^+K^-K^+K^-$ and $K^+K^-K_s^0 K_s^0$ with $K_s^0 \to \pi^+\pi^-$, using 4C kinematic fits. In order to increase the detection efficiency, which is affected by K decays, 1C kinematic fits are also performed, using the hypothesis $J/\psi \to K^+K^-K^\pm$ ($K_{missing}^\mp$). The resulting detection efficiency for the ϕK^+K^- mode after combining the 4C-fit and 1C-fit data sets is $\sim 37\%$, falling off at masses ≥ 1.75 GeV. For the $\phi K_s^0 K_s^0$ mode, the efficiency is

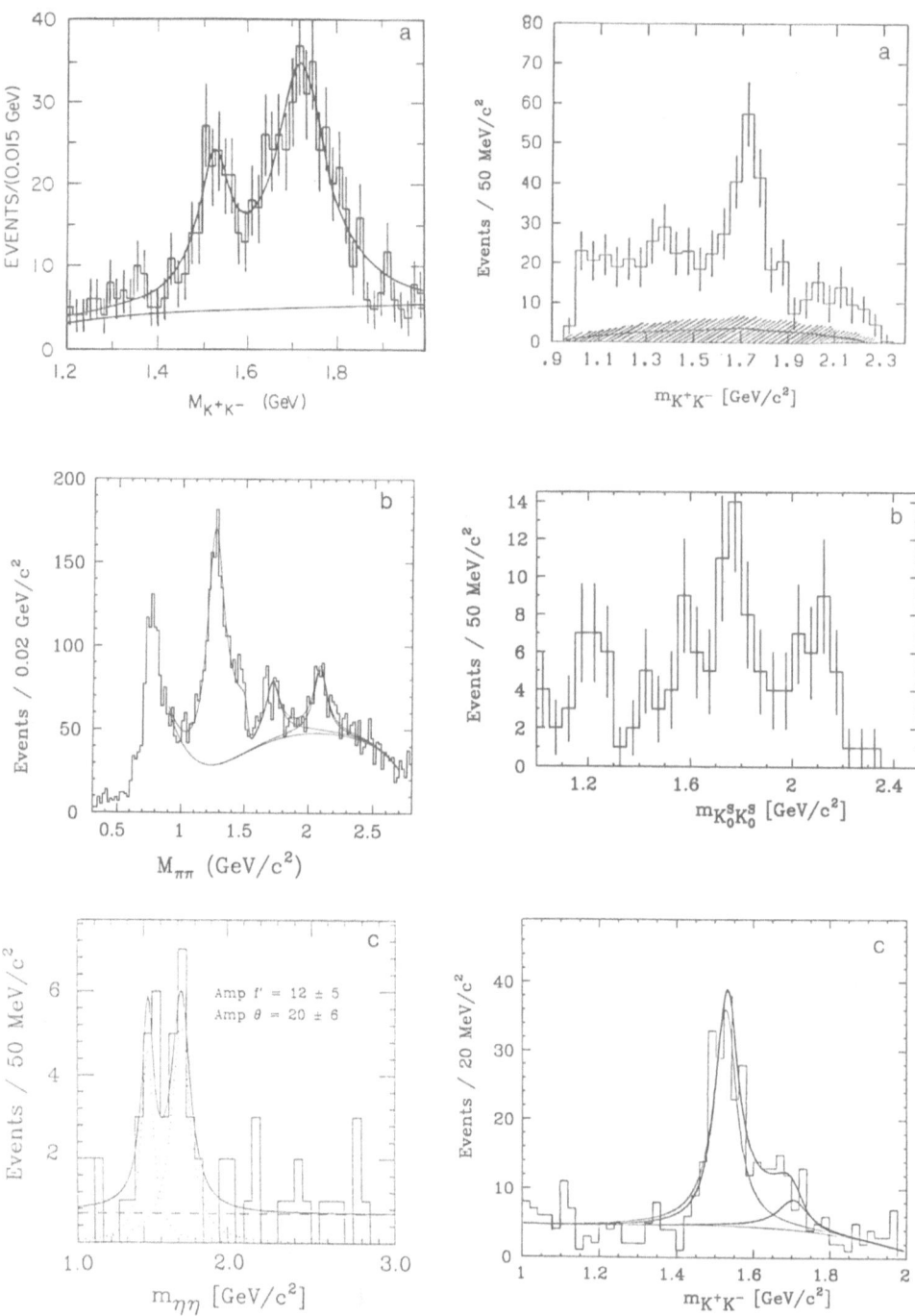

Fig.12. a) The Mark III K^+K^- invariant mass spectrum from $J/\psi \to \gamma K^+K^-$, b) The Mark III $\pi^+\pi^-$ invariant mass spectrum from $J/\psi \to \gamma\pi^+\pi^-$, c) The Chrystal Ball $\eta\eta$ invariant mass spectrum from $J/\psi \to \gamma\eta\eta$.

Fig.13. The Mark III $K\bar{K}$ invariant mass spectra from: a) $J/\psi \to \omega K^+K^-$, b) $J/\psi \to \omega K_s^0 K_s^0$ and c) $J/\psi \to \phi K^+K^-$. The solid curve shown in c) is a fit to 2 coherent Breit-Wigner amplitudes.

TABLE 1

Measured Branching Ratios in $J/\psi \rightarrow \gamma PP$ Decays

Mode $J/\psi \rightarrow$	Object	Mass MeV/c^2	Width MeV/c^2	$B(J/\psi \rightarrow \gamma X) \times$ $B(X \rightarrow PP)$ $[10^{-4}]^\ddagger$	Exp.
$\gamma K^+ K^-$	$f'_2(1525)$	$1525 \pm 10 \pm 10$	85 ± 35	$3.0 \pm 0.7 \pm 0.6$	MKIII†
		1531 ± 10	102.6 ± 29.7	$2.5 \pm 0.6 \pm 0.4$	DM2
	$f(1720)$	$1720 \pm 10 \pm 10$	130 ± 20	$4.8 \pm 0.6 \pm 0.9$	MKIII†
		1707 ± 10	166.4 ± 33.2	$4.6 \pm 0.7 \pm 0.7$	DM2
$\gamma K_0 \bar{K}_0$	$f'_2(1525)$	1525 fixed	85 fixed	$1.9 \pm 0.7 \pm 0.5$	MKIII†
		1525 fixed	70 fixed	$1.5 \pm 0.3 \pm 0.3$	DM2
	$f(1720)$	1720 fixed	130 fixed	$4.5 \pm 1.2 \pm 1.1$	MKIII†
		1711 ± 9	173 ± 22	$3.1 \pm 0.7 \pm 0.6$	DM2
$\gamma \pi^+ \pi^-$	$f_2(1270)$	1274 fixed	176 fixed	$6.7 \pm 0.5 \pm 0.5$	MKIII*
		1270 fixed	214 ± 5	$6.8 \pm 0.2 \pm 1.0$	DM2*
	$f'_2(1525)$	1525 fixed	70 fixed	$0.22 \pm 0.1 \pm 0.06$	MKIII*
		1525	41 ± 10	$0.11 \pm 0.11 \pm 0.02$	DM2*
	$f_2(1720)$	1720 fixed	130 fixed	$1.1 \pm 0.2 \pm 0.5$	MKIII*
		1695	220 ± 20	$0.9 \pm 0.2 \pm 0.2$	DM2*
	$X(2100)$	$2089 \pm 10 \pm 15$	$127 \pm 34^{+150}_{-50}$	$1.5 \pm 0.2 \pm 0.7$	MKIII
		2025 ± 21	312 ± 30	$2.2 \pm 0.3 \pm 0.4$	DM2
$\gamma \pi^0 \pi^0$	$f_2(1270)$	$1278 \pm 7 \pm 5$	$137 \pm 15 \pm 20$	$(17 \pm 1 \pm 5) \times B_{f \rightarrow \pi^0 \pi^0}$	CB
	$f_2(1720)$	1720 fixed	130 fixed	$0.78 \pm 0.22 \pm 0.27$	CB
	$X(2100)$	2086 fixed	210 fixed	$0.94 \pm 0.24 \pm 0.32$	CB
$\gamma \eta \eta$	$f_2(1525)$	1525 fixed	85 fixed	$1.9 \pm 0.8 \pm 0.5$	CB
	$f_2(1720)$	1720 fixed	130 fixed	$2.6 \pm 0.8 \pm 0.7$	CB

$\dagger$) Mark III results obtained from the initial 2.7×10^6 J/ψ data sample.

$*$) coherent fit with interference between $f_2(1270)$, $f'_2(1525)$ and $f_2(1720)$.

$\ddagger$) PP stands for either $K^+ K^-$, $K^0 \bar{K}^0$, $\pi^+ \pi^-$, $\pi^0 \pi^0$ or $\eta \eta$.

generally lower, and shows an explicit $K_s^0 K_s^0$ mass dependence. Details of the analysis are given in Ref. 41.

Figure 13c displays the $K^+ K^-$ invariant mass spectrum, which shows a peak at the $f'_2(1525)$ and a shoulder around 1.7 GeV. This shoulder probably results from $\theta / f_2(1720)$ production, since a spin-parity analysis in that mass region favors

TABLE 2

Measured Branching Ratios in Hadronic $\omega K\bar{K}$ and $\phi K\bar{K}$ Decays

Mode $J/\psi \to VX$	Object	Mass MeV/c^2	Width MeV/c^2	$B(J/\psi \to VX) \times B(X \to K\bar{K})^\dagger$ $[10^{-4}]$	Exp.
$\omega K^+ K^-$	$f_2'(1525)$	1520 fixed	75 fixed	< 1.2 at 90% C.L.	MKIII
	$f_2(1720)$	$1731 \pm 10 \pm 10$	$110^{+45}_{-35} \pm 15$	$2.25^{+0.6}_{-0.55} \pm 0.5$	MKIII
	$f_2(1720)$	1749 ± 17	165 ± 54	$2.0 \pm 0.25 \pm 0.3$	DM2
$\omega K^0 \bar{K}^0$	$f_2(1720)$	1730 fixed	100 fixed	$3.2 \pm 1.2 \pm 1.0$	MKIII
$\phi K^+ K^-$	$f_2'(1525)$	1525 fixed	70 fixed	$2.42 \pm 0.26 \pm 0.7$	MKIII
	$f_2'(1525)$	1515 fixed	85 fixed	$3.04 \pm 0.26 \pm 0.5$	DM2
	$f_2(1720)$	1730 ± 13	163 ± 30	$1.52 \pm 0.27 \pm 0.43$	MKIII
	$f_2(1720)$	1686 ± 14	162 ± 35	$1.81 \pm 0.14 \pm 0.3$	DM2
$\phi K^0 \bar{K}^0$	$f_2'(1525)$	1525 fixed	70 fixed	$2.6 \pm 0.86 \pm 0.74$	MKIII
	$f_2'(1525)^*$	1525 fixed	fixed ?	$2.15 \pm 0.35 \pm 0.45$	DM2
	$f_2'(1720)$	1730 fixed	163 fixed	$1.62 \pm 0.54 \pm 0.46$	MKIII
	$f_2(1720)^*$	1720 fixed	fixed ?	$1.8 \pm 0.35 \pm 0.35$	DM2

* Fit to 2 incoherent BW line shapes; DM2 results from Ref. 42.

$\dagger$ $K\bar{K}$ stands for either K^+K^- or $K^0\bar{K}^0$.

a $J^P = 2^+$ assignment. The K^+K^- mass spectrum is, therefore, fitted with two coherent Breit-Wigner amplitudes plus a smooth background term. The resulting fit, which is shown overplotted in Fig. 13c, yields the parameters in Table 2. For comparison, the corresponding DM2 results are also given.[42] The mass and width of the higher mass structure agree with the parameters of the $\theta/f_2(1720)$. The branching ratio for the $\theta/f_2(1720)$ is only a factor ~ 1.5 smaller than that for the $f_2'(1525)$. The analysis of $J/\psi \to \phi K_s^0 K_s^0$ yields similar results, shown in Table 2. In both the $\omega K\bar{K}$ and $\phi K\bar{K}$ hadronic modes, the results from Mark III and DM2 are consistent.

4.3 $J/\psi \to \omega\pi^+\pi^-$ and $J/\psi \to \phi\pi^+\pi^-$

The hadronic decay $J/\psi \to \omega\pi^+\pi^-$ is observed in the 5π final state. The data sample is obtained by performing 5C kinematic fits to the hypothesis $J/\psi \to \pi^+\pi^-\pi^0\pi^+\pi^-$.[40] Though the 5π channel is the largest single J/ψ decay mode, the 5π background is substantially reduced by requiring an ω candidate, since at the ω mass, the signal-to-background ratio is $\sim 8 : 1$. The $\omega\pi\pi$ channel also contains a contribution from $J/\psi \to b_1^\pm(1235)\pi^\mp$ with $b_1^\pm(1235) \to \omega\pi^\pm$, which has not been removed. The efficiency for $J/\psi \to \omega\pi^+\pi^-$ is of the order of 25%, and is constant in the $0.3 - 2.4$ GeV mass range.

Figure 14b shows the $\pi^+\pi^-$ invariant mass spectrum recoiling against the ω. A low mass structure at ~ 500 MeV and a prominent $f_2(1270)$ signal are visible. The spectrum shows no indication of an $f_0(975)$ or a $\theta/f_2(1720)$. The $f_2(1270)$ line shape is consistent with being a single resonance. Fixing the mass of the resonance at 1277 MeV, a Breit-Wigner parametrization yields a width of $\Gamma_f = 182 \pm 10$ MeV. The branching ratio is:

$$B(J/\psi \to \omega f) = (4.93 \pm 0.25 \pm 1.25) \times 10^{-3} .$$

The low mass structure is not yet understood. It may result from a threshold effect,[43] or be a quark mass effect due to spontaneous quark pair creation.[44]

The total branching ratio for the $\omega\pi\pi$ channel, including $J/\psi \to b_1(1235)\pi$ is:

$$B(J/\psi \to \omega\pi^+\pi^-) = (7.8 \pm 0.1 \pm 1.6) \times 10^{-3} .$$

The process $J/\psi \to \phi\pi\pi$ can only proceed through the doubly-disconnected diagram shown in Fig. 10c. This channel is studied in the $K^+K^-\pi^+\pi^-$ final state using 4C kinematic fits. This mode is nearly background free. The efficiency is of the order of 25%, falling off rapidly above 1.75 GeV due to kaon decays.

Figure 14c displays the $\pi^+\pi^-$ invariant mass spectrum recoiling against the ϕ. This doubly OZI-violating decay mode exhibits interesting features. A huge peak at the $f_0(975)$ is observed. Further structures appear in the $1.2 - 1.5$ GeV mass region and around 1.7 GeV. The peculiar line shape of the $f_0(975)$ has a simple explanation. In the quark model, the $f_0(975)$ is classified as the $I = 0$ $s\bar{s}$ member of the scalar nonet. Hence, the $f_0(975)$ is expected to decay preferentially to $K\bar{K}$. But, since the resonance lies below the $K\bar{K}$ threshold, it decays mostly to $\pi\pi$. Above threshold, where enough phase space is available, the $K\bar{K}$ decay becomes dominant, producing the sharp fall-off of the $\pi\pi$ signal at ~ 1 GeV. The $f_0(975)$ signal is fitted using the coupled-channel parametrization of Flatté.[45] The resulting branching ratio is:

$$B(J/\psi \to \phi f_0(975)) \cdot B(f_0(975) \to \pi^+\pi^-) = (2.3 \pm 0.3 \pm 0.6) \times 10^{-4} .$$

A relativistic parametrization of the $f_0(975)$ signal gives the same answer. This result is in good agreement with the DM2 measurement of $(2.38 \pm 0.2 \pm 0.4) \times 10^{-4}$.

The origin of the structure in the $1.2-1.5$ GeV mass range is not presently known. Considering the signal shape, it is likely that more than one resonance contribute. Candidates include the $f_2(1270)$, the $f_0(1300)$ and more exotic states, like the $X(1440)$[46] or a scalar glueball. To learn more, it will be necessary to perform a spin-parity analysis in that mass range. The enhancement at ~ 1.7 GeV is consistent with the mass and width of the $\theta/f_2(1720)$. The entire branching ratio for the $\phi\pi\pi$ channel is:

$$B(J/\psi \to \phi\pi^+\pi^-) = (9 \pm 0.4 \pm 2.3 \times 10^{-4} .$$

4.4 Comparison of KK and $\pi\pi$ Invariant Mass Spectra

Figures 14 show, for comparison, the $\pi^+\pi^-$ invariant mass spectra recoiling against a photon, an ω and a ϕ meson. The corresponding K^+K^- invariant mass spectra are presented in Figs. 15.

The dominant features are the $f_2(1270)$ and the $f_2'(1525)$, the $I = 0$ members of the 2^{++} tensor nonet. The $f_2(1270)$ is clearly seen in $\gamma\pi\pi$ and $\omega\pi\pi$, indicating a preferential coupling to u and d quarks. It may also contribute to the $\phi\pi\pi$ channel. No $f_2(1270)$ production is found in the $K\bar{K}$ mass spectra. The $f_2'(1525)$ is clearly observed in γKK and ϕKK. It is not visible in $\omega\pi\pi$, $\phi\pi\pi$ and ωKK, implying that it contains mainly $s\bar{s}$ quarks. However, its appearance in $\gamma\pi^+\pi^-$ indicates a small admixture of u and d quarks, of the order of 2%. Assuming that $K\bar{K}$, $\eta\eta$, and $\pi\pi$ are the main decay modes of the $f_2'(1525)$, one can determine the branching ratio for $f_2'(1525) \to K\bar{K}$ from the measurements in Table 1, yielding:

$$B(f_2'(1525) \to K\bar{K}) = 0.63 \pm 0.1 .$$

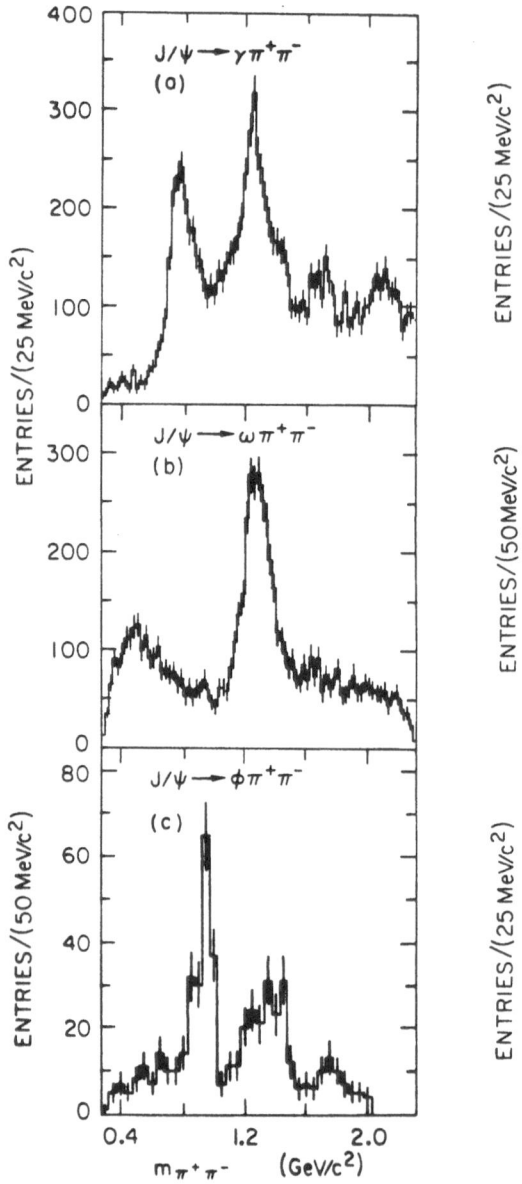

Fig.14. A comparison of the Mark III invariant mass spectra observed in the processes: a) $J/\psi \to \gamma K^+ K^-$, b) $J/\psi \to \omega K^+ K^-$, and c) $J/\psi \to \phi K^+ K^-$.

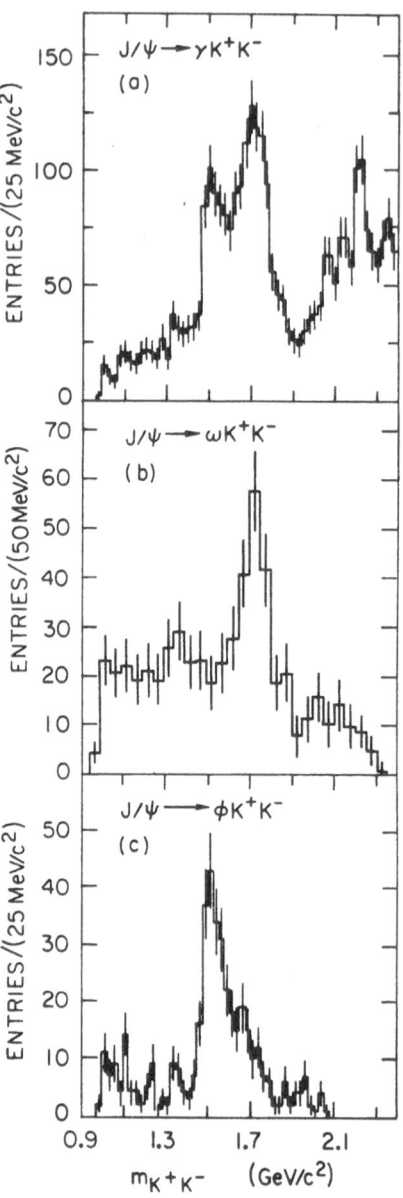

Fig.15. A comparison of the Mark III invariant mass spectra observed in the processes: a) $J/\psi \to \gamma \pi^+ \pi^-$, b) $J/\psi \to \omega \pi^+ \pi^-$, and c) $J/\psi \to \phi \pi^+ \pi^-$.

For this estimate and an S-wave phase space a comparison of the reduced branching ratios in the radiative decays to the $f_2(1270)$ and $f_2'(1525)$ yields:

$$\tilde{B}(J/\psi \to \gamma f_2(1270))/\tilde{B}(J/\psi \to \gamma f_2'(1525)) = 1.9 \pm 0.4\,,$$

which is consistent with the SU(3) prediction of two for ideally-mixed tensor states.

A comparison of the reduced branching ratios in the radiative and hadronic decays yields:

$$\tilde{B}(J/\psi \to \omega f_2(1270))/\tilde{B}(J/\psi \to \gamma f_2(1270)) = 4.4 \pm 1.2$$

and

$$\tilde{B}(J/\psi \to \phi f_2'(1525))/\tilde{B}(J/\psi \to \gamma f_2'(1525)) = 2.1 \pm 0.4\,,$$

which are consistent with the expectation for $q\bar{q}$ states.

The $\theta/f_2(1720)$ is clearly observed in $\gamma\pi\pi$, $\gamma K\bar{K}$, $\omega K\bar{K}$ and $\phi K\bar{K}$. A small signal is also seen in $\phi\pi\pi$. The $\omega\pi\pi$ mode shows no evidence for $\theta/f_2(1720)$ production. A detailed discussion on the $\theta/f_2(1720)$ is given in the next section.

In the Mark III data, the $f_0(975)$ appears clearly in $\phi\pi\pi$. In $\omega\pi\pi$, a small signal is observed near the $f_0(975)$ mass. In the Mark III data, therefore, the $f_0(975)$ is consistent with coupling preferentially to strange quarks. However, DM2 observes $J/\psi \to \omega f_0(975)$ with a branching ratio of $B(J/\psi \to \omega f_0(975)) \cdot B(f_0(975) \to \pi\pi) = (0.95 \pm 0.1 \pm 0.22) \times 10^{-5}$, which does not favor a $q\bar{q}$ classification of the $f_0(975)$. Besides the $q\bar{q}$ classification, other interpretations of the $f_0(975)$ exist, including a $q\bar{q}q\bar{q}$ state,[13] a loosely bound $K\bar{K}$ molecule[47] or two close states, one an $s\bar{s}$ meson and the other a glueball.[48] These interpretations predict different production rates in J/ψ decays and $\gamma\gamma$ collisions. For an $s\bar{s}$ state, one expects:[49] $\Gamma(f_0|(u\bar{u} + d\bar{d}))/\sqrt{2} \to \gamma\gamma) \simeq 4.5$ keV, $\Gamma(f_0|s\bar{s}) \to \gamma\gamma) \simeq 0.33$ keV, $B(J/\psi \to \omega f_0) = 0$ and $B(J/\psi \to \phi f_0) = B(J/\psi \to \rho^0 a_0)$. On the other hand, for a $K\bar{K}$ molecule, for example, one obtains: $\Gamma(f_0|K\bar{K}) \to \gamma\gamma) << \Gamma(f_0|s\bar{s}) \to \gamma\gamma)$ and $B(J/\psi \to \phi f_0) = 2 \cdot B(J/\psi \to \omega f_0) = 2 \cdot B(J/\psi \to \rho^0 a_0)$. The measurements of $\Gamma(f_0 \to \gamma\gamma) \cdot B(f_0 \to \pi\pi) < 0.8$ keV at 95% C.L.[50] and $B(J/\psi \to \rho^0 a_0) < 0.044$ at 90% C.L. (see Chap. 8) are not currently sensitive enough to draw a definite conclusion. On the other hand, the DM2 data suggest the interpretation of a $K\bar{K}$ molecule. For a final answer, further and more precise measurements are necessary.

The $X(2100)$, which is currently not classified, is only seen in $\gamma\pi\pi$ and $\gamma K\bar{K}$ (see Fig. 16). The mass region above 2 GeV is not accessible in $\phi\pi\pi$ and $\phi K\bar{K}$.

4.5 Status of the $\theta/f_2(1720)$

The current experimental status of the $\theta/f_2(1720)$ is:

- The $\theta/f_2(1720)$ is produced in radiative J/ψ decays with a total branching ratio of $B(J/\psi \to \gamma\theta) = (1.3\pm.14)\times10^{-3}$, assuming that $K\bar{K}$, $\pi\pi$ and $\eta\eta$ are the main decay modes. This branching ratio is a factor ~ 4 smaller than $B(J/\psi \to \gamma\iota)$. The reduced rates for $f_2(1270)$, $f_2'(1525)$ and $\theta/f_2(1720)$ production in radiative J/ψ decays are: $\tilde{\Gamma}_{\gamma f} : \tilde{\Gamma}_{\gamma f'} : \tilde{\Gamma}_{\gamma\theta} = 1.9 : 1 : 2.4$.
- The $\theta/f_2(1720)$ is observed in hadronic decays $J/\psi \to \omega\theta$ and $J/\psi \to \phi\theta$, which demonstrates that this state is not completely decoupled from $q\bar{q}$ states. The ratio $\tilde{B}(J/\psi \to \omega\theta)/\tilde{B}(J/\psi \to \phi\theta) = 1.0 \pm 0.2$ indicates that the $\theta/f_2(1720)$ couples more strongly to strange quarks than an SU(3) singlet. The ratio $\tilde{B}(J/\psi \to \omega\theta)/\tilde{B}(J/\psi \to \gamma\theta) = 0.54 \pm 0.1$, however, is smaller than the corresponding ratios for the tensor mesons.
- The $\theta/f_2(1720)$ is probably not an $s\bar{s}$ state, since it is not observed in $K^-p \to K_s^0 K_s^0 \Lambda$, whereas the $f_2'(1525)$ is clearly seen in that process.[51] However, at

Brookhaven, a group has recently claimed to observe the $\theta/f_2(1720)$ in the process $\pi^- p \to K_s^0 K_s^0 n$.[52] In a coupled channel analysis performed in the $K_s^0 K_s^0$ mass spectrum, a 2^{++} state with a mass of 1730^{+2}_{-10} MeV and a width of 122^{+74}_{-10} MeV has been found.

- The $\theta/f_2(1720)$ is not produced in $\gamma\gamma$ collisions. The upper limit at 95% C.L. is $\Gamma_{\gamma\gamma}(\theta) \cdot B(\theta \to K\bar{K}) < 0.2$ keV.[53]

- The stickiness ratios for the $f_2(1270)$, $f_2'(1525)$ and the $\theta/f_2(1720)$ are $S_f : S_{f'} : S_\theta = 1 : 9 : > 16$ for an S-wave phase space. This indicates that the $\theta/f_2(1720)$ is different from a $q\bar{q}$ state.

- The $\theta/f_2(1720)$ decays in a flavor non-symmetric way. The observed rates in radiative J/ψ decays are $\Gamma_{KK} : \Gamma_{\pi\pi} : \Gamma_{\eta\eta} = 3.6 \pm 1.6 : 0.5 \pm 0.3 : 1$, while the SU(3) predictions for a singlet after phase space correction are: $\Gamma_{K\bar{K}} : \Gamma_{\pi\pi} : \Gamma_{\eta\eta} = 6 : 12 : 1$. It has been suggested that final state interactions are responsible for this deviation.

- The $\theta/f_2(1720)$ is confirmed as an isoscalar by comparing the decay rates of $\theta \to \pi^+\pi^-$ with $\theta \to \pi^0\pi^0$ and $\theta \to K^+K^-$ with $\theta \to K_s^0 K_s^0$.

- Angular distribution measurements classify the $\theta/f_2(1720)$ as a $J^{PC} = 2^{++}$ state. All three helicity amplitudes are present with approximately equal strength. This shows that the $\theta/f_2(1720)$ is different from the tensor states $f_2(1270)$ and $f_2'(1525)$.

- There is no room for the $\theta/f_2(1720)$ in the lowest tensor nonet.

- The interpretation of the $\theta/f_2(1720)$ as a radial excitation of the $f_2'(1525)$ is problematic. First, the production rate is too large. Second, a lighter partner is missing. Third, the mass of the θ is too low, since the first radial excitation of the $f_2'(1525)$ is expected at ~ 2 GeV.

In conclusion, the $\theta/f_2(1720)$ is probably not a $q\bar{q}$ state, though it is seen in hadronic J/ψ decays. The hybrid and glueball hypotheses are possible interpretations. However, if the Brookhaven result is confirmed and the $\theta/f_2(1720)$ is observed in ordinary hadronic processes, the glueball interpretation becomes unlikely, unless the $\theta/f_2(1720)$ mixes with other $q\bar{q}$ states.

5. THE STATUS OF THE $\xi(2230)$

In 1983, Mark III presented evidence for a narrow state at 2.2 GeV in the channel $\gamma K^+ K^-$ by studying a sample of 2.7×10^6 J/ψ decays.[54] The narrowness of the structure ($\Gamma \leq 4.0$ MeV at 95% C.L.) motivated several theoretical interpretations including a high spin $J \geq 2$ $s\bar{s}$ state,[55] a glueball,[56] a hybrid,[57] a 4 quark state,[58] the Higgs boson,[59] and others.[60] Since DM2 did not see this state in an analysis of their first data,[61] Mark III recorded an additional data sample of 3.1×10^6 J/ψ decays, confirming the $\xi(2230)$ in $\gamma K^+ K^-$. In addition, a signal was found in $\gamma K_s^0 K_s^0$.[62] At the same time, DM2 finished the analysis of the full data sample of 8.6×10^6 J/ψ events and reported an upper limit for $\xi(2230)$ production.

Figures 16a,b show the $K^+ K^-$ and $K_s^0 K_s^0$ invariant mass spectra from Mark III. Besides the $f_2'(1525)$ and $f_2(1720)$, a clear signal at 2.2 GeV appears in both channels. In addition, a $X(2100)$ signal is visible in the $K_s^0 K_s^0$ mass spectrum. The $K_s^0 K_s^0$ spectrum is essentially background-free, since the direct hadronic decays $J/\psi \to K_s^0 K_s^0$ and $J/\psi \to \pi^0 K_s^0 K_s^0$ are forbidden by C-parity conservation. An estimate for the non-$K_s^0 K_s^0$ background is shown by the cross-hatched histogram in Fig. 16b. The $K^+ K^-$ spectrum contains background from radiative Bhabhas and from two hadronic channels: $J/\psi \to \rho\pi \to \pi^+\pi^-\pi^0$ and $J/\psi \to K^{*\pm}(892)K^\mp \to K^+K^-\pi^0$. Although all events with more than 1 photon are rejected if they fit to the hypotheses

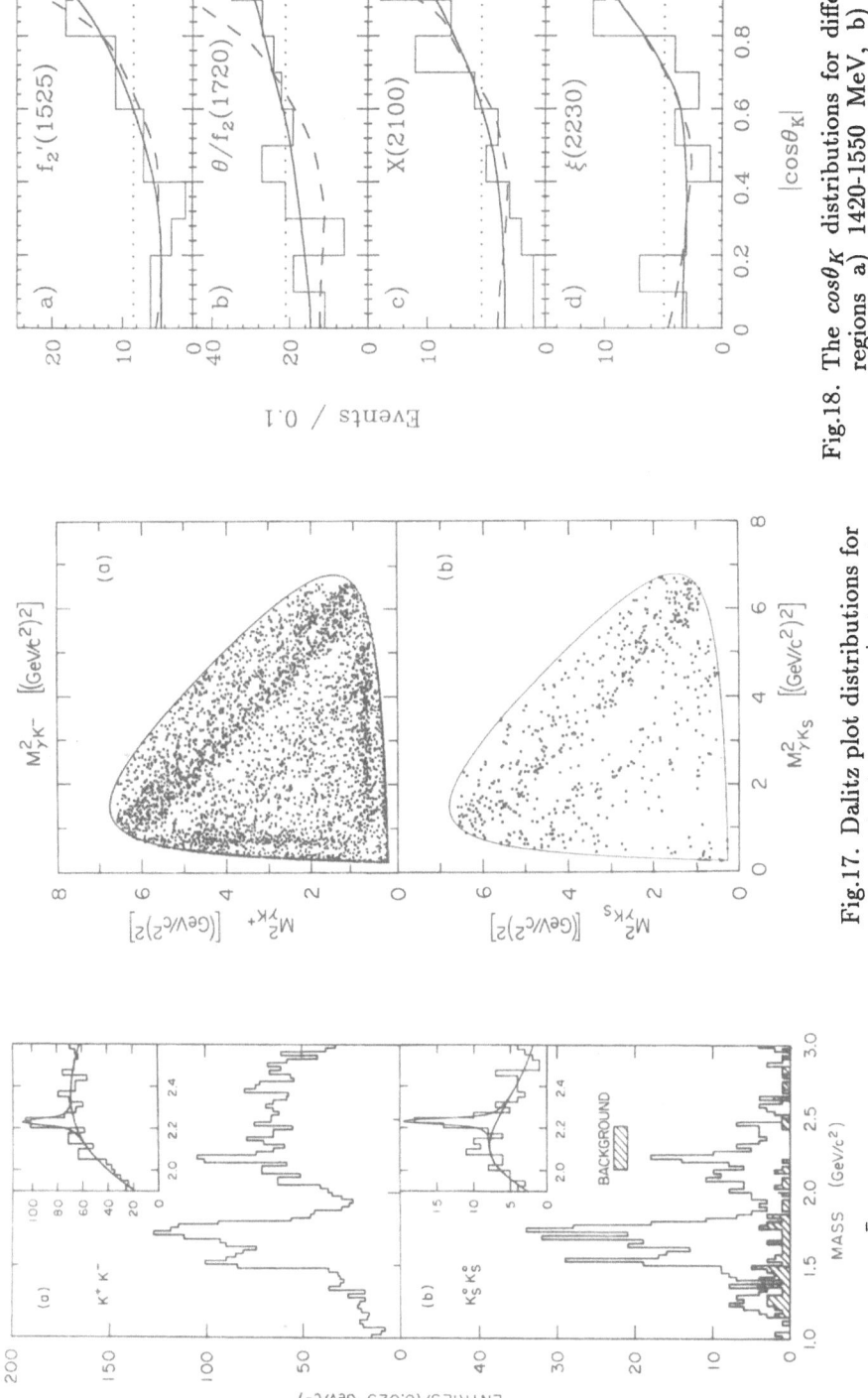

Fig.16. Mark III $K\bar{K}$ invariant mass spectra from a) $J/\psi \to \gamma K^+ K^-$, and b) $J/\psi \to \gamma K_s^0 K_s^0$. The insets show a fit in the 2-2.5 GeV mass range.

Fig.17. Dalitz plot distributions for a) $J/\psi \to \gamma K^+ K^-$, and b) $J/\psi \to \gamma K_s^0 K_s^0$.

Fig.18. The $\cos\theta_K$ distributions for different mass regions a) 1420-1550 MeV, b) 1620-1820 MeV, c) 2080-2180 MeV, and d) 2180-2280 MeV. Predictions for spin 0, 2 and 4 are shown by dotted, solid and dashed lines.

203

$J/\psi \to \pi^+\pi^-\pi^0$ or $J/\psi \to K^+K^-\pi^0$, those in which the π^0 decays asymmetrically and one photon is not detected, remain. While $\rho\pi^0$ events mainly populate the lower mass region, the substantial background above 2 GeV comes from $J/\psi \to K^+K^-\pi^0$. This is supported by the Dalitz plot shown in Fig. 17a. Two clear K^* bands are visible; these feed into the 2-3 GeV mass region. The $\rho\pi$ events populate the diagonal edge of the Dalitz plot, while radiative Bhabhas appear at the horizontal and vertical boundary. The Dalitz plot for $\gamma K_s^0 K_s^0$ (Fig. 17b) again demonstrates that the non-$K_s^0 K_s^0$ background is very small. The four resonances $f_2'(1525)$, $\theta/f_2(1720)$, $X(2100)$ and $\xi(2230)$ are clearly visible as diagonal bands in both plots.

To determine the resonance parameters, a maximum likelihood fit is performed in the 1.9 to 2.6 GeV mass region, including a smooth background parametrization plus a Breit-Wigner resonance convoluted with a gaussian resolution function. The mass resolution at 2 GeV is 10 MeV. The fits shown in the insets of Fig. 16 yield masses and widths of:

$$\gamma K^+K^- : \qquad m_\xi = 2230 \pm 6 \pm 14 \text{ MeV}, \qquad \Gamma_\xi = 26 \,{}^{+20}_{-16}\pm 17 \text{ MeV}$$

$$\gamma K_s^0 K_s^0 : \qquad m_\xi = 2232 \pm 7 \pm 7 \text{ MeV}, \qquad \Gamma_\xi = 18 \,{}^{+23}_{-15}\pm 10 \text{ MeV}$$

The systematic error includes an uncertainty in the parametrization of the background shape, as well as uncertainties in the absolute mass scale.

The significance of the signal determined from the likelihood fit is 4.5 s.d. for γK^+K^- and 3.6 s.d. for $\gamma K_s^0 K_s^0$. The product branching ratios are measured to be:

$$B(J/\psi \to \gamma\xi) \cdot B(\xi \to K^+K^-) = (4.2 \,{}^{+1.7}_{-1.4}\pm 0.8) \times 10^{-5}$$

$$B(J/\psi \to \gamma\xi) \cdot B(\xi \to K_s^0 K_s^0) = (3.1 \,{}^{+1.6}_{-1.3}\pm 0.7) \times 10^{-5}$$

Mark III has also performed a spin-parity analysis of the $\xi(2230)$ in the relatively background-free $\gamma K_s^0 K_s^0$ channel using the maximum likelihood technique.[63] The procedure is similar to that used for the $\theta/f_2(1720)$ (see Chap. 3.1), except that the ratios of helicity amplitudes x and y are assumed to be real. The angular distributions depend on three variables: θ_γ, the polar angle of the photon in the lab frame, θ_K, the polar angle of the K_s^0 in the $K_s^0 K_s^0$ rest frame, and ϕ_K, the angle between the event plane defined by the photon and the initial positron and the resonance decay plane defined by the direction of the two K_s^0's. The angle which bears the greatest sensitivity to the different spin hypotheses and which needs the smallest acceptance corrections is θ_K. Figure 18 displays the θ_K distributions for the $\xi(2230)$ region in comparison to the $f_2'(1525)$, $\theta/f_2(1720)$ and $X(2100)$. In addition, the predictions for spin hypotheses $J = 0, 2$ and 4 are plotted. The $J = 2$ hypothesis fits the data best, although spin 4 cannot be excluded. The angular distribution is affected by contamination from the broad structure at 2.1 GeV, which favors a spin 2 assignment. A detailed study considering different levels of contamination shows that the $\xi(2230)$ is not likely to be a spin 0 state. For $J = 2$ (4), the ratio of helicity amplitudes are $x = -0.67^{+0.14}_{-0.16}$ $(1.29^{+0.62}_{-0.30})$ and $y = 0.13^{+0.21}_{-0.19}$ $(0.4^{+0.76}_{-0.39})$. This indicates that, for the $J = 2$ hypothesis, the helicity-2 component is suppressed, as it is for the $f_2(1270)$ and the $f_2'(1525)$, whereas, for the $J = 4$ hypothesis, the results are inconclusive. In order to set a stringent limit on the spin 0 assignment and to distinguish between spin 2 and 4, more data are required.

DM2 has performed a re-analysis of $J/\psi \to \gamma K^+K^-$ and $J/\psi \to \gamma K_s^0 K_s^0$.[39] The resulting $K\bar{K}$ invariant mass spectra are shown in Fig. 19. DM2 still claims to observe no $\xi(2230)$ signal in its data. Assuming a zero width object, DM2 determines the following upper limits:

$$B(\psi \to \gamma\xi) \cdot B(\xi \to K^+K^-) < 1.7 \times 10^{-5} \text{ at } 90\% \text{ C.L.},$$

$$B(\psi \to \gamma\xi) \cdot B(\xi \to K_s^0 K_s^0) < 2.0 \times 10^{-5} \text{ at 95\% C.L.},$$

which disagree with the Mark III results. For a 50 MeV wide resonance, the DM2 limit on $\xi \to K^+K^-$ increases to 4.3×10^{-5} at 90% C.L. The measurements of the masses, widths, and branching ratios of the $f_2'(1525)$ and $f_2(1720)$ are consistent with the Mark III results. The DM2 detection efficiency in the K^+K^- channel, however, is more than a factor 2 smaller than the Mark III detection efficiency. The DM2 $K_s^0 K_s^0$ invariant mass spectrum looks very similar to the Mark III $K_s^0 K_s^0$ mass distribution, except that the spike at 2.23 GeV is moved downward by 1 bin ($\sim$ 30 MeV). Their K^+K^- mass spectrum shows no significant structure at 2.23 GeV. This discrepancy can perhaps be explained by different detector performances:[42]

- The DM2 TOF resolution is 540 ps, yielding a $\geq$ 3 s.d. K/π separation for momenta $\leq$ 450 MeV/c. The Mark III TOF resolution is on the average 205 ps, allowing a $\geq$ 3 s.d. K/π separation for momenta up to $\sim$950 MeV/c. The kaons originating from the two-body decay of a 2.2 GeV object have momenta between 650 and 1300 MeV/c.

- In the DM2 detector, the shower counter, covering 70% of 4π sr, lies outside the coil. For photon showers below 300 MeV, the energy resolution scales like $\Delta E/E = 19\%/\sqrt{E}$ and remains constant at $\sim$ 35% above 300 MeV. In the Mark III detector, the electromagnetic calorimeter lies inside the coil and covers 94% of 4π sr. The energy resolution is $17\%/\sqrt{E}$.

Two hadronic scattering experiments report the observation of a resonance at 2.2 GeV. The GAMS[65] collaboration has studied the process $\pi^-p \to \eta\eta'n$ at incident pion energies of 38 GeV and 100 GeV. The $\eta\eta'$ mass spectra plotted in Fig. 20 show a structure at $m = 2220 \pm 10$ MeV. The width of $\Gamma \sim 80$ MeV is consistent with the resolution. At 38 GeV, the cross section is found to be (50 ± 17) nb. The angular distributions favor a $J^{PC} = 2^{++}$ or 4^{++} assignment. The LASS group has studied the reactions $K^-p \to K_s^0 K_s^0 \Lambda$ and $K^-p \to K^+K^-\Lambda$[51]. Their $K_s^0 K_s^0$ invariant mass spectrum is shown, in Fig. 21, in comparison with the corresponding Mark III spectrum. A structure at 2.2 GeV is observed. A moment analysis favors a $J \geq 2$ hypothesis.

Mark III has also searched for other decay modes of the $\xi(2230)$. So far, none has been found, and the following upper limits on the product branching ratio at 90% C.L. have been set:

$\mu^+\mu^-$:	$B < 5.0 \times 10^{-6}$	$\pi^+\pi^-$:	$B < 2 \times 10^{-5}$
K^*K :	$B < 2.5 \times 10^{-4}$	K^*K^* :	$B < 3 \times 10^{-4}$
$\eta\eta$:	$B < 7.0 \times 10^{-5}$	$p\bar{p}$:	$B < 2 \times 10^{-5}$
$\omega\phi$:	$B < 5.9 \times 10^{-5}$	$\phi\phi$:	$B < 5.5 \times 10^{-5}$

For all modes except $\mu^+\mu^-$, $\omega\phi$ and $\phi\phi$ the limits are based on a data sample of 2.7×10^6 J/ψ decays. The $\xi(2230)$ has also been searched for in other processes. In Υ decays, CLEO[66] has set limits on $\xi(2230)$ production of $B(\Upsilon \to \gamma\xi) \cdot B(\xi \to K_s^0 K_s^0) < 2 \times 10^{-4}$ and of $B(\Upsilon' \to \gamma\xi) \cdot B(\xi \to K^+K^-) < 9 \times 10^{-5}$. In $p\bar{p}$ annihilation, two experiments, at BNL and LEAR, have determined limits (3 s.d.) on $B(\xi \to p\bar{p}) \times B(\xi \to K^+K^-)$ for different spins and widths. At BNL,[67] the limits for a 35 (7) MeV wide state are $B < 5.6 \times 10^{-4}$ (12×10^{-4}) for $J^{PC} = 0^{++}$ and $B < 1.8 \times 10^{-4}$ (3.8×10^{-4}) for $J^{PC} = 2^{++}$. At LEAR,[68] the results are similar, yielding for a width of 30 (3) MeV, limits of $B < 5.4 \times 10^{-4}$ (18×10^{-4}) for $J^{PC} = 0^{++}$ and $B < 1.3 \times 10^{-4}$ (4.6×10^{-4}) for $J^{PC} = 2^{++}$.

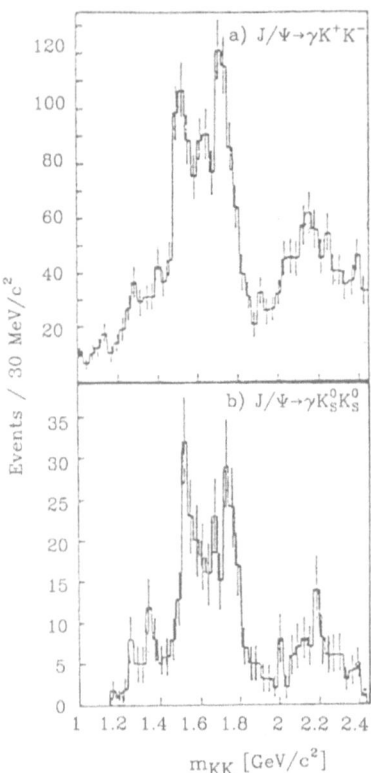

Fig.19. The DM2 $K\bar{K}$ mass spectra for a) $J/\psi \to \gamma K^+ K^-$ and b) $J/\psi \to \gamma K^0_s K^0_s$.

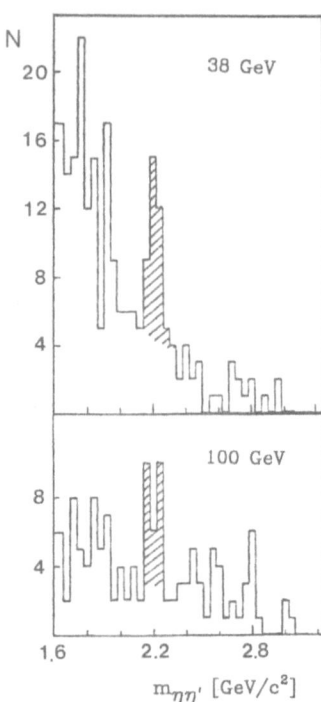

Fig.20. The GAMS $\eta\eta'$ invariant mass spectra from $\pi^- p \to \eta\eta' n$ for 38 GeV and 100 GeV incident π energies.

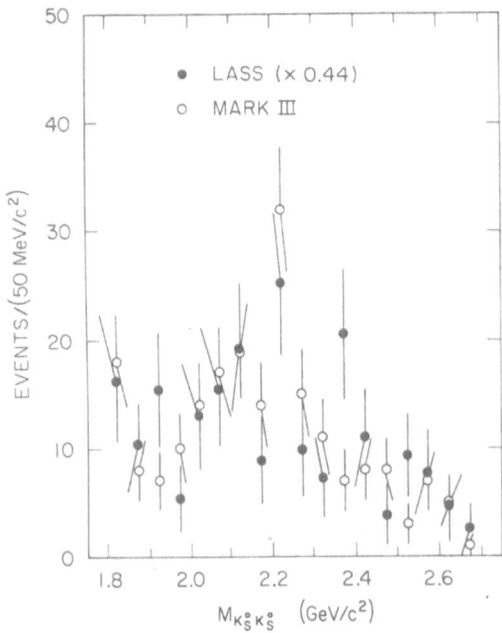

Fig.21. The LASS $K^0_s K^0_s$ mass spectrum in the 1.8-2.8 GeV region from $K^- p \to K^0_s K^0_s \Lambda$.

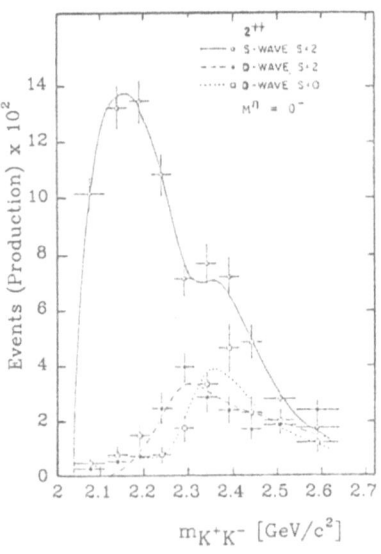

Fig.22. A partial wave analysis of the $\phi\phi$ mass spectrum observed at BNL in $\pi^- p \to \phi\phi n$.

For a classification of the $\xi(2230)$, more information is needed. However, the Higgs boson hypothesis has become rather unlikely. For the standard Higgs,[59] the predicted total branching ratio, $B(\psi \to \gamma H^0) \simeq 2.9 \times 10^{-5}$, is smaller than the measured value for $B(\psi \to \gamma\xi) \cdot B(\xi \to K\bar{K}) = (9.3 \pm 1.2) \times 10^{-5}$. In addition, the ξ mass lies below the Linde-Weinberg bound of 7 GeV.[69] In models with 2 or more Higgs doublets one expects a large rate for the decay $b \to s\xi$, yielding $B(b \to s\xi) \approx 50\%$. In a search for $B \to HX$, CLEO[66] set an upper limit on $\xi(2230)$ production of $B(B \to \xi X) \cdot B(\xi \to K^+K^-) < 0.5\%$ at 90\% C.L., thereby excluding this interpretation. Finally, a spin assignment $J \geq 2$ also contradicts the Higgs hypothesis. The glueball interpretation also becomes unlikely if one assumes that the state seen by LASS and GAMS is the $\xi(2230)$. In addition, the flavor non-symmetric decays of the $\xi(2230)$ disfavor a glueball classification. The hybrid hypothesis looks promising. A crucial test would be to find the ξ in the doubly OZI suppressed decay $\xi \to \omega\phi$, which is discussed in Chapter 6.5. The most plausible explanation is an $L = 3$ $s\bar{s}$ $J^{PC} = 2^{++}$ or $J^{PC} = 4^{++}$ $q\bar{q}$ state. For a 2^{++} state, the only rate comparable to Γ_{KK} is Γ_{K^*K}, while the rates $\Gamma_{K^*K^*}$ and $\Gamma_{\eta\eta}$ are a factor ~ 3 smaller than Γ_{KK}. For a 4^{++} state the dominant rate, $\Gamma_{K^*K^*}$, would be a factor ~ 4 larger than Γ_{KK}, whereas the rates Γ_{K^*K} and $\Gamma_{\phi\phi}$ would be of the order of Γ_{KK}. Mark III is currently searching for $J/\psi \to \gamma K^*K^*$ and $J/\psi \to \gamma K^*K$ to test these predictions.

6. $J/\psi \to \gamma VV$ and $J/\psi \to \gamma\gamma V$

The study of the γVV decays is motivated by several interesting issues. Besides studying η_c decays and looking for the $\xi(2230)$, these channels provide an excellent tool to search for glueballs, hybrids or other new states. Glueballs have already been claimed in a VV channel. In a BNL experiment, three broad 2^{++} resonances, named g_T,[70] have been observed in the hadronic process $\pi^-p \to \phi\phi n$. Figure 22 shows the result of a partial wave analysis. The masses and widths of the three states are measured to be:

$$g_T: \quad m = 2011 \pm^{62}_{76} \text{ MeV}, \quad \Gamma = 202 \pm^{67}_{62} \text{ MeV}$$

$$g_{T'}: \quad m = 2297 \pm 28 MeV, \quad \Gamma = 149 \pm 41 MeV$$

$$g_{T''}: \quad m = 2339 \pm 55 MeV, \quad \Gamma = 319 \pm^{81}_{69} MeV.$$

If these states are really glueballs, they should be produced in $J/\psi \to \gamma VV$ decays. In addition, the γVV channels have two nice properties. First, the spin-parity of a state can be easily determined from a decay plane analysis of the vector mesons. Second, SU(3) relations may be used to identify a given state. For an SU(3) singlet, the individual decay rates behave like:

$$\Gamma_{\rho\rho} : \Gamma_{K^*K^*} : \Gamma_{\phi\phi} : \Gamma_{\omega\omega} = 3 \times p_\rho^3 : 4 \times p_{K^*}^3 : 1 \times p_\phi^3 : 1 \times p_\omega^3. \tag{8}$$

There is, however, a caveat. Due to SU(3) breaking, the comparison may be more complex in some cases. For example, SU(3) relations do not work well in η_c decays, as discussed in Section 6.6. A signature for hybrids is provided by decays like $J/\psi \to \gamma\omega\phi$, $\gamma\omega\rho$ or $\gamma\rho\phi$.

The study of the $\gamma\gamma V$ modes is motivated by the important issue of whether the $\iota/\eta(1440)$ has an electromagnetic width. A comparison of $\gamma\gamma\rho$ with $\gamma\gamma\phi$ allows one to distinguish a glueball from a $q\bar{q}$ or a $q\bar{q}g$ state. [107] I will discuss the VV systems $\rho\rho$, $\omega\omega$, $\phi\phi$ and $\omega\phi$, as well as the γV system $\gamma\rho$. The analyses of $J/\psi \to \gamma K^*\bar{K}^*$, $\gamma\gamma\omega$ and $\gamma\gamma\phi$ are in progress.

6.1 $J/\psi \to \gamma\rho\rho$

Mark III has studied both decay modes $J/\psi \to \gamma\rho^0\rho^0$ and $J/\psi \to \gamma\rho^+\rho^-$ using a sample of 2.7×10^6 J/ψ decays.[71] Stringent selection criteria are necessary, since a

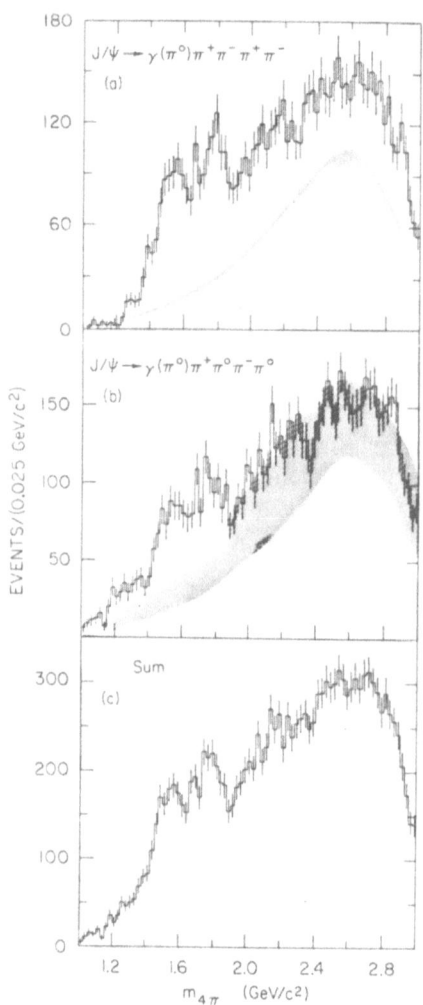

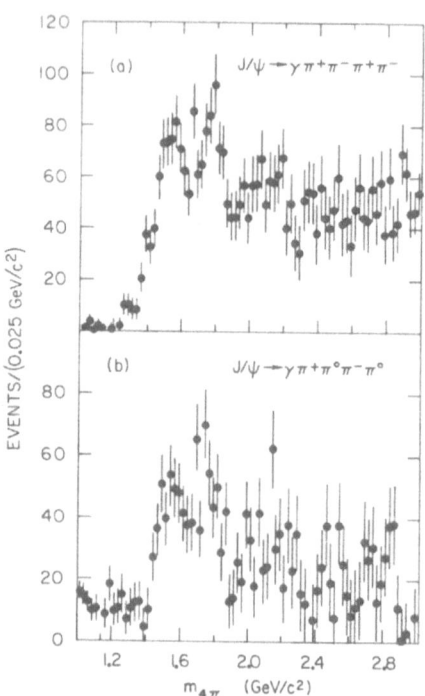

Fig.24. Background subtracted 4π invariant mass spectra from a) $J/\psi \to \gamma\pi^+\pi^-\pi^+\pi^-$ and b) $J/\psi \to \gamma\pi^+\pi^0\pi^-\pi^0$.

Fig.23. The 4π invariant mass spectra from a) $J/\psi \to \gamma\pi^+\pi^-\pi^+\pi^-$, b) $J/\psi \to \gamma\pi^+\pi^0\pi^-\pi^0$, and c) sum. The bands indicate the 5π-background estimate.

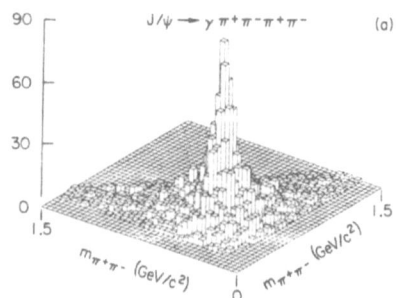

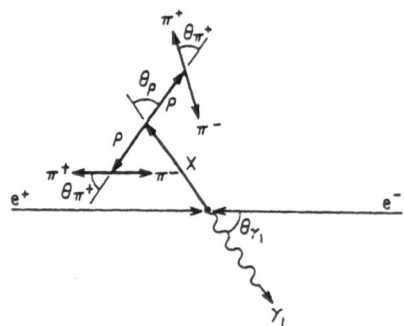

Fig.26. Decay schematics for $\psi \to \gamma\rho\rho$.

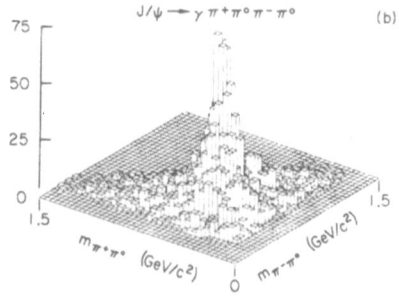

Fig.25. Mass correlations: a) $m_{\pi^+\pi^-}$ vs $m_{\pi^+\pi^-}$ and b) $m_{\pi^+\pi^0}$ vs $m_{\pi^-\pi^0}$

large background contribution from $J/\psi \to 5\pi$ feeds into both channels. The $\gamma\rho^0\rho^0$ channel is studied in the $\gamma\pi^+\pi^-\pi^+\pi^-$ final state using 4C kinematic fits. The detection efficiency is 21%. The $\gamma\rho^+\rho^-$ channel is analyzed in the $\gamma\pi^+\pi^0\pi^-\pi^0$ decay mode with $\pi^0 \to 2\gamma$ using 6C kinematic fits. The detection efficiency for $\gamma\pi^+\pi^0\pi^-\pi^0$ is only 5%. Further details of the selection criteria are given in Ref. 71.

Figure 23 shows the 4π invariant mass distributions for both modes. The 5π background estimate is indicated by the shaded bands. Subtracting the background yields the mass spectra shown in Fig. 24. Two bumps are visible at masses of ~ 1.55 GeV and ~ 1.8 GeV, both approximately 100 MeV wide. The $\pi\pi$ mass correlation plots displayed in Figs. 25 show clear $\rho\rho$ signals. Below 2 GeV, the branching ratios are:

$$B(J/\psi \to \gamma\pi^+\pi^-\pi^+\pi^-) = (3.05 \pm 0.08 \pm 0.45) \times 10^{-3}$$

$$B(J/\psi \to \gamma\pi^+\pi^0\pi^-\pi^0) = (8.3 \pm 0.2 \pm 3.1) \times 10^{-3}.$$

The ratio $r = B(J/\psi \to \gamma\pi^+\pi^0\pi^-\pi^0)/B(J/\psi \to \gamma\pi^+\pi^-\pi^+\pi^-) = 2.7 \pm 1.1$ is consistent with the expectation $r = 2$ for an isoscalar 4π system made from 2 isovector $\pi^+\pi^-$ or $\pi^\pm\pi^0$ subsystems. The entire branching ratio for the $\gamma\pi^+\pi^-\pi^+\pi^-$ channel below 3 GeV is:

$$B(J/\psi \to \gamma\pi^+\pi^-\pi^+\pi^-) = (6.4 \pm 0.2 \pm 0.8) \times 10^{-3}.$$

The spin-parity of the $\rho\rho$ system is determined using two independent techniques. The first consists of a measurement of χ, the angle between the ρ decay planes in the $\rho\rho$ center-of-mass system. Chang and Nelson,[72] as well as Trueman,[73] have pointed out that χ provides a unique signature for states with even spin and odd parity, in analogy with Yang's parity test for the π^0.[74] The second method is a full multi-channel maximum likelihood fit. The decay $J/\psi \to \gamma\rho\rho$, with $\rho \to \pi\pi$, which is sketched in Fig. 26, is characterized by 7 angles. These angles are: θ_γ, the polar angle of the photon in the lab frame; θ_ρ and ϕ_ρ, the polar and azimuth angles of the vector meson in the $\rho\rho$ helicity frames; $\theta_{\pi_1}, \phi_{\pi_1}, \theta_{\pi_3}, \phi_{\pi_3}$, the polar and azimuth angles of π_1 and π_3 in their ρ rest frames. In terms of the helicity angles, χ is given by $\chi = \phi_{\pi_1} + \phi_{\pi_2}$. Besides χ, the polar angles θ_{π_i} are also sensitive to determine the spin-parity of the resonance.

The χ distribution is defined by:[72]

$$R(\chi) = 1 + \beta \cos 2\chi. \tag{9}$$

with:

$$\beta = \frac{2\eta|A_{11}|^2}{2|A_{11}|^2 + |A_{00}|^2 + 4|A_{10}|^2 + 2|A_{1-1}|^2}, \tag{10}$$

where A_{ij} are the helicity amplitudes for the decay $X \to \rho\rho$ with helicities $i = 0, \pm 1$ and $j = 0, \pm 1$, and η is the parity of the state X. The helicity amplitudes depend on the relative orbital angular momentum of the $\rho\rho$ system $L_{\rho\rho}$, its total spin $S_{\rho\rho}$, and the total angular momentum $J = L_{\rho\rho} + S_{\rho\rho}$. Table 3 summarizes the helicity amplitudes and the parameter β for different spin-parity hypotheses. Equation (10) implies that the sign of β is defined by the parity and that $0 \leq \beta \cdot \eta \leq 1$. Thus, for even spin and for odd (even) parities, $R(\chi)$ has a $\sin^2 \chi$ ($\cos^2 \chi$) component. For a pseudoscalar state $\beta = -1$, while β vanishes for odd values of J.

The distribution for θ_π is given by the expression:

$$R(\theta_\pi) = 1 + \frac{\xi}{2}(3\cos^2\theta_\pi - 1), \tag{11}$$

with:

$$\xi = 2\frac{|A_{00}|^2 - |A_{11}|^2 - |A_{1-1}|^2 + |A_{10}|^2}{2|A_{11}|^2 + |A_{00}|^2 + 4|A_{10}|^2 + 2|A_{1-1}|^2}. \tag{12}$$

TABLE 3

Projections of Helicity States and Predictions for the Amplitudes
β and ξ for Different J^P Hypotheses

J^P	L	S	A_{11}	A_{00}	A_{10}	A_{1-1}	β	ξ	J^P	L	S	A_{11}	A_{00}	A_{10}	A_{1-1}	β	ξ
0^+	0	0	$\frac{1}{3}$	$-\frac{1}{3}$	—	—	$\frac{2}{3}$	0	2^+	0	2	$\frac{1}{30}$	$\frac{2}{15}$	$\frac{1}{10}$	$\frac{1}{5}$	$\frac{1}{15}$	0
0^+	2	2	$\frac{1}{6}$	$\frac{2}{3}$	—	—	$\frac{1}{3}$	1	2^+	2	0	$\frac{1}{3}$	$-\frac{1}{3}$	—	—	$\frac{2}{3}$	0
0^-	1	1	$-\frac{1}{2}$	—	—	—	-1	-1	2^+	2	2	$-\frac{1}{21}$	$-\frac{4}{21}$	$-\frac{1}{28}$	$\frac{2}{7}$	$\frac{2}{21}$	$-\frac{3}{14}$
1^-	1	1	—	—	$-\frac{1}{4}$	—	0	$\frac{1}{2}$	2^-	1	1	$\frac{1}{5}$	—	$\frac{3}{20}$	—	$-\frac{2}{5}$	$-\frac{1}{10}$
1^+	2	2	—	—	$\frac{1}{4}$	—	0	$\frac{1}{2}$	2^-	3	1	$-\frac{3}{10}$	—	$\frac{1}{10}$	—	$-\frac{3}{5}$	$-\frac{2}{5}$

A $\sqrt{}$ is implied over each coefficient A_{ij}.

Values of the amplitude ξ for different spin-parity hypotheses are listed in Table 3. For a pseudoscalar state, one obtains $\xi = -1$, yielding $R(\theta_\pi) \propto \sin^2\theta_\pi$. Hence, $R(\chi)$ and $R(\theta_\pi)$ distribution allow a unique identification of a pseudoscalar state.

For the $\gamma\rho\rho$ channel, however, the discrimination power of these distributions is reduced because of the 5π and combinatorial backgrounds (2π pair combinations per event). Figure 27a shows the $\chi_{\rho\rho}$ distribution of the $\gamma\pi^+\pi^-\pi^+\pi^-$ mode for masses below 2 GeV. A $\sin^2\chi$ component is clearly visible, providing evidence for structures with even spin and odd parity. The only significant contribution, in fact, comes from $J^P = 0^-$. The background contribution in the $\gamma4\pi$ channel is relatively large, but it remains constant as shown in Fig. 27b. The $\chi_{\rho\rho}$ distribution is fitted to a constant plus a $\sin^2\chi$ term. The fit, shown overplotted in Fig. 27a, yields a pseudoscalar fraction of $37 \pm 3\%$ in $\gamma4\pi$ below 2 GeV.

The multi-channel spin-parity analysis makes use of all the information contained in the $\gamma\rho\rho$ channel, which consists of 9 variables, in addition to the 4π mass: 7 angles as discussed above and the two π-pair masses, m_{12} and m_{34}. A weight is assigned to each event in a defined mass interval for each hypothesis included in the maximum likelihood fit. The log likelihood function is maximized, varying the channel fractions and all other parameters upon which the weight depends. For example, for $J > 0$, these parameters include the helicity ratios x and y. It is important to note that no significant channel has been excluded. The result, then, provides an estimate of the fraction of events from each source. Further details are given in Ref. 71.

For the $\gamma\rho\rho$ analysis, 10 channels are included in the fit: 6 channels for $J/\psi \to \gamma X$, $X \to \rho\rho$ with spin-parities $J^P = 0^\pm, 1^\pm$ and $2^\pm$ and 4 channels consisting of $\rho\rho$ phase space, $\rho\pi\pi$, 4π and $a_2(1320)\pi$, for which no angular correlations are taken into account. The $a_2\pi$ channel accounts for a possible feedthrough from $J/\psi \to a_2\rho$, with $a_2 \to \rho\pi$ and also absorbs possible contributions from $J/\psi \to \gamma a_1(1270)\pi$.

In order to reduce the number of free parameters, three additional assumptions are made:

i) for a given spin-parity, only the lowest value of $L_{\rho\rho}$ is considered
ii) the ratios of helicity amplitudes x and y are assumed to be real
iii) no interference between amplitudes of individual channels exist.

The first assumption is certainly plausible for $\rho\rho$ masses close to threshold. The second assumption holds for other resonances produced in J/ψ decays.[36]

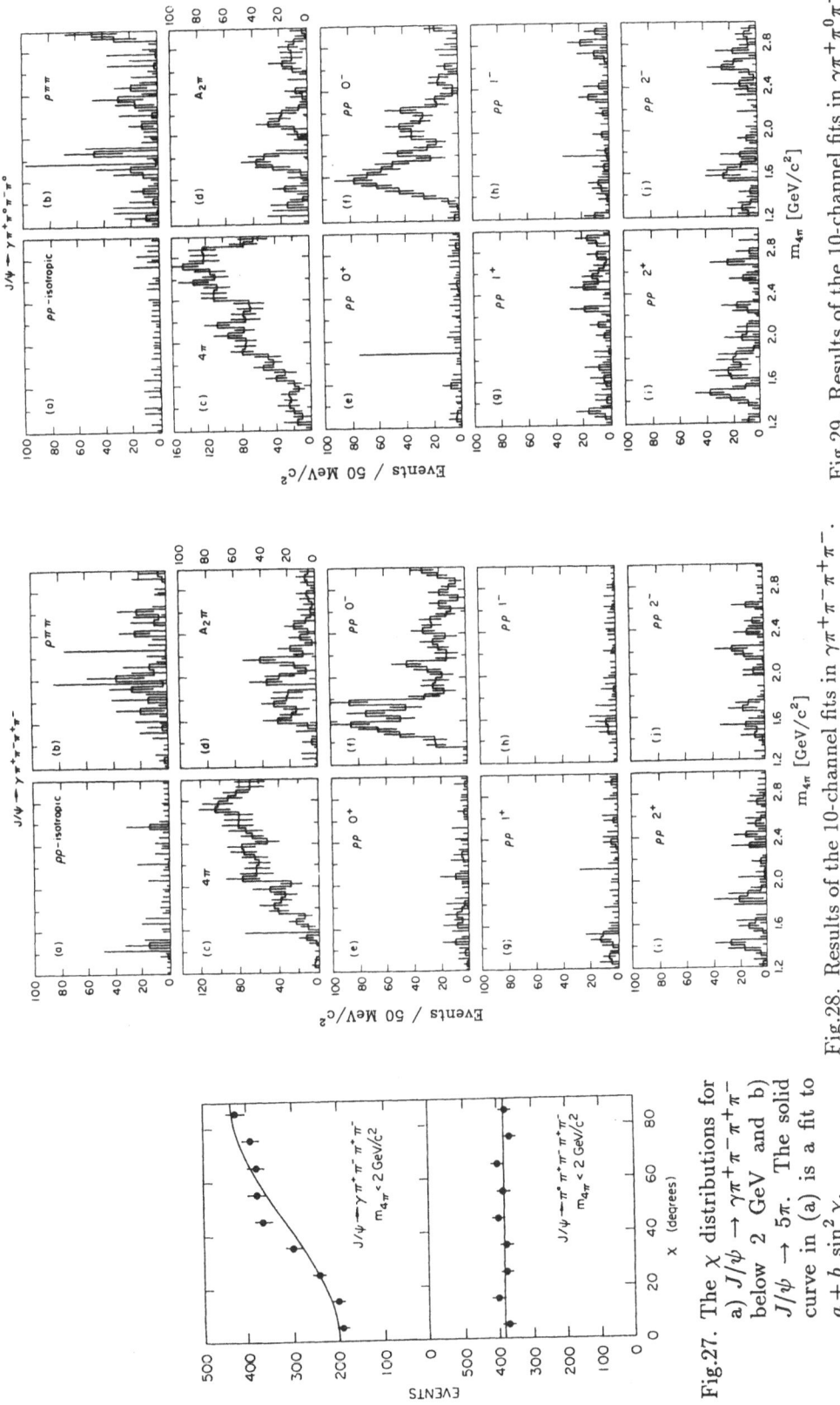

Fig.27. The χ distributions for a) $J/\psi \rightarrow \gamma \pi^+ \pi^- \pi^+ \pi^-$ below 2 GeV and b) $J/\psi \rightarrow 5\pi$. The solid curve in (a) is a fit to $a + b \sin^2 \chi$.

Fig.28. Results of the 10-channel fits in $\gamma \pi^+ \pi^- \pi^+ \pi^-$.

Fig.29. Results of the 10-channel fits in $\gamma \pi^+ \pi^0 \pi^- \pi^0$.

211

Figures 28 and 29 show the 4π invariant mass distributions from the maximum likelihood fit in each channel for the radiative modes $\gamma\pi^+\pi^-\pi^+\pi^-$ and $\gamma\pi^+\pi^0\pi^-\pi^0$, respectively. The errors plotted are statistical. The most prominent contributions are observed in the 4π and $\rho\rho$ 0^- channels, where the $\rho\rho$ 0^- channel exhibits the most significant structure below 2 GeV. Though two clear peaks are not observed as in Figs. 24, the shape of the $0^-\rho\rho$ mass is not inconsistent with 2 states. The data samples also contain a small amount of $\rho\pi\pi$ events, $\rho a_2(1320)$ feedthroughs, and a small $\rho\rho$ 2^+ component below 1.6 GeV; other channels show no significant contributions. A similar 10-channel analysis, which is performed in the 5π mode as a cross check, indicates that the background feeds predominantly into the 4π phase space channel. In addition, small contributions to the $\rho\pi\pi$ and $a_2\pi$ channels are found, but no significant contribution is seen in any $\rho\rho$ channel.

For the combined $\gamma 4\pi$ data samples, the fit yields a fraction of $(46 \pm 8)\%$ of all $\gamma 4\pi$ events due to the $0^-\rho\rho$ channel. Above 2 GeV, the fraction is 20%. This is consistent with the result from the decay plane analysis. After background correction, the $\rho\rho$ 0^- contribution in $\gamma 4\pi$ below 2 GeV increases to $(51 \pm 9)\%$. The fraction of the $\rho\rho$ 0^- channel contributing to $\gamma\rho\rho$ below 2 GeV is $(76 \pm 11)\%$.

The product branching ratio for the pseudoscalar $\rho\rho$ component below 2 GeV is

$$B(J/\psi \to \gamma X_{0-}) \cdot B(X_{0-} \to \rho\rho) = (4.7 \pm 0.3 \pm 0.9) \times 10^{-3}.$$

This is consistent with the DM2 result of $(3.6 \pm 0.12 \pm 0.54) \times 10^{-3}$, [40] if the large systematic errors are taken into account. In a recent multi-channel analysis carried out by DM2, three 0^- states were found in the $\rho\rho$ invariant mass spectrum at 1497 ± 3 MeV, 1812 ± 4 MeV and 2107 ± 9 MeV.[72] The widths of these states have been measured to be 126 ± 2 MeV, 110 ± 8 MeV and 244 ± 23 MeV, respectively. The results for the first and second state agree with the Mark III observations. In addition, the Mark III $\rho\rho$ mass spectra are consistent with a third state around 2.1 GeV.

Since no significant signals are observed in the 2^+ $\rho\rho$ channel above 1.6 GeV, upper limits at 90% C.L. for the $\theta/f_2(1720)$ and the g_T states are derived from the events in the 1.6-1.85 GeV and 2.1-2.4 GeV mass intervals, yielding:

$$B(J/\psi \to \gamma\theta) \cdot B(\theta \to \rho\rho) < 5.5 \times 10^{-4}$$

$$B(J/\psi \to \gamma g_T) \cdot B(g_T \to \rho\rho) < 6.0 \times 10^{-4}.$$

The DM2 upper limit for g_T production is $B(J/\psi \to \gamma g_T) \cdot B(g_T \to \rho\rho) < 6.8 \times 10^{-4}$.[75]

6.2 $J/\psi \to \gamma\omega\omega$

Mark III has analyzed the radiative decay $J/\psi \to \gamma\omega\omega$ in the $\gamma\pi^+\pi^-\pi^0\pi^+\pi^-\pi^0$ final state. The data are selected with 6C kinematic fits using the initial 2.7×10^6 J/ψ event sample.[76] Figure 30 shows the 3π mass correlation plot with 4 entries per event. A prominent $\omega\omega$ signal is found. Since the processes $J/\psi \to \omega\omega$ and $J/\psi \to \pi^0\omega\omega$ are forbidden by C-parity conservation, the $\omega\omega$ signal proves the existence of $J/\psi \to \gamma\omega\omega$. The $\omega\omega$ invariant mass spectrum, which is obtained by selecting ω candidates with $|m_{3\pi} - m_\omega| \leq 30$ MeV, is depicted in Fig. 31a. The shaded band represents an estimate of the residual background which feeds in from processes like $J/\psi \to 7\pi$, $\gamma 6\pi$ and $\omega 4\pi$. The width of the band includes both statistical and systematic uncertainties. The $\omega\omega$ mass distribution shows a prominent structure at 1.8 GeV, which is approximately 250 MeV wide. The spectrum exhibits no other significant features. The detection efficiency is $(5.3 \pm 1.1)\%$, independent of $m_{\omega\omega}$. Below 2 GeV, the branching ratio is measured to be:

$$B(J/\psi \to \gamma\omega\omega) = (1.22 \pm 0.07 \pm 0.31) \times 10^{-3},$$

where the systematic error includes uncertainties from the background subtraction.

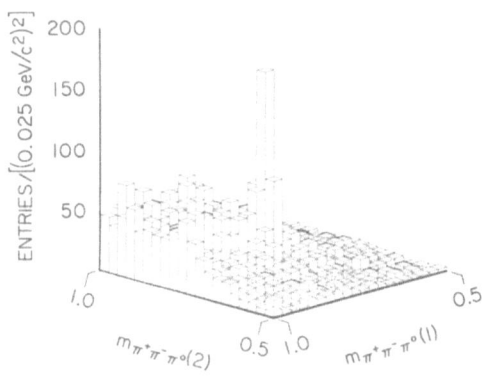

Fig.30. The $\pi^+\pi^-\pi^0$ vs $\pi^+\pi^-\pi^0$ mass correlation for $\psi \to \gamma 2(\pi^+\pi^-\pi^0)$.

Fig.31. a) The $\omega\omega$ invariant mass spectrum. The shaded band indicates the background estimate. The inset shows the η_c region. b) The branching ratio $B(\psi \to \gamma\omega\omega)$ as a function of mass. The dashed (dotted) lines are the phase space predictions for an S-wave (P-wave).

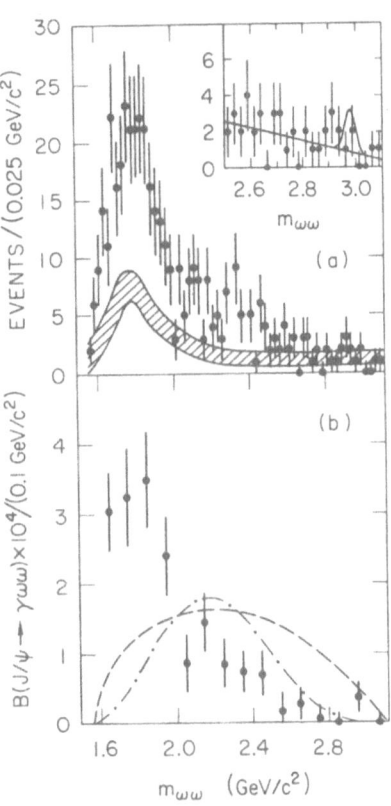

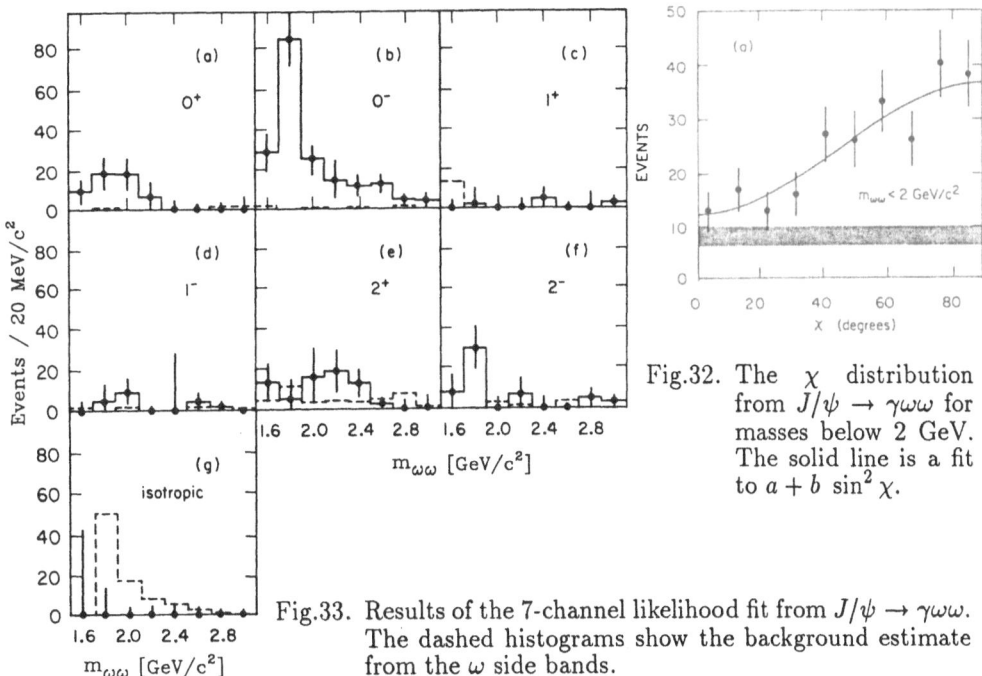

Fig.32. The χ distribution from $J/\psi \to \gamma\omega\omega$ for masses below 2 GeV. The solid line is a fit to $a + b\,\sin^2\chi$.

Fig.33. Results of the 7-channel likelihood fit from $J/\psi \to \gamma\omega\omega$. The dashed histograms show the background estimate from the ω side bands.

213

DM2 has recently reported on the observation of a 0^- state above the $\omega\omega$ threshold.[72) DM2 measures a branching ratio of $(1.06 \pm 0.16 \pm 0.32) \times 10^{-3}$, which is consistent with the Mark III result. The total branching ratio for $J/\psi \to \gamma\omega\omega$ from Mark III is:

$$B(J/\psi \to \gamma\omega\omega) = (1.76 \pm 0.09 \pm 0.45) \times 10^{-3}.$$

The branching ratio for $J/\psi \to \gamma\omega\omega$ as a function of $m_{\omega\omega}$ is plotted in Fig. 31b. For comparison, the 3 body phase space predictions for S-wave and P-wave $\omega\omega$ systems are also shown. Both predictions provide a poor description of the data.

To determine the spin-parity of the low mass structure, both an ω decay plane analysis and a multi-channel maximum likelihood fit have been performed. Figure 32 shows the χ distribution of the $\gamma\omega\omega$ signal events. A prominent $\sin^2 \chi$ component is found, providing evidence that the structure has even spin and odd parity. The background, which is determined from a χ angle analysis of the ω side bands $(0.723 \le m_{3\pi} \le 0.753 \text{ GeV}/c^2$ and $0.813 \le m_{3\pi} \le 0.843 \text{ GeV}/c^2)$, produces a constant distribution. The background here is much smaller than in the $\gamma 4\pi$ channel, as indicated by the shaded band in Fig. 32. The χ distribution is fitted to the parametrization $a + b\sin^2 \chi$. The fit, which is represented by the solid curve in Fig. 32, yields a fraction of $(75 \pm 16)\%$ for the (even)$^-$ component after background correction. As will be shown below, the main contribution to the mass spectrum below 2 GeV/c^2 is the 0^- component.

The multi-channel maximum likelihood fit of the $\omega\omega$ final state includes 7 channels: isotropic $\omega\omega$ for $J/\psi \to \gamma\omega\omega$ phase space and six channels for $J/\psi \to \gamma X$, $X \to \omega\omega$ with spin-parities $J^P = 0^\pm, 1^\pm$ and $2^\pm$. To reduce the number of free parameters, the same three assumptions used in the $\gamma\rho\rho$ analysis, are employed.

The resulting $\omega\omega$ invariant mass distributions for individual channels are displayed in Fig. 33. The dashed histograms indicate the background estimate resulting from a maximum likelihood fit performed in the ω side bands. Below 2 GeV, the dominant contribution is the pseudoscalar channel. A small 2^- component may also be present. No other channel shows a significant contribution. The background primarily contributes to the isotropic $\omega\omega$ channel. The $\omega\omega$ 0^- channel is essentially background-free. Below 2 GeV/c^2, the pseudoscalar component amounts to $(85 \pm 19)\%$ of the entire $\gamma\omega\omega$ mode after background correction. From the events in the 2^+ channel, upper limits at 90% C.L. can be estimated for $\theta/f_2(1720)$ and g_T production $(2.1 \le m_{\omega\omega} \le 2.4 \text{ GeV})$:

$$B(J/\psi \to \gamma\theta) \cdot B(\theta \to \omega\omega) < 2.4 \times 10^{-4}$$

$$B(J/\psi \to \gamma g_T) \cdot B(g_T \to \omega\omega) < 2.6 \times 10^{-4}.$$

6.3 $J/\psi \to \gamma\phi\phi$

Mark III has performed a new analysis of $J/\psi \to \gamma\phi\phi$ in the two final states $\gamma K^+K^-K^+K^-$ and $\gamma K^+K^-K_s^0 K_L^0$, with $K_s^0 \to \pi^+\pi^-$ using a sample of 4.9×10^6 J/ψ decays.[77) The $\gamma K^+K^-K^+K^-$ mode is strongly affected by K decays, especially in the low $\phi\phi$ mass region where kaons have low momenta: below 2.4 GeV, about 60% of the events have a mis-measured track. Therefore, 4C kinematic fits very often fail, resulting in a low detection efficiency. However, a substantial gain in efficiency is obtained by performing 1C kinematic fits, in which the mis-measured track is excluded from the fit. In order to be consistent for events with four well-measured tracks, all three-track combinations are fitted and the "best" fit is selected. Since 1C kinematic fits are not as effective in background suppression as 4C fits, stringent selection criteria are imposed.

214

Figure 34a shows the scatter plot of the K^+K^- invariant masses $m_{K_1^+K_1^-}$ versus $m_{K_2^+K_2^-}$, where $K_2^+K_2^-$ is the pair containing the mismeasured track. A clear $\phi\phi$ signal is observed providing evidence for the process $J/\psi \to \gamma\phi\phi$, since the hadronic decays $J/\psi \to \phi\phi$ and $J/\psi \to \phi\phi\pi^0$ are forbidden by C-parity invariance. Both K^+K^- mass distributions peak at the nominal ϕ mass. The resolutions are $\sigma_1 = 4.2 \pm 0.3$ MeV for $m_{K_1^+K_1^-}$ and $\sigma_2 = 5.6 \pm 0.5$ MeV for $m_{K_2^+K_2^-}$. The scatter plot indicates that the background level is small. Potential background sources are decay modes containing $K^+K^-\pi^+\pi^- + n\gamma$ events and the channels $\gamma\phi K^+K^-$, ϕK^+K^- and $\phi K^+K^-\pi^0$.

The $\phi\phi$ invariant mass spectrum shown in Fig. 35a is obtained after selecting all events with $\delta \leq 3$, where δ is defined by:

$$\delta^2 = ((m_{K_1K_2} - m_\phi)/\sigma_1)^2 + ((m_{K_2K_2} - m_\phi)/\sigma_2)^2. \tag{13}$$

The η_c and a resonance around 2.2 GeV are visible. The spectrum shows no other significant structures. The background is estimated from the elliptical band $5 \leq \delta \leq 6$. The resulting distribution shown by the shaded histogram demonstrates that the background is small and uniform in $m_{\phi\phi}$. In addition, side bands below each ϕ, covering the same area as the signal region, have been examined. This estimate yields a similar background spectrum. According to Monte Carlo studies, $\sim 40\%$ of these events are in fact $\phi\phi$ signal events, which indicates that the background from $\gamma\phi K^+K^-$, ϕK^+K^- and $\phi K^+K^-\pi^0$ is negligible. The detection efficiency obtained from a Monte Carlo simulation is shown overplotted in Fig. 35a. The efficiency varies almost linearly from $\sim 7\%$ to $\sim 19\%$ in the 2.1-2.4 GeV mass region. While the efficiency loss at low $\phi\phi$ masses is mainly due to kaon decays, the drop above 3 GeV is caused by the decreasing photon-finding efficiency. The mass resolution increases from 12 MeV at 2.2 GeV to 20 MeV at 3.0 GeV.

In the analysis of the $\gamma K^+K^-K_s^0K_L^0$ channel, the K_s^0 is detected in the $\pi^+\pi^-$ decay mode, while the K_L^0 is not required to be observed since the K_L^0 detection efficiency in the Mark III detector is only $\sim 50\%$. To improve the $K_s^0K_L^0$ mass resolution 2C kinematic fits are performed to the hypothesis $J/\psi \to \gamma K^+K^-K_s^0(K_{L missing}^0)$. To suppress background from $K^+K^-\pi^+\pi^- + n\gamma$ events, stringent selection criteria are again imposed. Figure 34b shows the resulting scatter plot of $m_{K^+K^-}$ versus $m_{K_s^0K_L^0}$, which again exhibits a clear $\phi\phi$ signal. The $m_{K^+K^-}$ and $m_{K_s^0K_L^0}$ projections peak at the correct ϕ mass. The mass resolutions are $\sigma_1 = 4 \pm 0.4$ MeV for $m_{K^+K^-}$ and $\sigma_2 = 6.4 \pm 0.8$ MeV for $m_{K_s^0K_L^0}$. The scatter plot indicates that the background level for this mode is also small.

The final data sample is extracted by selecting all events with $\delta \leq 3$ where δ is defined by an expression equivalent to Eq. 13 for the $\gamma K^+K^-K_s^0K_L^0$ mode. The resulting $\phi\phi$ invariant mass spectrum depicted in Fig. 35b confirms the structure at 2.2 GeV. The η_c signal is much broader, as the mass resolution is worse : 15 MeV at 2.2 GeV and ~ 40 MeV at 3 GeV. The residual background is again estimated by analyzing the elliptical band $5 \leq \delta \leq 6$. The result, which is shown by the shaded histogram in Fig. 34b, demonstrates that the background is small and uniform in $m_{\phi\phi}$. In addition, a side-band analysis below each ϕ confirms that the background from modes such as $\gamma\phi K^+K^-$, ϕK^+K^-, and $\phi K^+K^-\pi^0$, with $\phi \to K_s^0K_L^0$ is negligible as before. Finally, Monte Carlo studies show that the hadronic mode $\phi K_s^0K_s^0$ is not a significant background source. The detection efficiency, shown by the solid curve in Fig. 35b, exhibits characteristics similar to that for $\gamma K^+K^-K^+K^-$.

The $\phi\phi$ invariant mass spectra for both modes, after efficiency correction, are displayed in Figs. 36a,b. The main features are the prominent low mass structure and the η_c. To measure the resonance parameters, both mass spectra are fitted to two Breit Wigner line shapes plus a phase space term. The resulting fits are shown overplotted in Figs. 36. The resonance parameters of the low mass state obtained from the fit in

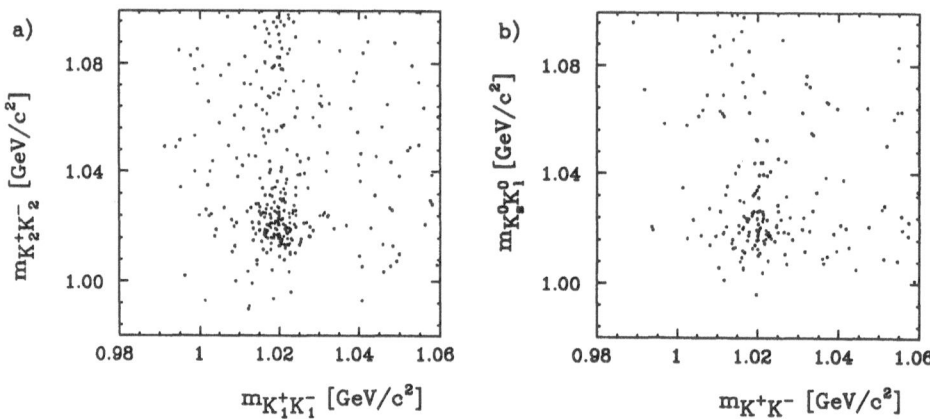

Fig.34. Scatter plots of a) $m_{K_1^+ K_1^-}$ vs $m_{K_2^+ K_2^-}$ and b) $m_{K^+ K^-}$ vs $m_{K_s^0 K_L^0}$ from the reaction $J/\psi \to \gamma 4K$.

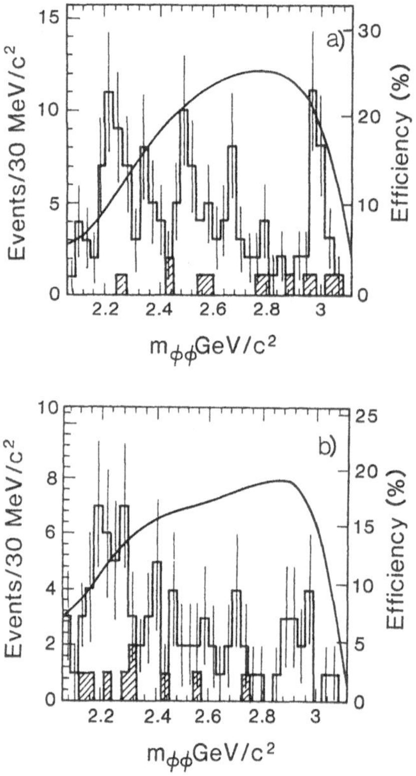

Fig.35. The $\phi\phi$ invariant mass spectra from a) $J/\psi \to \gamma K^+ K^- K^+ K^-$ and b) $J/\psi \to K^+ K^- K_s^0 K_L^0$. The shaded histograms are background estimates. The solid curves show the detection efficiencies.

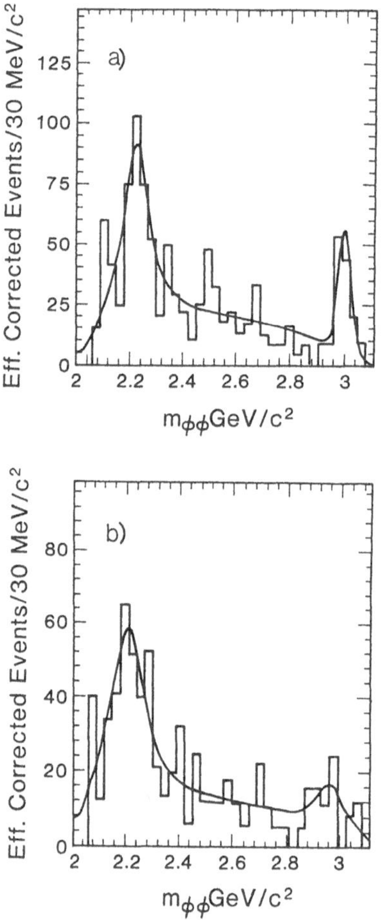

Fig.36. The $\phi\phi$ mass spectra after efficiency correction from a) $J/\psi \to \gamma K^+ K^- K^+ K^-$ and b) $J/\psi \to \gamma K^+ K^- K_s^0 K_L^0$. The solid curves are fits to 2 Breit-Wigner line shapes plus phase space.

the $\gamma K^+K^-K^+K^-$ and $\gamma K^+K^-K_s^0 K_L^0$ final states are:

$$m_X = 2220 \pm 15 \pm 12 \text{ MeV}, \qquad \Gamma_X = 114 \pm 45 \pm 35 \text{ MeV}$$
$$m_X = 2206 \pm 20 \pm 13 \text{ MeV}, \qquad \Gamma_X = 150 \pm 46 \pm 35 \text{ MeV},$$

respectively. The corresponding product branching ratios of the two modes are:

$$B(J/\psi \to \gamma X_{2.2}) \cdot B(X_{2.2} \to \phi\phi) = (3.4 \pm 0.6 \pm 0.5) \times 10^{-4}$$
$$B(J/\psi \to \gamma X_{2.2}) \cdot B(X_{2.2} \to \phi\phi) = (3.0 \pm 0.6 \pm 0.5) \times 10^{-4},$$

respectively. For the η_c, the fit yields masses of $m_{\eta_c} = 2998 \pm 7 \pm 24$ MeV for $J/\psi \to \gamma K^+K^-K^+K^-$ and $m_{\eta_c} = 2961 \pm 36 \pm 25$ MeV for $J/\psi \to \gamma K^+K^-K_s^0 K_L^0$. Both values are consistent with the nominal η_c mass. The product branching ratios for η_c production in the two modes are:

$$B(J/\psi \to \gamma\eta_c) \cdot B(\eta_c \to \phi\phi) = (0.85 \pm 0.2 \pm 0.2) \times 10^{-4}$$
$$B(J/\psi \to \gamma\eta_c) \cdot B(\eta_c \to \phi\phi) = (0.50 \pm 0.2 \pm 0.2) \times 10^{-4}.$$

The first value is consistent with Mark III's published result of $(1.02 \pm 0.25 \pm 0.14) \times 10^{-4}$.[78] The second value, though lower, is still reasonably consistent. This measurement, however, is less reliable, because of low statistics, a worse mass resolution, and uncertainties in the phase space parametrization. Mark III has also performed an analysis of the $\gamma K^+K^-K^+K^-$ channel with 4C kinematic fits to cross check the η_c results. The resulting mass of $m_{\eta_c} = 2978 \pm 9 \pm 10$ MeV is close to the nominal value and the product branching ratio of

$$B(J/\psi \to \gamma\eta_c) \cdot B(\eta_c \to \phi\phi) = (0.85 \pm 0.25 \pm 0.2) \times 10^{-4},$$

is in excellent agreement with the result from the 1C kinematic fit. The total branching ratio for the $\gamma\phi\phi$ mode is measured to be $B(J/\psi \to \gamma\phi\phi) = (8 \pm 0.8 \pm 1.2) \times 10^{-4}$.

To determine the spin-parity of the low mass resonance, the distribution of the helicity angles χ and θ_K are analyzed, where χ is the angle between the ϕ decay planes and θ_k is the polar angle of the K^+ or the K_s^0 in its ϕ rest frame. Figures 37a,b show the χ and $\cos\theta_K$ distributions in the $2.14 \leq m_{\phi\phi} \leq 2.3$ GeV mass range for the $\gamma K^+K^-K^+K^-$ mode. The corresponding distributions for $\gamma K^+K^-K_s^0 K_L^0$ are depicted in Figs. 37c,d. The solid curves shown overplotted are the predictions for different spin-parity hypotheses. The χ angle distributions peak at large angles. Since the acceptance is flat, as Monte Carlo studies show, the spin-parity of the low mass state is restricted to $J^P = (\text{even})^-$. The $\cos\theta_K$ distributions exhibit a strong $\sin^2\theta_K$ dependence, which is characteristic of $J^P = 0^-$. Since the $\cos\theta_K$ acceptances are also nearly uniform, the $X_{2.2}$ can be identified as a pseudoscalar state. To prove the reliability of this technique the χ and $\cos\theta_K$ distributions are examined in the η_c region defined by $2.92 \leq m_{\phi\phi} \leq 3.04$ GeV. Figures 37e,f show the resulting plots for the $\gamma K^+K^-K^+K^-$ mode. Both distributions exhibit the characteristic shape of a 0^- state, thus confirming the validity of this technique. A study of the entire mass spectrum indicates that the region below 2.4 GeV is dominantly 0^-. By fitting the χ distributions to $a + b\sin^2\chi$ Mark III has determined upper limits at 90% C.L. for the g_T tensor states ($2 \leq m_{\phi\phi} \leq 2.4$ GeV) and the $\xi(2230)$ ($2.18 \leq m_{\phi\phi} \leq 2.3$ GeV):

$$B(J/\psi \to \gamma g_T) \cdot B(g_T \to \phi\phi) < 8.6 \times 10^{-5},$$
$$B(J/\psi \to \gamma\xi(2230)) \cdot B(\xi(2230) \to \phi\phi) < 5.5 \times 10^{-5}.$$

DM2 has also analyzed the $\gamma\phi\phi$ mode in the $\gamma K^+K^-K^+K^-$ final state.[79] A low mass enhancement is found in their spectrum at 2.23 GeV which favors a $J^P = 0^-$ assignment. This observation is in agreement with the Mark III results. However, the claim that the remaining spectrum below 2.9 GeV is consistent with 2^+ is not confirmed by Mark III. In addition, the results for the η_c are not consistent in both experiments. DM2 observes a smaller η_c signal resulting in a product branching ratio

TABLE 4

Results for $J/\psi \to \gamma\gamma\rho$ **from Different Experiment**

Mass [MeV/c²]	Width [MeV/c²]	$B(J/\psi \to \gamma X) \cdot B(X \to \gamma\rho) \times 10^4$	Experiment
$1420 \pm 15 \pm 20$	$133 \pm 55 \pm 30$	$1.1 \pm 0.24 \pm 0.25$	Mark III
1401 ± 18	174 ± 44	$0.9 \pm 0.2 \pm 0.14$	DM2
1390 ± 25	185^{+110}_{-80}	$1.9 \pm 0.5 \pm 0.4$	CB

of $B(J/\psi \to \gamma\eta_c) \cdot B(\eta_c \to \phi\phi) = (0.31 \pm 0.07 \pm 0.04) \times 10^{-4}$ which is substantially lower than the Mark III measurement.

6.4 $J/\psi \to \gamma\gamma\rho$

The main interest in studying the channel $J/\psi \to \gamma\gamma\rho$ is the issue of whether the $\iota/\eta(1440)$ decays to $\gamma\rho$. The results of this study may provide important clues for the interpretation of the $\iota/\eta(1440)$ as discussed in Chapter 7.4. Mark III has performed 4C kinematic fits to the hypothesis $J/\psi \to \gamma\gamma\pi^+\pi^-$, using a sample of 2.7×10^6 J/ψ decays.[80] The detection efficiency in the 1400 MeV mass region is 35 %, the mass resolution is 11 MeV. Figure 38 shows the $\pi^+\pi^-\gamma_{low}$ invariant mass spectrum. Besides the η', a broad structure at 1.4 GeV is visible. Fitting the mass spectrum to a Breit Wigner amplitude yields the resonance parameters summarized in Table 4. For comparison, the results from DM2[42] and Crystal Ball[81] are also listed. All 3 experiments show that the width is consistent with the $\eta(1440)$ while the mass is systematically too low by 1 to 2σ, being closer to the $f_1(1420)$. A preliminary spin-parity analysis, in which the polar angle of the ρ in the $\gamma\rho$ rest frame has been measured, gives nearly equal probabilities for the 0^- and 1^+ hypotheses. This suggests that the observed structure may contain more than one resonance. A possible interpretation is that the main peak is due to the $f_1(1420)$, but contributions from the $\iota/\eta(1440)$ and the $f_1(1285)$ are also present. However, without a partial wave analysis no definite conclusion is possible. Currently, a new Mark III analysis on the full data sample of 5.8×10^6 J/ψ decays is in progress, including a multi-channel spin-parity analysis in the 1.4 GeV mass region.

6.5 $J/\psi \to \gamma\omega\phi$

The process $J/\psi \to \gamma\omega\phi$ can only proceed through a doubly-disconnected diagram as shown in Fig. 39. Besides providing a signature for hybrid states this channel plays an important role in our understanding of the decay mechanisms for $\eta_c \to VV$, as discussed in the next section.

Mark III has studied the radiative decay $J/\psi \to \gamma\omega\phi$ in the $\phi \to K^+K^-$ and $\omega \to \pi^+\pi^-\pi^0$ final state, using 5.8×10^6 J/ψ events.[82] The data sample is selected by 5C kinematic fits to the hypothesis $J/\psi \to K^+K^-\pi^+\pi^-\pi^0$. The detection efficiency is constant at $\sim 9\%$. The mass resolution increases from 12 MeV at 2.2 GeV to 17 MeV at the η_c.

The resulting $\omega\phi$ invariant mass spectrum is plotted in Fig. 40. A total of 32 ± 6 events is observed after subtracting the background, which is uniform at ~ 0.3 events per 20 MeV bin. This proves the existence of doubly-disconnected diagrams. The branching ratio for the entire spectrum is :

$$B(J/\psi \to \gamma\omega\phi) = (1.4 \pm 0.25 \pm 0.28) \times 10^{-4},$$

The $\omega\phi$ spectrum shows no significant structures. The spectrum in the 2.2 GeV region

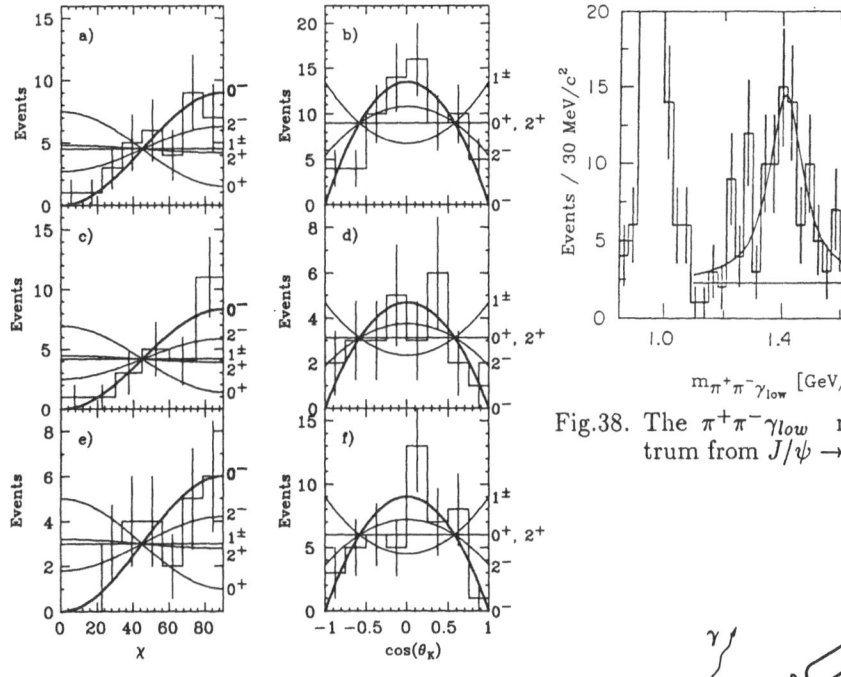

Fig.37. The χ and $\cos\theta_K$ distributions from $J/\psi \rightarrow \gamma\phi\phi$: a,b) the 2.14-2.3 GeV mass region in the $\gamma K^+K^-K^+K^-$ final state; c,d) the 2.14-2.3 GeV mass region in the $\gamma K^+K^-K^0_s K^0_L$ final state; e,f) the 2.92-3.04 GeV mass region in the $\gamma K^+K^-K^+K^-$ final state.

Fig.38. The $\pi^+\pi^-\gamma_{low}$ mass spectrum from $J/\psi \rightarrow \gamma\gamma\rho$.

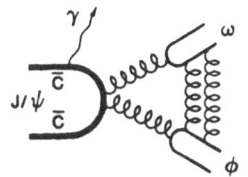

Fig.39. Lowest order diagram for $J/\psi \rightarrow \gamma\omega\phi$.

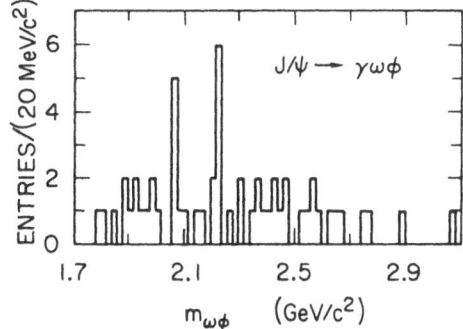

Fig.40. The $\omega\phi$ spectrum from the reaction $J/\psi \rightarrow \gamma\omega\phi$

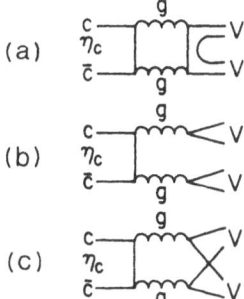

Fig.41. Decay mechanisms of $\eta_c \rightarrow VV$.

allows us to determine an upper limit for $\xi(2230)$ production:

$$B(J/\psi \to \gamma\xi(2230)) \cdot B(\xi(2230) \to \omega\phi) < 5.9 \times 10^{-5} \quad \text{at 90\% C.L.}$$

No events are observed in the η_c region, yielding an upper limit at 90% C.L. of:

$$B(J/\psi \to \gamma\eta_c) \cdot B(\eta_c \to \omega\phi) < 1.3 \times 10^{-5}.$$

6.6 η_c Decays

The η_c is the $c\bar{c}$ $^1S_0$ ground state lying ~ 115 MeV below the J/ψ. The first observation of the η_c claimed by DASP in $J/\psi \to 3\gamma$ was rather controversial.[83] According to theoretical expectations, the mass observed at 2.82 ± 0.04 GeV was too low and the product branching ratio of $B(J\psi \to \gamma\eta_c) \cdot B(\eta_c \to \gamma\gamma) = 1.4 \times 10^{-4}$ was too high. In addition, Crystal Ball [84] could not confirm the DASP results, setting a limit at 90% C.L. of $B(J/\psi \to \gamma\eta_c) \cdot B(\eta_c \to \gamma\gamma) < 1 \times 10^{-4}$. Later, the η_c was established at 2.98 GeV both by the Crystal Ball [85] in measurements of the inclusive photon spectrum and by Mark II [86] in studies of hadronic final states. The most precise mass measurement has been obtained by the gas jet experiment in $p\bar{p}$ collisions at the ISR,[87] yielding $m_{\eta_c} = 2982 \pm^{2.7}_{2.3}$ MeV. The η_c masses measured by Mark III average at 2982 MeV, while the values from DM2 are about 10 MeV lower. The spin-parity assignment of $J^P = 0^-$ for the η_c obtained by Mark III has been confirmed in several experiments.

At the J/ψ, η_c's are copiously produced through an M1 photon transition. The observed branching ratio for this process is $B(J/\psi \to \gamma\eta_c) = 1.27 \pm 0.36\%$.[6] The η_c decays dominantly through 2 gluons into hadrons, which is also an OZI-suppressed mode. The 2 photon rate is expected to be very small. Including first order QCD corrections, one obtains, with $\alpha_s(m_\psi^2) = 0.21$:[88]

$$\frac{\Gamma(\eta_c \to gg)}{\Gamma(\eta_c \to \gamma\gamma)} = \frac{2}{9}\frac{1}{e_c^4}\frac{\alpha_s^2(q^2)}{\alpha^2}[1 + \frac{14}{\pi}\alpha_s(q^2) + O(\alpha_s^2)] = 1.8 \times 10^3. \tag{14}$$

Recent measurements of $\eta_c \to \gamma\gamma$, which average at $\Gamma_{\gamma\gamma}(\eta_c) = 6 \pm 3$keV,[89] are consistent with the prediction of Eq. 14. A study of hadronic final states provides some insight into the coupling of the 2 gluon system to quark flavors. The branching ratios for several hadronic modes measured by Mark III are shown in Table 5. It seems that the η_c preferentially decays to $\eta\pi\pi$, $\eta'\pi\pi$ and modes containing strange quarks. The large decay rates to $\eta\pi\pi$ and $\eta'\pi\pi$ are not yet understood. One could explain them by assuming that the η_c contains a small admixture of the η and η'.[90] The fraction of light quarks in the η_c would then decay through an OZI-allowed mode, thus yielding such enhancements. On the other hand, if this hypothesis were true, one would expect to observe mixing effects in other decays, such as $J/\psi \to \gamma\eta$, $J/\psi \to \gamma\eta'$, $\eta \to \gamma\gamma$ and $\eta' \to \gamma\gamma$. The J/ψ measurements and the 2 photon results, however, indicate no such mixing. For a better understanding of the 3 and 4 body decay modes, higher statistics are necessary, and all 2 body intermediate states need to be separated.

Further clues about the η_c decay mechanisms and the coupling of the 2 gluon system to quarks can be obtained from a systematic study of $\eta_c \to VV$ decays. The lowest order diagrams for these decays are depicted in Fig. 41. For the decays into $\phi\phi$, $\rho\rho$ and $\omega\omega$, all three diagrams can, in principle, contribute. For $\eta_c \to K^*\bar{K}^*$ the diagrams in Figs. 41a and 41c are possible while for $\eta_c \to \omega\phi$, only the diagram in Fig. 41b is allowed. For an SU(3) singlet, one expects the reduced VV decay rates $\tilde{\Gamma}_{\phi\phi} : \frac{1}{2}\tilde{\Gamma}_{K^{*0}\bar{K}^{*0}} : \tilde{\Gamma}_{\rho^0\rho^0} : \tilde{\Gamma}_{\omega\omega}$ to be equal, while $\tilde{\Gamma}_{\omega\phi}$ is zero. After phase space corrections, the Mark III branching ratios from Table 5 yield:

$$\tilde{\Gamma}_{\phi\phi} : \frac{1}{2}\tilde{\Gamma}_{K^{*0}\bar{K}^{*0}} : \tilde{\Gamma}_{\rho^0\rho^0} : \tilde{\Gamma}_{\omega\omega} : \tilde{\Gamma}_{\omega\phi} = 1 : 0.26 \pm 0.14 :< 0.70 :< 0.47 :< 0.16,$$

indicating large SU(3) violations. The reduced rates increase with the number of

TABLE 5

Hadronic Decays of the η_c

	MARK III		MARK III	DM2
Mode	$B(\eta_c \to X)$ [%]	Mode	$B(\eta_c \to VV)$ [%]	$B(\eta_c \to VV)$ [%]
$\eta\pi\pi$	5.4 ± 1.3	$K^{*0}\bar{K}^{*0}$	0.47 ± 0.25	0.43 ± 0.19
$KK\pi$	4.8 ± 1.1	$\phi\phi$ a)	0.67 ± 0.16	$0.31 \pm 0.07 \pm 0.04$
$\eta'\pi\pi$	4.1 ± 1.3	$\rho^0\rho^0$	< 0.47 at 90% C.L.	$0.87 \pm 0.26 \pm 0.13$
$K^+K^-\pi^+\pi^-$	2.1 ± 0.3	$\omega\omega$	< 0.31 at 90% C.L.	< 0.7 at 90% C.L.
$K^{*0}\pi^+\pi^- +$ c.c.	2.0 ± 0.5	$\omega\phi$ b)	< 0.1 at 90% C.L.	—
$\pi^+\pi^-\pi^+\pi^-$	1.3 ± 0.5			

All Mark III data are based on 2.7×10^6 J/ψ's, except a) 4.9×10^6 J/ψ's and b) 5.8×10^6 J/ψ's. A 28% normalization uncertainty is not included in the errors. Limits on $\eta \to a_0\pi$, $a_2\pi$ and $f\eta$ are found in Ref. 78.

strange quarks, as predicted for a 0^- state. The DM2[75] results, which are also shown in Table 5, are totally different. While the K^*K^* rate is consistent with the Mark III value, the $\phi\phi$ rate is a factor 2 lower. The $\rho\rho$ rate measured by DM2 is a factor ~ 2 larger than the Mark III limit. For the $\omega\omega$ mode, DM2 reports a factor ~ 2 larger limit than Mark III. From these measurements DM2 determines a ratio of reduced rates of:

$$\tilde{\Gamma}_{\phi\phi} : \tfrac{1}{2}\tilde{\Gamma}_{K^{*0}\bar{K}^{*0}} : \tilde{\Gamma}_{\rho^0\rho^0} : \tilde{\Gamma}_{\omega\omega} = 1 : 0.6 \pm 0.4 : 1.73 \pm 0.72 :< 1.4 \,,$$

which, due to the large errors, is consistent with the SU(3) predictions. Before drawing a definite conclusion on $\eta_c \to VV$ decays, Mark III must cross check the results for $\eta_c \to \rho^0\rho^0$ and $\eta_c \to \omega\omega$ with the full data sample.

From the limit on $\eta_c \to \omega\phi$, one can conclude that the diagram in Fig. 41b is small. If it were dominant, one could relate the reduced rates to the decay rates for $V \to e^+e^-$, yielding:

$$\frac{\tilde{\Gamma}_{\omega\phi}}{\tilde{\Gamma}_{\phi\phi}} = \frac{|\psi_\omega(0)|^2}{|\psi_\phi(0)|^2} = 2\frac{m_\omega^2}{m_\phi^2}\frac{\Gamma(\omega \to e^+e^-)}{\Gamma(\phi \to e^+e^-)} = 0.6 \pm 0.05 \,,$$

which is clearly ruled out by the observed limit.

If the above SU(3) breaking pattern remains valid, one could explain the smaller $K^{*0}\bar{K}^{*0}$ rate in a simple way. If the gluons couple preferentially to strange quarks, the diagram in Fig. 41c is always reduced, because it contains one coupling to a d quark. The diagram in Fig. 41a, on the other hand, is also reduced if the 2 gluons couple to a $d\bar{d}$ instead of an $s\bar{s}$ pair. Since the two diagrams interfere positively, the $\phi\phi$ rate is enhanced relative to the $K^{*0}\bar{K}^{*0}$ rate.

6.7 Conclusion on $J/\psi \to \gamma VV$

The pseudoscalar states observed in $\gamma\rho\rho$, $\gamma\omega\omega$ and $\gamma\phi\phi$ are currently not understood. Since the branching ratios are rather large, these states may be indications of new physics. Another possibility is that they are just the second radial excitations of the pseudoscalar mesons. In order to obtain further clues about the nature of these states a comparison of the individual VV mass spectra is very helpful. For identical states found in different VV channels SU(3) relations could be used to check if they are singlets. However, for such a comparison more information is required. First, the channel $J/\psi \to \gamma K^* K^*$ needs to be analyzed. Second, the $\gamma\rho\rho$ and $\gamma\omega\omega$ modes have to be reanalyzed with the full statistics. Third, a multi-spin-parity analysis is mandatory for all VV modes.

7. SYSTEMATIC STUDY OF $J/\psi \to 1^{--}\eta\pi\pi$ and $J/\psi \to 1^{--}K\bar{K}\pi$

Mark III has studied the $\eta\pi\pi$ and $K\bar{K}\pi$ systems recoiling against a photon, an ω meson and a ϕ meson, using a sample of 5.8×10^6 J/ψ decays. While the radiative decays and the hadronic modes $\omega K\bar{K}\pi$, $\phi K\bar{K}\pi$ and $\omega\eta\pi\pi$ proceed through the regular annihilation diagram (Figs. 4), the decay $J/\psi \to \phi\eta\pi\pi$ can only proceed through doubly-disconnected diagrams (Fig. 10c).

The main interest of the $K\bar{K}\pi$ system lies in the study of the 1400 to 1500 MeV mass region,[91] where the $\iota/\eta(1440)$ and the $f_1(1420)$ have been observed. The $\iota/\eta(1440)$ resonance was first seen by Mark II in $J/\psi \to \gamma K_s K^{\pm}\pi^{\mp}$[92], who reported a mass and width of $m_\iota = 1440 \pm_{15}^{10}$ MeV and $\Gamma_i = 50 \pm_{20}^{30}$ MeV. It was mistaken for the $f_1(1420)$, since the parameters were similar. Crystal Ball [93] later performed an isobar analysis of the $\iota/\eta(1440)$ in the channel $J/\psi \to \gamma K^+ K^- \pi^0$. They found that the $\iota/\eta(1440)$ was a $J^{PC} = 0^{-+}$ state which decayed dominantly through quasi-two-body decays: $J/\psi \to \gamma\iota$, $\iota \to a_0(980)\pi$, $a_0(980) \to K^+ K^-$. To emphasize that this state was not the $f_1(1420)$, it was named $\iota(1440)$. This state has been considered as a candidate for the pseudoscalar glueball.

The $f_1(1420)$ meson, originally called the E meson, was first seen in $\pi^- p \to K\bar{K}\pi n$ in 1967,[94] and later confirmed by several other experiments. It has been classified as a $J^{PC} = 1^{++}$ $s\bar{s}$ state belonging to the axial vector nonet. Its mass and width are well measured: $m_E = 1422.3 \pm 2.1$ MeV and $\Gamma_E = 55.9 \pm 3.4$ MeV. However, several experiments have reported that the state at 1420 MeV is a 0^{-+} resonance, causing an experimental controversy about the spin-parity assignment of the $f_1(1420)$ meson. The situation is sketched in Table 6, which summarizes several results on $f_1(1420)$ production in $p\bar{p}$ annihilations, hadronic scattering and 2 photon collisions. On the other hand the $f_1(1420)$ has not been observed in the following processes: $\gamma\gamma \to f_1 \to K\bar{K}\pi$, $\gamma\gamma \to f_1 \to \eta\pi\pi$, $\gamma\gamma^* \to f_1 \to \eta\pi\pi$ and $K^- p \to f_1\Lambda \to K\bar{K}\pi\Lambda$.[95]

The results from 2 photon collisions yield conclusive evidence that a 1^{++} meson at 1.42 GeV exists [100]. A signal is found in $\gamma\gamma^*$, but not in $\gamma\gamma$ collisions. This is expected as Yang's theorem states that two massless photons cannot produce a $J = 1$ resonance. On the other hand, the BNL-E771[101] data, which show a 0^{-+} state, look convincing. Perhaps this structure corresponds to the same state seen at KEK in $\eta\pi\pi$ with a mass of 1390 MeV. [101] The BNL-E771 group also observes an enhancement at ~ 1400 MeV in the 1^{+-} channel. A similar 1^{+-} object is seen by LASS in $K^- p \to K\bar{K}\pi n$[102]. Under the assumption that all results are correct, this may indicate that there are three or more states in the $1400 - 1460$ MeV mass region. Above 1500 MeV, another 1^{++} resonance has been reported by two experiments. LASS[102] sees an ~ 80 MeV wide structure at ~ 1530 MeV, while the BNL-E771 group[103] observes a state at $m = 1512 \pm 10$ MeV with a width of $\Gamma = 40 \pm 20$ MeV.

The $\eta\pi\pi$ system is of great interest because it provides another access to the $f_1(1420)$, the $\iota/\eta(1440)$ and the $f_1(1285)$. Partial wave analyses have suggested that the $f_1(1420)$ and the $\iota/\eta(1440)$ have large quasi-two-body decays to $a_0(980)\pi$ with subsequent decays $a_0(980) \to K\bar{K}$. Since the $a_0(980)$ also decays to $\eta\pi$,[94] one expects to observe an $\iota/\eta(1440)$ signal in the $\eta\pi\pi$ mass spectrum produced in radiative J/ψ

TABLE 6

Production of $f_1(1420)$ Meson in Hadronic Scattering Experiments and $\gamma\gamma^*$ Collisions

Process	Mode	Mass	Width	J^{PC}	Reference
$\bar{p}p \to f_1\pi\pi,^{a)}$	$a_0\pi, K^*K$	1425 ± 7	80 ± 10	0^{-+}	Baillion[94]
$\pi^- p \to f_1 n,^{a)}$	K^*K	1426 ± 6	40 ± 15	1^{++}	Dionisi[96]
$\pi^+(p)p \to \pi^+(p)f_1 p,^{a)}$	K^*K	1425 ± 2	62 ± 5	1^{++}	Armstrong[97]
$\pi^- p \to f_1 n^{a)}$	$a_0\pi, K^*K$	1421 ± 3	70 ± 8	0^{-+}	Chung[98]
$p\bar{p} \to f_1 X^{a)}$	$a_0\pi$	1424 ± 3	60 ± 10	0^{-+}	Reeves[99]
$\gamma\gamma^* \to f_1^{a)}$	K^*K ?	1417 ± 13	35^{+47}_{-20}	1^{++}	TPC, MKII[100]
$\pi^- p \to X n^{b)}$	$a_0\pi$	1390 ± 10	42 ± 16	0^{-+}	KEK[101]

a) with observation of f_1 in $K_s K^{\pm}\pi^{\mp}$, b) with observation of X in $\eta\pi^+\pi^-$

decays, while the $f_1(1420)$ is expected to appear in the $\eta\pi\pi$ spectra from hadronic J/ψ decays.

The $f_1(1285)$, originally named the D-meson, has only been observed in hadron interactions[6] and is classified as the 1^{++} $(u\bar{u}+d\bar{d})$ partner of the $f_1(1420)$ in the axial vector nonet. Its parameters are well measured, yielding $m_D = 1283.4 \pm 1.3$ MeV and $\Gamma_D = 24.9 \pm 1.6$ MeV. The observed modes consist of $\eta\pi\pi$ (49%), of which 75% is due to $a_0(980)\pi$, 4π (40%) and $K\bar{K}\pi$ (11%).[6] Therefore, an $f_1(1285)$ signal is expected to appear in the $\eta\pi\pi$, 4π and $K\bar{K}\pi$ mass spectra produced in hadronic J/ψ decays. The current limit in radiative J/ψ decays is $B(J/\psi \to \gamma f_1(1285)) < 0.6\%$ at 90% C.L.

The $\eta(1275)$ has been observed in partial wave analyses of the $\eta\pi\pi$ system by 2 hadronic scattering experiments.[104] It is interpreted as the radial excitation of the η. Since the radiative decays $J/\psi \to \gamma\eta$ and $J/\psi \to \gamma\eta'$ have large branching ratios one also expects to observe the radial excitations of the η and η' in the $\gamma\eta\pi\pi$ channel.

7.1 Study of the $K\bar{K}\pi$ Mass Spectra

7.1.1 $J/\psi \to \gamma K\bar{K}\pi$

Using 5.8×10^6 J/ψ events, Mark III has analyzed the radiative decay $J/\psi \to \gamma K\bar{K}\pi$ in three independent modes: $\gamma K_s^0 K^{\pm}\pi^{\mp}$, $\gamma K^+ K^- \pi^0$ and $\gamma K_s^0 K_s^0 \pi^0$ with $K_s^0 \to \pi^+\pi^-$.[105] For a $K\bar{K}\pi$ isoscalar, the corresponding relative branching ratios are 4 : 2 : 1. The individual data samples are selected by performing 4C kinematic fits to the hypothesis $J/\psi \to \gamma\pi^+\pi^- K^{\pm}\pi^{\mp}$ and 5C kinematic fits to the hypotheses $J/\psi \to \gamma K^+ K^- \pi^0$ and $J/\psi \to \gamma\pi^+\pi^-\pi^+\pi^-\pi^0$. The reconstruction efficiencies in the $1.3 - 1.6$ GeV mass range are 21% for $\gamma K_s^0 K^{\pm}\pi^{\mp}$ and 18% for both $\gamma K^+ K^- \pi^0$ and $\gamma K_s^0 K_s^0 \pi^0$. The mass resolution is 7-9 MeV, depending on the decay mode.

The resulting $K\bar{K}\pi$ invariant mass spectra between 1 and 2 GeV are plotted in Figs. 42. The main feature is a prominent $\iota/\eta(1440)$ signal and a hint for an $f_1(1285)$. Table 7 summarizes the parameters of the $\iota/\eta(1440)$ obtained in individual decay modes by different experiments. The Mark III $K\bar{K}\pi$ invariant mass spectrum after combining the three individual modes is displayed in Fig. 43a. The $K\bar{K}\pi$ mass

223

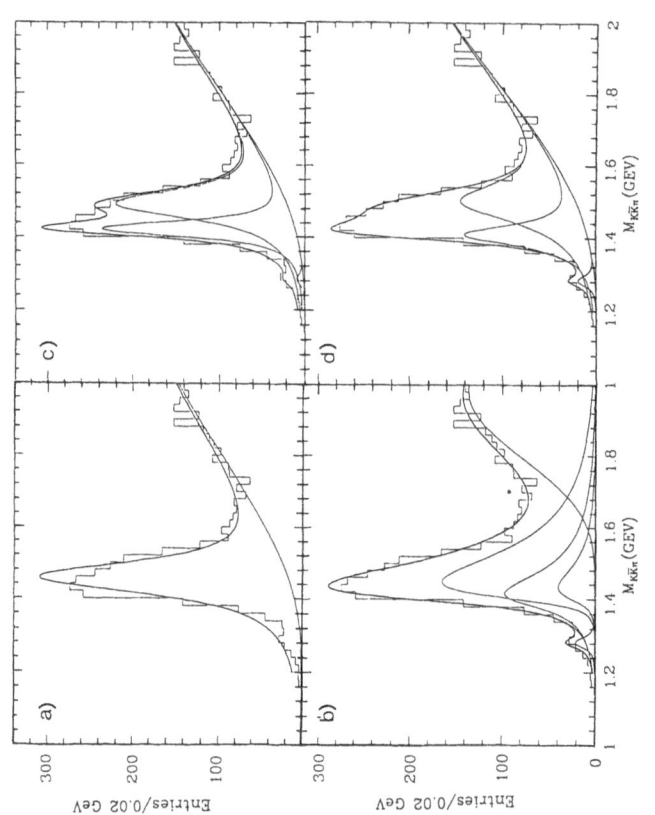

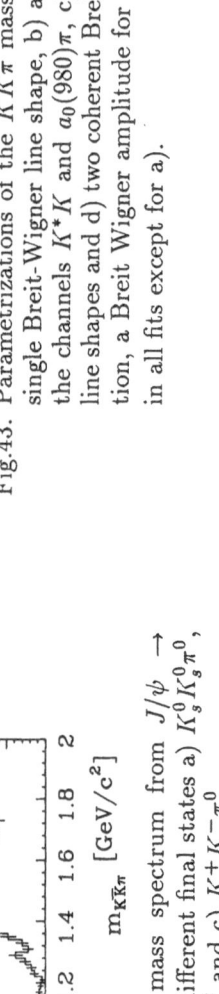

Fig. 43. Parametrizations of the $K\bar{K}\pi$ mass in the $\iota/\eta(1440)$ region; a) a single Breit-Wigner line shape, b) a single state decaying through the channels K^*K and $a_0(980)\pi$, c) two incoherent Breit-Wigner line shapes and d) two coherent Breit-Wigner line shapes. In addition, a Breit Wigner amplitude for the $f_1(1285)$ has been included in all fits except for a).

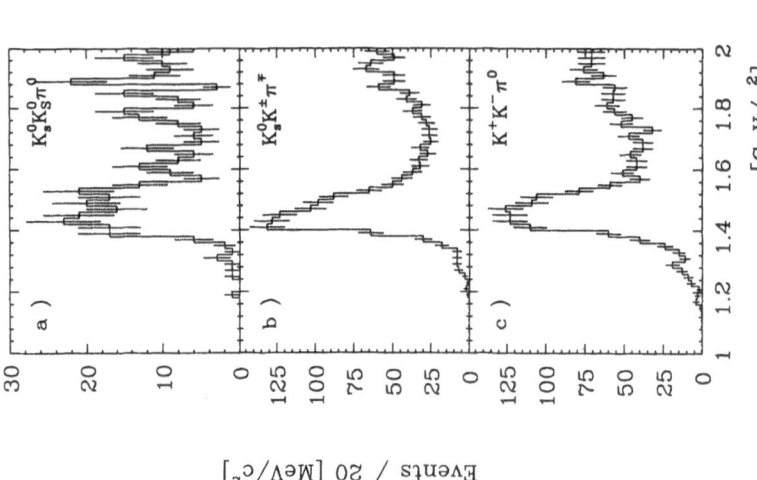

Fig. 42. The $K\bar{K}\pi$ mass spectrum from $J/\psi \rightarrow \gamma K\bar{K}\pi$ for different final states a) $K_s^0 K_s^0 \pi^0$, b) $K_s^0 K^{\pm}\pi^{\mp}$ and c) $K^+K^-\pi^0$.

TABLE 7

Production of the $\eta(1440)$ in Radiative J/ψ Decays

Mass	Γ	$B(J/\psi \rightarrow \gamma\iota) \times 10^{-3}$	Mode	Group
$1440.0 \pm 10.0 \pm 15.0$	$50 \pm 30 \pm 20$	4.3 ± 1.7	$K_s^0 K^\pm \pi^\mp$	MKII[91]
$1440.0 \pm 20.0 \pm 15.0$	$55 \pm 20 \pm 30$	4.3 ± 1.2	$K^+ K^- \pi^0$	CB[92]
1461 ± 5	101 ± 10	$4.9 \pm 0.2 \pm 0.8$	$K^+ K^- \pi^0$	MKIII[†] [105]
$1456 \pm 5 \pm 6$	$95 \pm 10 \pm 15$	$5.0 \pm 0.3 \pm 0.8$	$K_s K^\mp \pi^\pm$	MKIII[†]
1451 ± 3	96.6 ± 10	$4.1 \pm 0.5 \pm 6$	$K^+ K^- \pi^0$	DM2[61]
$1460 \pm 3 \pm 8$	$100 \pm 12 \pm 15$	$4.1 \pm 0.6 \pm 0.9$	$K_s^0 K^\mp \pi^\pm$	DM2

†) based on the initial 2.7×10^6 J/ψ decays

spectrum is fitted to a single relativistic Breit-Wigner amplitude plus the background function $(m-m_0)^\alpha \cdot e^{-\beta m}$, where m_0 is the $K\bar{K}\pi$ threshold and α, β are free parameters. The fit is shown overplotted in Fig. 43a. The resulting resonance parameters are listed in Table 8. From Fig. 43a, it is already evident that the fit is very poor. The fit probability is only $P(\chi^2) = 1.4 \times 10^{-3}$. This raises the question of whether the distorted line shape of the $\iota/\eta(1440)$ is caused by the opening of a new channel. In order to study this issue, Mark III has performed a Dalitz plot analysis in the $\iota/\eta(1440)$ mass region.[106] Monte Carlo samples were generated for $\iota \rightarrow K^*K$, $K^* \rightarrow K\pi$ and $\iota \rightarrow a_0(980)\pi$, $a_0 \rightarrow K\bar{K}$ where the ι was produced as a pseudoscalar in radiative J/ψ decays. Figures 44 display the Monte Carlo generated Dalitz plots of $M^2_{K_1\pi}$ versus $M^2_{K_2\pi}$ for the K^*K and $a_0\pi$ modes separately in the low $(1.35 - 1.45$ GeV) and high $(1.45-1.55$ GeV) $K\bar{K}\pi$ mass range, respectively. The Dalitz plots for K^*K show nodes in the middle of the K^* bands and constructive interference in the overlap region of the K^* bands. The corresponding plots for $a_0\pi$ display a broad bandlike structure parallel to the diagonal boundary, which is due to clustering in the low $K\bar{K}$ mass region. These different behaviors have a simple explanation. The decay $\iota \rightarrow K^*K$ proceeds through a P-wave due to parity conservation. Angular momentum conservation requires the helicity of the K^* to be zero. This leads to a $\cos^2\theta$ distribution of the K^* decay products, producing nodes and constructive interference at the K^* band crossing. The decay $\iota \rightarrow a_0\pi$ proceeds in an S-wave, which produces a uniform distribution in the region close to the boundary due to the small a_0 mass.

Figure 45 shows the Dalitz plot for the combined data sample in the 1.35-1.4 GeV and 1.45-1.55 GeV mass regions. The corresponding $K\pi$ and $K\bar{K}$ projections of the Dalitz plot are displayed in Fig. 46. Below 1450 MeV, the data cluster mainly near the diagonal boundary. The $K\bar{K}$ projection indicates a peak near the $a_0(980)$, while the $K\pi$ projection is inconclusive. A broad distribution is observed, peaking below the K^*. Though the $a_0\pi$ mode is clearly present, K^*K production cannot be ruled out. For most events, the $K\bar{K}\pi$ mass lies below the K^*K threshold. Therefore, events originating from $\iota \rightarrow K^*K$ can only contribute if the $K\pi$ mass lies in the lower tail, thus producing a distribution which peaks below the K^*. Above 1450 MeV, the Dalitz plot shows nodes in the middle of the K^* bands and a large cluster at the crossing of the K^* bands. The $K\bar{K}$ projections exhibit a clear K^* signal while no $a_0(980)$ is visible in the $K\bar{K}$ projection. This is the first evidence for $\iota \rightarrow K^*K$ in radiative J/ψ decays. Assuming that the $\iota/\eta(1440)$ is a single resonance, two decay modes contribute to the $K\bar{K}\pi$ spectrum. The decay to K^*K should be affected by threshold effects,

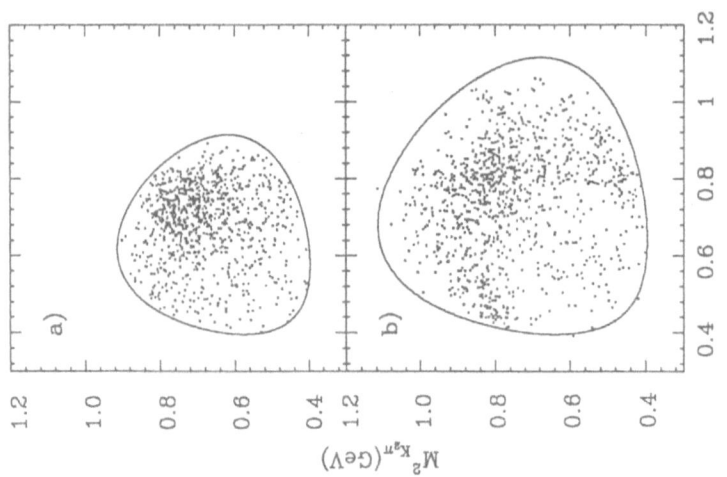

Fig.45. Dalitz plots of the $J/\psi \to K\bar{K}\pi$ data sample for a) the 1350-1450 MeV mass range and b) the 1450-1550 MeV mass range.

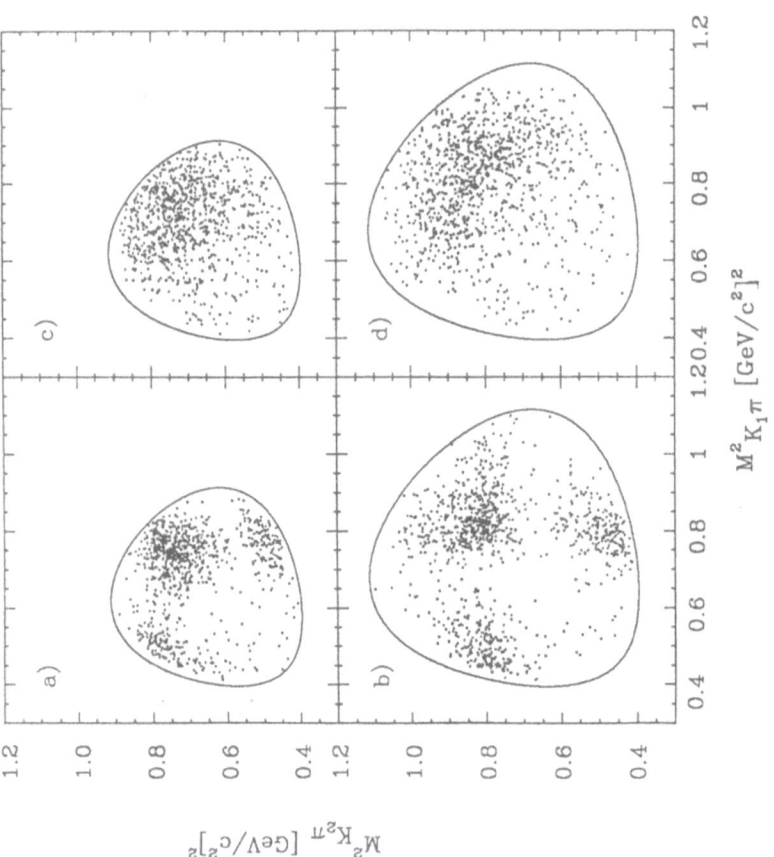

Fig.44. Dalitz plots from a Monte Carlo simulation of $J/\psi \to K\bar{K}\pi$ with intermediate K^*K and $a_0\pi$ productions; a) 1350-1450 MeV region for K^*K, b) 1450-1550 MeV region for K^*K, c) 1350-1450 MeV region for $a_0\pi$, d) 1450-1550 MeV region for $a_0\pi$.

since the ι peak lies only ~ 60 MeV above the K^*K threshold. In addition, phase space is limited, and there is a P-wave momentum barrier in this channel. Therefore, a coupled channel analysis of the $\iota/\eta(1440)$ has been performed to determine whether these effects can account for the distorted $K\bar{K}\pi$ mass spectrum.

In a coupled channel model, the line shape for a pseudoscalar $\iota/\eta(1440)$ produced in radiative J/ψ decays is given by:

$$\frac{dN}{dm} = C\frac{mk^3\Gamma_{K\bar{K}\pi}(m)}{|m_p^2 - m^2 + \Pi(m^2) - Re\Pi(m_p^2)|^2}, \tag{15}$$

where C is the normalization constant, m the $K\bar{K}\pi$ mass, m_p the pole position and k the photon energy. The function $\Pi(m^2)$ describes loop corrections to the bare propagator, which introduce a mass shift. The unitarity condition relates the imaginary part of $\Pi(m^2)$ to the total width by

$$Im\Pi(m^2) = -m_p\Gamma_\iota(m) = -m_p\{\Gamma_{K\bar{K}\pi}(m) + \Gamma_0\}, \tag{16}$$

where Γ_0 denotes a sum over partial widths accounting for all other ι decay modes such as $\iota \to \eta\pi\pi$. Assuming analyticity of $\Pi(m^2)$, the real and imaginary part of $\Pi(m^2)$ are connected by a dispersion relation

$$Re\Pi(s) = \frac{1}{\pi}\wp\int\frac{Im\Pi(s')}{s - s'}ds'. \tag{17}$$

Assuming that the $K\bar{K}\pi$ channel has contributions to both the K^*K and $a_0\pi$ modes, the $K\bar{K}\pi$ partial width is given by:

$$\Gamma_{K\bar{K}\pi}(m) = \frac{G}{m^3}\int\int |B_{K^*}(m_{K_1\pi})q_1\sqrt{D_1(q_1R_\iota)}\cos\theta_1$$
$$+ B_{K^*}(m_{K_2\pi})q_2\sqrt{D_1(q_2R_\iota)}\cos\theta_2 + \delta e^{i\phi}B_{a_0}(m_{K\bar{K}})|^2 dm_{K_1\pi}^2 dm_{K_2\pi}^2, \tag{18}$$

where G is a constant, $\cos\theta_i$ is the angle of the π in the $K_i\pi$ rest frame, q_i is the momentum of the $K_i\pi$ system in the $K\bar{K}\pi$ rest frame, and $D_1(q_iR_\iota) = \{1 + (q_iR_\iota)^2\}^{-1}$ is the $L = 1$ Blatt-Weiskopf penetration factor[107] with interaction radius R_ι. For the K^* two relativistic P-wave Breit-Wigner amplitudes $B_{K^*}(m_{K_i\pi})$ are used, corresponding to the two $K\pi$ combinations in the event, which interfere constructively. The $a_0(980)$ amplitude, $B_{a_0}(m_{K\bar{K}})$, is expressed by a Flatté parametrization[43]. The parameters δ and ϕ denote the magnitude and phase of the a_0 amplitude relative to the K^* amplitudes.

The $K\bar{K}\pi$ mass distribution for the combined event sample is fitted with this coupled channel line shape by varying m_p, $\Gamma_\iota(m_p)$, δ, ϕ, the number of signal events and the background shape. In addition, an amplitude for the $f_1(1285)$ with fixed mass and width is included. The fits are performed for different ratios $r = \Gamma_{K\bar{K}\pi}(m_p)/\Gamma_\iota(m_p)$ and values of R_ι. Acceptable fits result for values of $0.6 \leq r \leq 1$ and $R_\iota \geq 0.4$ fm. The best fit, obtained for $r = 0.8$ and $R_\iota = 2fm$, is shown in Fig. 43b. The corresponding resonance parameters are listed in Table 8. The fit probability is 47%. The rate of the $a_0\pi$ channel relative to the K^*K channel is approximately $1 : 2$. The phase is $\phi = -0.8^{+1.0}_{-1.7}$. As a cross check, the individual $K\bar{K}\pi$ decay modes are also fitted by the same technique. The resulting product branching ratios are consistent with the hypothesis of an isoscalar ι.

An alternative explanation for the distorted $\iota/\eta(1440)$ line shape is the presence of more than one resonance. Therefore, the $K\bar{K}\pi$ mass spectrum is also fitted both to two incoherent Breit-Wigner amplitudes and to two coherent Breit-Wigner resonances, as shown in Figs. 43c,d. A non-interfering Breit-Wigner amplitude for the $f_1(1285)$ is also included in these fits. In both cases, good fits are obtained with probabilities of 15% and 58%, respectively. Table 8 summarizes the resulting resonance parameters. In the incoherent fit, the narrower state is consistent with the $f_1(1420)$, while the

227

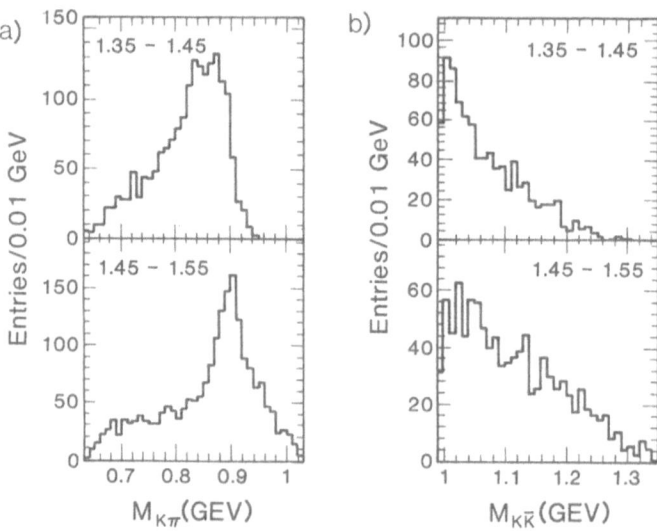

Fig.46. a) The $K\pi$ mass projections from the Dalitz plot and b) the $K\bar{K}$ mass projections from the Dalitz plot shown for the 1350-1450 MeV and 1450-1550 MeV $K\bar{K}\pi$ mass regions.

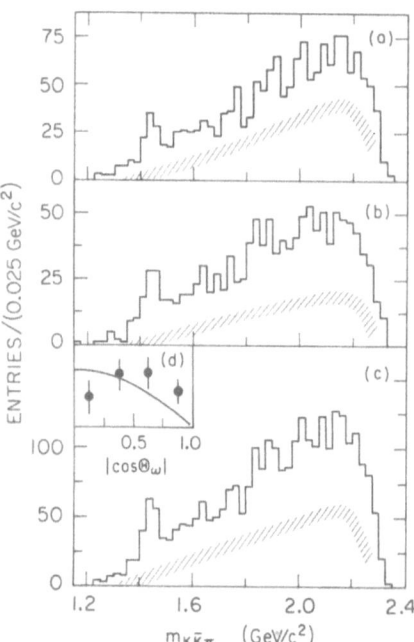

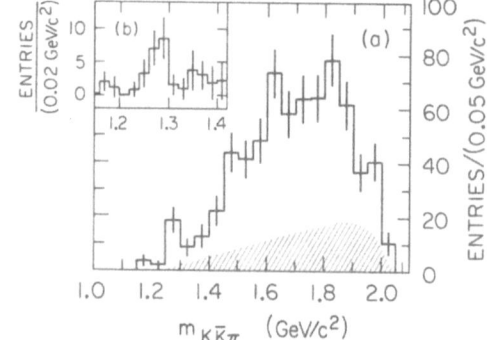

Fig.48. Summed $K^+K^-\pi^0$ and $K^0_s K^\pm\pi^\mp$ invariant mass spectra for $J/\psi \rightarrow \phi\, K\bar{K}\pi$, b) The 1200 MeV mass region after selecting $m_{K\bar{K}} < 1.15$ GeV. the shaded bands shows the background estimate.

Fig.47. a) $K^0_s K^\pm\pi^\mp$ invariant mass spectrum from $J/\psi \rightarrow \omega K^0_s K^\pm\pi^\mp$, b) $K^+K^-\pi^0$ invariant mass spectrum from $J/\psi \rightarrow \omega K^+K^-\pi^0$, c) combined spectrum, d) distribution of $\cos\Theta_\omega$ with prediction for $J^P = 0^-$ (solid curve).

228

TABLE 8

Fit Results for Different Hypotheses of the

$\iota/\eta(1440)$ Line Shape and $f_1(1285)$ Branching Ratios

Fit	m_ι (MeV)	Γ_ι (MeV)	$B(J/\psi \to \gamma\iota)$ $\times B(\iota \to K\bar{K}\pi)$	$B(J/\psi \to \gamma f_1(1285))$	$P(\chi^2)$ (%)
1 B-W	1457 ± 2	104 ± 5	$(5.1 \pm 1.2) \times 10^{-3}$	$< 2.8 \times 10^{-5}$	$\sim 10^{-13}$
C.Ch.	1454 ± 3	160 ± 11	$(6.3 \pm 1.4) \times 10^{-3}$	$11 \pm 2 \times 1 \times 10^{-5}$	47
3 Incoh.	1422 ± 3	43 ± 7	$(1.8 \pm 0.5) \times 10^{-3}$	$4.8 \pm 2.4 \times 10^{-5}$	15
BW	1490 ± 5	95 ± 5	$(3.2 \pm 0.8) \times 10^{-3}$		
2 coh. BW +	1409 ± 5	69 ± 11	$(1.5 \pm 0.4) \times 10^{-3}$	$8.5 \pm 3.1 \times 10^{-5}$	58
1 incoh. BW	1499 ± 9	138 ± 25	$(2.7 \pm 0.8) \times 10^{-3}$		

C.Ch. stands for coupled channel analysis, BW for Breit-Wigner line shapes.

All data are based on 5.8×10^6 J/ψ decays.

broader state is moved up to 1490 MeV. The widths are consistent with the $f_1(1420$ and $\iota/\eta(1440)$. In the coherent case, the narrower state lies lower than the $f_1(1420)$, while the broader state is moved to 1500 MeV. Both widths increase by 40-50%. A possible interpretation of these results is that either the $\iota/\eta(1440)$ and the $f_1(1420)$ or the $\iota/\eta(1440)$ and the $X(1390)$ are observed.

In order to obtain further clues about the $f_1(1420) - \iota/\eta(1440)$ mass region, it is important to perform a multi-channel spin-parity analysis for different $K\bar{K}\pi$ mass intervals. Previous spin-parity analyses by Mark III and DM2 have been performed on the entire resonance, studying directly the decays to three pseudoscalars and thus were independent of quasi-two-body decays. Both groups have measured $J^{PC} = 0^{-+}$ for the entire resonance.

7.1.2 $J/\psi \to \omega K\bar{K}\pi$

The $K\bar{K}\pi$ final state recoiling against an ω meson is studied in the two reactions: $J/\psi \to \omega K_s^0 K^\pm \pi^\mp$ and $J/\psi \to \omega K^+K^-\pi^0$, with $\omega \to \pi^+\pi^-\pi^0$ and $K_s^0 \to \pi^+\pi^-$.[108] The data samples are selected by performing 5C kinematic fits to the hypothesis $J/\psi \to K^\pm \pi^\mp \pi^+\pi^-\pi^+\pi^-\pi^0$ and 6C kinematic fits to the hypothesis $J/\psi \to K^+K^-\pi^+\pi^-\pi^0\pi^0$. Figures 47a,b show the $K_s^0 K^\pm \pi^\mp$ and $K^+K^-\pi^0$ invariant mass spectra. The combined $K\bar{K}\pi$ mass spectrum is depicted in Fig. 47c. The shaded band represents background estimates obtained from sidebands above and below the ω signal. A resonance near 1.4 GeV is observed. Fitting the low $K\bar{K}\pi$ mass region by a single Breit-Wigner line shape plus a quadratic background function yields:

$$m = 1442 \pm 5 \, ^{+10}_{-17} \text{ MeV}, \qquad \Gamma = 40 \, ^{+17}_{-13} \pm 5 \text{ MeV}.$$

The systematic errors include uncertainties in the absolute mass scale (10 MeV) and account for possible mass shifts due to K^*K or $a_0\pi$ substructures. A fit which includes K^*K production yields a 7 MeV mass decrease due to the rapidly rising K^*K phase space above threshold. The product branching ratio for $X(1440)$ production is

$$B(J/\psi \to \omega X(1440)) \cdot B(X(1440) \to K\bar{K}\pi) = (6.8 \, ^{+1.9}_{-1.6} \pm 1.7) \times 10^{-4}.$$

The $f_1(1285)$ is not seen in the $K\bar{K}\pi$ mass spectrum, yielding an upper limit of:

$$B(J/\psi \to \omega f_1(1285)) \cdot B(f_1(1285) \to K\bar{K}\pi) < 1.1 \times 10^{-4} \text{ at 90\% C.L.}.$$

Angular distributions have been analyzed to determine the spin-parity of the $X(1440)$. Figure 47d shows the measured distribution of the normal to the ω-decay plane in the ω helicity frame θ_ω, corrected for acceptance effects. The solid line represents the prediction for $J^P = 0^-$; a uniform distribution is expected for $J^P = 1^+$. The data are inconsistent with the pseudoscalar prediction; a fit yields only a probability of 6%. In addition, a multi-channel spin-parity analysis has been performed on the $K\bar{K}\pi$ mass system consisting of a $J^P = 0^-$ intermediate state decaying to a $a_0(980)\pi$, a $J^P = 1^+$ state decaying to K^*K and an isotropic distribution. The axial vector hypothesis is favored by the fit. Hence, the $X(1440)$ very likely corresponds to the $f_1(1420)$. In addition, the width of the $X(1440)$ is not consistent with the $\iota/\eta(1440)$. The limits at 90% C.L. yield $24 < \Gamma < 84$ MeV. The total branching ratios in the $\omega K^0_s K^\pm \pi^\mp$ channel is:

$$B(J/\psi \to \omega K_s K^\pm \pi^\mp) = (2.95 \pm 0.14 \pm 0.7) \times 10^{-3}.$$

Requiring intermediate K^* production, the total branching ratio amounts to:

$$B(J/\psi \to \omega K^* \bar{K} + c.c.) = (5.3 \pm 1.4 \pm 1.4) \times 10^{-3}.$$

7.1.3 $J/\psi \to \phi K\bar{K}\pi$

To study the $K\bar{K}\pi$ mass spectrum recoiling against a ϕ meson, the following modes are examined: $J/\psi \to \phi K^+ K^- \pi^0$ with $\phi \to K^+ K^-$, $J/\psi \to \phi K^0_s K^\pm \pi^\mp$ with $\phi \to K^+ K^-$ and $J/\psi \to \phi K^0_s K^\pm \pi^\mp$ with $\phi \to K^0_s K^0_L$.[108] The data samples for the first mode are selected by 5C kinematic fits to the hypothesis $J/\psi \to K^+ K^- K^+ K^- \pi^0$. For the second mode, two samples are selected to gain statistics. Events with 6 observed charged tracks are subjected to 4C kinematic fits to the hypothesis $J/\psi \to K^+ K^- \pi^+ \pi^- K^\pm \pi^\mp$, while for events with 5 observed charged tracks, 1C kinematic fits are applied to the hypothesis $J/\psi \to \pi^+ \pi^- \pi^\mp K^\pm K^\pm (K^\mp_{missing})$. In case of the third mode, the sample is selected by using 1C kinematic fits with the assumption that the K^0_L is missing. Figure 48 shows the combined $K\bar{K}\pi$ invariant mass spectrum recoiling against a ϕ. The shaded region represents the background estimate from a sideband above the ϕ mass. Above 1.3 GeV, the spectrum is consistent with phase space for $J/\psi \to \phi K^*K$. A small signal near the $f_1(1285)$ is visible. Requiring that the $K\bar{K}$ invariant mass of the $K\bar{K}\pi$ system lies below 1.15 GeV enhances the signal, as shown in Fig. 48b.

A fit of the $K\bar{K}\pi$ mass spectrum to a Breit-Wigner line shape plus the $J/\psi \to \phi K^*K$ phase space distribution yields the resonance parameters:

$$m_{f_1} = 1279 \pm 6 \pm 10 \text{ MeV}, \qquad \Gamma_{f_1} = 14 \,{}^{+20}_{-14} \pm 10 \text{ MeV},$$

$$B(J/\psi \to \phi f_1(1285)) \cdot B(f_1(1285) \to K\bar{K}\pi) = (0.6 \pm 0.2 \pm 0.1) \times 10^{-4},$$

which are consistent with the $f_1(1285)$. The $K\bar{K}\pi$ mass spectrum shows no signals at the $f_1(1420)$ or the $\iota/\eta(1440)$, translating into upper limits at 90%C.L. of:

$$B(J/\psi \to \phi E) \cdot B(E \to K\bar{K}\pi) < 1.2 \times 10^{-4} \qquad \text{for } \Gamma = 40 - 60 \text{ MeV}$$
$$B(J/\psi \to \phi \iota) \cdot B(\iota \to K\bar{K}\pi) < 2.1 \times 10^{-4} \qquad \text{for } \Gamma = 92 \text{ MeV}.$$

The total branching ratio in the $\phi K^0_s K^\pm \pi^\mp$ channel is:

$$B(J/\psi \to \phi K_s K^\pm \pi^\mp) = (7.0 \pm 0.6 \pm 1.0) \times 10^{-4}.$$

Requiring intermediate K^* production in the $\phi K \bar{K} \pi$ mode yields a branching ratio of:

$$B(J/\psi \rightarrow \phi K^* \bar{K} + c.c.) = (2.0 \pm 0.3 \pm 0.3) \times 10^{-3}.$$

The hadronic decays $J/\psi \rightarrow \omega K \bar{K} \pi$ and $J/\psi \rightarrow \phi K \bar{K} \pi$ are dominated by phase space distributed $\omega K^* K$ and $\phi K^* K$ intermediate states. The ratio $B(J/\psi \rightarrow \omega K^* K + c.c.)/B(J/\psi \rightarrow \phi K^* K + c.c.) = 2.7 \pm 1.0$ is not consistent with the expectation from $SU(3)$ symmetry, which is 0.93 for an S-wave phase space.[109]

7.2 Study of the $\eta \pi \pi$ Mass Spectra

7.2.1 $J/\psi \rightarrow \gamma \eta \pi \pi$

The radiative decay $J/\psi \rightarrow \gamma \eta \pi \pi$ is observed in two final states resulting from $\eta \rightarrow \gamma \gamma$ and $\eta \rightarrow \pi^+ \pi^- \pi^0$, with $\pi^0 \rightarrow \gamma \gamma$. The data are selected by 5C and 6C kinematic fits which include the η and π^0 mass constraints. Background from the prominent hadronic decay $J/\psi \rightarrow \rho \pi \rightarrow \pi^+ \pi^- \pi^0$, which feeds into the $\gamma \pi^+ \pi^- (\eta \rightarrow \gamma \gamma)$ mode because of spurious neutral showers, is reduced by rejecting events that contain a π^0. Background feeding into the $\gamma (\eta \rightarrow \pi^+ \pi^- \pi^0) \pi^+ \pi^-$ mode from $\gamma 6\pi$ and 7π decays of the J/ψ is suppressed by allowing no additional showers with energies greater 20 MeV. The detection efficiencies for both modes in the 1.2-1.5 GeV mass region determined from a phase space Monte Carlo simulation are $\epsilon(\eta \rightarrow \gamma \gamma) = 40\%$ and $\epsilon(\eta \rightarrow \pi^+ \pi^- \pi^0) = 15\%$. Further details are given in Ref. 110.

Figure 49 displays $\eta \pi^+ \pi^-$ invariant mass spectrum from the $\gamma (\eta \rightarrow \gamma \gamma) \pi^+ \pi^-$ channel, showing a prominent η' signal. The branching ratios for $J/\psi \rightarrow \gamma \eta'$, which are listed for both modes in Table 9, are consistent with previous measurements.[106] In order to examine the mass spectrum for other structures Fig. 50a shows $m_{\eta \pi^+ \pi^-}$ for both modes combined in the 1-3 GeV mass region. Besides an η_c signal, two peaks around 1.28 and 1.39 GeV and broader structures in the 1.5-2 GeV mass region are evident. The $\eta \pi$ submasses indicate copious $a_0(980)$ production, as shown in Fig. 51. In the $\eta \rightarrow \gamma \gamma$ final state, for example, a fit including a Breit-Wigner line shape plus a polynomial background term yields:

$$m_{a_0} = 987 \pm 2 \text{ MeV}, \qquad \Gamma_{a_0} = 54 \pm 9 \text{ MeV},$$

which is consistent with the nominal values of the $a_0(980)$. In the other final state similar results are obtained. Requiring intermediate $a_0(980)$ production in the $\eta \pi \pi$ sample by the criterion $|m_{\eta \pi} - m_\delta| < 50$ MeV produces the spectrum in Fig. 50b. The peaks at 1.28 GeV and 1.39 GeV are now clearly visible, whereas the η_c signal disappeared and the structures in the 1.5-2 GeV mass region are reduced but remain significant. The $\eta \pi \pi$ mass spectra in Figs. 50a,b are fitted in the 1.1-1.54 GeV mass region to 2 incoherent Breit-Wigner curves plus a linear background term. Since the signal at 1.28 GeV is consistent with the $f_1(1285)$, its mass and width have been fixed in the fit at the nominal values of the $f_1(1285)$. For the higher mass state all parameters are kept free. The fit results obtained from both modes for both states are listed in Table 9.

If the lower mass state is the $f_1(1285)$, the fit yields $B(J/\psi \rightarrow \gamma f_1(1285)) = (1.2 \pm 0.4) \times 10^{-3} \times B(a_0(980) \rightarrow \eta \pi)$. For an estimated branching ratio of $B(a_0(980) \rightarrow \eta \pi) = 0.67$,[6] one obtains $B(J/\psi \rightarrow \gamma f_1(1285)) = (0.18 \pm 0.06)\%$, which is consistent with the previous limit of 0.6%. A contribution from the $\eta(1275)$ cannot be excluded without a multi-channel spin-parity analysis. Because of this uncertainty, the signal will be referred to as "$f_1(1285)$", in quotes. The higher mass state is probably not the $f_1(1420)$, since it lies several s.d. below the $f_1(1420)$ mass. This observation is confirmed by a DM2 result also shown in Table 9. It is more likely that this object corresponds to the 0^{-+} state observed at KEK and will, therefore, be called the $X(1390)$. Perhaps, this state is the radial excitation of the η'. The $\iota/\eta(1440)$ is not

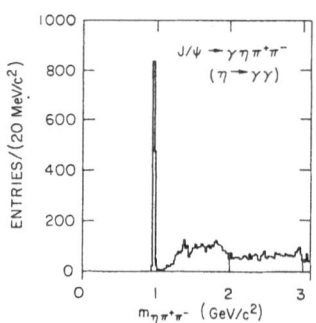

Fig.49. The $\eta\pi^+\pi^-$ mass spectrum from $J/\psi \to \gamma\eta\pi^+\pi^-$ with $\eta \to \gamma\gamma$.

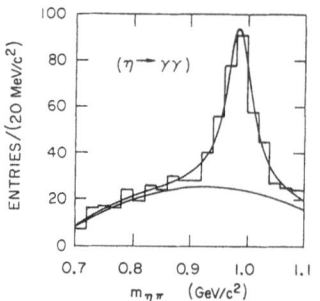

Fig.51. The $\eta\pi$ submass indicating $a_0(980)$ production.

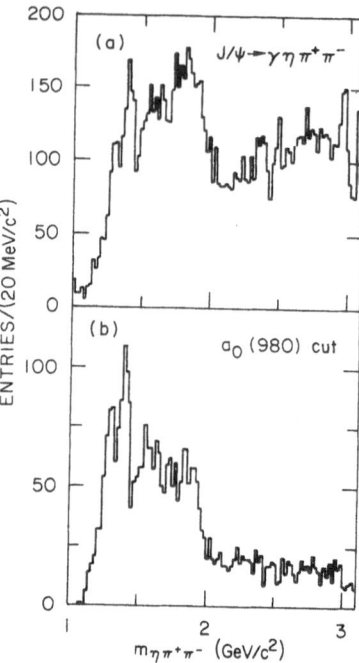

Fig.50. a) The $\eta\pi^+\pi^-$ mass spectrum from $J/\psi \to \gamma\eta\pi^+\pi^-$ for the $\eta \to \gamma\gamma$ and $\eta \to \pi^+\pi^-\pi^0$ final states combined, b) The $\eta\pi^+\pi^-$ mass spectrum with intermediate $a_0(980)$ production.

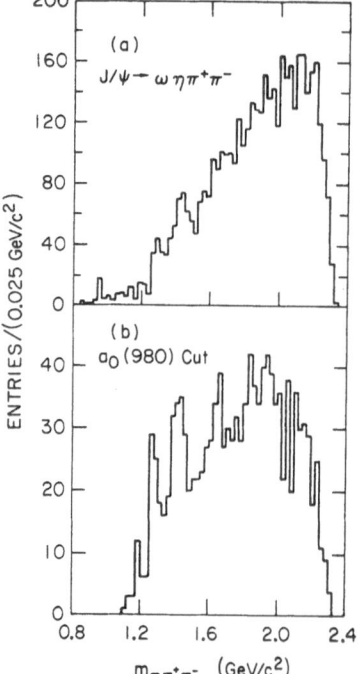

Fig.52. a) The $\eta\pi^+\pi^-$ mass spectrum for $\eta \to \gamma\gamma$ from $J/\psi \to \omega\eta\pi^+\pi^-$, b) the $\eta\pi^+\pi^-$ mass spectrum with intermediate $a_0(980)$ production.

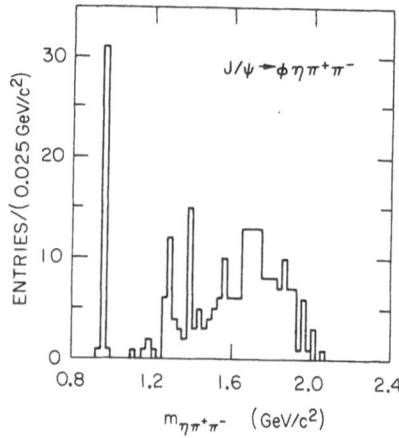

Fig.53. The $\eta\pi^+\pi^-$ invariant mass spectrum from $J/\psi \to \phi\eta\pi^+\pi^-$.

232

TABLE 9

Resonance Parameters and Branching Ratios

for $J/\psi \rightarrow \{\gamma, \omega, \phi\}\eta\pi\pi$ Measured by Mark III

Object	Observed in $J/\psi \rightarrow$	Mass [MeV/c^2]	Width [MeV/c^2]	BR [$\times 10^{-4}$]
η'	$\gamma\eta\pi^+\pi^-$ (a)	958 (fixed)	0.24 (fixed)	$43 \pm 2 \pm 11$
	$\gamma\eta\pi^+\pi^-$ (b)	958 (fixed)	0.24 (fixed)	$33 \pm 2 \pm 7$
	$\omega\eta\pi^+\pi^-$ (a)	958 (fixed)	0.24 (fixed)	$1.7 \pm 0.3 \pm 0.2$
	$\phi\eta\pi^+\pi^-$ (a)	958 (fixed)	0.24 (fixed)	$2.9 \pm 0.5 \pm 0.4$
$f_1(1285)$?	$\gamma\eta\pi^+\pi^-$ (a)*	1283 (fixed)	26 (fixed)	$4.1 \pm 1.2 \pm 0.3$
$f_1(1285)$?	$\gamma\eta\pi^+\pi^-$ (b)*	1283 (fixed)	26 (fixed)	$4.8 \pm 1.7 \pm 0.4$
$f_1(1285)$	$\omega\eta\pi^+\pi^-$ (a)	$1283 \pm 6 \pm 10$	$14^{+19}_{-14} \pm 10$	$4.3 \pm 1.2 \pm 1.3$
$f_1(1285)$?	$\phi\eta\pi^+\pi^-$ (a)	$1283 \pm 6 \pm 10$	$24^{+20}_{-14} \pm 10$	$1.6^{+0.6}_{-0.5} \pm 0.4$
$f_1(1285)$	$\phi\eta\pi^+\pi^-$ (a)	—	—	1.83 ± 0.6 †
$X(1390)$	$\gamma\eta\pi^+\pi^-$ (a)*	1382 ± 6	69 ± 23	$7.8 \pm 1.8 \pm 0.7$
	$\gamma\eta\pi^+\pi^-$ (b)*	1400 ± 7	62 ± 16	$7.8 \pm 2.7 \pm 0.7$
	$\gamma\eta\pi^+\pi^-$ (a)	1391.5 ± 3	52 ± 9	$6.2 \pm 0.5 \pm 2.0$ †
$f_1(1420)$	$\omega\eta\pi^+\pi^-$ (a)	$1421 \pm 8 \pm 10$	$45^{+32}_{-23} \pm 15$	$9.2 \pm 2.4 \pm 2.8$

(a) with $\eta \rightarrow \gamma\gamma$, (b) with $\eta \rightarrow \pi^+\pi^-\pi^0$, *) through $a_0(980)\pi$, †) DM2 results.
For the η', BR refers to $B(J/\psi \rightarrow \{\gamma, \omega, \phi\}\eta')$; for the other states, BR refers to $B(J/\psi \rightarrow \{\gamma, \omega, \phi\}X) \cdot B(X \rightarrow \eta\pi\pi)$

observed in the $\eta\pi\pi$ invariant mass spectrum. At the $\iota/\eta(1440)$ mass, a sharp dip is seen. One interpretation[112] is that this dip is caused by destructive interference between the $\iota/\eta(1440)$ or the $f_1(1420)$ with other resonances. The nature of the structures above 1.5 GeV is presently unknown. For a complete understanding of the different features observed in the $\gamma\eta\pi\pi$ channel, it is necessary to perform a partial wave analysis.

7.2.2 $J/\psi \rightarrow \omega\eta\pi\pi$

The hadronic decay $J/\psi \rightarrow \omega\eta\pi^+\pi^-$ is studied in final states with $\omega \rightarrow \pi^+\pi^-\pi^0$ and $\eta \rightarrow \gamma\gamma$.[40] The data are selected by performing 6C kinematic fits to the hypotheses $J/\psi \rightarrow \pi^+\pi^-\pi^0\eta\pi^+\pi^-$. The dominant background is from $J/\psi \rightarrow 7\pi$. Though a large amount of this background is eliminated by rejecting events with 2 π^0's or events which contain an additional photon with an energy greater than 60 MeV, the signal-to-background ratio is still ~ 0.7. Figure 52a displays the resulting $\eta\pi^+\pi^-$ invariant mass spectrum. Besides an η' signal, peaks at the $f_1(1285)$, and the $f_1(1420)$ resonances are visible. The higher mass region is not very conclusive. It is dominated by the 7π background which exhibits a typical phase space distribution. To extract the resonance parameters, the $\eta\pi\pi$ mass spectrum is fitted in the 1.-1.5 GeV region to 2 Breit-Wigner amplitudes plus a quadratic background curve. The resonance parameters are listed in Table 9. The resulting masses and widths are consistent with the parameters of the $f_1(1285)$ and $f_1(1420)$. The detection efficiency above 1.2 GeV is $\sim 11\%$.

A study of the $\eta\pi$ subsystems again indicates copious $a_0(980)$ production. In fact, the decay $f_1(1285) \rightarrow \eta\pi\pi$ proceeds almost entirely through an intermediate $a_0(980)$ state. For the $f_1(1420)$, the quasi-two-body decay into $a_0(980)\pi$ is dominant, while in the higher mass region the $\eta\pi$ submass shows no evidence for an $a_0(980)$ resonance. Requiring intermediate a_0 production by the condition $|m_{\eta\pi}-m_{ao}| < 50$ MeV produces the $\eta\pi\pi$ mass spectrum in Fig. 52b. Both the $f_1(1285)$ and the $f_1(1420)$ structures stand out more clearly. New structures may also be present in the higher mass region. However, only a partial wave analysis can decide this matter conclusively.

7.2.3 $J/\psi \rightarrow \phi\eta\pi\pi$

The decay $J/\psi \rightarrow \phi\eta\pi^+\pi^-$ is observed in the $\eta \rightarrow \gamma\gamma$ and $\phi \rightarrow K^+K^-$ final state.[40] Candidates are selected by performing 4C kinematic fits to the hypothesis $J/\psi \rightarrow K^+K^-\pi^+\pi^-\gamma\gamma$ and the mass condition $|m_{\gamma\gamma} - m_\eta| \leq 55$MeV. The detection efficiency for this mode is $\sim 20\%$. Figure 53 shows the resulting $\eta\pi^+\pi^-$ invariant mass spectrum. Besides a prominent η' signal, a peak at the $f_1(1285)$ is observed. The resonance parameters obtained from a Breit-Wigner parametrization plus a quadratic background shape are listed in Table 9. The mass and width are consistent with the $f_1(1285)$. A study of the $\eta\pi$ subsystem indicates that this resonance is entirely consistent with the decay through an intermediate $a_0(980)$ state. The $\eta\pi\pi$ spectrum shows no evidence for the $f_1(1420)$ or the $\iota/\eta(1440)$. A high single bin fluctuation is observed at 1390 MeV. The higher mass region exhibits no significant structures.

7.3 Comparison of $K\bar{K}\pi$ and $\eta\pi\pi$ Invariant Mass Spectra

Figures 54 show a comparison of the $K\bar{K}\pi$ invariant mass spectra recoiling against a photon, an ω, and a ϕ meson. The corresponding comparison of the $\eta\pi\pi$ mass spectra is presented in Figs. 55. The mass distributions from the $\gamma\eta\pi\pi$ and $\omega\eta\pi\pi$ channels include the requirement of intermediate $a_0(980)$ production.

The $f_1(1285)$-like structure appears in all $\eta\pi\pi$ recoil spectra. A much weaker "$f_1(1285)$" signal is also found in the $\gamma K\bar{K}\pi$ and $\phi K\bar{K}\pi$ channels. If the state seen in these processes is always the $f_1(1285)$, as suggested by the narrow width, the observed production mechanisms are rather puzzling. Classified as a pure $(u\bar{u} + d\bar{d})$ state, the $f_1(1285)$ is not expected to appear in $\phi\eta\pi\pi$ and $\phi K\bar{K}\pi$, but significant signals are found in both channels. A comparison of the branching ratios in the $\omega\eta\pi\pi$ and $\phi\eta\pi\pi$ channels yields:

$$B(J/\psi \rightarrow \omega\text{"}f_1(1285)\text{"})/B(J/\psi \rightarrow \phi\text{"}f_1(1285)\text{"}) = 2.7 \pm 1.6,$$

which is consistent with the prediction for an SU(3) singlet. This suggests three possibilities: 1) the $f_1(1285)$ contains $s\bar{s}$ quarks, 2) contributions from doubly-disconnected diagrams are large, or 3) a different state, such as the $\eta(1275)$, is observed in $\phi\eta\pi\pi$ and $\phi K\bar{K}\pi$, though the $\eta(1275)$ has a factor ~ 3 larger width than observed. In the radiative decays, the $f_1(1285)$ assignment is again preferred, which would imply that the gluons in the two gluon system have a non-zero mass, because a spin $J = 1$ state cannot be formed from massless gluons. On the other hand, the $\eta(1275)$ is also expected to appear in $\gamma\eta\pi\pi$. In order to obtain further clues on both issues, it is necessary to perform a multi-channel spin-parity analysis.

The decay pattern of the $f_1(1285)$-like structure is consistent with a $(u\bar{u} + d\bar{d})$ state. The "$f_1(1285)$" decays preferentially to $\eta\pi\pi$. In addition, the $\eta\pi\pi$ and $K\bar{K}\pi$ mass spectra show dominant intermediate $a_0(980)$ production in the $f_1(1285)$ region.

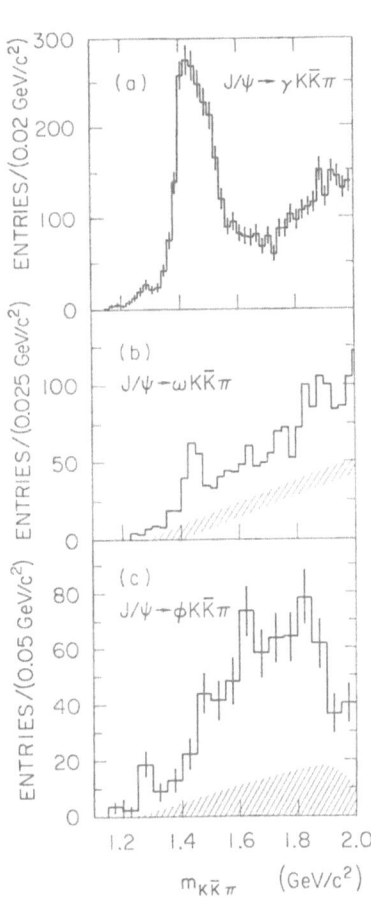

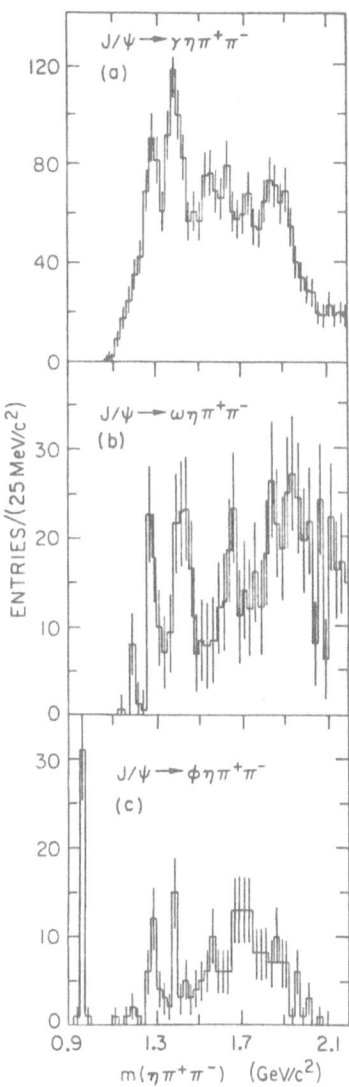

Fig.54. A comparison of Mark III $K\bar{K}\pi$ invariant mass spectra observed in the reactions: a) $J/\psi \rightarrow \gamma K\bar{K}\pi$, b) $J/\psi \rightarrow \omega K\bar{K}\pi$ and c) $J/\psi \rightarrow \phi K\bar{K}\pi$.

Fig.55. A comparison of Mark III $\eta\pi\pi$ invariant mass spectra observed in the reactions: a) $J/\psi \rightarrow \gamma\eta\pi^+\pi^-$, b) $J/\psi \rightarrow \omega\eta\pi^+\pi^-$ and c) $J/\psi \rightarrow \phi\eta\pi^+\pi^-$.

A comparison of the $\eta\pi\pi$ and $K\bar{K}\pi$ decay modes yields:

$$\frac{B(J/\psi \to \gamma\text{“}f_1(1285)\text{”}) \cdot B(\text{“}f_1(1285)\text{”} \to \eta\pi\pi)}{B(J/\psi \to \gamma\text{“}f_1(1285)\text{”}) \cdot B(\text{“}f_1(1285)\text{”} \to K\bar{K}\pi)} = 3.6 \pm 1.1$$

and

$$\frac{B(J/\psi \to \phi\text{“}f_1(1285)\text{”}) \cdot B(\text{“}f_1(1285)\text{”} \to \eta\pi\pi)}{B(J/\psi \to \phi\text{“}f_1(1285)\text{”}) \cdot B(\text{“}f_1(1285)\text{”} \to K\bar{K}\pi)} = 2.7 \pm 1.6\,,$$

where the branching ratios of the individual $\gamma\eta\pi\pi$ modes have been averaged and the result of the coupled channel fit for $\gamma K\bar{K}\pi$ has been used. Similar results are obtained for the other parametrizations of the $K\bar{K}\pi$ mass spectrum. Both ratios are consistent with the expected $f_1(1285)$ decay pattern, as the value from the Particle Data Group,[6] $B(f_1(1285) \to a_0(980)\pi)/B(f_1(1285) \to K\bar{K}\pi) = 3.3 \pm 1.1$, shows. These results , however, are not of sufficient precision to exclude a possible $\eta(1275)$ contribution. The classification of the $f_1(1285)$ as a $(u\bar{u} + d\bar{d})$ state is also confirmed by results from two-photon collisions. The $f_1(1285)$ is observed in $\gamma\gamma^*$ collisions with a reduced rate of $\tilde{\Gamma}(f_1(1285) \to \gamma\gamma^*) = 4.7 \pm 1.25 \pm 0.85$ keV,[89] which is typical for a $(u\bar{u} + d\bar{d})$ state. For an $s\bar{s}$ state, the rate would be an order of magnitude lower, because $\Gamma_{\gamma\gamma^*} \sim e_q^4$, where e_q is the quark charge. In conclusion, unless different states are observed in the $\eta\pi\pi$ and $K\bar{K}\pi$ recoil mass spectra, the production and decay patterns of the $f_1(1285)$ are not consistent.

The $f_1(1420)$ appears clearly in the $K\bar{K}\pi$ and $\eta\pi\pi$ mass spectra recoiling against the ω meson. No evidence for $f_1(1420)$ production is found in $\phi\eta\pi\pi$ and $\phi K\bar{K}\pi$. In $\gamma\eta\pi^+\pi^-$ a different state is observed, while in $\gamma K\bar{K}\pi$ the $f_1(1420)$ is possibly seen. These measurements suggest that the $f_1(1420)$ is mainly a $u\bar{u} + d\bar{d}$ state rather than a pure $s\bar{s}$ state. Further support for this hypothesis is given by the results from hadron collisions, which started the whole discussion, and by observations from two photon collisions. The $f_1(1420)$ is produced in π^-p and pp scattering but not in K^-p scattering. It appears in the tagged events $\gamma\gamma^* \to K\bar{K}\pi$. The production rate corrected for phase space

$$\tilde{\Gamma}(f_1(1420) \to \gamma\gamma^*) \cdot B(f_1(1420) \to K_s^0 K^{\pm}\pi^{\mp}) = (1.9 \pm 0.6) \text{ keV}^{[89]}$$

is larger than the expected rate of 0.4 keV for an $s\bar{s}$ state.[31]

The classification of the $f_1(1420)$ as an $s\bar{s}$ state was based on several arguments. If the $f_1(1285)$ and $f_1(1420)$ are 1^{++} isoscalars, the higher mass state should contain mainly $s\bar{s}$ quarks. This assignment is supported by the SU(3) mass relations for the axial vector nonet. Other supports come from the decay patterns observed in hadron production and $\gamma\gamma^*$ collisions. The $f_1(1420)$ mainly decays to $K\bar{K}\pi$, which proceeds dominantly through K^*K and not through $a_0\pi$. In $\gamma\gamma^*$ collisions the $f_1(1420)$ is observed in $K\bar{K}\pi$ but not in $\eta\pi\pi$. On the other hand, if the state seen in $\omega\eta\pi\pi$ is the $f_1(1420)$, the decay pattern observed in J/ψ decays is different; the decay to $\eta\pi\pi$ is larger than to $K\bar{K}\pi$:

$$\frac{B(J/\psi \to \omega f_1(1420)) \cdot B(f_1(1420) \to \eta\pi\pi)}{B(J/\psi \to \omega f_1(1420)) \cdot B(f_1(1420) \to K\bar{K}\pi)} = 1.35 \pm 0.66\,.$$

Though the observed resonance parameters are consistent with the $f_1(1420)$, a spin-parity measurement has not yet been performed.

In summary, the experimental data cannot be explained by the naïve quark model. The quark content of the $f_1(1420)$ is inconsistent in production and decay. A recent model by Seiden et al.,[112] which parametrizes the $f_1(1285)$ and $f_1(1420)$ wave functions by a particular mixture of $u\bar{u}$, $d\bar{d}$, and $s\bar{s}$ quark pairs, and includes contributions from doubly-disconnected diagrams, can explain the J/ψ results. On the other hand, if one assumes that the 1^{++} state at 1530 MeV is the $s\bar{s}$ partner of the $f_1(1285)$, the quark model would leave no room for the $f_1(1420)$. Chanowitz suggests that the $f_1(1420)$ could actually be a $g(u\bar{u} + d\bar{d})$ hybrid.[113] This would explain the observed production and decay patterns, since the hybrid can be produced in $\gamma\gamma^*$ collisions and decay to $s\bar{s}$ states. This idea is interesting because the lightest $J = 1$ hybrid is a 1^{-+} exotic. However, this explanation is inconsistent with the spin-parity measurements of the $f_1(1420)$. Another model, which assumes mixing, suggests a $q\bar{q}$ assignment with $J^{PC} = 1^{+-}$.[114]

The $\iota/\eta(1440)$ has been observed only in $J/\psi \rightarrow \gamma K \bar{K} \pi$. Its status is discussed in the next section. The $K \bar{K} \pi$ systems recoiling against an ω and a ϕ show no signals above 1.5 GeV. In the radiative decay, the higher mass region of the $\eta \pi^+ \pi^-$ spectrum shows interesting structures. In order to identify these states one needs to perform a partial wave analysis. While the $\phi\eta\pi\pi$ channel shows no structures in the higher mass region, the $\eta\pi\pi$ mass spectrum from $J/\psi \rightarrow \omega\eta\pi\pi$ may contain resonances above 1.5 GeV.

7.4 Status of the $\iota/\eta(1440)$

The current experimental status of the $\iota/\eta(1440)$ is:

- The process $J/\psi \rightarrow \gamma\iota$, $\iota \rightarrow K \bar{K} \pi$ is the largest single mode in radiative J/ψ decays. For a single state decaying to K^*K and $a_0(980)\pi$, the branching ratio is $(6.3 \pm 1.4) \times 10^{-3}$. The reduced rates for η, η' and ι scale as $\tilde{\Gamma}_{\gamma\eta} : \tilde{\Gamma}_{\gamma\eta'} : \tilde{\Gamma}_{\gamma\iota} = 1 : 6 : 14$.

- The $\iota/\eta(1440)$ is an isoscalar, since $B(\iota \rightarrow K_s K^{\pm} \pi^{\mp})/B(\iota \rightarrow K^+ K^- \pi^0) = 2 \pm 0.5$. Its spin-parity has been measured to be $J^P = 0^-$.

- The $\iota/\eta(1440)$ is not observed in hadronic decays $J/\psi \rightarrow \omega X$ and $J/\psi \rightarrow \phi X$. The upper limit on ι production in $\phi K \bar{K} \pi$ is $B(J/\psi \rightarrow \phi\iota) \cdot B(\iota \rightarrow K \bar{K} \pi) < 1.2 \times 10^{-4}$ at 90% C.L.

- The $\iota/\eta(1440)$ is not produced in ordinary hadronic processes. This provides strong evidence that the $\iota/\eta(1440)$ is decoupled from $q\bar{q}$ states.

- The $\iota/\eta(1440)$ is suppressed in $\gamma\gamma$ collisions. The upper limit at 95% C.L. is $\Gamma_{\gamma\gamma}(\iota) \cdot B(\iota \rightarrow K \bar{K} \pi) < 1.6$ keV. For comparison, the η and η' are produced with rates of $\Gamma_{\gamma\gamma}(\eta) = 0.514 \pm 0.031$ keV and $\Gamma_{\gamma\gamma}(\eta') = 4.25 \pm 0.19$ keV.

- The stickiness ratios for the η, η' and ι are $S_\eta : S_{\eta'} : S_\iota = 1 :\sim 4 :> 83$. This indicates that the $\iota/\eta(1440)$ is different from $q\bar{q}$ states.

- The $\iota/\eta(1440)$ decays are not flavor symmetric. The only mode observed is $\iota \rightarrow K \bar{K} \pi$, with dominant intermediate K^* production indicating a preferential coupling to strange quarks as expected for a 0^- glueball.

- The $\iota/\eta(1440)$ line shape does not fit a single simple BW parametrization. The distorted line shape is equally well explained by a single state decaying through the quasi-two-body decays $\iota \rightarrow K^*K$ and $\iota \rightarrow a_0\pi$ or by a superposition of two states. Without a mass dependent partial wave analysis, one cannot decide between the parametrizations of two coherent or two incoherent Breit-Wigner line shapes.

- The interpretation of the $\iota/\eta(1440)$ as a radial excitation of the η' is problematic. First, the production rate is too large compared to $J/\psi \rightarrow \gamma\eta$ and $J/\psi \rightarrow \gamma\eta'$. Second, under the assumption that the $\eta(1275)$ and the $X(1390)$ are the radial excitations, too many pseudoscalars are found in the 1400 MeV mass region to fit within the quark model.

In conclusion, the $\iota/\eta(1440)$ is very likely not a $q\bar{q}$ state. The interpretation of a glueball or at least a state containing a strong glueball component is very promising.[115] However, a unique identification is not possible before several issues are settled. First, the limits on ι production in $\gamma\gamma$ have to become more stringent. Second, the radial excitations of the η and η' have to be established. Third, a mass dependent partial wave analysis must be performed in the ι region. Finally, the structure in $J/\psi \to \gamma\gamma\rho$ has to be understood. The observation of a small $\iota/\eta(1440)$ signal in $\gamma\rho$ would not rule out the glueball interpretation. Several models predict rates for $G \to \gamma V$ of up to a few MeV.[116] For example, a bag model prediction by Donoghue, which considers glueball-$\eta - \eta'$ mixing, as well as additional mechanisms for $G \to \gamma V$, yields $\Gamma_{\gamma\rho}(G) = 0.4 - 1.6$ MeV. A pole model by Palmer and Pinsky obtains a rate of 3.5 MeV. If the signal observed in Fig. 38 were entirely due to $\iota/\eta(1440)$, one would estimate $\Gamma(\iota \to \gamma\rho) = 1.9 \pm 0.7$ MeV for $\Gamma_i = 95$ MeV and $B(\iota \to K\bar{K}\pi) = 1$, which is in the right range. On the other hand, if the $\iota/\eta(1440)$ were a hybrid, Barnes and Close predict $\Gamma_{\gamma\rho}(H) \approx 1.5$ MeV, using $q\bar{q}$-$q\bar{q}g$ mixing in the bag model. As pointed out by Barnes and Close, a comparison of the $\gamma\gamma\rho$ and $\gamma\gamma\phi$ channels allows one to distinguish between a glueball and a $q\bar{q}$ or $q\bar{q}g$ state. For a glueball the peak should be at the same mass in both modes, while for a $q\bar{q}$ or $q\bar{q}g$ state the peak in $\gamma\phi$ is expected to be ~ 300 MeV higher than in $\gamma\rho$.

8. HADRONIC TWO BODY DECAYS

In this last chapter, I will briefly summarize the Mark III measurements on the two body hadronic modes, $J/\psi \to VP$, $J/\psi \to VT$, $J/\psi \to VS$ and $J/\psi \to VV$. The points of interest include quark-gluon couplings, the quark content of the η and η' and the magnitude of doubly-disconnected diagrams.

Mark III has updated its systematic study on VP decays.[117] The measured branching ratios for 11 different modes are listed in Table 10. These results are confirmed by DM2 measurements, except for $J/\psi \to \omega\eta'$. DM2 yields a $\sim 2\sigma$ higher branching ratio than Mark III. For a comparison of the VP decays, Table 11 lists the reduced branching ratios $\tilde{B}(J/\psi \to VP) = B(J/\psi \to VP)/P_V^3$, where P_V is the center-of-mass momentum of the vector boson. Two features are immediately observed. First, SU(3) breaking is visible. For example, the rates $\frac{1}{3}\tilde{\Gamma}_{\rho\pi}$ and $\frac{1}{2}\tilde{\Gamma}_{K^{*\pm}K^\pm}$ are expected to be equal in the SU(3) symmetric limit. The measurements, however, yield $\frac{1}{3}\tilde{B}_{\rho\pi} : \frac{1}{2}\tilde{B}_{K^{*\pm}K^\mp} = 1 : 0.66 \pm 0.09$, indicating a weaker coupling to s quarks. A detailed analysis shows that the SU(3) breaking is proportional to the number of $s\bar{s}$ pairs in the final state. Second, a comparison of $J/\psi \to K^{*\pm}K^\mp$ and $J/\psi \to K^{*0}\bar{K}^0 + c.c.$, indicates possible isospin violation. The measurements yield $\tilde{B}_{K^{*\pm}K^\mp}/\tilde{B}_{K^{*0}\bar{K}^0+c.c.} = 0.82 \pm 0.05 \pm 0.09$, which is ~ 1.5 s.d. lower than the prediction of equal rates. However, for $J/\psi \to \rho\pi$, the ratio $\tilde{B}_{\rho^0\pi^0}/\tilde{B}_{\rho^\pm\pi^\mp} = 0.49 \pm 0.01 \pm 0.03$ is consistent with the isospin prediction of $1/2$.

In order to describe the VP decays quantitatively, a phenomenological model has been suggested[118] which includes contributions from doubly disconnected diagrams. The parametrization shown in Table 11 for each channel is based on the following facts:

a) The decays $\rho\eta$, $\rho\eta'$ and $\omega\pi^0$ are isospin-violating and, therefore, proceed only through the EM diagram.

b) The amplitude for the isospin-violating decay $\phi\pi^0$ vanishes, as the ϕ is a pure $s\bar{s}$ state, while the π^0 is a $(u\bar{u} - d\bar{d})$ state.

c) Doubly-disconnected diagrams can only contribute to the $\omega\eta$, $\phi\eta$ and $\phi\eta'$ channels, because both mesons have to be isoscalars.

TABLE 10

Mark III Branching Ratios for Hadronic 2 Body Decays

Mode	MARK III $B(J/\psi \to VP)$ [10^{-3}]	DM2 $B(J/\psi \to VP)$ [10^{-3}]	Mode	MARK III $B(J/\psi \to VT)$ [10^{-3}]
$\rho\pi$	$14.2 \pm 0.1 \pm 1.9$	12.7 ± 9	$\rho\, a_2$	$11.8 \pm 0.8 \pm 2.9$
$K^{*+}K^- + c.c.$ [a]	$5.3 \pm 0.1 \pm 0.6$	5.4 ± 4.4	$K^{*+}K^{**-} + c.c.$	$8.1 \pm 1.2 \pm 1.2$
$K^{*0}\bar{K}^0 \pm c.c.$	$4.3 \pm 0.1 \pm 0.5$	3.9 ± 0.6	$K^{*0}\bar{K}^{**0} + c.c.$	$7.6 \pm 1.1 \pm 1.3$
$\omega\eta$	$1.65 \pm 0.06 \pm 0.21$	1.33 ± 0.17	ωf	$4.93 \pm 0.25 \pm 1.3$
$\omega\eta'$ [a]	$0.16 \pm 0.02 \pm 0.02$	0.4 ± 0.11	$\omega f'$	< 0.12
$\phi\eta$ [a]	$0.66 \pm 0.04 \pm 0.09$	0.64 ± 0.1	ϕf	< 0.37
$\phi\eta'$ [a]	$0.30 \pm 0.03 \pm 0.04$	0.4 ± 0.05	$\phi f'$ [b]	0.81 ± 0.1
$\omega\pi^0$	$0.48 \pm 0.02 \pm 0.06$	0.51 ± 0.8		
$\rho^0\eta$	$0.19 \pm 0.01 \pm 0.03$	0.19 ± 0.5	a) averages over several modes	
$\rho^0\eta'$	$0.11 \pm 0.01 \pm 0.02$	0.08 ± 0.03	b) with $B(f' \to K\bar{K}) = 0.63$	
$\phi\pi^0$	< 0.0068	< 0.013		

In addition, the wave functions of the η and η' have been parametrized by:[119]

$$|\eta\rangle = X_\eta |u\bar{u} + d\bar{d}\rangle/\sqrt{2} + Y_\eta |s\bar{s}\rangle\,, \quad |\eta'\rangle = X_{\eta'}|u\bar{u} + d\bar{d}\rangle/\sqrt{2} + Y_{\eta'}|s\bar{s}\rangle\,, \quad (19)$$

where X_i and Y_i indicate the fraction of light quarks and strange quarks, respectively. Assuming no mixing with other states implies $X_\eta^2 + Y_\eta^2 = X_{\eta'}^2 + Y_{\eta'}^2 = 1$. The pseudoscalar mixing angle is determined by:

$$\tan\theta_p = -(\sqrt{2}X_\eta + Y_\eta)/(X_\eta - \sqrt{2}Y_\eta) = (X_{\eta'} - \sqrt{2}Y_{\eta'})/(\sqrt{2}X_{\eta'} + Y_{\eta'})\,. \quad (20)$$

Using this model, Mark III has performed global fits to the ten measurements in Table 11. If the additional constraints $X_\eta^2 + Y_\eta^2 = 1$ and $X_{\eta'} = -Y_\eta$ are imposed, the fit yields:

$$g = 1.1 \pm 0.04 \qquad e = 0.12 \pm 0.05 \qquad r = -0.15 \pm 0.01$$

$$s = 0.11 \pm 0.03 \qquad \theta_e = 1.23 \pm 0.14 \qquad X_\eta = 0.81 \pm 0.02.$$

The probability of this fit is $P(\chi^2) = 5.1\%$. From Eq. 20, one obtains a mixing angle of $\theta_P = (-19 \pm 2)°$, in good agreement with a recent estimate of $\theta_P \approx -20°$, which is based on a quadratic mass matrix,[119] as well as results from $\gamma\gamma$ collisions[120] and radiative[121] J/ψ decays yielding $\theta_P = (-18.4 \pm 1.1)°$ and $\theta_P = (-22 \pm 2)°$, respectively. For a fit without the above constraints on X_i and Y_i, the probability increases to 98.6%, yielding $X_\eta^2 + Y_\eta^2 = 1.0 \pm 0.16$ and $X_{\eta'}^2 + Y_{\eta'}^2 = 1.44 \pm 0.25$. The other parameters remain within the errors of the above results. This observation demonstrates that the η and η' are completely understood in terms of u, d and s quarks, thus excluding mixing with other states. Furthermore, doubly-disconnected diagrams are relevant. An unconstrained fit with $r = 0$ yields similar results, but the probability is only 0.03%, while, for a constrained fit, the probability is negligibly small.

TABLE 11

Parametrization of Amplitudes for $J/\psi \to VP$

and the Reduced Branching Ratios

Process	Amplitude A_{VP}	$\tilde{B} \times 10^3$
$\rho^+\pi^-, \rho^0\pi^0, \rho^-\pi^+$	$g + e$	1.56 ± 0.16
$K^{*+}K^-, K^{*-}K^+$	$g(1-s) + e$	1.026 ± 0.089
$K^{*0}\bar{K}^0, \bar{K}^{*0}K^0$	$g(1-s) - 2e$	0.836 ± 0.055
$\omega\eta$	$(g+e)X_\eta + \sqrt{2}rg(\sqrt{2}X_\eta + Y_\eta)$	0.61 ± 0.064
$\omega\eta'$	$(g+e)X'_\eta + \sqrt{2}rg(\sqrt{2}X_{\eta'} + Y_{\eta'})$	0.078 ± 0.01
$\phi\eta$	$(g(1-2s) - 2e)Y_\eta + rg(\sqrt{2}X_\eta + Y_\eta)$	0.288 ± 0.034
$\phi\eta'$	$(g(1-2s) - 2e)Y_{\eta'} + rg(\sqrt{2}X_{\eta'} + Y_{\eta'})$	0.180 ± 0.026
$\rho^0\eta$	$3eX_\eta$	0.071 ± 0.010
$\rho^0\eta'$	$3eX_{\eta'}$	0.052 ± 0.009
$\omega\pi^0$	$3e$	0.159 ± 0.017
$\phi\pi^0$	0	< 0.0026

g: 3 gluon decay amplitude; e: EM decay amplitude; r: fraction of doubly disconnected diagrams in strong decays; s: SU(3) violation. In addition, a phase θe between
g and e is included.

In the VT decay sector, the modes $\rho a_2(1320)$, $K^*(892)K_2^*(1430)$, $\omega f_2(1270)$,
$\omega f_2'(1525)$, $\phi f_2(1270)$ and $\phi f_2'(1525)$ have thus far been studied. Since the decays
including the $f_2(1270)$ and $f_2'(1525)$ have already been discussed in Chapter 3, I will
focus here on $K^*K_2^*(1430)$ and $\rho a_2(1320)$. Using 4C kinematic fits, Mark III has analyzed both the neutral and charged modes of $J/\psi \to K^*K_2^*(1430)$ in the $K^+\pi^-K^-\pi^+$
and $K^+\pi^0K^-\pi^0$ final states, respectively.[122] Figure 56 shows the $K\pi$ invariant mass
spectra opposite a $K^{*0}(892)$. Besides a prominent $K_2^*(1430)$ signal, a significant peak
at the $K^{*0}(892)$ mass is observed, which results from $J/\psi \to K^{*0}(892)\bar{K}^{*0}(892)$, as
discussed later. The other $K\pi$ mass spectra show the same features. The reconstruction efficiencies for the neutral and charged $K^*K_2^*(1430)$ modes are 15% and 8%,
respectively. The resulting branching ratios are given in Table 10. Within the errors,
both values are equal, implying isospin symmetric decays.

The hadronic modes $J/\psi \to \rho^0 a_2^0(1320)$ and $J/\psi \to \rho^\pm a_2^\mp(1320)$ are studied with
6C kinematic fits in the $\eta\pi^+\pi^-\pi^0$ final state with $\eta \to \gamma\gamma$ resulting from the decays
$\rho^\pm \to \pi^\pm\pi^0$ ($\rho^0 \to \pi^+\pi^-$) and $a_1^\pm \to \eta\pi^\pm$ ($a_2^0 \to \eta\pi^0$).[123] The $\eta\pi$ invariant mass
spectrum, after combining the charged and neutral modes, is shown in Fig. 57. Besides
a large $a_2(1320)$ signal, no other significant features are observed in the $\eta\pi$ mass
distribution. The detection efficiencies for the neutral and charged ρa_2 mode are 35%
and 28%, respectively. The branching ratio is given in Table 10 together with the
Mark III measurements for the other VT decays. After phase space corrections, the
comparison of the reduced branching ratios yields:

$$\tfrac{1}{3}\tilde{B}_{\rho a_2} : \tfrac{1}{4}\tilde{B}_{K^* K_2^*} : \tilde{B}_{\omega f} : \tilde{B}_{\phi f'} = 1 : 1.1 \pm 0.5 : 1.2 \pm 0.3 : 0.27 \pm 0.08 \,.$$

Except for $\phi f'$, all values are consistent with the SU(3) prediction.

In the $J/\psi \to VS$ decays, so far only the $\phi f_0(975)$ mode has been observed. The $f_0(1300)$ is expected in the $\omega\pi\pi$ channel. Since this state is close to the $f_1(1270)$ and rather broad, it can only be found by a partial wave analysis. As discussed earlier, the $\phi\pi\pi$ spectrum shows some structure in the 1200-1500 MeV mass region which may have contributions from an $f_0(1300)$. A similar argument holds for the $K_0^*(1350)$, which is close to the $K_2^*(1430)$ and rather broad. Since it decays dominantly to $K\pi$, a multi-channel analysis in the $K^* K\pi$ mode may provide further clues. The $a_0(980)$ should be seen in the $\eta\pi$ spectrum in Fig. 57. A small enhancement may be seen in the 1 GeV mass region, resulting in an upper limit at 90% C.L. of

$$B(J/\psi \to \rho a_0(980)) < 4.4 \times 10^{-4} \,.$$

In the naïve quark model, the $\rho a_0(980)$ decay rate is expected to be of the same order of magnitude as the $\rho\pi$ decay rate. Since this limit is an order of magnitude lower after phase space corrections, this may provide evidence that the $a_0(980)$ is not a $q\bar{q}$ state. Other interpretations of the $a_0(980)$ include a $K\bar{K}$ molecule[47] or a $q\bar{q}q\bar{q}$ state.

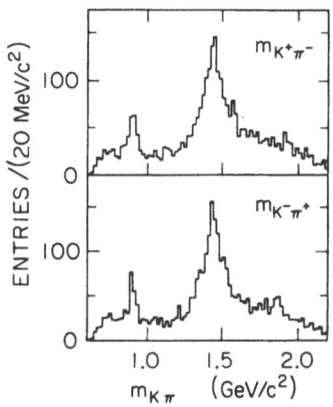

Fig.56. The $K\pi$ invariant mass spectra recoiling against a $K^*(982)$ in the process $J/\psi \to K^* K^{\pm} \pi^{\mp}$.

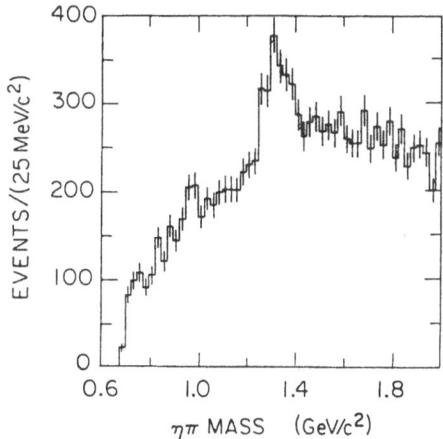

Fig.57. Sum of the $\eta\pi^+$, $\eta\pi^-$ and $\eta\pi^0$ mass spectra from $J/\psi \to \rho\eta\pi$.

The observation of a K^* signal in the $K\pi$ mass spectrum in Fig. 56 provides evidence for the SU(3) forbidden decay $J/\psi \to K^*(982)\bar{K}^*(982)$.[124] The branching ratios for the charged and neutral modes are relatively large:

$$B(J/\psi \to K^{*0}\bar{K}^{*0}) = (4.5 \pm 0.7 \pm 1.0) \times 10^{-4}$$

$$B(J/\psi \to K^{*+}K^{*-}) = (10.0 \pm 2.0 \pm 4.0) \times 10^{-4} \,.$$

This may indicate that either a large electromagnetic amplitude or SU(3) violating effects are present. The ratio of the two branching ratios is 2.25 ± 0.21 after phase space corrections. The same phenomena is found in the SU(3) forbidden decays $J/\psi \to K\bar{K}$:

$$\tilde{B}(J/\psi \to K^+ K^-)/\tilde{B}(J/\psi \to K_s^0 K_L^0) = 2.42 \pm 0.5. \quad [125]$$

Mark III also has found evidence for the SU(3) forbidden VV decay $J/\psi \to \rho^+\rho^-$, which is observed in the $\pi^+\pi^0\pi^-\pi^0$ final state.[124] The branching ratio is of the same order of magnitude as the one for $J/\psi \to K^{*+}K^{*-}$:

$$B(J/\psi \to \rho^+\rho^-) = (9.4 \pm 1.0 \pm 1.5) \times 10^{-4}.$$

It is worth noting that these branching ratios are a factor ~ 5 larger than the corresponding branching ratios for the $J/\psi \to PP$ decays.

9. CONCLUSION

The results from J/ψ decays, combined with recent measurements from two photon collisions and hadronic scattering experiments, have provided much new information, which is important for the understanding of $q\bar{q}$ states and the classification of the $\iota/\eta(1440)$ and the $\theta/f_2(1720)$. The data confirm that the $\iota/\eta(1440)$ and the $\theta/f_2(1720)$ are unlikely $q\bar{q}$ states. For the $\iota/\eta(1440)$ the glueball interpretation is the most promising. For the $\theta/f_2(1720)$, both the hybrid and glueball interpretations are possible. The $\xi(2230)$ has been observed in $J/\psi \to \gamma K^+K^-$ and $J/\psi \to \gamma K_s^0 K_s^0$. The $\xi(2230)$ is probably also seen in $K^-p \to K\bar{K}\Lambda$ and $\pi^-p \to \eta\eta'n$. The nature of this state is still unknown. In the γVV decays pseudoscalar states have been observed near threshold. Their nature is currently not understood. The $f_2(1270)$ and $f_2'(1525)$ are completely understood as nearly ideally-mixed tensor mesons, while the $a_0(980)$ and the $f_0(975)$ are probably not scalar mesons. The pseudoscalar states can be well described with a simple model, which includes doubly-disconnected diagrams. The presence of doubly-disconnected diagrams is observed in J/ψ decays. The η and η' are completely understood in terms of light and strange quarks. No other components are needed. The measured pseudoscalar mixing angle agrees well with predictions from a quadratic mass matrix. SU(3) breaking is seen in J/ψ decays.

On the other hand, the systematic studies of J/ψ decays have also raised new questions: for example, is the $f_1(1420)$ really a 1^{++} $q\bar{q}$ state, since it shows different properties in production and decay? Is the $\iota/\eta(1440)$ a single resonance or are there more states in that energy region? What is the nature of the 0^{-+} states produced in γVV decays? What is the $\xi(2230)$? Finally, several more fundamental questions remain open. For example, where is the scalar glueball. Do oddballs exist in the gg sector? Where are the hybrid states? These questions provide a great challenge to the field in the next few years.

ACKNOWLEDGEMENTS

I would like to thank the organizers for arranging a successful school in a nice environment. I enjoyed the interesting talks as well as many lively discussions. I also want to thank D. Hitlin, S. Meshkov, and W. Wisniewski for helpful discussions.

This work was supported in part by U.S. Department of Energy Contract No. DE-AC03-81-ER40050 and the Alexander von Humboldt Foundation.

REFERENCES

1. J. J. Aubert et al., Phys. Rev. Lett. **33**, 1404 (1974).
2. J. Augustin et al., Phys. Rev. Lett. **33**, 1406 (1974).
3. C. Bacci et al., Phys. Rev. Lett. **33**, 1408 (1974); W. Braunschweig et al., Phys. Lett. **53B**, 393 (1974); ibid. **53B**, 491 (1975).
4. J. Augustin et al., Phys. Rev. Lett. **33**, 1453 (1974).

5. T. Appelquist *et al.*, *Phys. Rev. Lett.* **34**, 365 (1975); E. Eichten *et al.*, *Phys. Rev. Lett.* **34**, 369 (1975).

6. M. Aguilar-Benitez, Particle Data Group, *et al.*, *Phys. Lett.* **170B**, 1 (1986).

7. F. Vanucci *et al.*, *Phys. Rev.* **D15**, 1814 (1977).

8. For other recent experimental reviews of J/ψ decays, see: W. Toki, Invited talk at the Charm Workshop, Beijing , SLAC-PUB-4410 (1987); U. Mallik, SLAC Summer Inst. of Part. Phys., Stanford (1986), SLAC-PUB-4238 (1987).

9. S. Okubo, *Phys. Lett.* **5**, 165 (1963); G. Zweig, CERN Preprints CERN-TH-401, 402, 412 (1964); J. Iizuka, *Prog. Theor. Phys. Suppl.* **37-38**, 21 (1966).

10. T. Appelquist and H. D. Politzer, *Phys. Rev. Lett.* **34**, 43 (1975); R. Barbieri, R. Gatto and R. Kögerler, *Phys. Lett.* **60B**, 183 (1976); R. Barbieri, R. Gatto and E. Remiddi, *ibid.* **61B**, 465 (1976).

11. R. van Royen and V. F. Weisskopf, *Nu. Cim.* **50A**, 617 (1967); R. Barbieri *et al.*, *Phys. Lett.* **57B**, 455 (1975); W. Celmaster, *Phys. Rev.* **D19**, 1517 (1979).

12. S. J. Brodsky *et al.*, *Phys. Lett.* **73B**, 203 (1978); *Phys. Rev.* **D28**, 228 (1983).

13. R. J. Jaffe, *Phys. Rev.* **D15**, 267 (1977); J. Weinstein and N. Isgur, *Phys. Rev.* **D27**, 588 (1983).

14. T. Barnes, these proceedings.

15. G. Godfrey and N. Isgur, *Phys. Rev.* **D32**, 189 (1985); N. Isgur, these procs.

16. S. Suzuki, *Proc. 2nd Int. Conf. on Hadron Spectroscopy at KEK*, p. 64 (1985).

17. F. Close, these proceedings.

18. S. Meshkov, *Proceedings of John Hopkins Workshop*, p. 185 (1982); P. M. Fishbane and S. Meshkov, *Comments Nucl. Part. Phys.* **13**, 325 (1984); S. Meshkov, *Proc. of the Aspen Winter Phys. Conf.*, p. 87, Aspen (1986).

19. C. N. Yang, *Phys. Rev.* **77**, 242 (1950).

20. H. W. Hamber and V. M. Heller, *Phys. Rev.* **D29**, 928 (1984); C. Michael and I. Teasdale, *Nucl. Phys.* **B215**, 433 (1983); K. Ishikawa, M. Teper and G. Schierholz, *Phys. Lett.* **116B**, 429 (1982); K. Ishikawa *et al.*, *Phys. Lett.* **120B**, 387 (1983); K. Seo, Univ. of Chicago preprint EF-82-10 (1982).

21. B. Berg, A. Billoire and C. Vohwinkel, *Phys. Rev. Lett.* **57**, 400 (1986); T. DeGrand, Univ. of Colorado preprint COLO-HEP-139 (1986); A. Patel *et al.*, *Phys. Rev. Lett.* **57**, 1288 (1986); P. de Forcrand, *et al.*, *Z. Phys.* **C31**, 87 (1986); *Phys. Lett.* **152B**, 107 (1985).

22. C. E. Carlson, T. H. Hansson and C. Peterson, *Phys. Rev.* **D30**, 1594 (1984); J. F. Donoghue, K. Johnson and B. Li, *Phys. Lett.* **99B**, 416 (1981); M. Chanowitz and S. Sharpe, *Nucl. Phys.* **B222**, 211 (1983).

23. J. M. Cornwall and A. Soni, *Phys. Lett.* **120B**, 431 (1983).

24. D. Robson, *Nucl. Phys.* **130B**, 328 (1977).

25. J. M. Cornwall and A. Soni, *Phys. Rev.* **D29**, 1424 (1984).

26. M. Chanowitz, *VI. Int. Workshop on Photon-Photon Collisions*, Tahoe, 1984, ed. R. Lander (World Scientific).

27. C. E. Carlson *et al.*, *Phys. Lett.* **99B**, 353 (1981).

28. M. Chanowitz and S. Sharpe, *Nucl. Phys.* **B222**, 211 (1983).

29. T. Barnes and F. Close, *Nucl. Phys.* **B224**, 241 (1983); T. Barnes, *SIN Spring School on Strong Interactions*, UPTPT 85-21 (1985).

30. J. Govaerts *et al.*, *Nucl. Phys.* **B284**, 674 (1987).

31. M. Chanowitz, *Proc. 2nd Int. Conf. on Had. Spectr. at KEK*, p. 269 (1987).

32. M. Chanowitz, *Proc. XIV Int. Conf. on Multiparticle Dynamics*, p. 716 (1983).

33. F. J. Gilman, in "High Energy Physics and Nuclear Structure —1975", *Proceedings of the Sixth International Conference*, Santa Fe and Los Alamos.

34. D. Bernstein *et al.*, *Nucl. Instrum. Meth.* **226**, 301 (1984).

35. C. Edwards *et al.*, *Phys. Lett.* **48**, 458 (1982).

36. R. M. Baltrusaitis *et al.*, *Phys. Rev.* **D35**, 2077 (1987).

37. J. Augustin *et al.*, contrib. to *Int. Symp. on Lepton Photon Interact. at High Energies, Kyoto, 1985*, Orsay Report No. LAL 85/27, 1986.

38. G. Dubois, *Proc. Int. Europhysics Conf. in HEP*, in Uppsala, 1987 (1988).

39. R. A. Lee, Ph.D. Thesis, SLAC-PUB-282 (1985, unpublished).

40. L. Köpke, *Proc. 23rd Int. Conf. on HEP*, ed. S. Loken (World Scient., 1986).

41. R. Xu, Ph.D. Thesis, U.C. Santa Cruz (1987, unpublished).

42. J. E. Augustin *et al.*, *Proc. Int. Europhysics Conf. on HEP*, Bari (1985).

43. T. Barnes, K. Dooley and N. Isgur, *Phys. Lett.* **183B**, 210 (1987).

44. H. G. Dosch and D. Gromes, *Z. Phys.* **C34**, 555 (1987).

45. S. M. Flatté, *Phys. Lett.* **63B**, 224 (1976).

46. V. Chabaud *et al.*, *Acta Phys. Pol.* **B12**, 575 (1981); M. Nikolic, *Phys. Rev.* **D26**, 3141 (1982).

47. J. Weinstein and N. Isgur, *Phys. Rev. Lett.* **48**, 659 (1982; J. Weinstein and N. Isgur, *Phys. Rev.* **D27**, 588 (1983).

48. K. L. Au, D. Morgan and M. R. Pennington, RAL-85-099 (1985); K. L. Au, D. Morgan and M. R. Pennington, RAL-86-076 (1986).

49. T. Barnes, Univ. of Toronto preprint UTPT-85-26 (1985), *Proc. VII Int. Workshop on Photon-Photon Collisions*, Collège de France (1986); A. Seiden, *Proc. VII Int. Workshop on Photon-Photon Collisions*, Collège de France (1986).

50. D. Antresyan *et al.*, *Phys. Rev.* **D33**, 1847 (1986).

51. D. Aston, SLAC-PUB-4279 (1987), submitted to *Nucl. Phys. B*; T. Suzuki, *Proc. 2nd Int. Conf. on Hadron Spectroscopy at KEK*, p. 64 (1987).

52. R. Longacre, *Proc. 2nd Int. Conf. on Had. Spectr. at KEK*, p. 46 (1987).

53. H. Aihara *et al.*, *Phys. Rev. Lett.* **57**, 404 (1986).

54. K. Einsweiler, SLAC-PUB-3702 (1983).

55. S. Godfrey, R. Kokoski and N. Isgur, *Phys. Lett.* **141B**, 439 (1984).

56. B. F. L. Ward, *Phys. Rev.* **D31**, 2849 (1985).

57. M. Chanowitz and S. R. Sharpe, *Phys. Lett.* **132B**, 413 (1983).

58. S. Pakvasa *et al.*, *Phys. Lett.* **145B**, 134 (1984); *Phys. Rev.* **D31**, 2378 (1985).

59. H. Haber and G. Kane, *Phys. Lett.* **135B**, 196 (1984); R. Wiley, *Phys. Rev. Lett.* **52**, 585 (1984); R. Barnett, G. Senjanovic and D. Wyler, *Phys. Rev.* **D30**, 1529 (1984).

60. H. Haber, SLAC Report No. SLAC-PUB-3193, August 1983; S. Ono, *Phys. Rev.* **D35**, 944 (1987); M. Shatz, *Phys. Lett.* **138B**, 209 (1984); K. Yamawaki, M. Bando and K. Matumoto, *Phys. Rev. Lett.* **56**, 1335 (1986).

61. J. Augustin *et al.*, contrib. to *Int. Symp. on Lepton-Photon Interact. at High Energies*, Leipzig, 1984, Orsay Report No. LAL 84/30, October 1984.

62. R. M. Baltrusaitis *et al., Phys. Rev. Lett.* **56**, 107 (1985).

63. J. J. Becker *et al.*, contrib. to *23rd Int. Conf. on HEP*, Berkeley, 1986.

64. G. Sklarz, DM2 Results presented at *Hadron '87*, KEK, April 1987.

65. D. Alde *et al., Phys. Lett.* **177B**, 120 (1986).

66. S. Behrends *et al., Phys. Lett.* **137B**, 277 (1984).

67. J. Sculli *et al., Phys. Rev.* **D58**, 1715 (1987).

68. G. Bardin *et al., Phys. Lett.* **195B**, 292 (1987).

69. A. D. Linde, *Phys. Lett.* **70B**, 306 (1977); S. Weinberg, *Phys. Rev. Lett.* **36**, 294 (1976).

70. A. Etkin *et al., Phys. Rev. Lett.* **49**, 1620 (1982).

71. R. M. Baltrusaitis *et al., Phys. Rev.* **D33**, 1222 (1986).

72. N. P. Chang and C. T. Nelson, *Phys. Rev. Lett.* **40**, 1617 (1978).

73. T. L. Trueman, *Phys. Rev.* **D18**, 3423 (1978).

74. C. N. Yang, *Phys. Rev.* **77**, 722 (1950).

75. L. Stanco, talk at *Top. Sem. on Heav. Flav.*, San Miniato (1987), Orsay LAL-87-42; D. Bisello *et al., Phys. Lett.* **192B**, 239 (1987); *ibid.* **179B**, 289 (1986).

76. R. M. Baltrusaitis *et al., Phys. Rev. Lett.* **55**, 1723 (1985).

77. G. Eigen, *Proc. XVII Int. Symp. on Multipart. Dyn. at Seewinkel* (1986); W. J. Wisniewski, *Proc. 2nd Int. Conf. on Had. Spectr. at KEK*, (1987).

78. R. M. Baltrusaitis *et al., Phys. Rev.* **D33**, 629 (1986).

79. D. Bisello *et al., Phys. Lett.* **179B**, 294 (1986).

80. J. Richman, Ph.D. Thesis, Caltech, CALT-68-1231 (1985, unpublished.)

81. C. Edwards, Ph.D. Thesis, Caltech, CALT-68-1165 (1985, unpublished.)

82. J. Becker *et al.*, contrib. to *23rd Int. Conf. on HEP*, Berkeley, SLAC-PUB-4242 (1986).

83. W. Braunschweig *et al., Phys. Lett.* **67B**, 1243 (1978).

84. R. Partridge *et al., Phys. Rev. Lett.* **44**, 712 (1980).

85. R. Partridge *et al., Phys. Rev. Lett.* **45**, 1150 (1980); J. Gaiser *et al., Phys. Rev.* **D43**, 711 (1986).

86. T. Himmel *et al., Phys. Rev. Lett.* **45**, 1146 (1980).

87. C. Baglin *et al., Phys. Lett.* **187B**, 191 (1987).

88. R. Barbieri *et al., Nucl. Phys.* **B154**, 535 (1979); K. Hagawara, C. Kim and T. Yoshino, *Nucl. Phys.* **B177**, 461 (1981).

89. J. Olsson, Invited talk at the *Int. Symp. on Lepton-Photon Interactions at High Energies*, Hamburg (1987), DESY report 87-136 (1987).

90. H. Harari, *Phys. Lett.* **60B**, 172 (1976).

91. For further discussions on this subject see: L. Montanet, these procs.; S. Cooper, *Proc. 23rd Int. Conf. on HEP at Berkeley*, p. 67 (1986), SLAC-PUB-4139.

92. D. Scharre *et al., Phys. Lett.* **97B**, 329 (1980).

93. C. Edwards *et al., Phys. Rev. Lett.* **49**, 259 (1982).

94. P. Ballion *et al., Nuovo Cim.* **50A**, 393 (1967).

95. D. Aston *et al.*, SLAC-PUB-4340, June 1987; H. Aihara *et al., Phys. Rev. Lett.* **57**, 51 (1986); G. Gidal *et al.*, SLAC-PUB-4274, April 1987; S. Cooper, *Proc. 2nd Int. Conf. on Hadron Spectroscopy at KEK* (1987).

96. C. Dionisi *et al.*, *Nucl. Phys.* **B169**, 1 (1980).

97. T. Armstrong *et al.*, *Phys. Lett.* **146B**, 273 (1984).

98. S. Chung *et al.*, *Phys. Rev. Lett.* **55**, 779 (1985).

99. D. F. Reeves *et al.*, *Phys. Rev.* **D34**, 1960 (1986).

100. H. Aihara *et al.*, *Phys. Rev. Lett.* **57**, 2500 (1986); G. Gidal *et al.*, SLAC Report No. SLAC-PUB-4275, April 1987.

101. A. Ando *et al.*, *Phys. Rev. Lett.* **57**, 1296 (1986); T. Tsuru, *Proc. 2nd Int. Conf. on Hadron Spectroscopy at KEK*, p. 233 (1987).

102. D. Aston *et al.*, SLAC-PUB-4394 (1987), subm. *Phys. Lett. B*; Ref. 51 (Suzuki).

103. S. D. Protopopescu, *Proc. 2nd Int. Conf. on Had. Spectr. at KEK* (1987).

104. M. Stanton *et al.*, *Phys. Rev. Lett.* **42**, 346 (1979); A. Ando *et al.*, *Phys. Rev. Lett.* **57**, 1296 (1986).

105. J. Richman, Ph.D. Thesis, Caltech CALT-68-1231 (1985, unpublished).

106. R. Partridge, *XII Rencontre de Moriond*, Les Arcs (1987); J. J. Becker *et al.*, SLAC-PUB-4225, submitted to *Phys. Rev. Lett.*

107. J. M. Blatt and V. F. Weisskopf, *Th. Nucl. Phys.*, p. 261 (Wiley, N.Y.) 1952.

108. J. J. Becker *et al.*, *Phys. Rev. Lett.* **59**, 186 (1987).

109. See Table 3 in H. E. Haber and J. Perrier, *Phys. Ref.* **D32**, 2961 (1985).

110. R. M. Baltrusaitis *et al.*, contrib. to *23rd Int. Conf. on HEP at Berkeley*, No. 3433 (1986).

111. W. F. Palmer and S. S. Pinsky, *Phys. Rev.* **D27**, 2219 (1983).

112. A. Seiden, H. Sadrozinski and H. E. Haber, paper submitted to the *Int. Europhysics Conf. on HEP*, Uppsala (1987).

113. M. Chanowitz, *Phys. Lett.* **187B**, 469 (1987).

114. J. Iizuka, F. Masuda and T. Miura, Univ. of Tsukuba, GK-U.TSUK/87-1 (1987).

115. An alternative interpretation of the $\iota/\eta(1440)$ and a discussion of the 1400-1500 MeV mass region is given in: S. Meshkov *et al.*, *Proc. XXII Rencontre de Moriond*, Les Arcs (1987).

116. J. Donoghue, *Phys. Rev.* **D30**, 114 (1984); W. F. Palmer and S. S. Pinsky, *Phys. Rev.* **D30**, 1002 (1984) and Ref. 111; T. Barnes and F. E. Close, RAL-84-055 (1984).

117. D. Coffman *et al.*, SLAC-PUB-4424 (1988), submitted to *Phys. Rev. D.*

118. A. Seiden *et al.*, UCSC preprint 87/73 (1988), submitted to *Phys. Rev. D.*

119. F. J. Gilman and R. Kauffman, *Phys. Rev.* **D36**, 2761 (1987).

120. A. Seiden, Ref. 49

121. From $J/\psi \to \gamma\eta$ and $J/\psi \to \gamma\eta'$, using an average from Ref. 6, Ref. 59 and Ref. 78. Consistent results are also obtained from $\phi \to \gamma\eta$, $\rho \to \gamma\eta$ and $\rho \to \gamma\eta'$.

122. Tim Bolton (personal communication).

123. J. Becker *et al.*, contrib. to the *23rd Int. Conf. on HEP at Berkeley* (1986) SLAC-PUB-4243 (1987).

124. M. Scarlatella, Ph.D. Thesis, U.C. Santa Cruz, SCIPP 87/83 (1987, unpub.)

125. R. Baltrusaitis *et al.*, *Phys. Rev.* **D32**, 566 (1985).

HADRON SPECTROSCOPY WITH GAMS

Freddy G. Binon*

Institut Interuniversitaire des Sciences Nucléaires
B-1050 Brussels, Belgium

INTRODUCTION

Strong interactions at high energy are characterized by the production of a large number of secondary particles. For example, about ten secondaries are produced in π^-p collisions at 300 GeV. Of these, four are neutral on the average. Resonances, which are rather readily produced at these energies, also have a tendency to decay through many-body channels and contribute to this multiplicity.

Reactions with neutral final states have been much less studied than the more usual charged ones due to major experimental difficulties. As Frank Close[1] puts it: "Neutral particles present more of a headache to experimenters". In particular, the special class of reactions with neutral final states which decay eventually to γ-rays has been only seldom considered. However, this kind of reaction has gained much interest during the last decade in the context of the revival of hadronic spectroscopy induced by the J/ψ "November revolution" of 1974 and by the emergence of QCD as the theory of strong interactions which has led to the search for exotic states (glueballs, hybrids, etc.)... but more about this later.

The experimental methodological difficulties have been overcome only during the seventies thanks to systematic studies and developments which have been undertaken in particular at the Institute for High Energy Physics (IHEP) in Serpukhov (USSR) since 1973 and which have led to the conception of the multigamma spectrometer GAMS. GAMS**-type spectrometers allow the simultaneous measurement of the coordinates and energies of a large number of photons and the reconstruction, with high accuracy, of the masses and momenta of the decaying particles. They offer the possibility to detect events with a complicated topology by means of a very compact setup and to cover effectively a major part of the phase space available to the reactions under study.

---

*   Senior Research Associate – NFSR (Belgium)
**   GAMS is the Russian abbreviation for hodoscope automatic multiphoton spectrometer.

The main object of study of the GAMS Collaboration[*] up to now, and the only one we will consider in some detail here, is the exclusive binary reaction with neutral final states

$$\pi^- + p \rightarrow M^\circ \underset{\overset{\llcorner}{}\; k\gamma}{} + n \qquad\qquad (1)$$

where M° stands for neutral particles or states decaying eventually into photons and n is a neutron.

Exclusive measurements offer the advantage that kinematic constraints can be applied to each event. This greatly reduces the errors on measured quantities such as the mass of M°. Furthermore the allowed angular momentum states of M° are often restricted, in particular in the small t region (OPE), which improves one's ability to make spin assignments to the observed particle states.

Only M° states which decay into some combination of π°, η, ω or η' have been studied up to now. Table 1 gives the relevant properties of these particles.

One should notice that only strong and electromagnetic interactions are involved in these decays, i.e. they take place in less than about 10^{-16}s. This implies that all photons in the final state are produced at the interaction point of the incident π^- inside the target.

The total cross section for π^-p interactions in the energy domain ~ 50 GeV to ~ 250 GeV is nearly constant and equal to 24 mb of which 20 mb correspond to inelastic reactions. The integrated differential cross sections for $\pi^-p \rightarrow (\pi^\circ, \eta, \omega)n$, which are by far the most copiously produced neutral states, in this energy domain give for the ratio $\sigma \, [\pi^-p \rightarrow (\pi^\circ + \eta + \omega)n] \, / \, \sigma_{TOT} \, (\pi^-p)$ a value smaller than 10^{-4}.

Table 1. Quantum numbers and main neutral decay modes of light neutral mesons[2].

M°(mass in MeV)	$I^G J^{PC}$	Main neutral decay modes	B.R.(%)
π°(135)	$1^- 0^{-+}$	2γ	98.8
η(549)	$0^+ 0^{-+}$	$\begin{cases} 2\gamma \\ 3\pi^\circ \rightarrow 6\gamma \end{cases}$	38.9 31.0
ω(783)	$0^- 1^{--}$	$\pi^\circ\gamma \rightarrow 3\gamma$	8.7
η'(958)	$0^+ 0^{-+}$	$\begin{cases} 2\gamma \\ \eta\pi^\circ\pi^\circ \begin{cases} \rightarrow 6\gamma \\ \rightarrow 10\gamma \end{cases} \end{cases}$	2.4 8.4 6.7

*   The Institutions participating in the GAMS Collaboration are:
        Institute for High Energy Physics, Serpukhov, USSR
        Institut Interuniversitaire des Sciences Nucléaires, Belgium
        Los Alamos National Laboratory, Los Alamos, USA
        Laboratoire d'Annecy de Physique des Particules, France

This means that neutral final states are produced at most once for 10,000 interacting pions in the target and that the reactions one is interested in are occuring at a rate at least one order of magnitude lower. Thus it is necessary to use a high intensity pion beam in order to get a reasonable number of interesting events. Also a very good anticoincidence system for charged particles is essential in order to reject as much as possible all non neutral final states.

The following section is devoted to a short description of the experimental setup, especially of GAMS spectrometers and their main characteristics. Then the essential steps of the data analysis, up to the reconstruction of the events, are explained. Next a survey of the main results obtained up to now by the GAMS Collaboration is given and two new results, not yet published, are studied in more detail in order to illustrate the power of the method and the physical interest of the results. Finally, some explanations are given about the actual extension of the GAMS program at CERN (NA12/2 Experiment) and of the plans of the Collaboration for the near future.

EXPERIMENTAL SETUP

A scheme of the experimental layout in the H8 beam line of the North Area at CERN (NA12 Experiment) is shown in Fig. 1. The central unit in the setup is the Cherenkov hodoscope spectrometer GAMS-4000. With it are

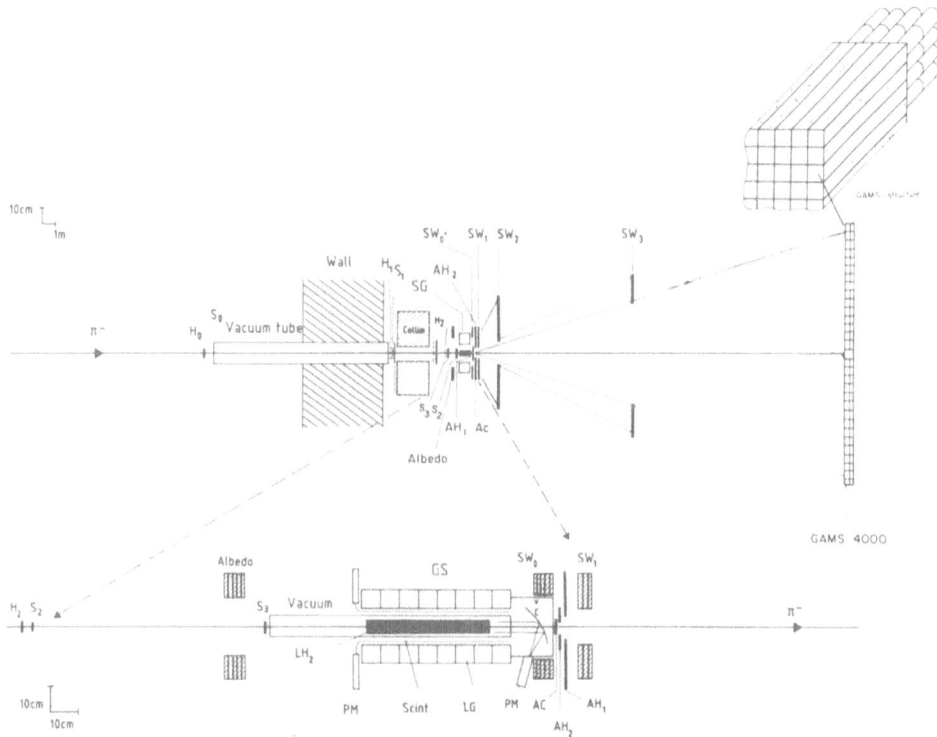

Fig. 1. Experimental layout. S_i: beam definition counters, H_i: beam hodoscopes; GS: guard system surrounding the 60 cm long liquid hydrogen target LH_2 (12 scintillation counters + 96 lead glass counters LG). AC and AH_i: scintillation counters; SW_i + Albedo: lead–scintillator sandwich counters.

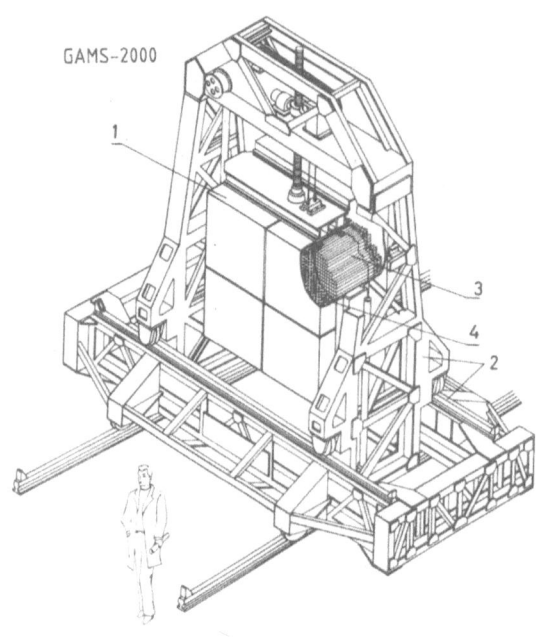

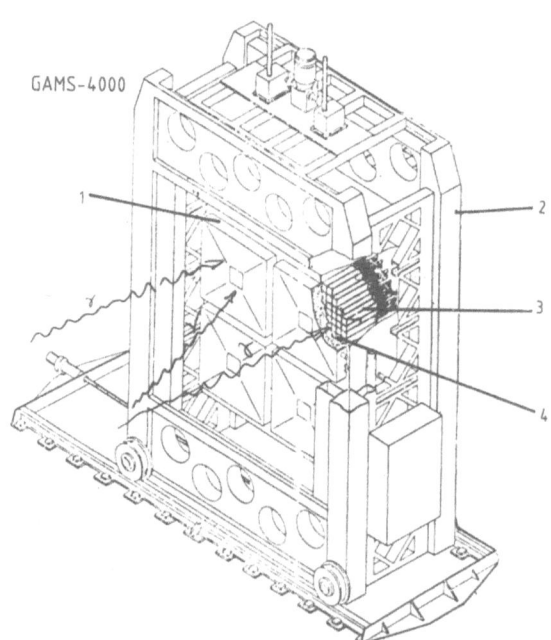

Fig. 2. General view of GAMS-2000 (above) and GAMS-4000 (below):
(1) lead-glass container; (2) moving frame;
(3) lead-glass counters; (4) light monitoring system.

250

measured the energy and the impact coordinates of each gamma in multigamma states. This, together with the knowledge of the origin of the interaction, allows the determination of the mass and of the momentum of these states. It should be underlined that no magnet is needed in the setup.

The GAMS Collaboration is running in parallel a second experiment installed at the IHEP 70 GeV proton synchrotron in Serpukhov (6th Joint CERN-IHEP Experiment). It is a smaller scale version of the CERN experiment; it is adapted to the lower energies available at Serpukhov, but otherwise both setups are quite similar. The central unit in this setup is GAMS-2000.

A summary of the main differences between the two detectors is given below (see Table 2). Unless otherwise stated, the examples given below concern the NA12 Experiment at CERN (GAMS-4000 detector).

GAMS Spectrometers

Detectors. GAMS-4000 is a matrix of 64 x 64 lead-glass cells with dimensions 38 x 38 x 450 mm³. The four central cells have been left out to provide a hole for the passage of particles which have not interacted in the target. The corresponding numbers for GAMS-2000 are 48 x 32 cells and a central hole corresponding to one cell. A general view of both detectors is shown in Fig.2. A detailed description of GAMS-2000 is given elsewhere[3].

A high energy photon (or electron) hitting the lead glass produces an electromagnetic shower. The electron component of the shower radiates Cherenkov light in the glass. The amount of Cherenkov light is proportional to the energy deposited in each counter which is viewed through its downstream end by a photomultiplier tube.

The lead-glass in GAMS-4000 represents ~ 16 radiation lengths and ~ 1 hadronic interaction length of material. However, the essential parameter of such a detector is the lateral dimention of the cell (2 Δ). If Δ is rather larger than b, the effective width of the electromagnetic showers in the glass, the coordinates of the photon (or electron) impact on the spectrometer can be measured only roughly, the accuracy being of the order of half the cell dimension. On the other hand, if Δ is similar or smaller than b (< 2cm in F-8 glass), the shower speads over a few cells. Comparing the amount of light produced in the different cells allows the determination of the photon (or electron) coordinates to an accuracy of ~ 1 mm (cf reference[3] and the references therein).

Because of their cellular structure, GAMS spectrometers are capable of measuring many simultaneous photons, contrary to the more conventional electromagnetic calorimeters consisting of sandwiches of scintillator strips and iron (or lead) sheets where the overlap of the different projections of the showers limits very much the number of simultaneous photons they can measure. This is an important feature at SPS energies, where events with ten or more photons can be easily detected by GAMS-4000 with reasonably high efficiency, which often allows the detection of two or more decay channels of many particles in the same experiment. For example, the η has comparable γγ and 3π° branching ratios. Thus, a resonance that decays into an ηη pair can be measured as a 4γ, 8γ, or 12γ final state, each having quite different topology. This provides a very powerful consistency test of the analysis procedures.

The minimum distance for which two overlapping photons may be identified and still be resolved is about 25 mm (i.e. even in the case when two gammas fall in the same cell) but this distance depends strongly on the relative energies of the photons and their absolute positions within a cell.

The measured energy resolution of the GAMS-4000 spectrometer for a single photon is $\sigma_E/E = 0.011 + 0.053 / \sqrt{E(GeV)}$.

Electronics. The pulse height analysis of each photomultiplier signal is made in parallel by 4096 individual analog to digital converters (ADC). In order to obtain the large dynamic range and the good linearity needed for GAMS, each photomultiplier tube is directly coupled to a gated charge to time converter (QTC). The time to digital conversion (TDC) is assured by a 50 MHz clock. The full dynamic range of the ADC channels is 12 bits, i.e. 4096 counts. The maximum conversion time is 250 μs per event.

Fast processing performs immediate pedestal subtraction and energy normalization (with coefficients from calibration runs). A wired pre-processor allows a determination of the effective mass, the total energy and the transverse momentum of the multigamma system being registered. This is done by evaluating the first and second moments of the signals in each cell relative to the axis of the spectrometer. In this way, a preselection of rare events can be done in presence of a huge flux of information before recording on magnetic tape (slow trigger). Taking into account the coding times, the system allows the recording of more than 500 events per second with a multiplicity of about 200 cells ($\sim$ 10 γ-rays).

A detailed description of the acquisition system of GAMS-4000 has been given elsewhere[4]. The main parameters of both GAMS detectors are given in Table 2.

Table 2.   Main parameters of GAMS-2000 and GAMS-4000

	GAMS – 2000	GAMS – 4000
Lead glass (made in USSR):		
– type-transparency class	TF1 – 000	F8 – 00
– PbO (%)	50	45
– density (g/cm³)	3.85	3.60
– radiation length X_0 (cm)	2.6	2.9
Detector:		
– cell dimensions (mm³)	38 x 38 x 450	
– number of cells	1535	4092
– matrix n x m – hole	48 x 32 – 1	64 x 64 – 4
– useful area (m²)	2.2	5.0
– weight of glass (t)	4	10
– PM type (made in USSR)	84–3	
– coordinate accuracy σ_x (mm)		
at 25 GeV	1.3	2
at 200 GeV	–	1
– σ_E/E for single γ (%)		
at 25 GeV	1.8	2.2
at 200 GeV	–	1.5
– mass resolution for particles decaying into γ	few %	
– max. number of simultaneous γ	10	20
Electronics:		
– QTC type	MQT 200 (Lecroy)	
– dynamic range in counts (bits)	4092 (12)	
– max. conversion time (μs)	250	
– max. number of events /s	$\sim$ 2000	
– gate width (ns)	60	

Calibration and monitoring. At the beginning and at the end of each run GAMS-4000 is calibrated in a beam of monoenergetic electrons. Such beams can readily be obtained at high energy accelerators thanks to the non negligible amount of energy radiated by fast electrons (synchrotron radiation) in the bending magnets of the beam transport systems[5]. For example, electrons with an initial energy of 300 GeV loose 12 GeV in transversing the H8 beam line at the CERN SPS, i.e. they are separated from the hadrons in the beam by more than 6 cm at the exit slit. Thus it is easy to select the electrons by carefully compensating the magnetic field in the bending magnets for the energy loss of the electrons. Typically, bursts of up to 50,000 electrons with a hadron contamination of less than 2% are obtained in this beam (Fig. 3).

All cells are successively irradiated by moving the detector across the defocused electron beam which irradiates about 2 x 5 cells at a time. This insures a uniform irradiation and also saves time as many cells are calibrated simultaneously. The speed of the detector displacement is adjusted so that each cell is hit by a sufficient number of electrons (300 to 500). It takes about 10 hours to perform the complete calibration of GAMS-4000.

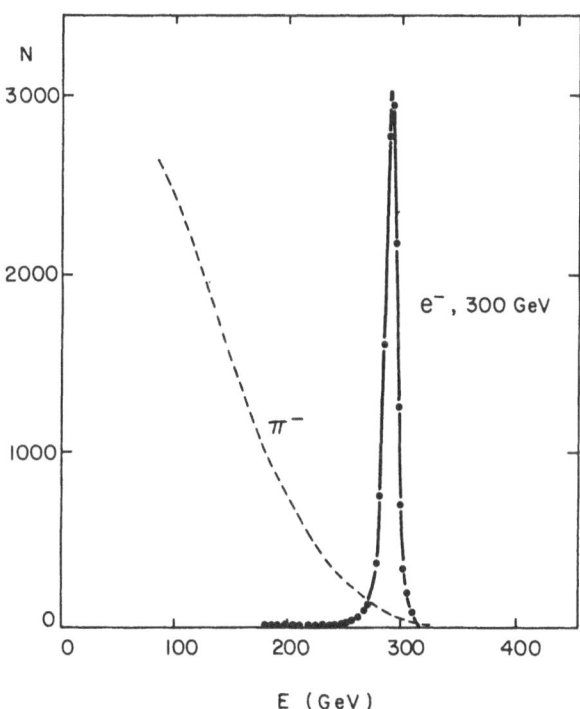

Fig. 3. Electrons separated from hadrons by synchroton radiation at 300 GeV.

The gain variations of the counters are monitored with a blue-light emitting flash-lamp which is triggered immediately after the end of every SPS burst. This light is distributed through optical fibres to each cell of GAMS. Because the light intensity fluctuations from flash to flash can be as large as 20%, the flash-lamp light output is normalized with respect to a stable reference PM which compares the flash-lamp light output with that of a pulsed light-emitting diode, stabilized in temperature.

Target, beam counters and triggering system. (cf Fig. 1)

The liquid hydrogen target is 60 cm long. The intensity of Cherenkov light emitted by the pions in the target is measured, allowing for the evaluation of the longitudinal coordinate of the vertex of the neutral state interaction with an accuracy of about 5 cm. One millimeter step scintillation hodoscopes (H_0, H_1, H_2) define the transverse coordinate and direction of the pions incident on the target.

The arrival time of a pion on the target is determined by the coincidence of signals from thin scintillators (S_0, S_1, S_2, S_3) placed upstream of the target. The latter is completely surrounded by thin scintillator counters (AC, AH_1, AH_2, GS-Scint) which detect the emission of charged particles. This guard system is completed by lead-glass Cherenkov counters (GS-LG) and lead-scintillator sandwiches (SW_0, SW_1, SW_2, SW_3, Albedo) which subtend the entire solid angle not covered by the GAMS-4000 spectrometer. These detectors make an unambiguous signal for exclusive neutral processes, given by the combination:

$$\prod_i S_i \cdot \overline{(AC + AH_1 + AH_2 + GS\text{-}Scint)} \cdot \overline{(\sum_j SW_j + Albedo)}$$

Only 6% of the incident π^- interact in the target of which a fraction 10^{-4} (cf introduction) give rise to neutral final states, i.e. a fraction 6.10^{-6} at most of the incident π^- give good events. In order to obtain a reasonable number of events per burst, the intensity of the pion beam has to be high, typically 3.10^7 π^-/burst in the CERN experiment. The most critical part in the trigger is the anticoincidence AC counter which must have an efficiency better than 99.99% in order to reject effectively all charged particles.

EVENT RECONSTRUCTION

In order to illustrate how particles decaying into photons can be reconstructed from the energy and the coordinates of the gammas measured in GAMS, one will consider here only the simplest case, namely events with an even number of well separated photons (non-overlapping showers) which may all be recombined into pairs corresponding to the decay of π°, η or η' mesons. A more comprehensive treatment has been given elsewhere[6].

The energy of the individual photons is in this case simply obtained by summing the amplitudes of the signals in the cells of GAMS touched by the shower multiplied by their respective calibration coefficient determined in the calibration with electrons. A first biased estimate of each photon impact coordinates is obtained from the center of gravity (first moment) of the shower in the cells. This estimate is corrected using an odd polynomial of the ninth degree which gives eventually the coordinates with an accuracy of the order of 1 mm. Moreover a shower profile function[6] is fit to the measured showers and a χ_γ^2 measuring the quality of the fit is associated with each gamma.

Several cuts are then applied, e.g. on the total energy deposited in GAMS which should be close to the incident pion energy as the recoiling neutron in reaction (1) carries only little energy away (< few GeV).

Next, events are classified according to their multiplicity m of gammas. Some problems may happen at this stage as the class of multiplicity m events can be contaminated by either the presence of a fake photon or by the escape of a low energy photon. The main source of fake gamma is the accidental occurence of a second interaction within the ADC gate which

depends on the incident beam intensity. This contamination is reduced
by one order of magnitude by applying a variable energy threshold on the
photons as a function of their distance to the central hole in GAMS. A
low energy gamma may escape detection if its falls either within the
central hole or outside the detector, or if its energy is lower than
the detection threshold value. Corresponding corrections are obtained
by Monte Carlo evaluations.

The knowledge of the energy and impact coordinates of the photons on
GAMS and of the vertex of the interaction inside the target, from which
the photons are emitted, determines completely the momentum of the photons.
If a particle of invariant mass M decays into m photons, their energies E_i
and momenta $\bar{p}_i$ in the laboratory are related to M by the following relation
(remember that $E_i^2 - p_i^2 = 0$ for photons):

$$M^2 = \left(\sum_{i=1}^{m} E_i \right)^2 - \left(\sum_{i=1}^{m} \bar{p}_i \right)^2 = \sum_{k=1}^{m} \sum_{\substack{\ell=1 \\ k<\ell}}^{m} m_{k\ell}^2$$

where   (2)

$$m_{k\ell}^2 = 4\, E_k E_\ell \sin^2 \frac{\alpha_{k\ell}}{2} \approx E_k E_\ell\, \alpha_{k\ell}^2$$

is the invariant mass squared of the pair of photons k and ℓ and $\alpha_{k\ell}$ is the
angle between $\bar{p}_k$ and $\bar{p}_\ell$.

Thus the way to proceed is simple, in principle. The invariant mass
of each γ-pair is evaluated. When for some pair it is found equal to
the mass of either a π°, a η or a η', within some specified interval
centered on the mass of the particle, this pair is associated with the
decay of the corresponding particle. In the simplest case one considers
here, the process stops when all γ have been univocally recombined into
pairs. The next step consists in trying higher order recombinations of
the photons. For example, if six photons at least are measured in GAMS
which can be recombined into 3 pairs corresponding to π° and if

$$\sum_{k'=1}^{6} \sum_{\ell'=1}^{6} m_{k'\ell'}^2 = (m_\eta \pm \Delta m)^2$$

where k'(ℓ') is a subset of indices corresponding to the paired photons,
then the 3 π° are associated with the decay of a η. Fig. 4 shows a typical
8-γ event measured at 100 GeV/c where one γ-pair corresponds to one η and
the other three with three π° which are further recombined into a η.

In practice, the identification of the particles is not always as
straightforward. For example, in the analysis of 4-γ events measured at
100 GeV/c, it happens that some can be recombined either into two η or
into two π°. The final identification is then based on a kinematical
fit of the specific reaction under study.

Remark. For a particle of mass M decaying into two photons, relation
(2) gives immediately:

$$\frac{\alpha_{12}}{2} \approx \sin \frac{\alpha_{12}}{2} = \frac{M}{2} / \sqrt{E_1 E_2}$$

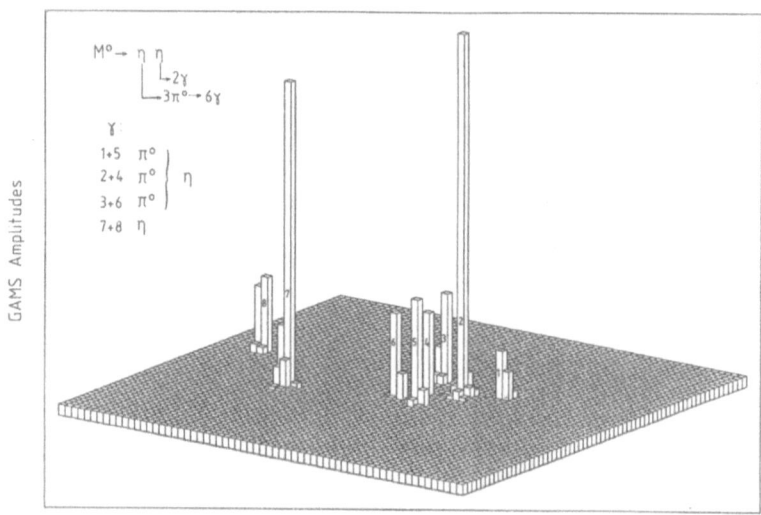

Fig. 4.   Isometric reconstruction of a typical 8-γ event
produced in reaction (1) at 100 GeV/c (L = 20 m).

In the case of symmetric decay (i.e. when the two photons are emitted in
the M rest frame perpendicularly to the direction of flight of M in the
laboratory), which implies $E_1 = E_2 = E_M/2$, E_M being the energy in the
laboratory of M, this relation becomes:

$$\alpha_{12}^{sym} \simeq 2 M / E_M.$$

For example, at 100 GeV α_{12}^{sym} for a $\pi°$ is 2.7 mr while for a η it is
11 mr. If the distance from the target to GAMS is 10 m, the separation of
the photons on GAMS is 2.7 cm and 11 cm, respectively. This means that the
photons from a decaying $\pi°$ are hardly separated at high energies unless
the distance L between the target and GAMS is taken very large in which
case the phase space coverage by the setup drops dramatically. Thus, L is
an important parameter which has to be optimized in relation with the
reaction under study.

SURVEY OF RESULTS OBTAINED WITH GAMS

Most physics results obtained up to now, both at IHEP and at CERN,
concern studies of exclusive production of neutral final states in the
charge exchange reaction (1) at 38, 100, and 230 GeV/c. Processes studied
include two-body cross sections of known mesons, rare decay modes of
established mesons, reconfirmation and measurement of branching ratios of
less well-established mesons, and the search for new states. Most of the
results presented in this section have been already published and the
other few are in press. A summary of these results is given in Table 3.

One of the first states discovered by this collaboration is the $J^{PC} =$
$= 6^{++}$ meson r(2510), decaying into 2 $\pi°$ [7]. Its mass and width are measured
to be (2510 ± 30) and (240 ± 30) MeV/c² respectively. It is seen as a
shoulder on the tail of the $J^{PC} = 4^{++}$ h(2030) in the charge exchange reaction
(1) at 38 GeV/c. This new state is a strong candidate for the $I^G=0^+$ member
of the $J^{PC}=6^{++}$ octet. This is supported by the fact that the mass of the
r(2510) places it right on the f and h trajectory of the Chew-Frautschi plot.

Table 3.   Summary of results obtained with GAMS

Decays of η

$\eta \rightarrow \gamma\gamma$	BR = 0.389 ± 0.004	38 GeV/c
$\eta \rightarrow 3\pi^\circ$	BR = 0.319 ± 0.004	38 GeV/c
$\eta \rightarrow \pi^\circ\gamma\gamma$	BR = $(7.1 \pm 1.7) \times 10^{-4}$	38 GeV/c
$\eta \rightarrow 3\gamma$	BR < 5×10^{-4}	38 GeV/c

Decays of η'

$\eta' \rightarrow 2\gamma$	BR = $(2.43 \pm 0.13) \times 10^{-2}$	38 GeV/c
$\eta' \rightarrow \eta\pi^\circ\pi^\circ$		38 GeV/c
$\eta' \rightarrow \omega\gamma$	BR = $(3.20 \pm 0.36) \times 10^{-2}$	38 GeV/c
$\eta' \rightarrow 3\pi^\circ$	BR = $(1.6 \pm 0.4) \times 10^{-3}$	38 GeV/c
$\eta' \rightarrow 3\gamma$	BR < 1×10^{-4}	38 GeV/c
$\eta' \rightarrow \pi^\circ\gamma\gamma$	BR < 8×10^{-4}	38 GeV/c
$\eta' \rightarrow 2\pi^\circ$	BR < 1×10^{-3}	38 GeV/c
$\eta' \rightarrow 4\pi^\circ$	BR < 5×10^{-4}	38 GeV/c

Recently Discovered States

$r(2510) \rightarrow 2\pi^\circ$	$J^{PC} = 6^{++}$	38 GeV/c
$G(1590) \rightarrow \eta\eta$ and $4\pi^\circ$	$J^{PC} = 0^{++}$	38/100/230 GeV/c
$G(1590) \rightarrow \eta\eta'$		38 GeV/c
$X(1750) \rightarrow \eta\eta$	$J^{PC} = 0^{++}$ or 2^{++}	38 GeV/c
$X(2220) \rightarrow \eta\eta'$	$J^{PC} = 2^{++}$ or 4^{++}	38/100 GeV/c
$X(1850) \rightarrow \eta\eta$ and $4\pi^\circ$	$J^{PC} = 2^{++}$	100 GeV/c
$X(1430) \rightarrow \eta\pi^\circ$	$J^{PC} = 1^{-+}$	100 GeV/c
$?(2100) \rightarrow \eta\eta$		100 GeV/c

$\eta\eta$ Decays of Known Mesons

$f(1270) \rightarrow \eta\eta$	BR = $(3.1 \pm 0.7) \times 10^{-3}$	38/100/230 GeV/c
$\varepsilon(1300) \rightarrow \eta\eta$	BR ≈ 2×10^{-2}	100/230 GeV/c
$f'(1525) \rightarrow \eta\eta$	σ.BR = (0.14 ± 0.05)nb	100/230 GeV/c
$h(2030) \rightarrow \eta\eta$	BR = $(2.2 \pm 1.0) \times 10^{-3}$	100/230 GeV/c

Other Decays

$B(1260) \rightarrow \omega\pi^\circ$ and $\omega\eta$	100 GeV/c
$g(1690) \rightarrow \omega\pi^\circ$	100 GeV/c
$D(1285) \rightarrow \eta\pi^\circ\pi^\circ$	100 GeV/c
$E(1420)? \rightarrow \eta\pi^\circ\pi^\circ$	100 GeV/c

Inclusive Measurements

$\pi^- p \rightarrow \chi X$	38 GeV/c

Exclusive Central Production Measurements

$\pi^- p \rightarrow \pi^- p\ M^\circ$	300 GeV/c

More unexpected has been the discovery of an isoscalar, $J^{PC} = 0^{++}$, G(1590)[8]. It was first observed in reaction (1) at 38 GeV/c decaying into $\eta\eta$, the η being observed in their 2γ decay mode. Results of the partial wave analysis of the $\eta\eta$ data at 38 GeV/c show clearly the G(1590) in the S-wave, with a mass measured to be (1592 ± 25) MeV/c² and a width of (210 ± 40) MeV/c². The f(1270) dominates the D-wave. The peak is slightly higher (1400 MeV/c²) than the tabulated value of the f-meson mass, due to angular momentum barrier effects. This is the first observation of the f(1270) decaying into $\eta\eta$. The branching ratio for this decay is measured to be (5.2 ± 1.7) x 10^{-3}.

A similar peak has been observed in the $\eta'\eta$ spectrum at 38 GeV/c with both the η and the η' being observed in their 2γ decay mode[9]. This peak is quite near threshold, but it cannot be accounted for by phase space or threshold effects. The $\eta'\eta$ final state is much harder to observe than $\eta\eta$ because the branching ratio for the 2γ decay of the η' is only 1.9%. Also, there is a large combinatorial background from the production of π° and η pairs which are three orders of magnitude more abundant. The branching ratio of this state into $\eta'\eta$ is measured to be (2.7 ± 0.8) times that into $\eta\eta$.

The $\eta\eta$ system produced in reaction (1) with 100 GeV/c incident pions has also been studied[10]. The η are observed in both their 2γ and $3\pi^{\circ}$ decay modes. A partial wave analysis was performed to extract the various spin contributions from the charge exchange cross-section. With three spins contributing there are ambiguities in the partial wave analysis – two possible solutions for the S and D-wave amplitudes with a unique solution for the G-wave amplitude. The mass region below 1.7 GeV/c², where the spin 4 wave is negligible, contains two resonances of spin zero. They are ident-ified with the ϵ(1300) and the G(1590) mesons. The mass and width of the G(1590) are measured to be (1580±30) MeV/c² and (280±40) MeV/c² respectively – in full agreement with the 38 GeV/c result. The spin two wave is charac-terized by a clean peak around 1.4 GeV/c² due to the f(1270) and f'(1525). This is the first observation of the decay f' → $\eta\eta$. The mass region above 1.7 GeV/c² is complicated by the ambiguity problem with no agreement between the two solutions. None of the spin zero or spin two structures in this region can be identified with a known meson (cf next section). The peak in the spin four wave can be identified with the h(2030) meson. Production cross-sections and branching ratios into $\eta\eta$ pairs have also been evaluated.

In both experiments, the most striking $\eta\eta$ resonance observed is the G(1590). It is not observed in either experiment to decay into $2\pi^{\circ}$ at a level of 0.3 of the 2η decay rate. It has also not been observed by other experiments[11] looking for the production of $K^{\circ}\overline{K^{\circ}}$ in reaction (1). The decay rate into $K^{\circ}\overline{K^{\circ}}$ is less than 0.6 of the 2η decay rate. The suppression of these decay modes makes it difficult to identify the G(1590) with a state composed of normal quarks. We have proposed it as a candidate for a glueball or state of high gluonic content.

It has been hypothesized by several authors[12,13] that the η and η' couple relatively strongly to gluon pairs. This is supported by the large branching ratios for the radiative decays of the J/ψ(3100) into the η and particularly into the η'. A mechanism of gluon decoloration has been pro-posed which would favor the decay of glueballs into the η and η' mesons preferably to either π° or K$^{\circ}$ mesons[14]. This mechanism gives branching ratios for G(1590), assumed to be a glueball, which are in agreement with the measured values.

A very nice series of results comes from the 38 GeV/c data with measurements of neutral decay widths of the η-meson[15]. A sample of 6 x 10^5 η-mesons were used to make precision measurements of four neutral decays of the η: $\gamma\gamma$, $3\pi^{\circ}$, $\pi^{\circ}\gamma\gamma$, and 3γ. The results, summarized in Table 4, improve

Table 4   Branching ratios and partial widths
of the neutral decays of the η

Channel	BR_i (%)[a]	Γ_i (eV)[b]
η → 2 γ	38.9 ± 0.4	470 ± 120
η → 3 π°	31.9 ± 0.4	380 ± 100
η → π°γγ	0.071 ± 0.017	0.85 ± 0.28
η → 3 γ	< 0.05	< 0.6

a) normalized to BR(η → neutrals) = (70.9 ± 0.7)% [2]
b) normalized to Γ(η → all) = (1.2 ± 0.3) KeV [16,17]

the error on the world average of the decays η → γγ and η → 3π° by a factor
of two. The measurement of the decay η → π°γγ confirms the only other
observation, made by the same apparatus, of this decay[18]. An upper limit is
placed on the charge conjugation violating decay η → 3γ, Γ(η→3γ)/Γ(η→2γ) <
< 1.2x10⁻³ at the 95% confidence level.

Studies have been made at IHEP of inclusive particle production in π⁻p
induced reactions. A unique study, at 38 GeV/c, of the production of χ-
particles states (the ³P_J charmonium states) has been made[19]. The χ are
detected through their radiative decay into J/ψ(3100) which is in turn
detected through its decay into e⁺e⁻. Only the χ(3510) and χ(3555) states
were seen in this experiment because of the small branching ratio (0.8%) for
the radiative decay of the χ(3415) into J/ψ(3100). The trigger was generated
by GAMS-2000 itself with the passive analog addition of the signals from the
last dynode of the photomultipliers in GAMS-2000. The cross-section for the
production of the J/ψ(3100) is in agreement with previous measurements[20].
The inclusive cross-section for χ production for x_F > 0 was measured to be
σ(π⁻p → χX) = (28 ± 11) nanobarns.

More recently, a closer investigation of the ηη and η'η spectrum from
reaction (1) at both 38 and 100 GeV/c have revealed two interesting structures. The first is a state of high spin at mass 2220 MeV/c² observed
decaying into η'η at both incident beam momenta[21]. The second is observed
at 1755 MeV/c² in a high statistics sample of data in the ηη decay spectrum
at 38 GeV/c [22].

The state at 2220 MeV/c² is observed in the sample of events that
contains four photons – both the η and η' being observed in their 2γ decay
mode. Because these events are heavily dominated by π°π° and ηπ°, all
events with any pair of photons consistent with π° or with two pairs of
photons consistent with ηη are rejected. All events with the second pair
of photons consistent with the η' are fitted with three constraints to
reaction (1). The resulting η'η mass spectrum in both data shows a small
peak at a mass of (2220 ± 10) MeV/c². The low statistics of the data sample
do not allow a determination of the width of the particle. Its observed width
is consistent with the instrumental resolution of the detectors (~ 5% at
2 GeV/c²). The statistical significance of these peaks are ~ 4.5σ in the
38 GeV/c data and ~ 3.5σ in the 100 GeV/c data.

The most striking feature of these data is the very anisotropic angular
distribution in Gottfried–Jackson frame of the η'η system. The angular
distributions in neighbouring mass bins are much flatter. The lack of statistics also renders a partial wave analysis of these data impractical.
However, the highly anisotropic angular distribution is a clear indication
of a spin J ≥ 2. If the state is a q̄q meson then it must have J^{PC}=(even)⁺⁺.

It is very tempting to identify this state with the $\xi(2230)$, observed by the MARK III group[23] in the radiative decays $J/\psi \to \gamma K\bar{K}$ and $J/\psi \to \gamma K_S^0 K_S^0$. They make a spin assignment of $J^{PC} = (\text{even})^{++}$, without excluding $J = 0$. There has been much speculation about the nature of the $\xi(2230)$ including its identification with gluonic or a hybrid state[24], a Higgs boson[25], or a high-spin $s\bar{s}$ meson[26]. If they are indeed the same state, it would be the first observation, in a hadronically induced reaction, of a newly discovered state produced in J/ψ radiative decay. This identification would also rule out the possibility that it is a Higgs boson, which must be a scalar.

The second narrow mesonic state was seen at a mass of 1755 MeV/c² in the $\eta\eta$ spectrum of reaction (1) at 38 GeV/c. The event selection was similar to that for the $\eta\eta$ events which discovered the G(1590), but with a roughly five times larger data sample. Because of the larger data sample it was possible to look for $\eta\eta$ pairs in the region of large momentum transfer, up to $|t| \sim 1$ (GeV/c)². The new state is not observed in the region $|t| < 0.2$ (GeV/c)² where the spectrum is dominated by other mesons such as G(1590) and h(2030). A clear narrow peak appears above a continuous background in the region $|t| > 0.35$ (GeV/c)². The statistical significance of this peak is $\sim 7\sigma$. The mass of the meson is (1755 ± 8) MeV/c². The observed width of the peak is consistent with the instrumental resolution of the detector so that the intrinsic width of the resonance is $\Gamma < 50$ MeV/c². The quantum numbers for this new state must be $J^{PC} = (\text{even})^{++}$ because of its $\eta\eta$ decay mode. The isotropic angular distribution in the Gottfried–Jackson frame of the $\eta\eta$ system indicates $J^{PC} = 0^{++}$, assuming spin alignment in the production process. A 2^{++} assignment is not ruled out.

The most striking features of the new state, which we call X(1750), are its small production cross-section and its relatively flat t-distribution. Its production cross-section is measured to be $\sigma(\pi^- p \to Xn) \cdot BR(X \to \eta\eta) = (3.5 \pm 1.5)$ nanobars. The t-distribution of the X(1750) can be described by an exponential function, $e^{-b|t|}$ with $b = (3.8 \pm 1.5)(\text{GeV/c})^{-2}$. This is significantly smaller than the value $b \approx 10$ $(\text{GeV/c})^{-2}$, typical at small $|t|$ for other mesons produced in reaction (1).

Several high precision studies of η' decays have been made in the 38 GeV/c data of reaction (1) The first has allowed to measure the isospin-violating decay $\eta' \to 3\pi°$ [27]. This rare process had not been observed before. The branching ratio of this decay is important as the knowledge of $\Gamma(\eta'\to3\pi°)/\Gamma(\eta'\to \eta\pi°\pi°)$ is related to the ratio of the masses of the u and d quarks[28]. The second is a measurement of the matrix element of the decay $\eta'\to \eta\pi°\pi°$ [29], observed in 6γ (BR = 8.4%). This decay has not been studied in the past because of the experimental difficulty in measuring large numbers of simultaneous photons. Data from the related charged decay $\eta' \to \eta\pi^+\pi^-$ have been obtained[30], but the global world statistics used to determine this decay is 1500 events. The present data sample contains about 5400 events. The matrix element of the neutral decay is shown to be approximately constant, with a weak dependence on one of the Dalitz variables.

The decay $\eta' \to \omega\gamma$ has also been studied[31]. This decay has been measured in only one previous experiment[32], where the $\omega \to \pi^+\pi^-\pi°$ channel was observed as a 70 event signal above an equal background. The present analysis uses a data sample of ~ 150 events over a ~ 100 events background. The branching ratio obtained for this decay is $BR(\eta' \to \omega\gamma) = (0.033 \pm 0.0036)$. The ratio of partial widths $R^{-1} = \Gamma(\eta'\to \rho\gamma)/\Gamma(\eta'\to \omega\gamma)$ is of particular interest because its measurement, depending only weakly on the dynamics of the interaction, provides directly the square of the d-quark charge. References to theoretical evaluations of this ratio are given in Ref.[31]. Using the tabulated value for the decay $\eta'\to \rho\gamma$, we obtain for this ratio, $R^{-1} = (9.1 \pm 0.9)$, in good agreement with predictions of the simple additive quark model.

Table 5. Branching ratios and partial widths
of the neutral decay modes of η'(958)[*])

Decay	BR (%)	Γ (KeV)
η' → 2γ	2.43 ± 0.13	4.25 ± 0.19 [35]
η' → ηπ°π°	21.7 ± 0.5 [2]	37.9 ± 2.7
η' → ωγ	3.20 ± 0.35	5.6 ± 0.7
η' → 3π°	0.160± 0.032	0.28 ± 0.06
η' → 3γ	< 0.01	< 0.02
η' → π°γγ	< 0.08	< 0.14
η' → 2π°	< 0.1	< 0.2
η' → 4π°	< 0.05	< 0.09

*) All upper limits are given for a 90% C.L.

Further extensive studies have been made at 38 GeV/c [33] of the η'
neutral decays into 2γ, ηπ°π°, ωγ and 3π°. Upper limits have also been
obtained for the η' decays into 3γ, π°γγ, 2π° and 4π°. The results
are summarized in Table 5. The fact that all decay channels have been
measured in the same set of data has allowed to reduce systematic errors to
a minimum. Former data on the 2-γ decay channel had only a low statistical
significance[2] (few tens of events) with the exception of one work[34] in
which, however, the rate of this decay was determined indirectly. The data
sample of 2-γ decays in this experiment exceeds eight thousand.

The measurement of BR(η' → 2γ) has allowed to determine the full width
of the η' using the partial width Γ(η' → 2γ) measured in e⁺e⁻ collisions[35]:
Γ(η' → all) = (176 ± 16)KeV. The matrix element of the G-parity violating
decay η' → 3π° has been determined and found approximately constant.

ONGOING STUDIES

The last section was devoted to a quick survey of the main results
obtained by the GAMS Collaboration since its beginning. No details were
given as these can easily be found in the quoted references. In order to
illustrate how physics results are obtained starting from the reconstructed
events in GAMS, one will consider here preliminary results concerning two
new channels which are under study, namely: M° → 4π° → 8γ and M° → π°η → 4γ
produced in reaction (1) at 100 GeV/c.

Study of M° → 4π° final states

Interest. For qq̄ mesons the 4π° decay mode is expected to be rare in
comparison with other 4π decay modes, due to ρ-dominance: M° → (ρρ) → 2π⁺2π⁻
or 2π°π⁺π⁻ but not to 4π° as ρ° cannot decay into 2π°. For mesons with an
exotic structure, e.g. glueballs, it has been shown[14,36] that BR(M°→4π°)/
/BR(M°→all 4π), can be as large as 1/5 if the gluon decoloration mechanism
is the main mechanism at work in this decay. Thus, ordinary qq̄ mesons like
f(1270), D(1235), h(2030) etc. are expected to decay rarely into 4π°
(0.1% to 1%) while this could become one of the main decay channels for
exotic mesons like G(1590), a scalar glueball candidate, for which
BR(G → 4π°) could be only twice lower than BR(G → ηη)[36].

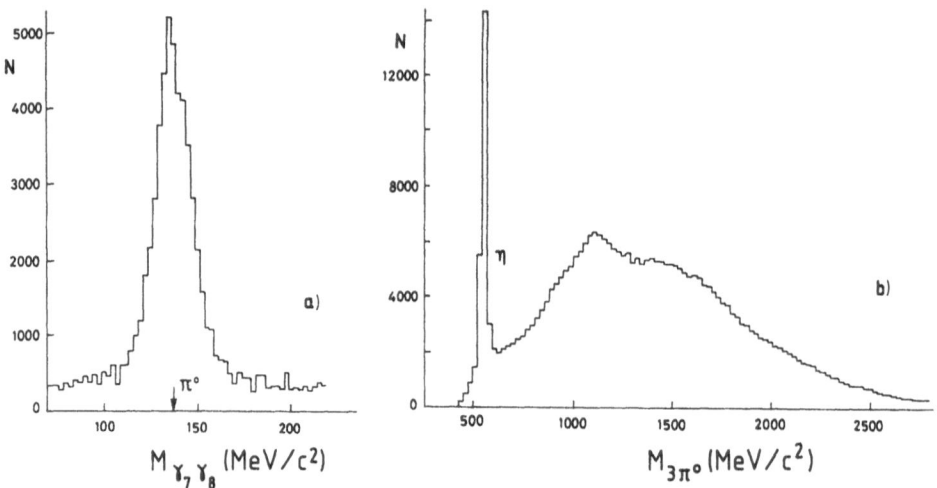

Fig. 5 a) Invariant mass of the fourth γ-pair after six out of eight
photons have been recombined into π°;
b) invariant mass spectrum of 3π° systems produced in the
reaction π⁻p → 4π°n at 100 GeV/c.

The study of the 4π° decay channel opens a new way to the search
for exotic states: glueballs, hydrids, etc. It had not been measured up
to now for a series of experimental difficulties including the need to
detect and measure momentum and coordinates of eight gammas and dis-
entangle the large combinatorial background (twenty-eight two-gamma
combinations for each decay event).

Mass spectra. The data presented here have been taken with a target
to GAMS distance of 20 m and concern only one third of the total
statistics. Amongst the 125,000 collected eight-gamma events, 40,000
correspond to four identified π° after a 5-C fit, fixing the mass of
each π° and the mass of the recoil neutron in reaction (1). Fig. 5a
shows the invariant mass spectrum of the fourth γ-pair after the first
three have been identified with π° (4-C fit). The background is small,
less than 10%. A search has been made for subsystems consisting of two
and three π°. Apart from a weak shoulder in the mass region of the
f(1270), no structure is observed in the 2π° mass spectrum (pure phase
space). On the contrary, the 3π° mass spectrum (Fig. 5b) shows a sharp
peak in the mass region of the η above a small combinatorial background.
Of the 4π° events, 30% correspond to ηπ° decays which are dominated by
the A₂(1320) meson. Appropriate cuts allow to eliminate these events.

The invariant mass spectrum of the remaining 4π° events is shown
in Fig. 6a. Three structures superposed on a continuous background
are visible at about 1.3 GeV/c², 1.6 GeV/c² and 1.8 GeV/c². The
t-distributions of the events in three mass bands around these values
(Fig. 6b) show a behaviour typical for one-pion exchange processes (OPE),
i.e. strong peaking in the forward direction. In order to isolate the OPE
contribution to the production process, only events with $|t| \leq 0.15$ (GeV/c)²
are further considered. This cut results in a loss of only 20% of the 4π°
events without changing much the shape of the mass spectrum (Fig. 6c).

Analysis. The allowed J^{PC} quantum numbers of 4π° systems are
$0^{++}, 2^{++}, 4^{++}$... As the production process is dominated by OPE, the
spin of the M° state is aligned i.e. its projection along the z-axis of
the Gottfried-Jackson system (t-frame) is zero (OPE without absorption).

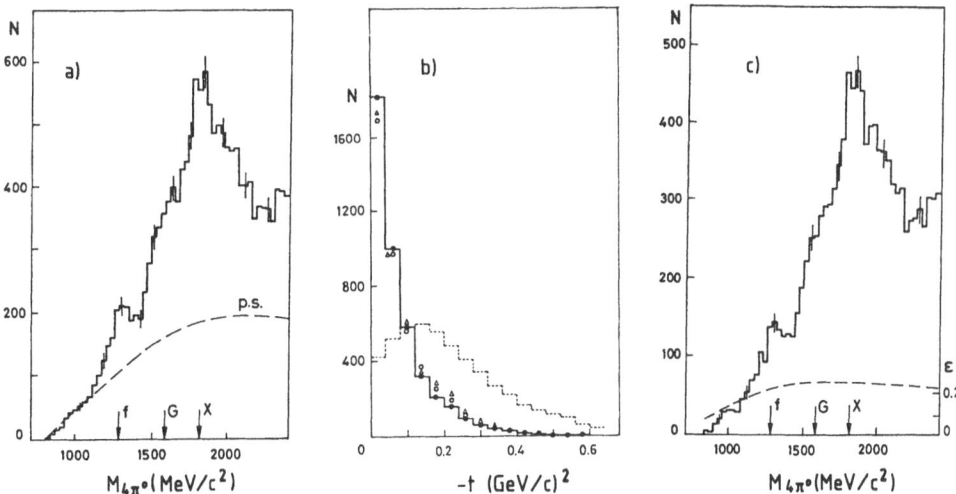

Fig. 6 a) Invariant mass spectrum of genuine $4\pi°$ systems (see text)
        produced in the reaction $\pi^-p \to 4\pi°n$ at 100 GeV/c
        (dashed line: pure phase space contribution);
     b) t-distribution of these events in the mass region around the
        f(1270) (Δ), the G(1590) (o) and the X(1810) ($\bullet$), respectively.
        The shape is typical of one pion exchange. The dashed histogram
        is the t-dependence of A_2(1320)$\to \eta\pi° \to 4\pi°$ events (ρ-exchange)
        measured in the same experiment;
     c) same as a) but for events with $|t| < 0.15$ (GeV/c)² (dashed line:
        detection efficiency).

In this case the matrix element for the decay of M° into $4\pi°$, considered
to consist of two $\pi°$-pairs in S-waves with a relative angular momentum
L = 2, symmetric under the exchange of any two pions (Bose-Einstein
statistics), is

$$M \div \sum_{i=2}^{4} [3 (\bar{e}_z \cdot \bar{p}_{1i})^2 - \bar{p}_{1i}^2]$$

which reduces to $M \div \sum_{i<j} \bar{p}_{ij}^2 \cdot (3 \cos^2\theta_{OB} - 1)$ if one introduces a
symmetrized cosine of the auxiliary angle θ_{OB} defined by

$$\cos^2 \theta_{OB} = \sum_{i<j} \bar{p}_{ij}^2 \cos^2\theta_{ij} / \sum_{i<j} \bar{p}_{ij}^2 \qquad (i,j = 1,2,3)$$

where $\bar{p}_{ij} = \bar{p}_i + \bar{p}_j$, $\bar{p}_i$ being the momentum of the $\underline{ith}$ $\pi°$ in the Gottfried-
Jackson frame of M°, and $\cos \theta_{ij} = \bar{e}_z \cdot \bar{p}_{ij}/|\bar{p}_{ij}|$, $\bar{e}_z$ being a unit vector
along the z-axis in this frame.

    Monte Carlo simulation show that $4\pi°$ decays of spin 2 states are
enhanced in an optimum way if $\cos \theta_{OB} < 0.4$ while for spin 0 states the
corresponding optimum cut is $\cos \theta_{OB} > 0.5$. These cuts applied to the
events of Fig. 6c give the invariant mass spectra shown in Figs. 7a and 7b
for $\cos \theta_{OB} < 0.4$ and $\cos \theta_{OB} > 0.5$, respectively. Two peaks are now clearly
showing up in Fig. 7a, one at ~ 1.3 GeV/c² and the other at ~ 1.8 GeV/c²
whilst the shoulder at 1.6 GeV/c² has nearly disappeared. In Fig. 7b, on the
contrary, the signal at 1.6 GeV/c² is much enhanced comparatively to the
other two which are nearly gone. The mass spectra in both figures have been

263

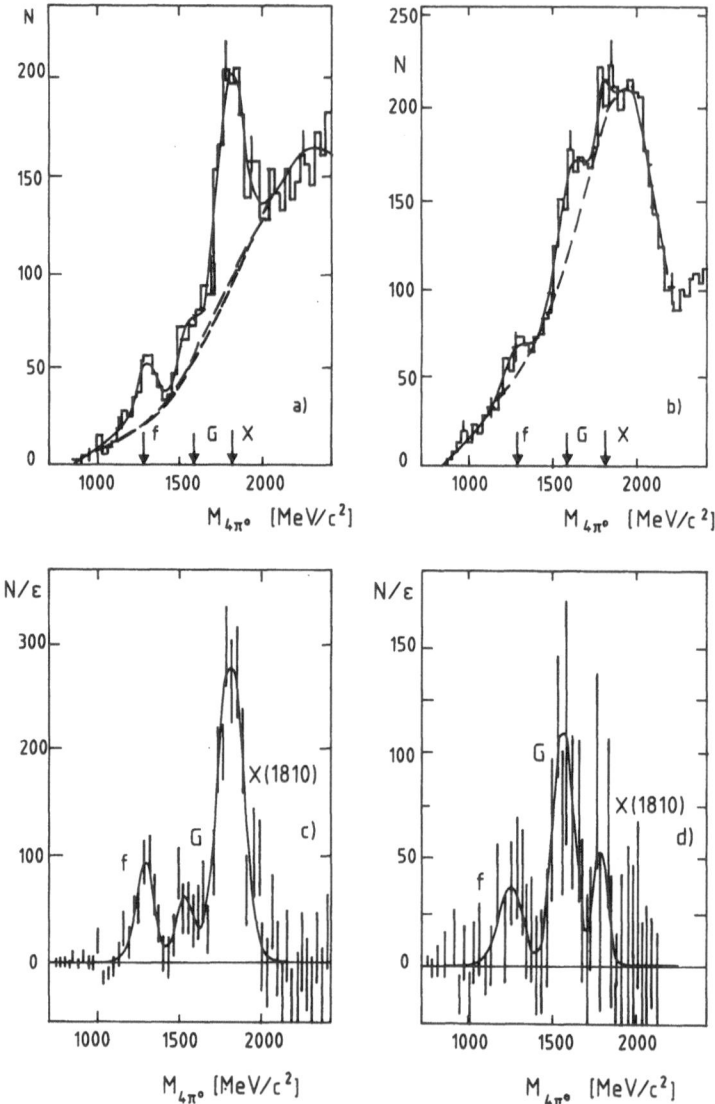

Fig. 7   a) and b): Invariant mass spectra of genuine $4\pi°$ events
of fig. 1c) with cos θ_{OB} < 0.4, favouring $J^P = 2^+$(a), and
with cos θ_{OB} > 0.5, favouring $J^P = 0^+$ (b). The curves
show the result of a fit with three Breit–Wigner curves (full
lines) superposed on a polynominal continuum (dashed lines);
c) and d): are the mass spectra corrected for efficiency
after subtraction of the continuum.

fit with three Breit-Wigner resonances superposed on a polynomial
continuum. The results of the fit are shown in Fig. 7c and Fig. 7d. A
comparison of the relative number of events in corresponding peaks in the
last two figures with Monte Carlo evaluations allows to fix unequivocally
the quantum numbers of those resonances: spin 2 for the 1.3 GeV/c² and
1.8 GeV/c² states, spin o for the 1.6 GeV/c² state.

The mass and width of the spin 0 state are M = (1570 ± 30) MeV/c² and
Γ = (145 ± 40) MeV/c². It is identified with the scalar meson G(1590), a
glueball candidate[14], observed previously in the ηη and ηη' decay modes[8-10].
After normalization of the 4π° data relatively to the reaction π⁻p → η'n
with η'→ ηπ°π°, measured simultaneously in the same experiment, and using
previous results on the ηη decay channel[10], one gets BR(G → 4π°)
/ BR(G → ηη) = 0.8 ± 0.3. This large branching ratio into 4π° strengthen
the interpretation of the G(1590) as a glueball candidate[36].

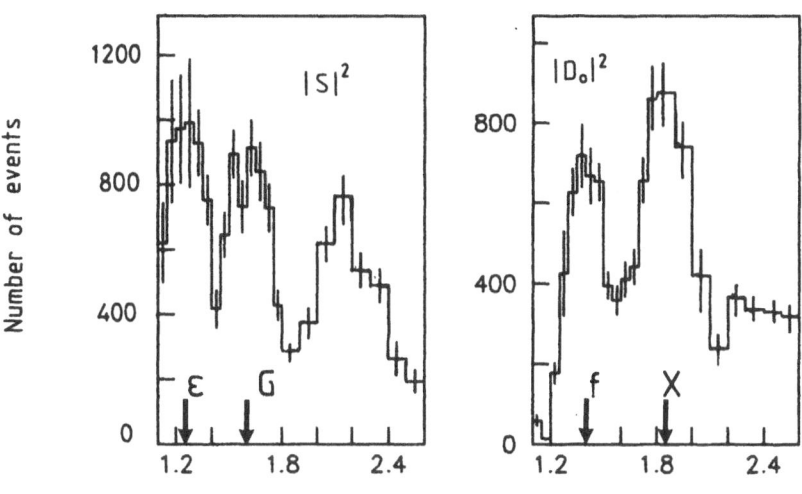

$$M_{\eta\eta} \text{ (GeV/c}^2)$$

Fig. 8   Result of a partial-wave analysis of the reaction π⁻p → ηηn
           measured at 100 GeV/c [10]. The solution shown here (which is one
           of two possible solutions) corresponds to "phase coherence". The
           D-wave shows cleary the presence of a state at (1850 ± 40) MeV/c²

The parameters of the first spin 2 state are M = (1286 ± 10) MeV/c² and
Γ = (140 ± 20) MeV/c². It is identified with the f(1270) meson. Using the
same normalisation procedure as above, one obtains BR(f→4π°)=(3±1).10⁻³ i.e.
the decay into 4π° is, as expected, a rare decay mode for the f(1270) meson.

The second spin 2 state has a mass M = (1810 ± 20) MeV/c² and a width
Γ = (170±30)MeV/c². This state cannot be identified with any known meson
decaying into ππ or K̄K but looks similar (Fig. 8) to a 2⁺⁺ state decaying
into ηη obtained as one of two possible solutions in a partial-wave analysis
of the reaction π⁻p → ηηn measured at 100 GeV/c [10]. Identification of
these two states leads to BR[X(1810)→ 4π°] / BR[X(1810) → ηη] = 0.8 ± 0.3.
On the other hand, the present experiment also gives the upper limit:
BR[X(1810) → 2π°] / BR[X(1810) → 4π°] < ¼. From Gershtein's argument[36]
one concludes that X(1810) is a good tensor glueball candidate.

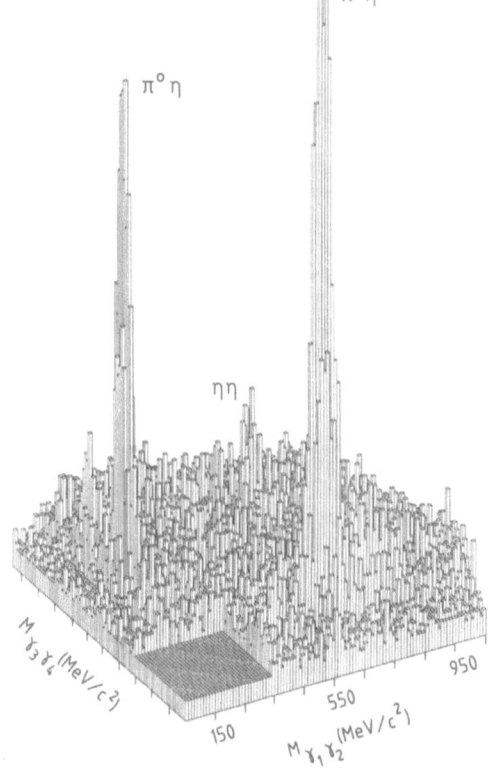

Fig. 9   Isometric view of all recombinations of 4γ into pairs. Even at
        this stage clear π°η and ηη peaks are seen over a small background.
        The π°π° events, which contribute to a peak about ten times higher
        than the π°η one, have been left out.

Study of ηπ° Final States

   Interest. The allowed J^P quantum numbers for ηπ° systems are given by
the natural series 0^+, 1^-, 2^+, 3^-, ... and their charge parity C is equal
to +1. However, ηπ° systems resulting from the decay of q$\bar{q}$ states can only
appear in even partial waves ($J^P = 0^+, 2^+, 4^+ ...$) as fermion-antifermion
pairs with an odd spin have C = −1. Thus the reaction π⁻p → ηπ°n is a good
channel to look for states with quantum numbers not accessible to q$\bar{q}$ states,
also called exotics.

   Mass spectrum. The data presented here have been obtained with a target
to GAMS distance of 20 m and comprise only ½ (14,000 ηπ° events) of the
total statistics. Another set of 24,000 events (taken at 15 m) is not yet
fully analysed. Fig. 9 shows a lego plot of all recombinations of the raw
4-γ data into pairs. The majority of 4-γ events detected in GAMS are π°π°
events. A first selection is made keeping only events that can be recombined
into two pairs, one having an invariant mass within 70 MeV/c² of the π° mass
and the other an invariant mass within 160 MeV/c² of the η mass. Then a fit
is performed with three constraints on the masses of the η, π° and n and the
ηπ° assignment is given on the basis of a minimal χ² criterium.

266

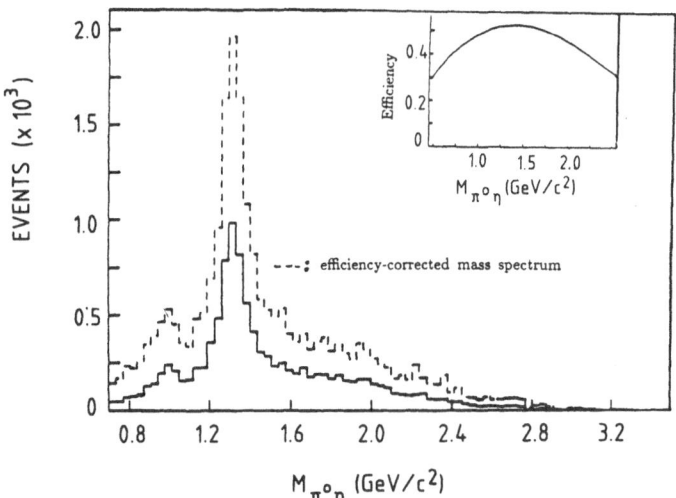

Fig. 10   Invariant mass spectrum in 35 MeV/c² bins of ηπ° events
produced in the reaction π⁻p → π°ηn at 100 GeV/c. Full histogram:
raw data; dashed histogram: mass spectrum corrected for efficiency.
The inset shows the detection efficiency versus the measured mass
of the ηπ° systems.

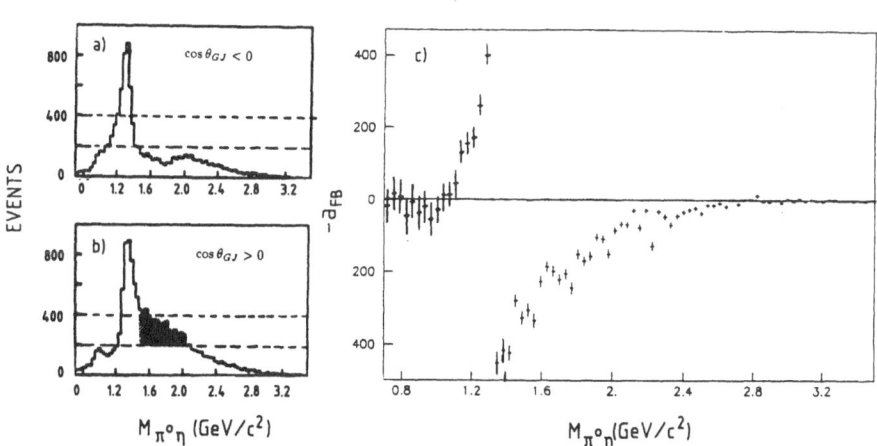

Fig. 11   a) and b) Invariant mass spectra for ηπ° events with cos θ$_{GJ}$< 0 (a)
and cos θ$_{GJ}$ > 0 (b), where θ$_{GJ}$ is the angle of the η in the
Gottfried–Jackson system;
(c) is the corresponding forward–backward asymmetry.

The raw invariant mass spectrum, integrated over the whole measured
range of four momentum $|t| \leq 0.8$ (GeV/c)², and the efficiency-corrected
spectrum are shown in Fig. 10. The correction is made for each individual
event as a function of its decay angles in the Gottfried–Jackson frame and
of the $\eta\pi°$ invariant mass. Possible contaminations from other channels
have been studied by Monte Carlo simulations. The $\eta\pi°\pi°$ channel, where two
soft photons produced by a low energy $\pi°$ escape detection and where the
remaining photons follow the constrained $\eta\pi°$ hypothesis with an acceptable
χ^2, is the largest source of background but its contribution is small,
about 5% in the low mass region (around the δ) and less elsewhere.

A dominant peak is observed in Fig. 10 at the mass of the well known
2^+ meson $A_2(1320)$ as well as a smaller one corresponding to the 0^+ meson
$\delta(980)$. Defining θ_{GJ} as the polar angle between the η and the incident beam
direction in the Gottfried–Jackson frame of the $\eta\pi°$ system, one expects a
symmetrical distribution in $\cos \theta_{GJ}$ of the events if only even partial waves
intervene in the decay. In reality (Fig.11) a quite significant asymmetry is
observed above ~ 1350 MeV/c² between events with $\cos \theta_{GJ} < 0$ and those with
$\cos \theta_{GJ} > 0$ (shaded area in Fig.11b) that leads to a strong forward-backward
asymmetry (Fig. 11c). This asymmetry had already been observed, but with a
lower statistics, in an experiment performed at 40 GeV/c [37]. Such an asymmetry can only be explained by the presence of an exotic odd-L amplitude
interfering with the $A_2(1320)$ in the D-wave [38].

Analysis (preliminary results). A first indication of the presence of
a P-wave interfering with the A_2 is already apparent in the decay angular
distributions of the $\eta\pi°$ systems in different mass bins (Fig. 12). An

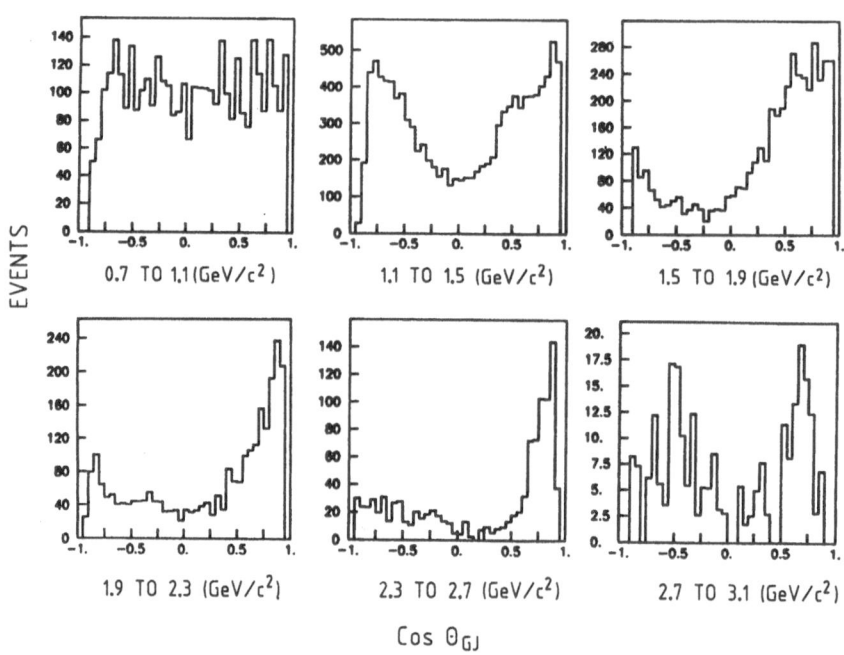

Fig. 12   Distribution of the $\eta\pi°$ events in 400 MeV/c²
mass bins versus $\cos \theta_{GJ}$.

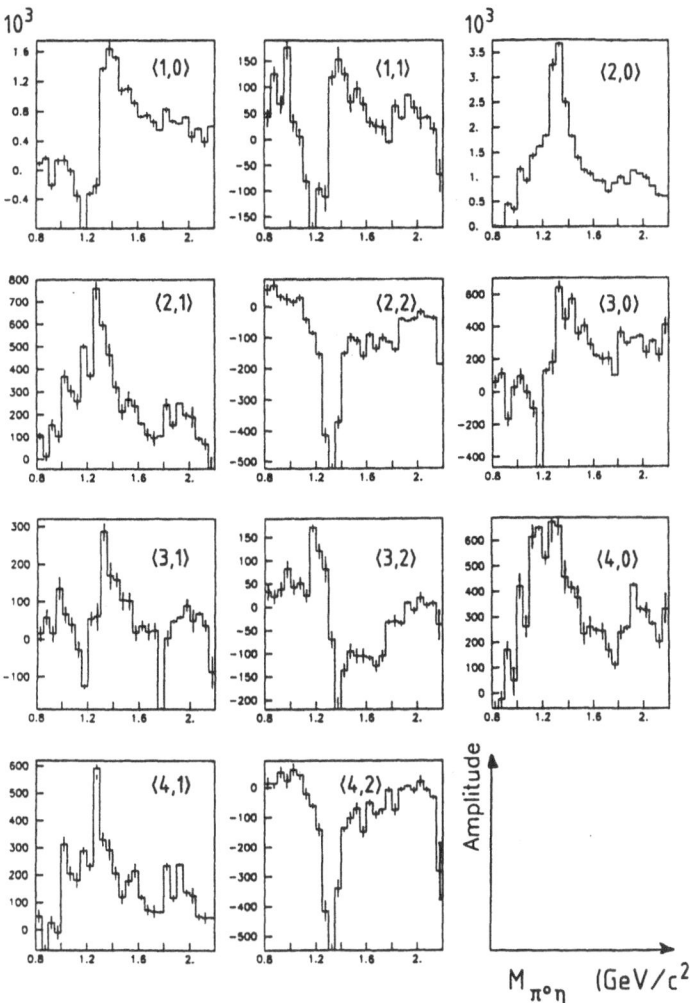

Fig. 13   Most significant unnormalized moments versus the mass of the $\eta\pi^\circ$ systems (<ℓ,m> is a shorthand for $\sqrt{4\pi}$ N <Y_ℓ^m>).

isotropic distribution, typical of S-wave behaviour, is observed in the $\delta(980)$ mass region, as expected. In the A_2 (1320) mass region, the D-wave behaviour is distorted in the forward direction in the angular domain ($0.6 \leq \cos\theta_{GJ} \leq 0.8$) where the contribution of $P_4(x)$ is large compared to those of $P_2(x)$ and $P_3(x)$, $P_i(x)$ being the usual Legendre polynomials with $x = \cos\theta_{GJ}$.

Having grouped the events in 35 MeV/c² mass bins, the unnormalized moments of spherical harmonics $\sqrt{4\pi}$ N < Y_ℓ^m > (N: number of events in a mass bin) of the experimental angular distributions are calculated for each mass bin using a χ^2 and a maximum likelihood method[39], both giving essentially the same results.   Moments with $\ell > 4$ and m > 2 are found everywhere consistent with zero, which means that only waves with $\ell \leq 2$, i.e. the S, P and D waves, are operating in this decay in the measured mass range.   The most significant moments are shown in Fig. 13.

The moments of the $\eta\pi^\circ$ angular distributions can be expressed as linear combinations of bilinear products of the productions amplitudes $L_{\lambda\pm}$ describing, to leading order in s, an $\eta\pi^\circ$ system of spin L and helicity λ produced by natural (+) or unnatural (−) parity exchange. Since the moments with m > 2 are consistent with zero, the amplitudes with $\lambda \geq 2$ can be ignored. A table of the relevant combinations for $\lambda = 0,1$ is given e.g. in ref.[40]. Solving numerically these equations with some reasonable assumptions concerning the phases (coherence), a preliminary solution has been obtained for the amplitudes which is shown in Fig. 14. It is found that A_2 (1320) production is dominated by natural parity exchange, as expected, and that a P-wave component with a resonant-like behaviour appears around 1.4 GeV/c². Before concluding that the latter is a $J^{PC} = 1^{-+}$ exotic state, a more detailed amplitude analysis is required. Such a study is in progress.

Addendum (August 87). This analysis has now been completed. It shows that the phase of the P-wave varies rapidly around 1430 MeV/c², as expected for a resonance.

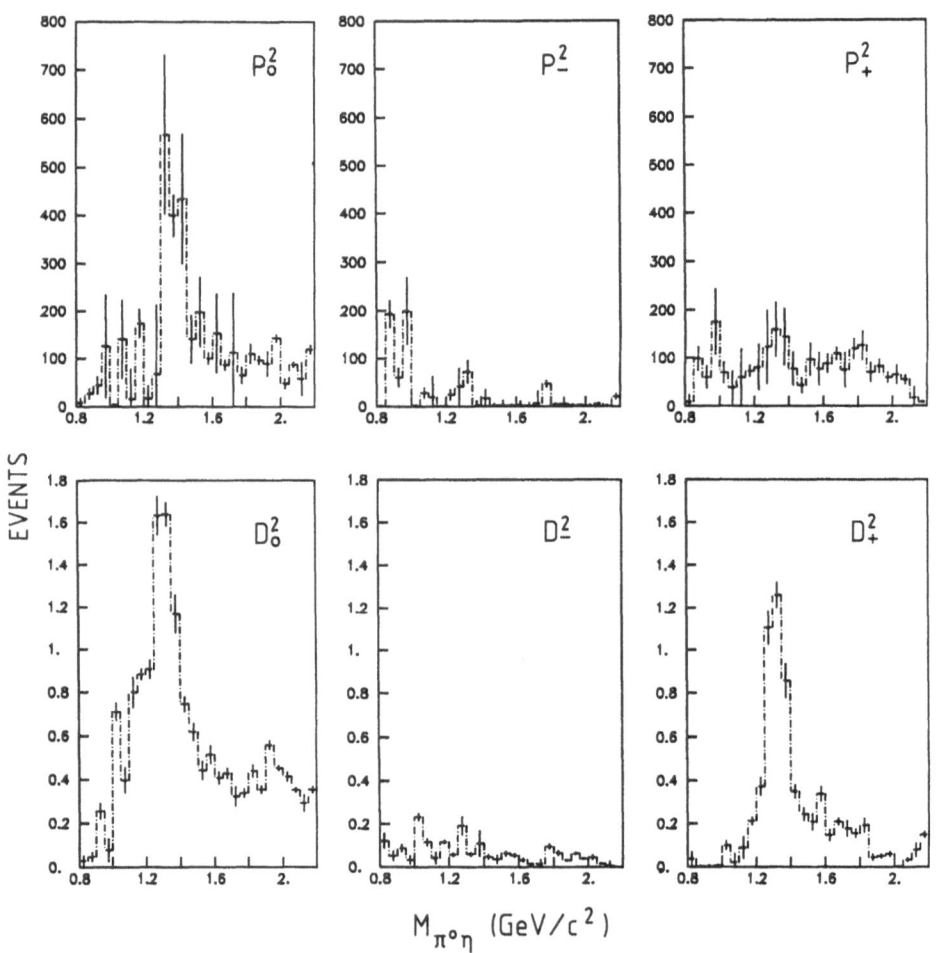

Fig. 14   Preliminary solution for the P² and D² amplitudes.
(number of events in D² waves: x10³)

270

The main line which has been pursued up to now by the GAMS Collaboration, both at CERN and at IHEP, is a systematic search for new neutral light meson resonances (including glueballs and exotic states) and for new rare neutral decay modes of known meson resonances produced in charge exchange reactions of π^- on protons. These resonances are detected as neutral final states with a high multiplicity of gamma rays.

Charge exchange reactions have produced a wealth of information on glueball candidates decaying into η and η' and other states. They show that the light meson spectrum is still not well understood. At the present stage a new division of tasks between the two setups is considered.

NA12/2*

The difficulty with identifying glueballs is that there is no clear-cut argument to prove the pure or quasi-pure gluonic nature of a meson. The problem can only be solved by a thorough study of the light mesons spectrum and of their many different decay channels on the one hand and to find most favourable conditions for the production of gluon-rich states on the other hand. J/ψ decay into ordinary light quark mesons is an OZI suppressed process and as such it is expected to occur through gluon hadronization. It should provide an ample source of gluonium. But so should the gluon sea in central hadron collisions.

Gluon rich states, if not pure gluoniun states, seem to play a more and more important role in hadron collisions with increasing energy. At the Sp$\bar{\text{p}}$S collider energies most jets appear to come from gluon hadronization[41]. A possible explanation of the universal rise of hadron total cross sections is the increasing contribution of central gg interactions and formation of gluonium states with a mass in the range of the G(1590) meson and above[42]. At energies above a few hundred GeV the collision of sea gluons inside interacting hadrons is expected to give rise to a substantial production of glueballs. Such states could be preferentially observed at rest in the center of mass of the collisions, with η' and η being an essential part of their decay products.

Thus, an extension of the NA12 program has been proposed and accepted at CERN under the name NA12/2. One of its goals is to study with 300 to 350 GeV hadrons (pions, protons) the exclusive reaction:

$$\text{hadron} + N \rightarrow \text{hadron} + N + M^\circ$$
$$\hookrightarrow \eta\eta, \eta'\eta, \eta'\eta' \rightarrow k\gamma$$

where M° are centrally produced ($x_F \simeq 0$) and are expected to have a large gluonic component. The number of final state photons, k, lies between 4 and 20, depending on the η and η' decay channel observed. In such a reaction the signal for a gluonic state should be enhanced in the decay into pairs made of η and η's because of the decoloration mechanism of gluons by gluons, which should be specific for glueball decay[14]. Cross sections of the order of ten microbarns may be expected with 350 GeV pions[43]. This reaction ends finally into multiphoton final states whose detection utilises

---

*   Since 1986, a group from Pisa University/INFN has joined the GAMS Collaboration.

the full performance capability of the GAMS-4000 spectrometer, in particular its self-triggering possibilities.

Although η and η' detection seems to be most promising to study gluonium states, the observation (or non-observation) of other decay channels may be decisive to resolve ambiguities. Long lived kaons and a fraction of the short lived kaons in neutral events can be measured with the hadron calorimeter HC-240 [44] installed just behind GAMS. HC-240 is also a valuable tool to distinguish hadronic showers from electromagnetic showers in complex events. Hadrons may be measured as such in events with moderate multiplicity.

Another goal of the experiment is the study of heavy quark meson production. GAMS is a good apparatus for the measurement of the inclusive production of high mass neutral mesons. As a matter of fact tests of the self-trigger capacity for GAMS have been made with GAMS-2000 at IHEP in which χ particles production has been measured through their radiative γJ/ψ decay (J/ψ → e⁺e⁻) [19]. Of particular interest in these studies is the $\eta_c(2980)$. Although a well confirmed state, it has only been seen in the radiative decays of the J/ψ(3100) and ψ(3685), and more recently at the ISR in pp scattering at threshold, seen decaying to γγ [44]. The observation of the $\eta_c(2980)$ and the evaluation of its production distribution as a function of x_F and $p_\perp$ would allow for a better understanding of how states composed of heavy quarks are produced in the hadronic interactions of the light quarks and gluons.

Calculations[45] show that the hadronic production of charmonium states should be dominated by the $\eta_c(2980)$. Searches have been made for its hadronic production, but only in the decay channels γγ, π°π°, and φφ. The branching ratios are small for all of these channels. The largest measured branching ratios for the $\eta_c(2980)$ are for the channels ηπ⁺π⁻ and η'π⁺π⁻ [46]. Its branching ratio into photons through the decays ηπ°π° and η'π°π° should be ~ 1.3%.

An evaluation based on the decay channels of c̄c observable by our apparatus along with their branching ratios and detection efficiencies shows that roughly 400 η_c should be observed in a 50 day running period. A few hundred χ(3510) and χ(3555) states should be observable through their decays χ → γ J/ψ and J/ψ → e⁺e⁻ during the same period of running.

To make these measurements, the NA12 set-up has been updated, as shown schematically in Fig. 15. A small aperture magnetic spectrometer has been added downstream of GAMS-4000. Five multi-wire proportional chambers (MWPC) are placed in this spectrometer, four of which to measure the trajectory of the diffractively scattered incident pion. The order of magnitude of the deflection on the last wire chamber is 0.02/(δp/p)mm or about 12mm for a 1.5 GeV/c² centrally produced resonance. An 8 x 8-cell GAMS-64 is placed downstream of the hole in GAMS-4000 to measure the forward produced photons. A small hadron calorimeter HC-9 is placed behind the last wire chamber to reject low energy hadron and muon background. Four planes of 2mm spaced MWPC cover the solid angle of GAMS-4000 as seen by the target. A live-target made of 16 pieces of 5mm thick scintillator allows for the determination of the longitudinal location of the interaction and the range of the recoil proton.

The fast trigger for exclusive central production is derived from the MWPCs and from GAMS-4000 itself. The wire chambers provide two trigger signals, one depending on the multiplicity of charged particles in the forward diffraction cone while the other depends on a predetermined threshold value of the deflection angle in the spectrometer. As for the GAMS self-

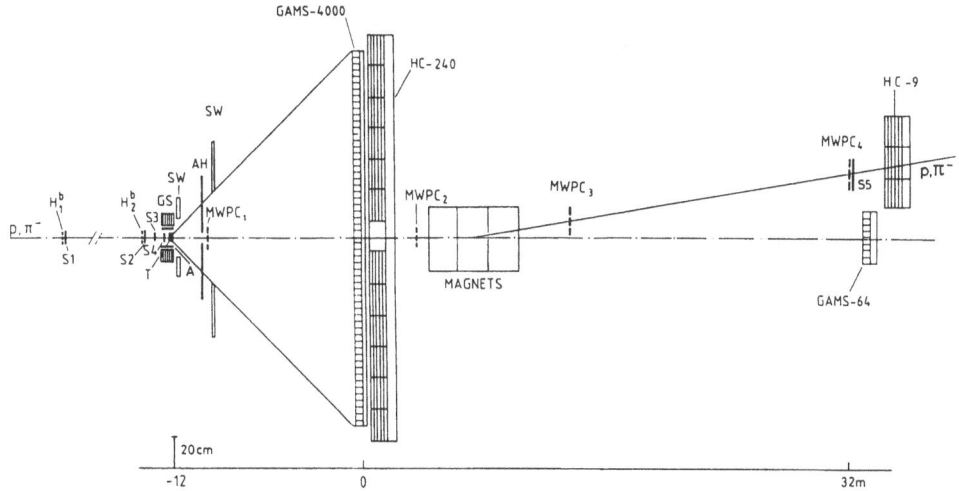

Fig. 15   NA12/2 experimental layout. In addition to the counters in
           Fig. 1 one finds CH$_i$: MWPCs; HC-240 and HC-9: hadron
           calorimeters; GAMS-64: electromagnetic calorimeter.

trigger, the detector is segmented into 256 4x4-cell blocks. The signals
from the last dynode of each photomultiplier tube in a 4x4 block are summed
in PM-adder circuits. The PM-adder signals are fanned out to a modular
electronics system to form analog signals proportional to the quantities:

$$M_x^{(n)} = \sum_{i,j} x_i^n E_{ij}$$

$$\left.\begin{array}{c}\\\\\\\\\\\end{array}\right\} \quad n = 0,1,2$$

$$M_y^{(n)} = \sum_{i,j} y_j^n E_{ij}$$

$$M_r = \sum_{i,j} r_{ij} E_{ij}$$

where $M^{(n)}$ is the nth moment of photons in GAMS-4000 and E_{ij} is the energy
deposited in the 4 x 4-cell block centered on (x_i,y_i), at a distance
$r_{ij} = \sqrt{x_i^2+y_j^2}$ from the central hole in GAMS. These signals may in turn
be used, either in an analog or digital manner, to form meaningful triggers.

     The proper functionning of the entire system has been checked during a
short data taking period in 1986. These preliminary data show clearly the
central production of G(1590) in the $\eta\eta$ channel at 300 GeV.

GAMS-4π

     Central collisions have significant cross sections at CERN SPS energies
only. This has been confirmed by the observation of a measurable production
of η-pairs at 300 GeV and their absence in a similar experiment performed at
40 GeV at IHEP. On the other hand, exclusive reactions may be conveniently
measured at IHEP as their cross-sections are quite larger at 40 GeV.

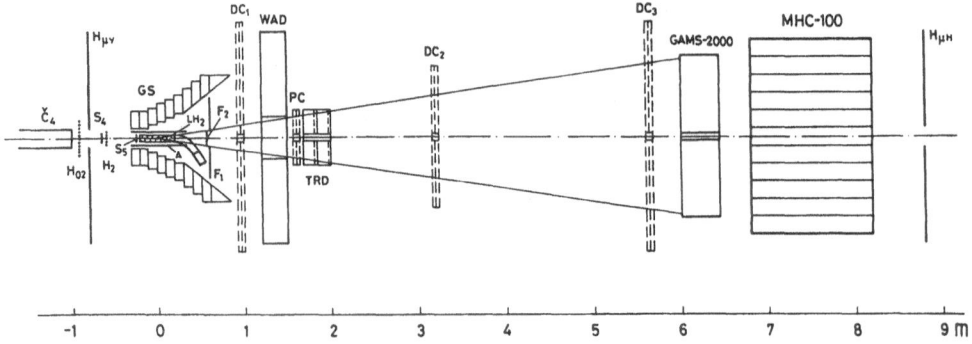

Fig. 16 GAMS-4π setup. GAMS-2000: 48 x 48 matrix of lead-glass cells;
         MHC-100 : modular hadron calorimeter; WAD : wide angle detector
         (lead-scintillator sandwiches); GS : wide angle guard system;
         TRD : transition radiation detector (electron identification);
         PC, DC_{1-3}: auxiliary wire and drift chambers; S,F,A:
         scintillation counters; H_i, H_μ: beam and muon hodoscopes;
         LH_2 : liquid hydrogen target.

        Substantial improvement of the setup at IHEP is foreseen. GAMS-2000,
which at present contains 48 x 32 = 1536 lead-glass cells, will be filled
to a square of 48 x 48 = 2304 cells in order to cover the solid angle of
acceptance in a symmetrical way. A wide angle calorimeter, WAD, consisting
of hodoscope lead-scintillator sandwiches, will considerably increase the
acceptance at a reasonable cost. A new guard system, with larger cells to
measure the energy of the soft gammas emitted at very large angle, will
cover most of the residual aperture making the setup as close as possible
to a 4π device (in the center of mass system).

        The general scheme of the new setup, called GAMS-4π, is given in
Fig. 16. Its main advantages over the actual setup are:

i)   a much larger geometrical acceptance without reducing the GAMS to
     target distance unduly and without increasing losses through the
     central hole. The half-aperture angle, which is ~ 0.2 rad. in
     GAMS-2000, will be increased to 0.7 rad. in GAMS 4π.

ii) a larger dynamical range for the measurement of gamma energies, i.e.
     a lesser number of small energy gammas will be lost. The efficiency
     for large multiplicity events is increased.

Acknowledgments

        This review would not have been possible were it not for the hard work
of all members of the GAMS Collaboration. To all my colleagues I want to
express my sincere thanks. I am grateful to the organizers of this school
for giving me the opportunity to present this review in such a pleasant
environment. I want to thank all participants in this school for many
stimulating discussions in a very congenial atmosphere. The help of Mrs.
M. d'Adhémar, V. Goidadin, C. Rigoni and M. Seinera in the preparation of
this paper is also gratefully acknowledged.

REFERENCES

1. F. Close, M. Marten and C. Sutton, "The Particle Explosion",p. 18, Oxford University Press, New York (1987).
2. Particle Data Group, Review of Particle Properties, Phys. Lett. 170B (1986) 1.
3. F.G. Binon et. al., Hodoscope multiphoton spectrometer GAMS-2000, Nucl. Instr. Meth. A248 (1986) 86. See also references therein.
4. D.M. Alde et. al., Acquisition system for the hodoscope spectrometer GAMS-4000, Nucl. Instr. Meth. A240 (1985) 343.
5. F.J. Farley et al., Separated high-energy electron beams using synchrotron radiation, Nucl. Instr. Meth. 103 (1972) 325.
   H.W. Atherton et al., Electron and photon beams in the SPS experimental area, Int. report CERN-SPS/85-43(EBS)(1985).
6. D.M.Alde et al., Data reduction procedures for large multicellular electromagnetic calorimeters, Int. report CERN-EP/87-28 (1987).
7. F.G. Binon et al., Observation of a spin-6 neutral meson r(2510), Lett. Nuovo Cim. 39 (1984) 41.
8. F.G. Binon et al., G(1590), a scalar meson decaying into two η-mesons, Nuovo Cim. 78A (1983) 313.
9. F.G. Binon et al., Study of $\pi^-p \to \eta'\eta\eta$ n in a search for glueballs, Nuovo Cim. 80A (1984) 363.
10. D.M. Alde et al., Production of G(1590) and other mesons decaying into η pairs by 100 GeV/c π^- on protons, Nucl. Phys. B269 (1986) 485.
11. B. Cohen et al., Amplitude analysis of the K^-K^+ system produced in the reactions $\pi^-p \to K^-K^+n$ and $\pi^+n \to K^-K^+p$ at 6 GeV/c, Phys. Rev. D22 (1980) 2595.
    A. Etkin et al., Amplitude analysis of the $K^\circ_s K^\circ_s$ system produced in the reaction $\pi^-p \to K^\circ_s K^\circ_s n$ at 23 GeV/c, Phys.Rev. D25 (1982) 1786.
12. J.F. Donoghue, Expectations for glueballs, AIP Conf. Proc. # 81, C.A. Heusch and W.T. Kirk Ed., p.97, Am. Inst. Phys., New-York (1982).
13. V. Novikov et al., Are all hadrons alike?, Nucl. Phys. B191 (1981) 301.
14. S.S. Gershtein, A.K. Likhoded, and Yu.D. Prokoshkin, G(1590)-meson and possible characteristic features of a glueball, Z. Phys. C24 (1984) 305.
15. D.M. Alde et al., Neutral decays of the η-meson, Z. Phys. C25 (1984) 225.
16. A. Weinstein et al., Observation of the production of η mesons in two photon collisions, Phys. Rev. D28 (1983) 2896.
17. A. Browman et al., Relative width of the η mesons, Phys. Rev. Lett. 32 (1974) 1067.
18. F.G. Binon et al., Observation of the decay $\eta \to \pi^\circ\gamma\gamma$, Nuovo Cim. 71A (1982) 497.
19. F.G. Binon et al., χ particle production in π^-p collisions at 38 GeV/c, Nucl. Phys. B239 (1984) 311.
20. G. Matthiae, Dilepton production in hadronic collisions, Rivista Nuovo Cim. 4 (1981) 3.
21. D.M. Alde et al., 2.22 GeV $\eta\eta'$ structure observed in 38 GeV/c and 100 GeV/c π^-p collisions, Phys. Lett. 177B (1986) 120.
22. D.M. Alde et al., Observation of a 1750 MeV narrow meson decaying into $\eta\eta$, Phys. Lett. 182B (1986) 105.
23. R.M. Baltrusaitis et al., Observation of a narrow $K\bar{K}$ state in J/ψ radiative decays, Phys. Rev. Lett. 56 (1986) 107.
24. M.S. Chanowitz and S.R. Sharpe, Glueballs and meiktons which decay to multi-kaon final states, Phys. Lett. 132B (1983) 413.
    B.F. Ward, Glueball theory of the ξ(2.22), Phys. Rev. D31 (1985) 2849.

25. H.E. Haber and G.L. Kane, Some tests whether a narrow neutral resonance can be a Higgs particle, Phys. Lett. 135B (1984) 196.
    R.S. Willey, Is the ξ(2.2) a Higgs boson?, Phys. Rev. Lett. 52 (1984) 585.
    R.M. Barnett et al., Tracking down Higgs scalars with enhanced couplings, Phys. Rev. D30 (1984) 1529.
26. S. Godfrey et al., ξ(2.22): an L = 3 $s\bar{s}$ Meson ?, Phys. Lett. 141B (1984) 439.
27. F.G. Binon et al., The isospin-violating decay $\eta' \to 3\pi°$, Phys. Lett. 140B (1984) 264.
28. D.J. Gross et al., Light-quark masses and isospin violation Phys. Rev. D19 (1979) 2188.
29. D.M. Alde et al., Matrix element of the $\eta'(958) \to \eta\pi°\pi°$ decay, Phys. Lett. 177B (1986) 155.
30. G.R. Kalbfleisch, η' (958) branching ratio, linear matrix element, and dipion phase shift, Phys. Rev. D10 (1974) 916. See also the references therein.
31. D.M. Alde et al., Decay $\eta' \to \omega\gamma$, Europhys. Lett. 3 (1987) 553.
32. C.J. Zanfino et al., Observation of the radiative decay $\eta' \to \omega\gamma$ in the reaction $\pi^- p \to \pi^+\pi^- 3\gamma n$, Phys. Rev. Lett. 38 (1977) 930.
33. D.M. Alde et al., Neutral decays of η' (958), submitted to Z. Phys. C.
34. W.D. Apel et al., Reaction $\pi^- p \to \eta' n$ in the 15-40 GeV/c momentum range, Phys. Lett. 83B (1979) 131.
35. J.E. Olsson, Photon photon interaction, Review talk given at the 1987 Intern. Conf. on Lepton and Photon Interactions at High Energies, Hamburg, July 1987 (to be published in the proceedings).
36. S.S. Gershtein, Exotic nature of scalar G (1590)-meson and possibilities for its further experimental study, Preprint IHEP 87-42, Serpukhov (1987).
37. W.D. Apel et al., Analysis of the reaction $\pi^- p \to \pi° \eta n$ at 40 GeV/c beam momentum, Nucl. Phys. B193 (1981) 269.
38. T. Barnes, The bag model and hydrid mesons, Preprint UTPT 85-21, Toronto Univ (1985).
39. G. Grayer et al., High statistics study of the reaction $\pi^- p \to \pi^-\pi^+ n$: apparatus, method of analysis, and general features of results at 17 GeV/c, Nucl. Phys. B75 (1974) 189.
40. G. Costa et al., An amplitude analysis of the K^+K^- system produced in the reaction $\pi^- p \to K^+ K^- n$ at 10 GeV/c, Nucl. Phys. B175 (1980)402 - see p. 406-7.
41. G. Pancheri, "Hidden" parton-parton scattering effects in very high energy hadron-hadron collisions, Proc. of the Santa Fe Meeting, T. Goldman and M.M. Nieto Ed., World Scientific, Philadelphia (1984).
42. S.S. Gershtein and A.A. Logunov, Growth of the hadron-hadron cross sections and its possible relation to the existence of glueballs Sov. J. Nucl. Phys. 39 (1984) 960.
    Yu.D. Prokoshkin, Growth of the total cross sections for hadron interactions, Sov. J. Nucl. Phys. 40 (1984) 1002.
43. Yu.D. Prokoshkin, Exclusive production and decay of glueballs, Sov. J. Part. and Nuclei 16 (1984) 253.
44. C. Baglin et al., Formation of charmonium states in antiproton-proton annihilation, Preprint CERN-EP/84-145 (1984).
45. R. Baier and R. Ruckl, Hadronic collisions : a quarkonium factory, Z. Phys. C19 (1983) 251.
46. A.L. Spadafora, Recent results from Mark III : The η_c and hadronic J/ψ decays, AIP Conf. Proc. #121, R.S. Pansini and G.B. Word Ed., p.12, Am. Inst. Phys., New-York (1984).
47. F.G. Binon et al., Modular hadron calorimeter, Nucl. Instr. Meth. A256 (1987) 444.

THE CERN Ω-WA76 EXPERIMENT: LIGHT MESON SPECTROSCOPY IN

THE CENTRAL REGION

Antimo Palano

CERN, European Organization for Nuclear Research, Geneva
Switzerland   and
Dipartimento di Fisica, University of Bari, Italy

ABSTRACT

The main results coming from the Ω WA76 experiment of the exclusive study of mesons produced centrally in a search for glueballs and exotic states are reviewed.

1. PHYSICS MOTIVATION

One of the basic expectations of the present theory of the strong interactions, QCD, is the existence of hadrons composed of gluons rather than quarks. The search for these new states, the glueballs, has been carried out in several experiments and has led to the discovery of a few candidates. However, the still undefined situation in the field of light meson spectroscopy makes it difficult to establish them as glueball states since the possibility that they are 4-quark states or radial excitations cannot be ruled out.

For this reason glueball hunting is a rather complex matter and the understanding of the low meson spectroscopy can only come from an overall comparison between meson production from different dynamical sources, i.e. J/ψ radiative and hadronic decays, $p\bar{p}$ annihilation, hadron induced reactions, $\gamma\gamma$ collisions, high p_T physics, central production etc.

Experiment WA76 at CERN was designed to study centrally produced exclusive final states, and searches for signs of new mesonic states. As the centre of mass energy increases the so-called "Double Pomeron Exchange"

(DPE) is one of the production mechanisms which is expected to become relatively more important and, if the Pomeron is composed of gluons, DPE can be expected to be a source of gluonium states [1].

A relatively high statistics exclusive study of meson production in all the possible final states in the reactions

$$\pi/p \; p \rightarrow \pi/p(X^o)p$$

is the physics goal of the WA76 experiment and is carried out here for the first time.

2. THE EXPERIMENT

The WA76 experiment started in 1982 with a run with π^+ and p at 85 GeV/c incident momentum and collected 10^7 events. The physics results presented in this paper come from this run.

A second run has been performed in 1986 with incident protons at 300 GeV/c and collected $1.2 \; 10^7$ events. Analysis of this set of data is in progress.

2.1 The Run at 85 GeV/c

What are the problems in studying the central production of mesons? The major problem is the large cross sections of diffractive reactions as compared with the DPE reactions. An unbiased study of the reaction $pp \rightarrow p_f$ ($\pi^+\pi^-$) p_s has been carried out with a bubble chamber exposed to a proton beam of 69 GeV/c [2]. Some results are shown in fig. 1(a) where the scatter plot of $M^2(p_f\pi^+\pi^-)$ vs. $M^2(p_s\pi^+\pi^-)$ is shown together with its projections. The data clearly show accumulation of events along the two axes indicating the contributions from the forward and backward diffractive production of the $p\pi^+\pi^-$ system with little evidence of a DPE contribution. The cross section for forward diffractive production of the $p\pi^+\pi^-$ system at this energy has been estimated to be ~ 300 μb.

On the other hand, theoretical calculations of the DPE contributions from π and p incident reactions from D.M. Chew and G.F. Chew (fig. 1(b)) [3], indicate 50 μb for the production of the ($\pi^+\pi^-$) system via DPE at 85 GeV/c.

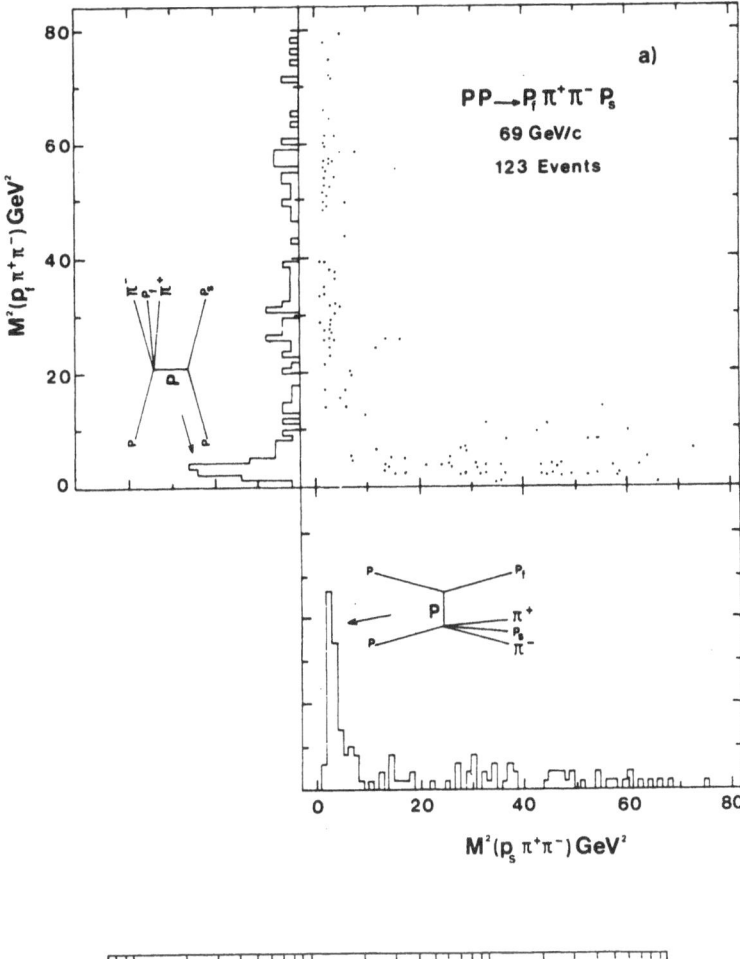

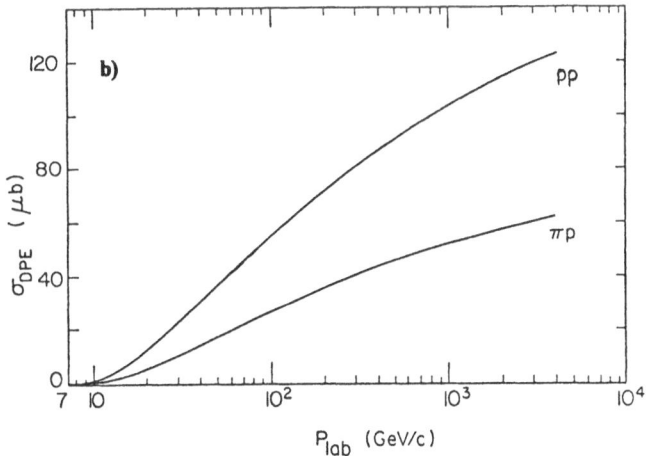

Figure 1: a) Squared $(p_f\pi^+\pi^-)$ effective mass vs. squared $(p_s\pi^+\pi^-)$ effective mass in the reaction $pp \to pp\pi^+\pi^-$ at 69 GeV/c from ref.[2]. b) Cross section variation of the DPE cross section as a function of the incident beam momentum for π and p beams, from ref. [3].

This means that, without any selection, the relative rate for the production of diffractive (forward and backward) $(\pi^+\pi^-)$ system with respect to the DPE production, at this energy, is

$$\sigma(\text{diffractive})/\sigma(\text{DPE}) \sim 12/1.$$

This calculation does not include other competitive channels other than DPE, i.e. Regge exchanges, Pomeron-Regge exchanges etc.

It is clear that strong selection criteria are needed in order to have a chance to select a really centrally produced system against the large abundance of unwanted diffractive or Regge exchanges contributions. In the following we explain how we solved these problems.

The layout of the Ω spectrometer at CERN used for this experiment is shown in fig. 2.

The positively charged H1 beam (45% p, 47% π^+) in the West Area was incident on a 60 cm long hydrogen target and two Cherenkov counters in the beam allowed positive identification of the pions and protons.

The trigger required:

(a) A fast particle defined by a hit in the counter A (54 $\times$ 60 cm^2) placed after Cherenkov C2, $\geq$ 1 hit in a single plane of the forward MWPCs T_2 and T_3 and no hit in the beam veto counter A_0;

(b) A slow particle defined by demanding one hit on any of the fourteen horizontal slabs of the Slow Proton Counter (SPC), (56 $\times$ 88 cm^2), and $\geq$ 1 hit on a single plane of the MWPCs situated on one side of the target. In addition, a hit was demanded in the side of the box counter (TS), surrounding the target, which was nearest the SPC.

(c) In order to reduce the backward diffraction or excitation we requested no hit in the other three sides of the box TS counter which was left open at its front end to allow particles produced centrally to escape downstream (fig. 3(a)). The effect of such a selection is visible in fig. 3(b) where the effective mass $p_s\pi^+$ is shown with no evidence of production of Δ^{++} which is the evidence that the backward diffractive contribution has been almost completely antiselected by the trigger conditions. In addition, the slow particle was identified by means of the pulse-height momentum correlation as shown in fig. 3(c).

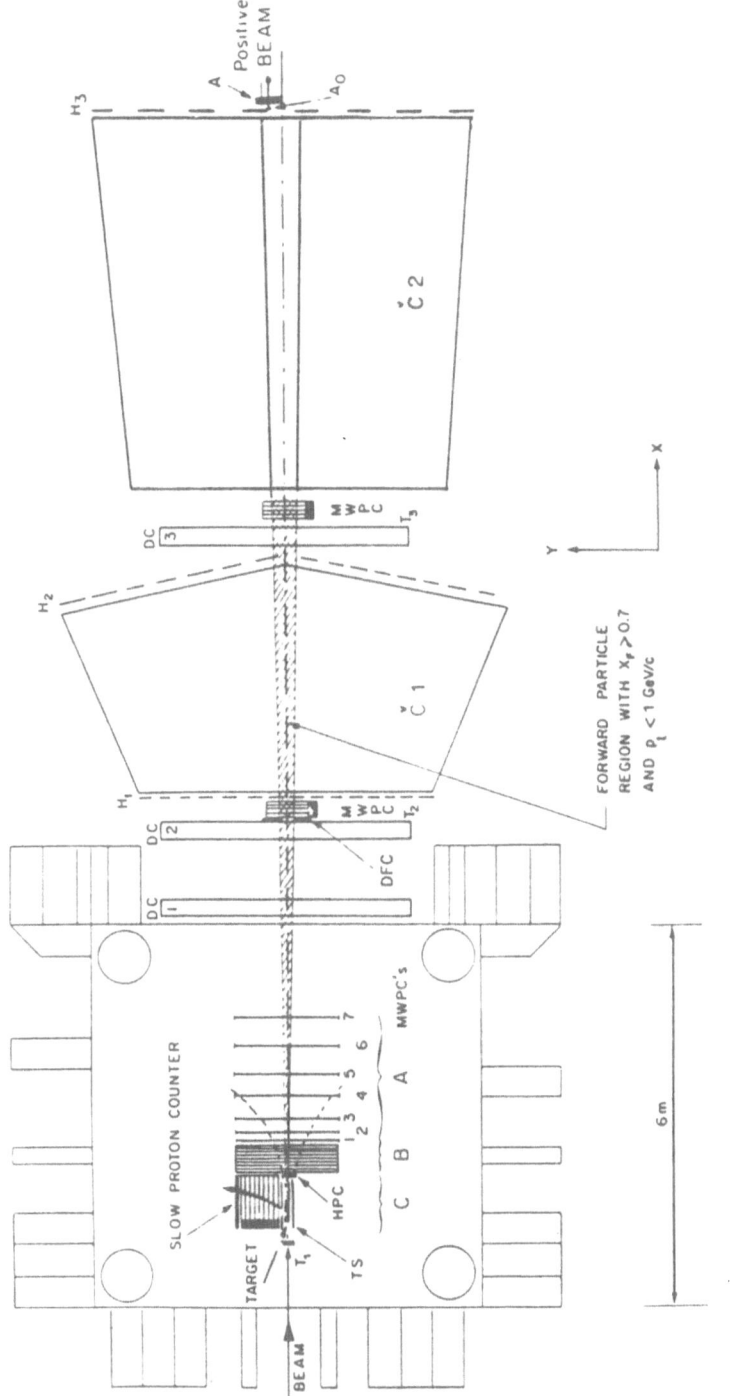

Figure 2. Layout of the CERN Ω spectrometer as used in the WA76 run at 85 GeV/c incident momentum.

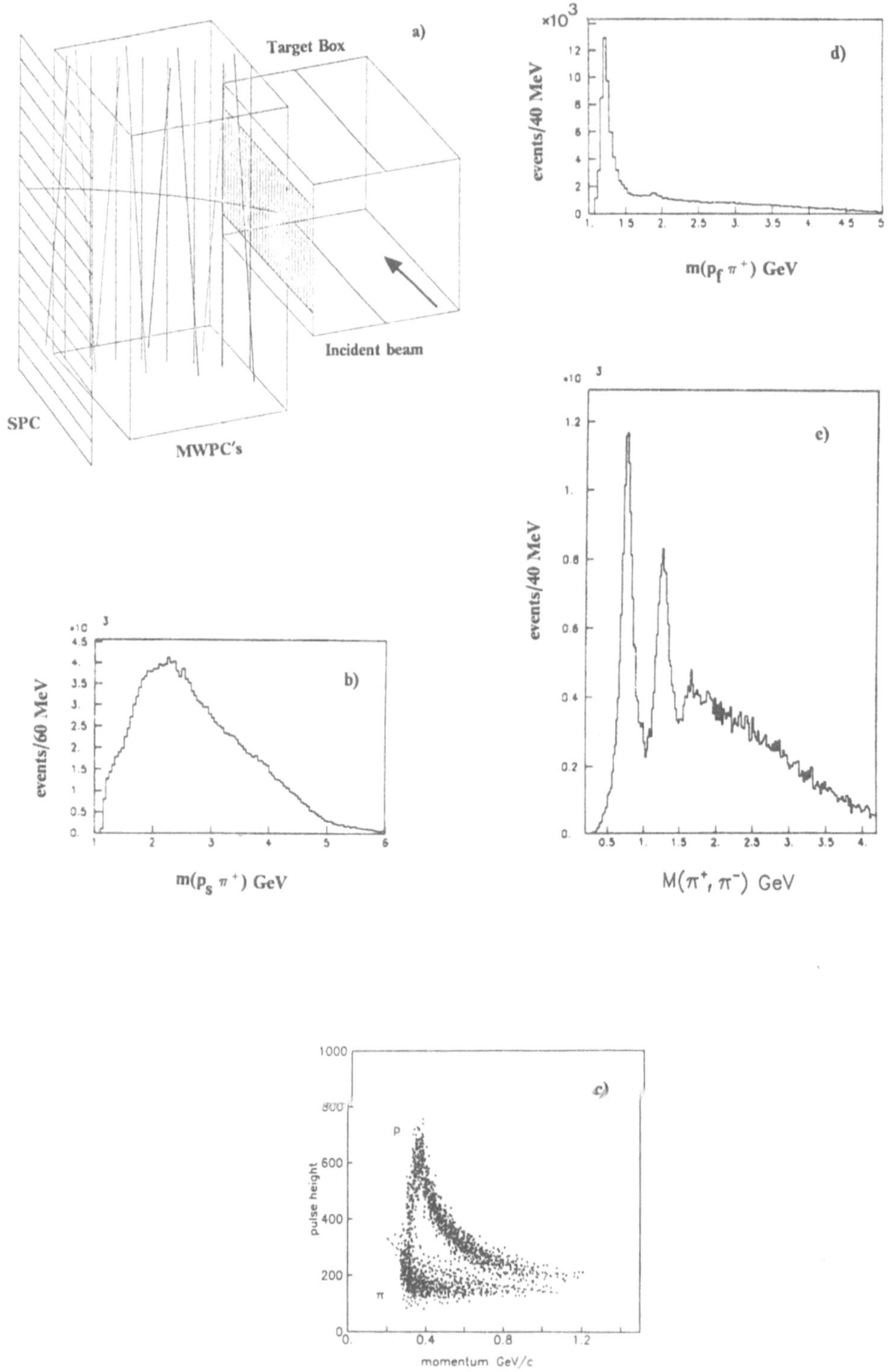

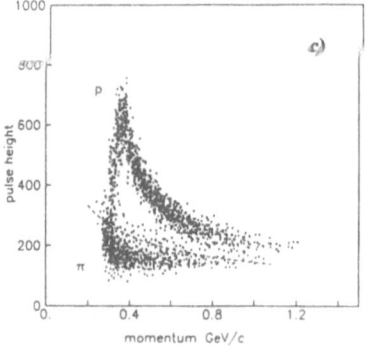

Figure 3. a) WA76 layout in the target region showing respectively, from left to right, the Slow Proton Counter (SPC), the side MWPC's and the target box. The curve indicates a typical slow proton trajectory. b) $p\pi^+$ effective mass; c) Pulse-momentum correlation for the slow particles. The dashed line indicates the p and π regions. d) $p_f\pi^+$ effective mass for incident p beam; e) $\pi^+_f\pi^-$ effective mass for π^+ incident beam.

(d) To reduce the forward diffraction or excitation we requested no hit in
two counters (DFC) of dimensions (30 × 60 cm^2) which were placed on
either side of the beam and just downstream of drift chamber 2 (DC2).
The effectiveness in reducing the forward diffractive contributions was
not complete as in the backward side as it can be seen from fig. 3(d)
where a clear Δ^{++} can be seen in the $p_f\pi^+$ effective mass for the p
incident data, and $\rho(770)$ and $f_2(1270)$ can be seen in the $\pi_f^+\pi^-$
effective mass (fig. 3(e)) for the π^+ incident data. The diffractive
contributions for the reactions $(\pi^+/p)p \rightarrow (\pi^+/p)_f\pi^+\pi^-p_s$, still present
in the data and antiselected by software cuts on these forward
resonances, resulted in ~ 45% of the data.

(e) Elastic events were vetoed by demanding a forward multiplicity of $\geq$ 2
(outside the beam region) in one plane of the "A$_3$" MWPC.

(f) In addition to the above conditions (called a π trigger) a K/p trigger
was imposed to select events having a negatively or positively charged
kaon or proton by demanding correlated hits in the hodoscopes H$_1$, H$_2$
and H$_3$ along with no correlated light in the appropriate region of the
multicell Cherenkov counters C1 and C2. An air bag was placed inside C1
in the beam region in order to reduce the multiple scattering on the
fast track.

The fast particle momentum, in the 85 GeV/c run, was measured by means
of three drift chambers which resulted in a momentum accuracy of
± 1.5 GeV/c at 85 GeV/c.

2.2 The run at 300 GeV/c

The layout used for the run at 300 GeV/c is essentially the same as the
one used in the 85 GeV/c run with a few important differences.

(a) The beam direction was measured by means of 2 doublets (y and z) of
50 µ pitch µ-strips detectors and 2 doublets (y and z) 1/3 mm
spacing scintillation counters. The beam momentum bite was held at
0.25%. This allows a very accurate determination of the beam parameters.

(b) The outgoing fast particle was measured by means of two sets of
4 planes of 50 µ pitch µ-strips detectors placed respectively at 6 m
and 10 m from the Ω centre as shown in fig. 4(a). The system was set
up in such a way that the beam was crossing only two offset y planes in
order to have a constant calibration of the exact position of the
µ-strips planes. An example of the raw hit distribution on the offset
plane of the 6 m µ-strips is shown in fig. 4(b). This layout was also

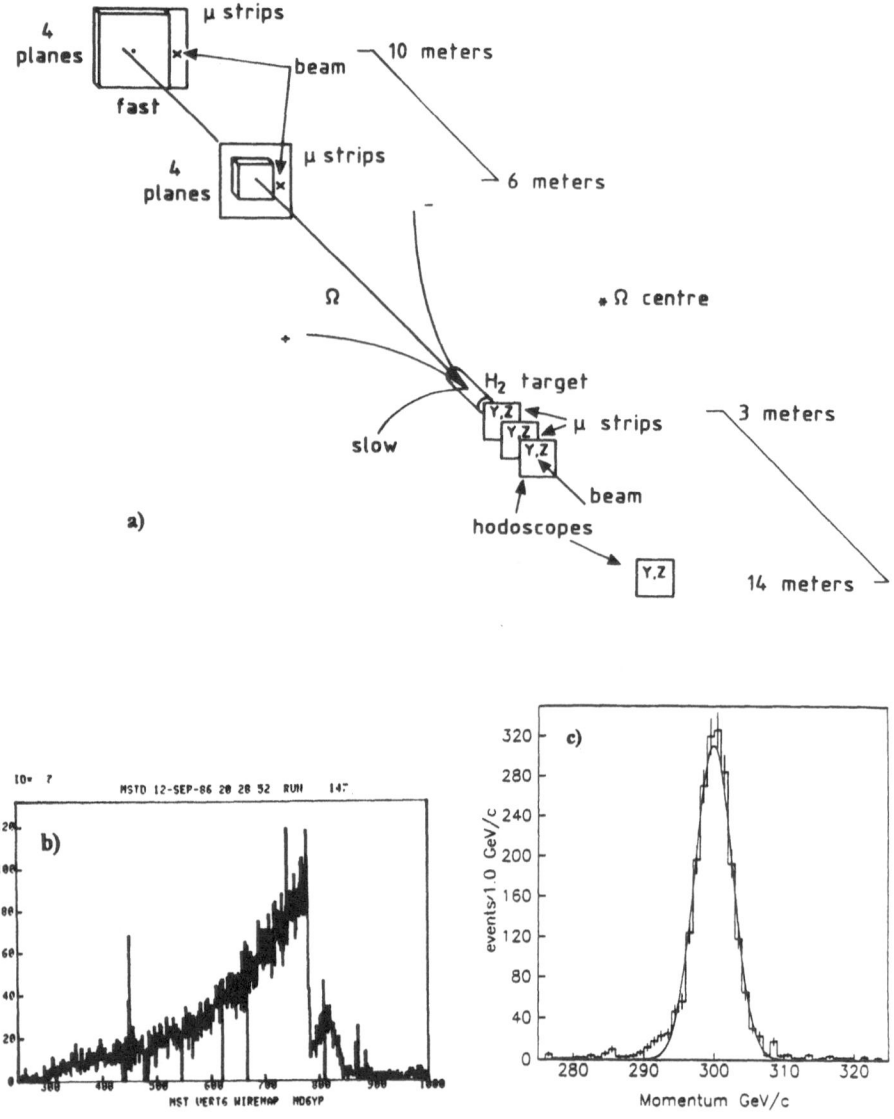

Figure 4. a) Layout of the μ-strips and scintillators used to measure the beam and the fast particle in the WA76 run at 300 GeV/c. b) Raw hits on the shifted μ-strip at 6m from the Ω centre; c) Beam momentum determination for the calibration of the alignment of the whole system. The curve represents a gaussian fit to the data.

automatically antiselecting the elastic events. The precision obtained
with this system has been estimated by measuring the momentum of beam
tracks by using of the first beam μ-strips and the forward μ-strips
planes and is ± 2.7 GeV/c at 300 GeV/c (see fig. 4(c)).

A typical event is shown in fig. 5(a) (charged) and 5(b) (neutral). The
fast track, which is not seen by the MWPCs and DC's, is measured by
using the vertex coordinates (determined very accurately in y and z by
the beam) and the two μ-strips points at 6 m and 10 m.

(c) The $\Omega\gamma$ detector was placed after Cherenkov 2 to detect γ, π^0 and η.

3. PHYSICS AT 85 GeV/c

The 85 GeV/c data have been analyzed, several results have been
published and the study of a few channels is still in progress. As
explained in chapter 2, the trigger selected the class of graphs shown in
fig. 6(a), i.e. double exchanges graphs with no possibility, due to the
limited rapidity gap between the fast and slow particles and the central
system, at this energy, to isolate DPE from other exchanges mechanisms. A
typical distribution of x-Feynman for the slow, the central system and the
fast particle are shown in fig. 6(b).

Isolation of the reaction $\pi^+/pp \rightarrow \pi^+/p\ (X^+X^-)p$

The reactions

$$\pi^+/pp \rightarrow \pi^+/p\ (X^+X^-)p, \quad \text{where } X = \pi, K, p$$

have been isolated from the sample of events having four outgoing tracks.
Fig. 7(a) shows the total missing P_t where a clear signal is evident from
candidate "4C" events. Fig. 7(b) shows the resulting missing P_x
distribution after requiring P_t cuts. The above plots show that the
momentum balance events can be clearly isolated at this energy. The energy
balance can be written as:

$$(\pi^+/p)_i\ p \rightarrow (\pi^+/p)_f\ P_s\ X_1^+\ X_2^-$$

$$E_i + m_p = E_f + E_s + \sqrt{(p_1^2 + m^2)} + \sqrt{(p_2^2 + m^2)}$$

equation which can be solved directly for m^2. The m^2 distribution for all
the data is shown in fig. 7(c) and is dominated by a large peak at the
squared π mass and is the signal of the reaction having the $\pi^+\pi^-$ system in
the final state. Two smaller peaks are also seen at the squared K and p
masses which indicate the presence of the K^+K^- and $p\bar{p}$ final states.

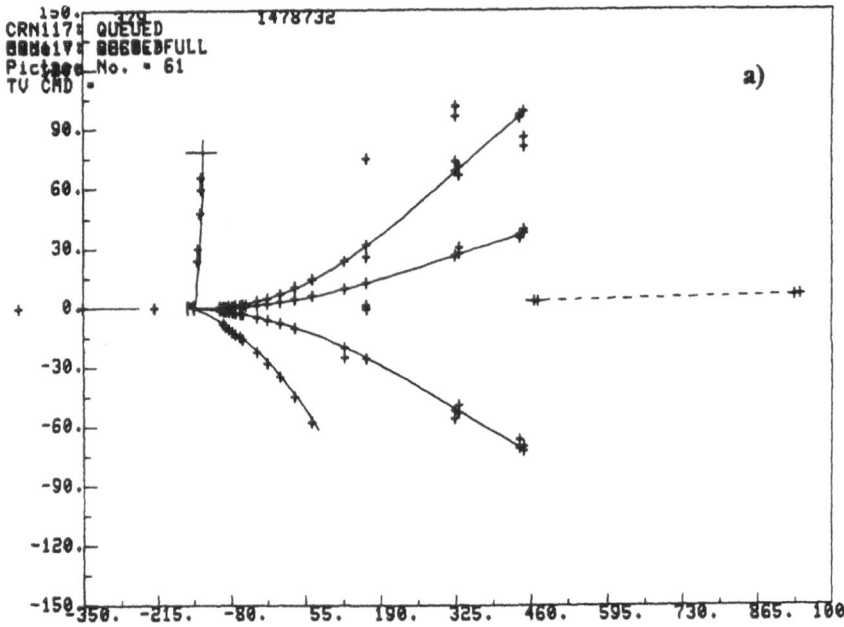

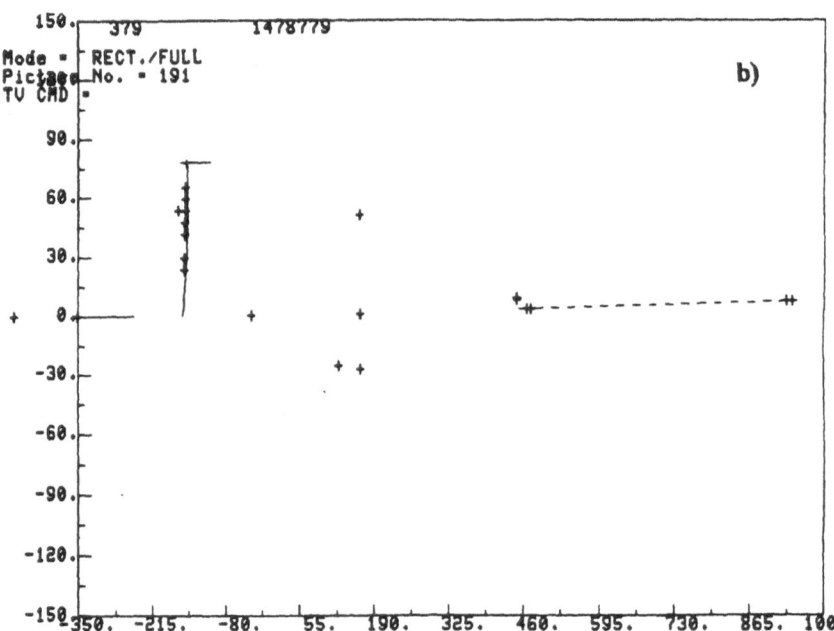

Figure 5: Typical events at 300 GeV/c. From left to wright are shown the beam μ-strips, the MWPC's, the Drift Chambers and the forward μ-strips. The dashed line indicates the reconstructed track in the forward μ-strips. a) Six outgoing charged tracks event; b) two charged and neutrals.

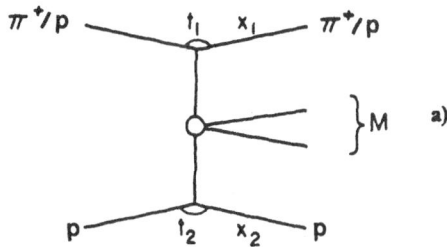

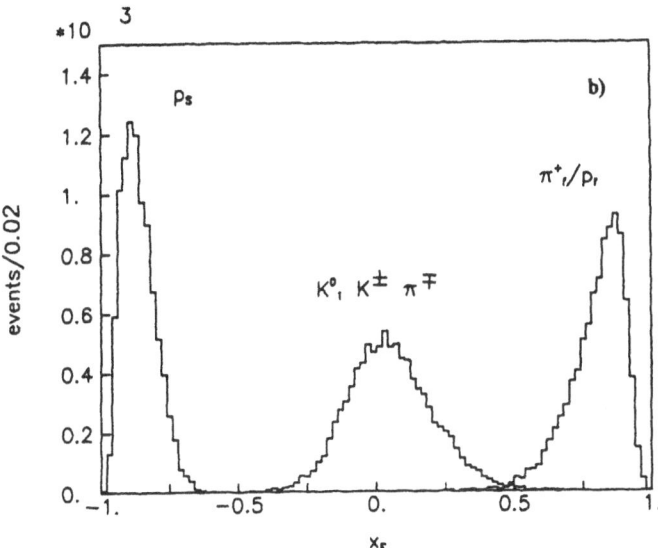

Figure 6. a) Feynman graph selected by the WA76 trigger conditions; b) Typical $x_{Feynman}$ distributions as selected by the trigger conditions. The x_F of the slow proton, the central system and the fast particle respectively are shown.

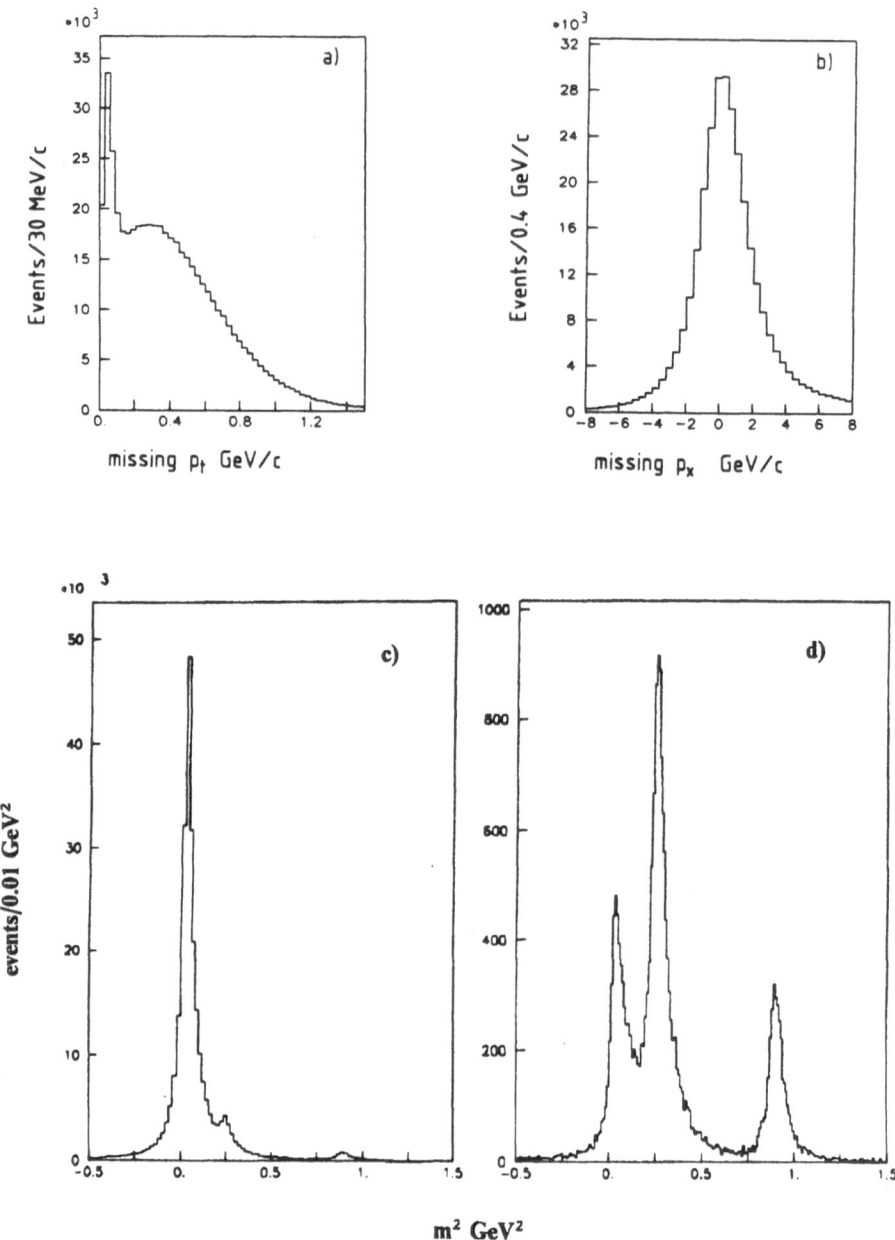

Figure 7. a) Missing transverse momentum distribution for events with four outgoing tracks; b) Missing longitudinal momentum after P_t cut. c) Ehrlich mass distribution for all the events and d) after having used the Cherenkov information to select the centrally produced $K^+ K^-$ final state.

A different spectrum is obtained by requiring one particle to be identified as a K or K/p by the Cherenkov system and the other, if reaching the Cherenkov system, to have a mass identification compatible with the K hypothesis (see fig. 7(d)). In this case the π peak is considerably reduced and clear signals of the K^+K^- and $p\bar{p}$ reactions can be seen. The latter final states can also be isolated with little background.

4.   SOME DYNAMICS FROM THE CENTRAL REGION

The dynamics and the processes underlying the central production of mesons are largely unknown and it is of interest to find possible hints of the dynamical sources of meson production. Some interesting features have been observed.

(a) Fig. 8(c) shows the centrally produced K^+K^- mass spectrum as obtained in the WA76 experiment at 85 GeV/c [4]. It is interesting to compare this spectrum with the K^+K^- mass spectrum coming from two different sources: incident π^- [5] (fig. 8(a)) and incident K^- [6] (fig. 8(b)). These spectra clearly show resonance production influenced by the nature of the incident beam, i.e. the π^- spectrum shows resonances coupled to non-strange quarks ($f_0(975)$, $f_2(1270)/a_2(1320)$, etc.), while the K^- induced spectrum shows resonances coupled to strange quarks ($\phi(1020)$, $f_2(1525)$). It can be seen that the spectrum of fig. 8(c) shows, in the low mass region, all these resonances together, indicating a "flavour blind" production mechanism, i.e. resonances coupled to strange and non-strange quarks are produced in the central region with similar strength.

(b) The study of channels with 4 or 5 particles in the final states enables a search for associated resonance production to be made. Evidence has been found for $\phi\phi$ production in the study of $2K^+2K^-$ final state [7] (fig. 9(a)), $K^{*0}\bar{K}^{*0}$ in the $K^+K^-\pi^+\pi^-$ final state [8] (fig. 9(b)) and probably $\phi\omega$ in the $K^+K^-\pi^+\pi^-\pi^0$ final state (fig. 9(c)). It should be noticed that in the 85 GeV/c run the π^0 where not detected so they were reconstructed only as missing particles. This implies that channels using π^0's are affected by uncertainties due to the bad experimental resolution. The $\phi\phi$ and $K^{*}\bar{K}^{*}$ effective mass distributions are shown in fig. 9(d) and 9(e) and show accumulations near threshold.

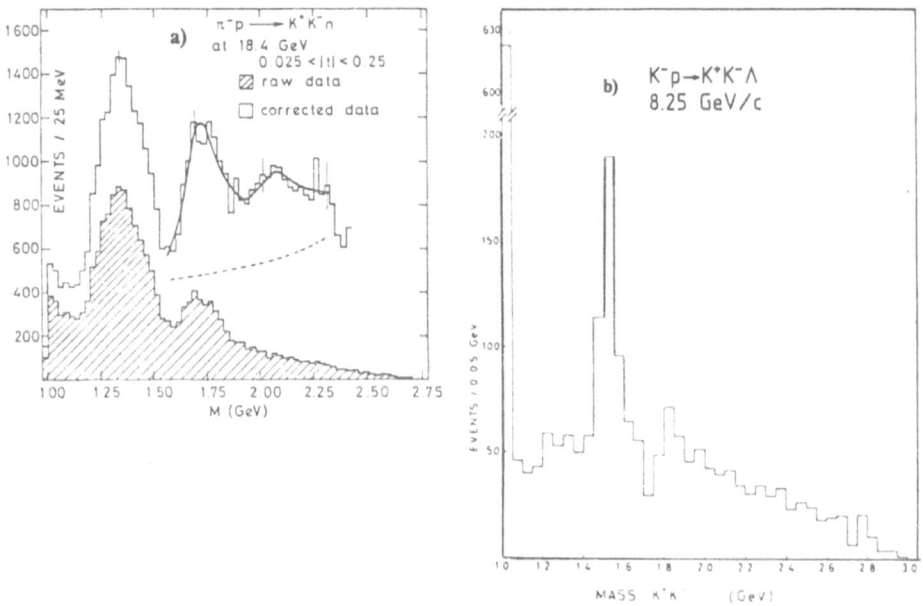

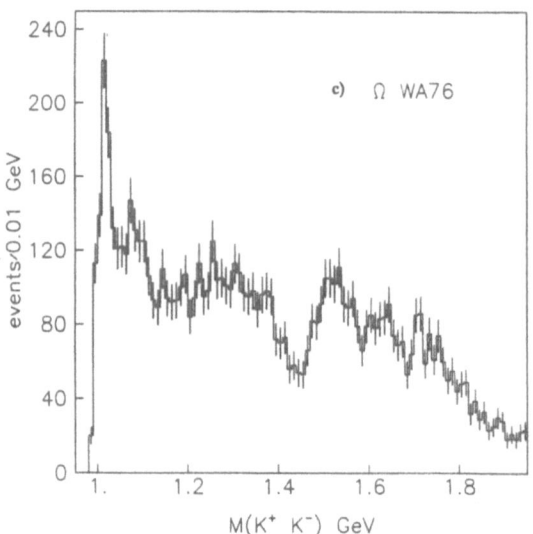

Figure 8. a) K^+K^- effective mass from incident π^- beam [5]. b) K^+K^- effective mass from incident K^- beam [6]. c) Centrally produced K^+K^- effective mass from WA76.

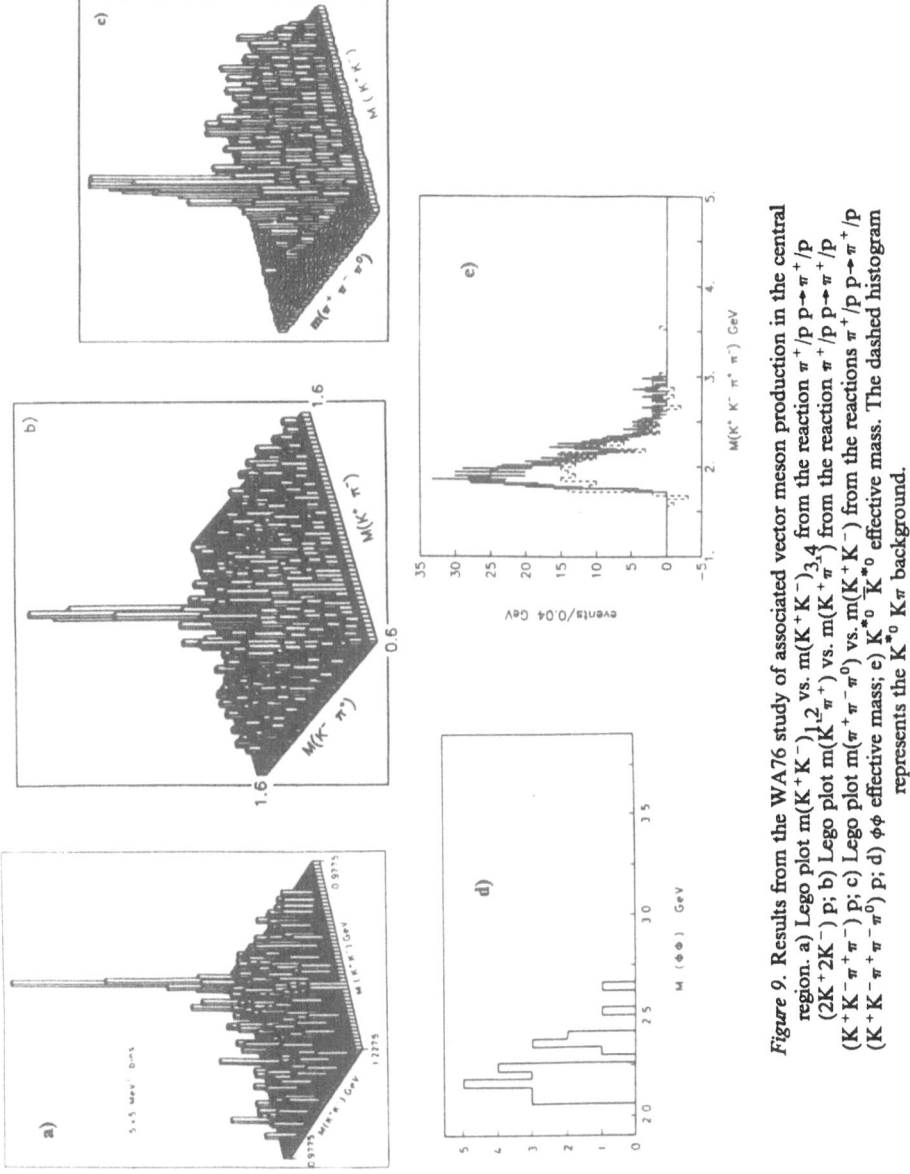

Figure 9. Results from the WA76 study of associated vector meson production in the central region. a) Lego plot $m(K^+K^-)_{1,2}$ vs. $m(K^+K^-)_{3,4}$ from the reaction $\pi^+/p \; p \to \pi^+/p$ $(2K^+2K^-)$ p; b) Lego plot $m(K^+_{\pi^+})$ vs. $m(K^+_{\pi^-})$ from the reaction $\pi^+/p \; p \to \pi^+/p$ $(K^+K^-\pi^+\pi^-)$ p; c) Lego plot $m(\pi^+\pi^-\pi^0)$ vs. $m(K^+K^-)$ from the reactions $\pi^+/p \; p \to \pi^+/p$ $(K^+K^-\pi^+\pi^-\pi^0)$ p; d) $\phi\phi$ effective mass; e) $K^{*0}\bar{K}^{*0}$ effective mass. The dashed histogram represents the $K^{*0} \; K\pi$ background.

291

(c) Ref. [9] has found an anomalous high $\phi\phi$ production in π^- induced reactions at 23 GeV/c: $\sigma(\phi\phi)/\sigma(\phi K^+ K^-) \sim 0.2$ and it has been suggested [10] that a large $\phi\phi$ production with respect to ϕKK production in π induced reactions is evidence of intermediate gluonic bound states. The study of the centrally produced $2K^+2K^-$ gives the largest ever observed production of $\phi\phi$: $\sigma(\phi\phi)/\sigma(\phi K^+ K^-) \sim 0.7$.

It is not clear, at present, how to interpret these experimental observations. One possibility is that gluons, as well as quarks, are interacting in the central region in order to produce the observed features.

5. THE SEARCH FOR THE 0^{++} GLUEBALL

Calculations of the glueball spectrum in QCD indicate the possibility that the lightest glueball should have $J^{PC} = 0^{++}$ and have a mass around 1 GeV [11]. If it is below the $K\bar{K}$ threshold it can only decay into pions so it should be present in $\pi\pi$ scattering data.

Recently, a multichannel analysis has been performed on different $\pi\pi$ and $K\bar{K}$ data but mostly based on CERN ISR (Axial Field Spectrometer (AFS) Collaboration [12]) data coming from the DPE reactions $pp \to pp\pi^+\pi^-$ and $pp \to ppK^+K^-$. The $\pi^+\pi^-$ mass spectrum is shown in fig. 10(a) and is characterized by a threshold enhancement and a sharp drop around the $K\bar{K}$ threshold, while the $K\bar{K}$ mass spectrum from the same experiment is shown in fig. 10(b). The analysis of ref. [13] interpreted the so-called "S^* effect" as due to the presence of two states,

$S_1(993)$ with $E_R = 0.993-0.023i$ GeV, $g_\pi = 0.23$, $g_K = 0.28$

$S_2(998)$ with $E_R = 0.998$ GeV, $g_K = 0.35$

where E_R is the position of the complex pole, g_π and g_K describing the $\pi\pi$ and $K\bar{K}$ couplings of the two resonances. The $S_1(993)$ is roughly equally coupled to $\pi\pi$ and $K\bar{K}$ and is interpreted as a possible scalar gluonium state. The $S_2(988)$, coupled only to $K\bar{K}$, is interpreted as a pure $s\bar{s}$ state or a $K\bar{K}$ molecule.

The WA76 data, summed over the two incident (π^+ and p) beams but split into two different t regions (t is the sum of t_1 and t_2, the four momentum transfer at the upper and the lower vertex respectively in the diagram shown in fig. 6(a) are shown in fig. 10(c,d). The low t region shows similar features to the AFS data, i.e. a threshold enhancement and a drop

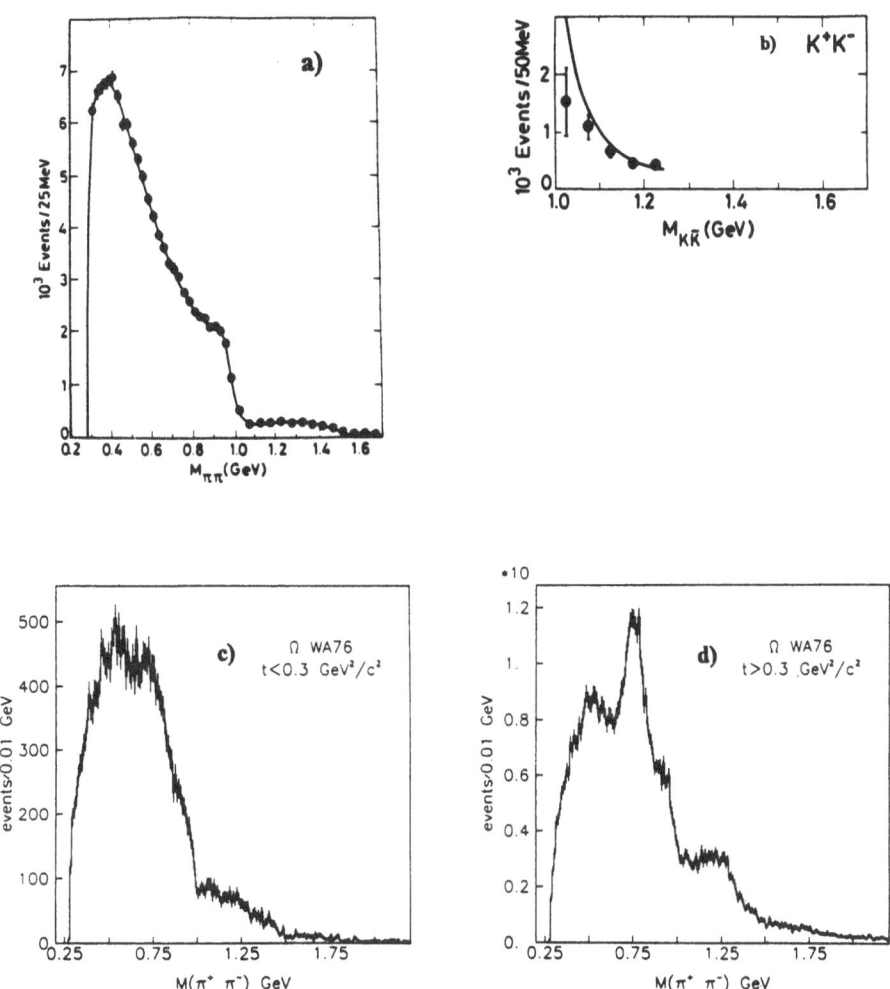

Figure 10: a),b) $\pi^+\pi^-$ and K^+K^- effective mass distributions from ref. [13]. The curves show projection of the multichannel fit. c),d) $\pi^+\pi^-$ effective mass from WA76 for $t < 0.3$ GeV^2/c^2 and $t < 0.3$ GeV^2/c^2 respectively.

at the $K\bar{K}$ threshold. The high t region, on the other hand, shows clear ρ and f_2 structures and a shoulder in the 0.94 GeV region. The total spectrum is shown in fig. 11(a).

One way to parametrize an object coupled to different decay channels is the Flatte' formulation [14] which has a very simple mathematical expression

$$\sigma_{\pi,K} \propto \frac{m_s \Gamma_\pi}{m_s^2 - m^2 - im_s(\Gamma_\pi + \Gamma_K)}$$

where $\Gamma_\pi = g_\pi q_\pi$ and $\Gamma_K = g_K q_K$. The fit of the π^+ incident data, in the WA76 experiment, with the above parametrization is shown in fig. 11(b) after background subtraction. The K^+K^- and $K^-\bar{K}^0$ mass spectra are shown in fig. 11(c,d). The two spectra show similar features with evidence of a large number of structures, in particular a threshold enhancement is visible in both spectra. In order to visualize the phase space contribution in both spectra the combinatorial phase space obtained by taking kaons from different events is plotted. Whether the structure seen in the $\pi\pi$ mass spectrum around 0.94 GeV and the threshold enhancement observed in the $K\bar{K}$ projection belong to a single or two resonances is not clear. The analysis of these channels is still in progress.

An inspection of the high mass region of both the K^+K^- and $K^0\bar{K}^0$ mass spectra show structure above the $f_2'(1525)$ which have been shown [4] not associated to the g meson (in the K^+K^- mass spectrum that is forbidden because only even spins are allowed). The K^+K^- mass spectrum, in addition, suggests the presence of two resonances in this region. The fit is shown in fig. 12a and gives the following parameters,

$m_1 = 1629 \pm 10$ MeV $\qquad \Gamma_1 = 82 \pm 30$ MeV

and

$m_2 = 1742 \pm 10$ MeV $\qquad \Gamma_2 = 82 \pm 30$ MeV

For comparison, fig. 12(b,c,d) shows the K^+K^- mass spectra recoiling against γ, θ and ϕ in the J/ψ decay [15]. The first two mass spectra show structure in the region of the θ with parameters $m(\theta) = 1720 \pm 10\pm 10$ MeV and $\Gamma(\theta) = 130 \pm 20$ MeV while an incoherent fit of the K^+K^- mass spectrum recoiling against the ϕ gives the values $m = 1671 \pm 15 \pm 10$ MeV and $\Gamma = 126 \pm 50 \pm 15$ MeV. If we are observing the same structures, the mass and widths of the structures observed in the J/ψ decay are consistent with the ones measured in the WA76 experiment.

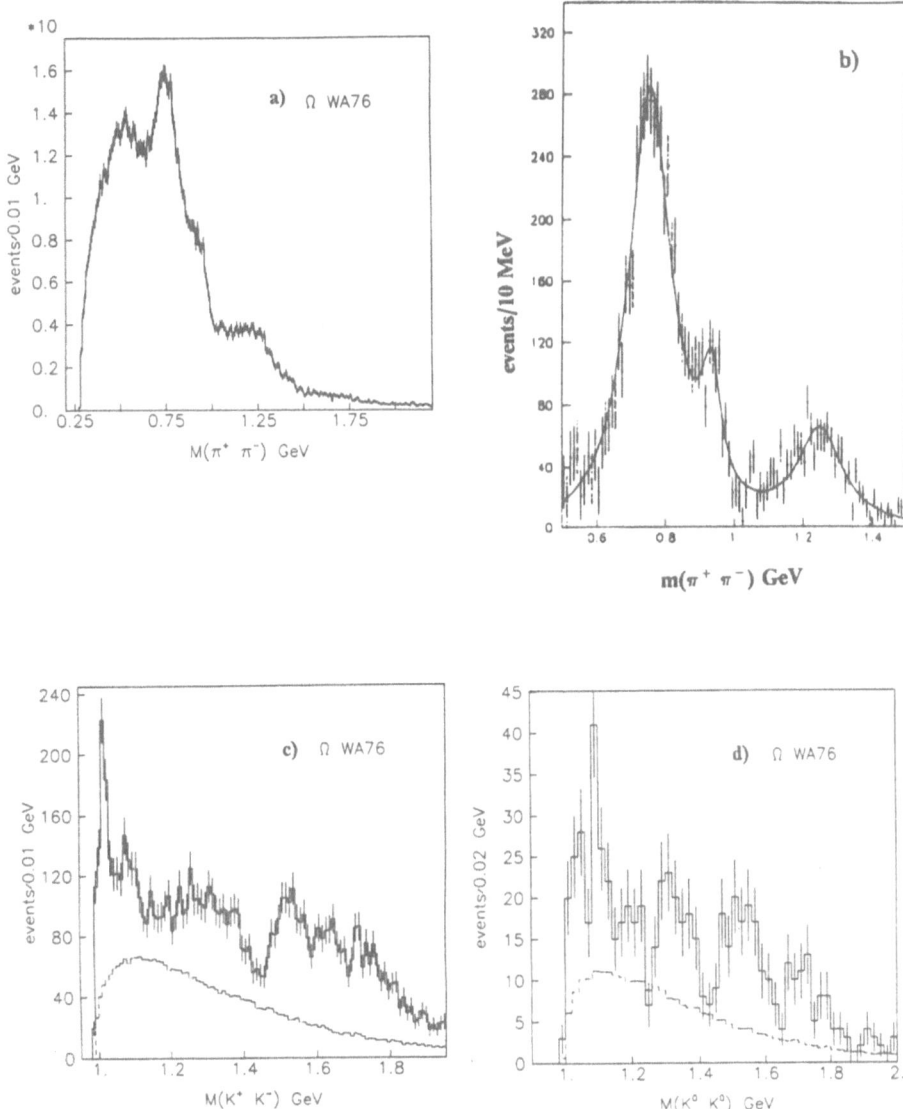

Figure 11. a) Total $\pi^+\pi^-$ effective mass from WA76; b) Preliminary fit of the π^+ incident data after background subtraction. c),d) K^+K^- and $K^0\overline{K}^0$ effective mass from WA76. The dashed histograms shows the estimated phase space from mixed events, the normalization is arbitrary.

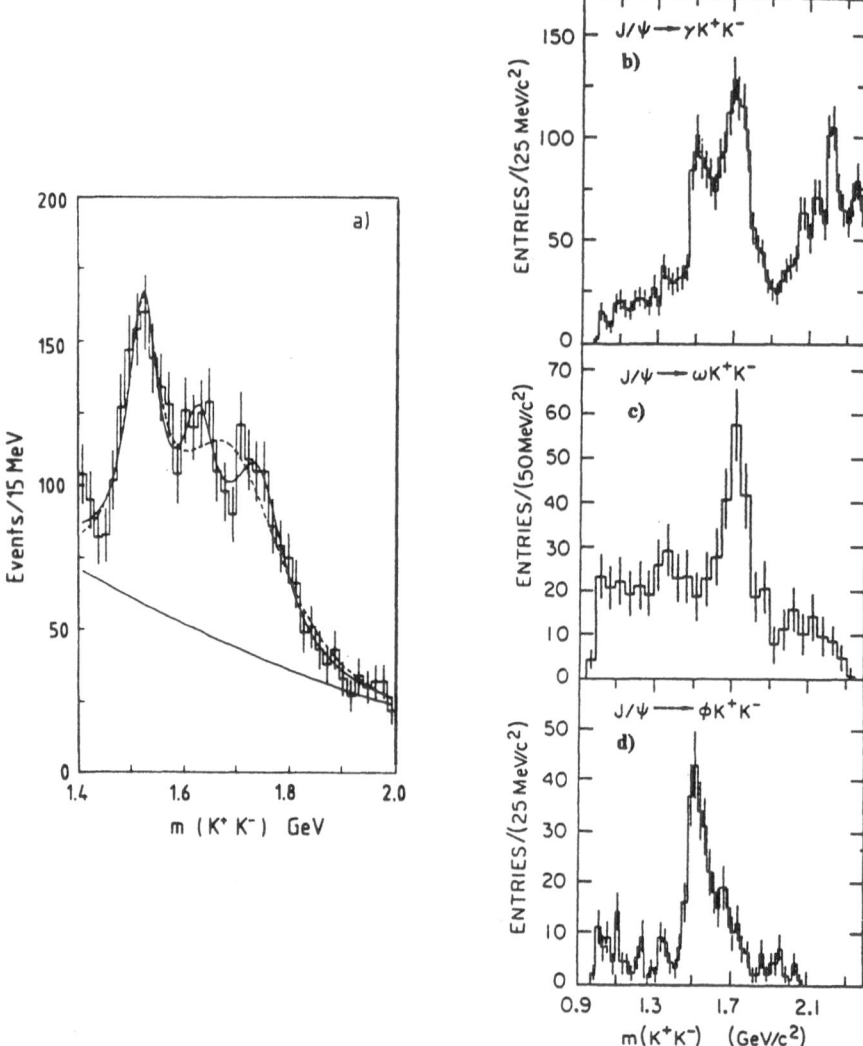

Figure 12. a) Fit of the WA76 K⁺K⁻ mass spectrum in the high mass region. b),c),d) K⁺K⁻ effective mass distribution recoiling against γ, ω and φ respectively in the J/ψ decay.

6. THE E/ι PUZZLE

The E/ι puzzle can be summarized as follows. One, two, three or even more states are observed decaying to $K\bar{K}\pi$ and/or $\eta\pi\pi$ in the mass region around 1.4 GeV which some experiments find to have $J^{PC} = 0^{-+}$ (ι) while others find $J^{PC} = 1^{++}$ (E). The interest in this region is that, if $J^{PC} = 0^{-+}$, the resonance could be a glueball candidate while, if $J^{PC} = 1^{++}$, it can be a normal $q\bar{q}$ object.

For a recent review of the subject see ref. [16]. The experimental situation can be summarized as follows:

(a) A pseudoscalar state, $J^{PC} = 0^{-+}$, is observed in the radiative J/ψ decay [17] and $p\bar{p}$ annihilation at rest [18] and is not seen together with the $f_1(1285)$, an established axial vector state. In both reactions the $\delta\pi$ decay mode seems to be important. The state seen in the J/ψ decay has a mass of ~ 1460 MeV and a width of ~ 100 MeV. In contrast the state seen in $p\bar{p}$ annihilation has a mass of 1425± 7 and a width of 80 ± 10 MeV. However the state seen in the radiative J/ψ decay looks more like a superposition of more than one state since a simple Breit-Wigner is not enough to describe the mass spectrum. In addition the existence of a $\delta\pi$ decay mode is not supported by the analysis of radiative decay of the J/ψ to $\eta\pi\pi$.

(b) Two experiments [19,20] find evidence for a pseudoscalar state at the E mass (~ 1420 MeV) in the study of the $K\bar{K}\pi$ and $\eta\pi\pi$ systems in π^- induced reactions at 8 GeV/c. However, this results is in contradiction with a previous, lower statistics, bubble chamber experiment [21] which found 1^{++} as quantum numbers of the same object. The study of the $\eta\pi\pi$ system revealed also the existence of a new pseudoscalar particle in the $f_1(1285)$ region, the $\eta(1275)$. These observations are not supported by the study of $\gamma\gamma$ collisions where these states, if belonging to the radial excitation sector, should clearly appear [22].

(c) The axial vector state E (1^{++}) is generally observed together with the $f_1(1285)$, has $K^*\bar{K}$ as dominant decay mode, a mass of 1425 MeV and a width of ~ 60 MeV. This state has been seen in π^- induced reactions [21], in the WA76 experiment [23], in $\gamma\gamma$ collisions [24] and in the hadronic decay of the J/ψ to $\omega K\bar{K}\pi$ [25]. The data from the WA76 experiment are shown in fig. 13 which contains the mass spectrum, the Dalitz plot showing clear K^* bands from the low background E signal and the results of the Dalitz plot analysis which shows a large (~ 86%) 1^{++} component and a small 0^{-+} component.

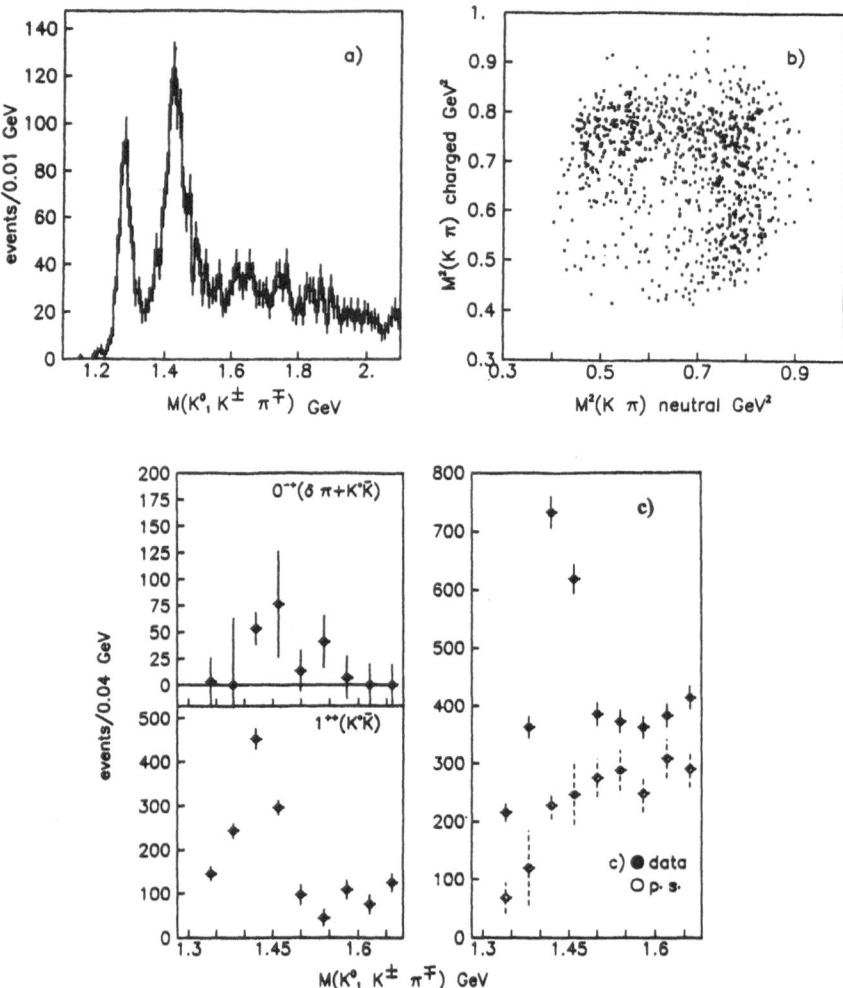

Figure 13. a) $K^0_1 K^\pm \pi^\mp$ centrally produced mass spectrum from incident π^+ and p at 85 GeV/c by using the CERN Ω spectrometer b) Dalitz plot, c) Results from Dalitz plot analysis.

(d) A new axial vector state, D'(1526) having a mass of 1526 ± 6 MeV and
a width of 107 ± 15 MeV has been observed by two experiments [26,27]
in the analysis of the $K\bar{K}\pi$ system from K^- induced reactions.

Too many pseudoscalars and axial vector mesons are present in this mass
region to be understood in terms of a simple quark model. As usual more
high statistics data are needed in order to help clarify the situation.

7. THE a_1 PROBLEM

The a_1 resonance, having $J^{PC} = 1^{++}$, is still a puzzle for light meson
spectroscopy. Its existence seem settled by two large experiments which
studied the 3π system in diffractive and charge exchange reactions [28,29]
and extracted the signal from a large Deck background. The mass and width
were determined to be respectively $m(a_1) = 1275 \pm 28$ MeV and $\Gamma(a_1)$ =
316 ± 45 MeV.

A different a_1 has been found in backward production of the charged 3π
system in the reaction $K^-p \to \Sigma^- \pi^+ \pi^+ \pi^-$ at 4.15 GeV/c [30]. Here mass and
width were found to be 1041 ± 13 MeV and 230 ± 50 MeV respectively.

Recent data from τ decay $\tau \to 3\pi\nu_\tau$ where the full 3π mass spectrum is
assumed to belong to the a_1 resonance find quite different results. Three
experiments find respectively 1056 ± 20 MeV [31], 1194 ± 14 MeV [32] and
1046 ± 11 MeV [33] for the a_1 mass and a quite large width of the order of
500 MeV.

A contribution to the a_1 problem also comes from the WA76 experiment
via the study of the centrally produced $2\pi^+ 2\pi^-$ system. The 4π mass spectrum
is shown in fig. 14(a) and shows a clear structure at the $f_1(1285)$ mass. A
fit which includes the reflections coming from the η' and D from the $\eta\pi\pi$
system where a slow π^o is not detected gives the following paramaters for
the mass and width:

m = 1280 ± 2 Γ = 39 ± 15 MeV.

The study of the 3π system in the 4π final state is made complex by the
presence of a combinatorial background, i.e. in the three pion mass
spectrum 3 entries are combinatorials and only one entry is physics. In
order to remove the combinatorial background the method of mixed events has
been used. Artificial events have been generated where each of the four
pions was taken from a different event. The resulting 3π mass spectrum

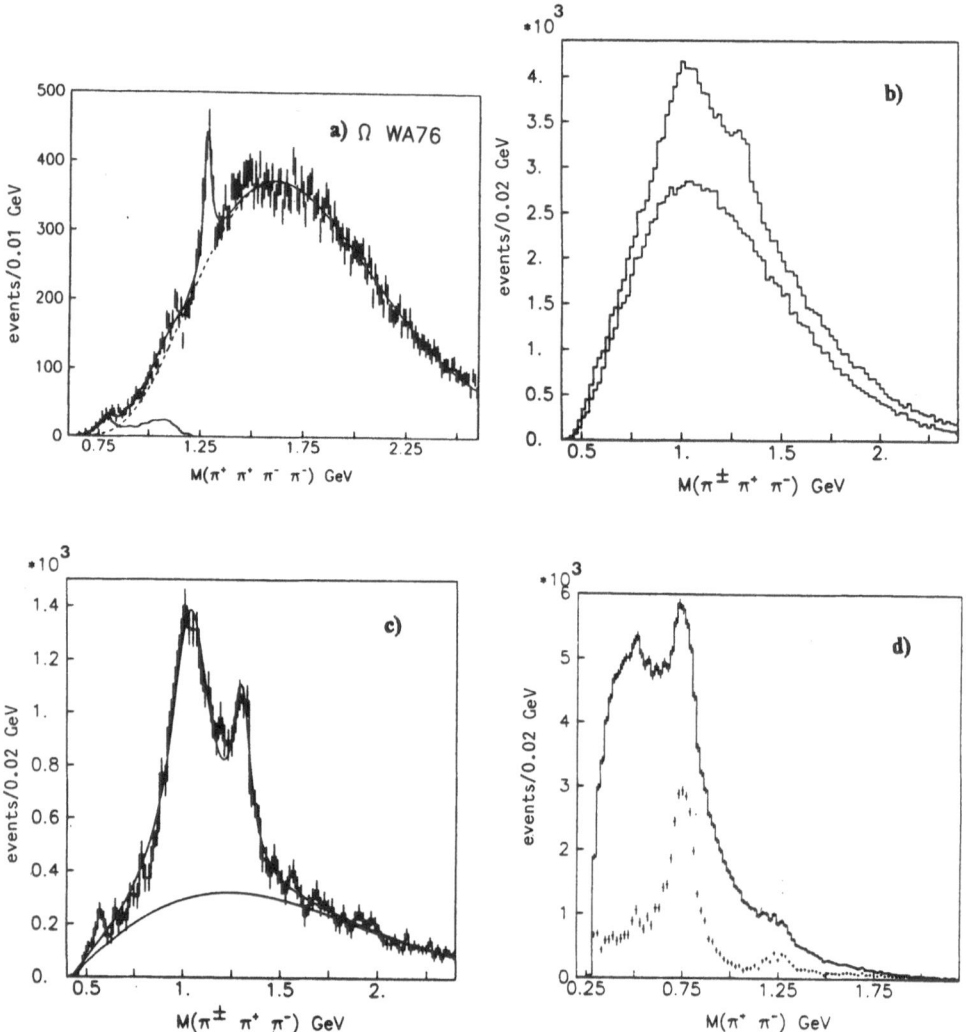

Figure 14. a) Centrally produced $2\pi^+2\pi^-$ effective mass distribution from WA76. The line is the result of the fit described in the text. b) 3π effective mass distribution (4 entries per event). The lower histogram shows the combinatorial background estimated by using the method of mixed events and normalized to 3/4 of the total entries. c) 3π effective mass after subtraction of the combinatorials. The line is the result of the fit described in the text. d) $\pi^+\pi^-$ effective mass in the same reaction. The same figure shows the result of subtracting the combinatorial background.

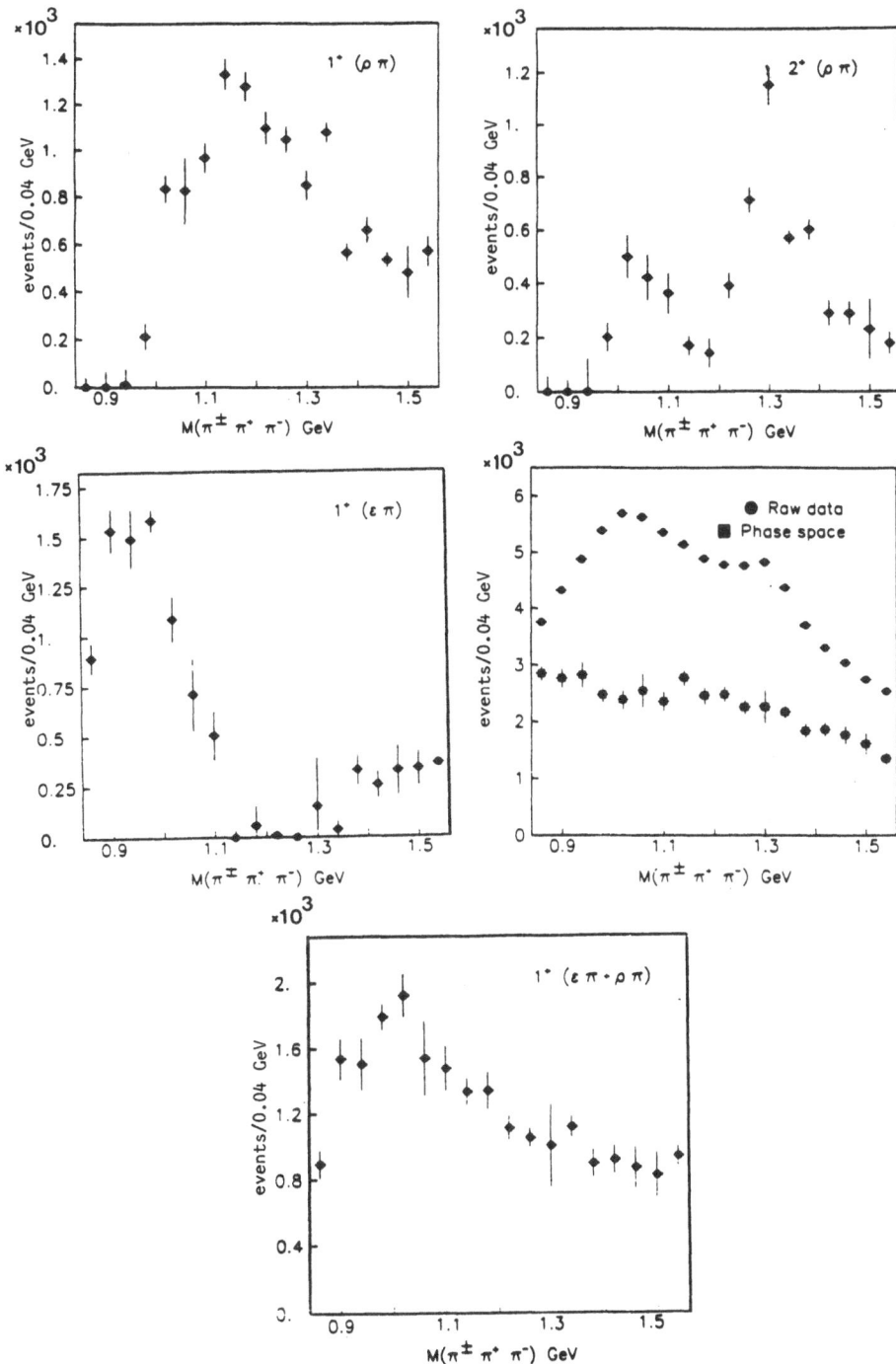

Figure 15: Results from the Dalitz plot analysis of the 3π system out of 4π centrally produced in WA76.

from this kind of Monte-Carlo events was normalized to be 3/4 of the total. The true spectrum as well as the combinatorial one is shown in fig. 14(b). The data show structure in the a_2 region and suggest a wide structure peaking above 1.0 GeV. The subtracted spectrum is shown in fig. 14(c) and clearly suggest the presence of two resonance structures. The subtracted spectrum has been fitted by using two simple Breit-Wigner shapes one being the standard a_2(1320) the other giving the following values:

   m = 1022 ± 20 MeV   Γ = 277± 50 MeV

in good agreement with the values quoted in ref. [30] for the backward a_1 production by incident K^-.

   In order to test if the method used to estimate the combinatorial background is biasing the data or generating false peaks, the same method has been applied to the $\pi^+\pi^-$ mass distribution which also suffers from combinatorial problems af the sime type as the 3π spectrum. The results are shown in fig. 14(d) where the raw 2π mass spectrum is shown together with the subtracted one. Clear structure can be seen at the ρ and f_2 masses with some indication of K^o. No artificial structure seems to be generated in this mass spectrum.

   We have attempted a Dalitz plot analysis of the 3π spectrum as a function of the mass. Preliminary results of this analysis are shown in fig. 15. The a_2(1320) signal is clearly seen in the 2^+ $\rho\pi$ amplitude. The 1^+ amplitudes show different behaviour: the 1^+ $\epsilon\pi$ amplitude has a threshold enhancement and gradually decreases as the $\rho\pi$ amplitude increases. Their sum is also shown in fig. 15 and shows a large structure having a maximum in the same region as the mass spectrum. The other waves (not shown) have only a phase space behaviour.

   Whether these data show evidence for an a_1 resonance is a matter for further investigation.

8. PHYSICS WITH NEUTRALS

   In the 85 GeV/c run no γ detector was available to the WA76 experiment so that it was possible to reconstruct π^o's only as missing particles. The π^o was reconstructed by assigning to it the missing momentum and the π^o mass. The experimental resolution in the effective mass which makes use of such a π^o has been estimated to be in the range 30–40 MeV so that the quality of these data is not so high as in the channels with only charged tracks.

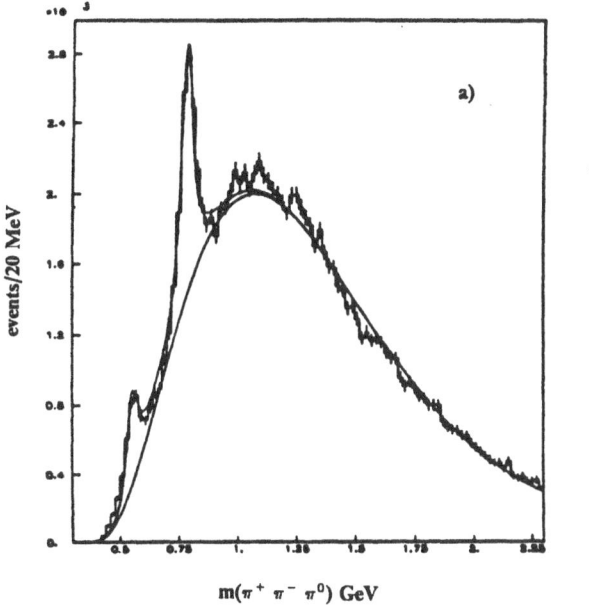

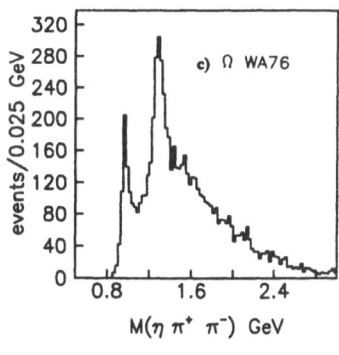

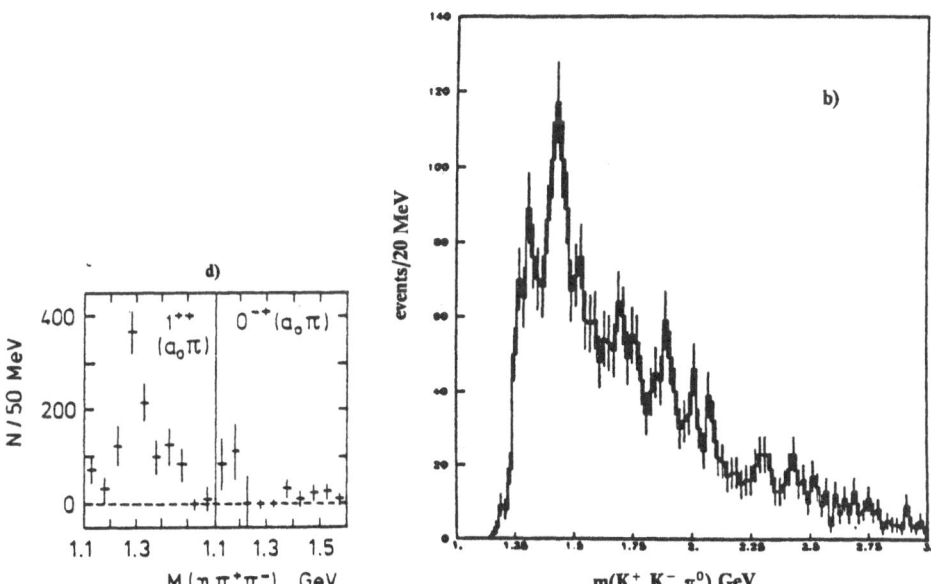

Figure 16. Physics with missing π^0's in WA76 at 85 GeV/c. a) $\pi^+\pi^-\pi^0$ effective mass, b) $K^+K^-\pi^0$ effective mass, c) $\eta\pi^+\pi^-$ effective mass, d) results from the Dalitz plot analysis of the $\eta\pi^+\pi^-$ system on the $J^{PC}=1^{++}$ and 0^{-+} $a_0\pi$ projections.

The $\pi^+\pi^-\pi^0$ effective mass for the events with two central charged
tracks after having removed the momentum balance events is shown in
fig. 16(a). The η and ω signals can be clearly seen and this plot shows
that the experimental resolution is good enough to detect narrow structures,
at least in the low mass region. The $K^+K^-\pi^0$ effective mass after having
required that two particles be identified by the Cherenkov system as K or
ambiguous K/p is shown in fig. 16(b). A clear E signal can be seen with a
possible hint of a D signal.

The $\eta\pi^+\pi^-$ system has been studied in the channel with 4 charged central
pions and a reconstructed π^0. After having plotted all the $\pi^+\pi^-\pi^0$ combinations,
the η region was selected and to the three particles in this region the η
mass was assigned. The resulting $\eta\pi^+\pi^-$ effective mass summed over the two
incident beams in shown in fig. 16(c) and shows clear $\eta'(958)$ and $f_1(1285)$
signals in spite of the large combinatorial background.

Since the physics in this channel is particularly interesting due to
the problematics related to the existence of two pseudoscalar states
observed to decay to $\eta\pi^+\pi^-$, a Dalitz plot analysis has been performed as a
function of the $\eta\pi\pi$ mass. The results are summarized in fig. 16(d) on the
1^{++} and 0^{-+} projections and can be summarized as follows:

The 1^{++} wave shows a large structure at the $f_1(1285)$ mass with a
possible hint of the presence of a 1^{++} $f_1(1420)$ signal. No evidence, on
the other hand, is observed for pseudoscalar states. It is difficult, with
these data to understand the effect of the experimental resolution on the
results of the Dalitz plot analysis. Better data with a γ detector are
needed in order to establish or not the existence of pseudoscalar states in
this channel.

9. CONCLUSIONS

A first look at the physics of the central region has been given by a
run in Ω at 85 GeV/c. The results are encouraging, several states,
including those on which discussions are actually going on in order to try
to establish their quark-gluon composition, are clearly observed in
particularly good conditions of signal to background. Further studies are
needed in order to understand the dynamics which is at work to produce
these states in the central region. A second run at 300 GeV/c has been
performed to study the energy variation of the observed states and to

search for new particles. In particular the resonances whose decays involve γ's should be detected.

REFERENCES

[1]   D. Robson et al., Nucl. Phys. B130 (1977) 328.

[2]   D. Denegri et al., Nucl. Phys. B98 (1975) 189.

[3]   D.M. Chew and G.F. Chew, Phys. Lett. 53B (1974) 191.

[4]   T.A. Armstrong et al., Phys. Lett. 167B (1986) 133.

[5]   W. Blum et al., Phys. Lett. 57B (1975) 403.

[6]   M. Baubillier et al., Z. Phys. C17 (1983) 309.

[7]   T.A. Armstrong et al., Phys. Lett. 66B (1986) 245.

[8]   T.A. Armstrong et al., Z. Phys. C34 (1987) 33.

[9]   A. Etkin et al., Phys. Rev. Lett. 40 (1978) 422;
       A. Etkin et al., Phys. Rev. Lett. 41 (1978) 784.

[10]   S.J. Lindenbaum, Comments Nucl. Part. Phys. 13 (1984) 285.

[11]   T.H. Hansson et al., Phys. Rev. D26 (1982) 2069.

[12]   T. Akesson et al., Nucl. Phys. B264 (1986) 154.

[13]   K.L. Au et al., Phys. Rev. D35 (1987) 1633.

[14]   S.M. Flatte' et al., Phys. Lett. 63 (1976) 224.

[15]   U. Mallik, preprint SLAC-PUB-4238 (1987).

[16]   A. Palano, CERN-EP/87-92, 18 May 1987.

[17]   D.L. Scharre et al., Phys. Lett. 97B (1980) 329;
       C. Edwards et al., Phys. Rev. Lett. 49 (1982) 259.

[18]   R. Armenteros et al., Int. Conf. on Elementary Particle Physics,
       Siena, Italy (1963);
       P. Baillon et al., Nuovo Cim. 50A (1967) 393.

[19]   S.U. Chung et al., Phys. Rev. Lett. 55 (1985) 779.

[20]   A. Ando et al., Phys. Rev. Lett. 57 (1986) 1296.

[21]   C. Dionisi et al., Nucl. Phys. B169 (1980) 1.

[22]   D. Antreasyan et al., SLAC-PUB-4305, May 1987.

[23]   T.A. Armstrong et al., Phys. Lett. 146B (1984) 273;
       T.A. Armstrong et al., Z. Phys. C34 (1987) 23.

[24] H. Aihara et al., Phys. Rev. Lett. 57 (1986) 51;
     H. Aihara et al., Phys. Rev. Lett. 57 (1986) 2500.
     G. Gidal et al., SLAC-PUB-4275, April 1987.

[25] J.J. Becker et al., Phys. Rev. Lett. 59 (1987) 186.

[26] Ph. Gavillet at al., Z. Phys. C16 (1982) 119.

[27] D. Aston et al., SLAC-PUB-4340, June 1987.

[28] C. Daum et al., Nucl. Phys. B182 (1981) 269.

[29] J. Dankowich et al., Phys. Rev. Lett. 46 (1981) 580.

[30] Ph. Gavillet et al., Phys. Lett. 69B (1977) 119.

[31] W. Ruckstuhl et al., Phys. Rev. Lett. 56 (1986) 2132.

[32] W.B. Schmidke et al., Phys. Rev. Lett. 57 (1986) 527.

[33] H. Albreckt et al., DESY 86-060 (1986) 527.

THE CRYSTAL BARREL EXPERIMENT (PS 197) AT LEAR

Crystal Barrel Collaboration: UC Berkeley, CERN
Univ. Hamburg, Univ. Karlsruhe, Queen Mary College London
UC Los Angeles, Univ. Mainz, Univ. München
Rutherford Appleton Lab., CRN Srasbourg
Univ. Zürich

Presented by Kersten Braune
CERN, CH-1211 Geneva 23, Switzerland

INTRODUCTION

The Crystal Barrel is a new detector which will combine good tracking and electromagnetic calorimetry over nearly the full solid angle. The detector will be located at the C2-beam line of the Low Energy Antiproton Ring (LEAR) at CERN. We will take data at cm-energies between $\sqrt{s} = 1.88$ GeV (i.e. stopping antiprotons) and $\sqrt{s} = 2.43$ GeV. In the corresponding mass range almost fifty mesons are known, most of them well understood as bound states of a quark q and an antiquark $\bar{q}$. Despite the experimental efforts in meson spectroscopy during the past 20 years many SU(3) $q\bar{q}$ nonets remain to be filled and some are overpopulated. One of our goals is to find the missing particles and to clarify existing ambiguities. A complete knowledge of the meson spectrum is essential for the identification of exotic particles like glueballs and hybrids. Once thought to rule out the quark model of hadrons, these exotics are predicted by QCD-inspired quark models and now their discovery would support the quark structure of matter and the underlying dynamical theory of quantum chromodynamics. Detailed experimental information about exclusive channels in $\bar{p}p$ annihilation can serve as an important test of QCD dynamics and new mathematical methods in non-perturbative QCD might lead to an understanding of hadron dynamics from first principles. Perhaps a light Higgs awaits discovery in $\bar{p}p$ annihilation as well as all the *sparticles* and *-inos* so much longed for.

THE EXPERIMENT

The Crystal Barrel detector is shown in Figure 1. The antiprotons enter from the left and interact in a liquid hydrogen, or an atmospheric pressure, hydrogen gas target. The interaction products which leave the target enter two layers of proportional wire chambers or, optionally, an X-ray drift chamber. Then they cross a jet drift chamber where the momentum of charged particles is determined and finally enter the electromagnetic calorimeter made of 1380 Cesium Iodide crystals. These detectors are inside a coil which produces a homogenous solenoidal magnetic field of 1.5 Tesla strength. Various types of fast hardware triggers on final state multiplicities and high level triggers on invariant masses will be available.

Liquid Hydrogen Target

Antiproton-proton interactions at rest with initial angular momentum $L = 0$ (i.e. annihilation from atomic S states) and in flight will be studied with a liquid hydrogen target. The target is 4 cm long with a diameter of 15 mm. Several windows and beam defining counters lead to multiple scattering and we expect a beam size of 1 cm FWHM at the target centre.

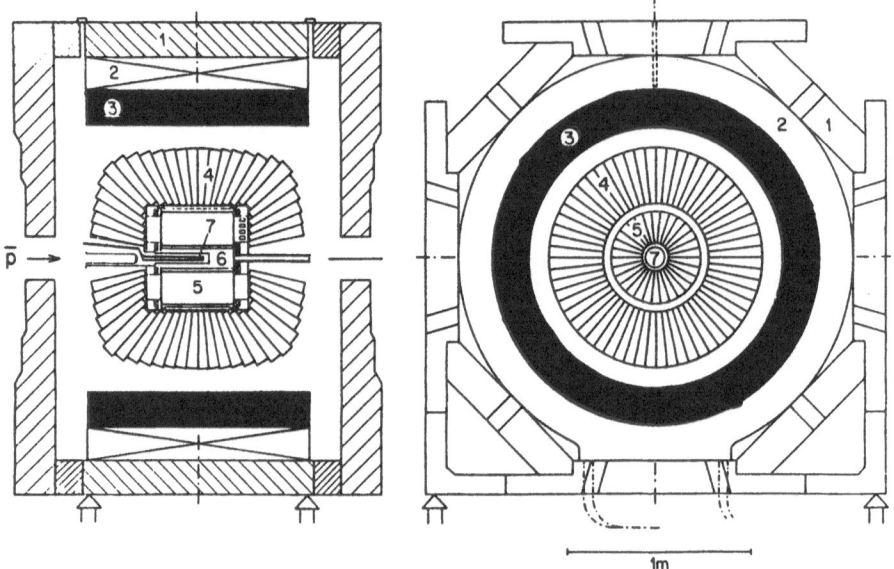

Figure 1.   The Crystal Barrel Detector.. Two perpendicular cuts of the detector can be seen. The yoke (1) and the coil (2&3) of the magnet, the CsI barrel (4), the jet drift chamber (5), the proportional chamber (6) and the liquid hydrogen target (7).

Proportional Wire Chambers

These cylindrical chambers of 35 cm active length have their anode wire layers at radii of 2.5 cm and 4.3 cm. The inner chamber has 120 and the outer chamber has 180 anode wires, leading to sectors in Φ of 2° and 3° The information from the proportional chamber will be available for the fast level zero trigger and the multiplicity of charged tracks will be known within 500 ns.

X-Ray Drift Chamber

Antiproton-proton annihilation from atomic P states (initial angular momentum L = 1) will be tagged by detecting the L X-rays. For these studies, the liquid hydrogen target and the proportional wire chambers will be replaced by an H_2 gas target and a cylindrical X-ray drift chamber. The chamber is 30 cm long, has one layer of sense wires at a radius of 4.3 cm and an outer radius of 5 cm. Chamber gas and H_2 gas target are separated by a 6 μm thin aluminized mylar membrane which is transparent to X-rays with energies greater than 500 eV. The X-rays convert in the argon/ethan gas of the chamber and the resulting signal on the sense wires is digitized by flash ADCs, giving drift time and pulse height information.

Jet Drift Chamber

This cylindrical drift chamber has a sensitive length of 40 cm, an inner radius of 5 cm and an outer radius of 27 cm. It is divided into 30 Φ sectors with 23 layers of staggered (±200 μm) sense wires. The drift cell is of the jet chamber type. The chamber gas is a 90:10 CO_2/Isobutane mixture with an average drift velocity of 8.5 μm/ns. The spatial resolution for each of the 23 measurements of a charged particle should be $\sigma \approx 100$ μm which leads to a momentum resolution $\sigma/p \approx 1.5\%$ to 3% for momenta between 100 MeV/c and 1 GeV/c. Charge division will be used to obtain the z coordinate of tracks with a precision of 4 mm. The pulse height on the sense wires will be measured by flash ADCs and should enable us to distinguish pions from kaons up to momenta of 500 MeV/c by their ionization energy loss.

CsI Barrel

The electromagnetic calorimeter is made of 1380 CsI(Tl) crystals of 16 radiation lengths and covers polar angles between $\Theta \geq 12°$ and $\Theta \leq 168°$ with complete coverage in azimuth. The crystals have a trapezoidal shape and are geometrically arranged to point to the interaction region at the centre of the detector. Each crystal covers 6° in Θ and 6° in Φ (12° in Φ for $\Theta \geq 150°$ and $\Theta \leq 30°$). The light from a crystal is collected in a wavelength shifter and read out by a photodiode. This type of readout imposes the use of charge sensitive preamplifiers and pulse shaping electronics which results in a noise equivalent of about $\sigma \approx 400$ KeV. The noise will contribute only at energies below ~ 100 MeV and will be negligible at higher energies. The expected energy resolution at 100 MeV is $\sigma/E \approx 4\%$ and 2% at 1 GeV. The reconstruction of π^0 and η mesons with a momentum of 500 MeV/c can then be achieved with a mass resolution of $\sigma \approx 8$ MeV/c^2, resp. $\sigma \approx 12$ MeV/c^2.

Magnet

All of the above detectors are located in a coil of 70 cm radius and 125 cm length. The coil produces a homogenous solenoidal magnetic field of 15 kGauss strength.

Trigger an Data Acquisition

Fast hardware triggers will be available for the total multiplicity and the charged multiplicity in the event. The number of charged particles as seen by the proportional wire chambers or the X-ray drift chamber will be known in less than 500 ns and used in the level 0 trigger. The total multiplicity (i.e. the number of clusters in the CsI barrel) will be determined by a purpose build fast cluster encoder (FACE), which also finds the coordinates of the clusters. A second check on the charged multiplicity comes from the jet drift chamber. This level 1 trigger on cluster and charged multiplicities is ready after about 10 μs. The high level 2 software triggers, based on VME processors, take over at this point. Decisions on cluster energies and invariant masses as well as some track finding is foreseen at this stage which takes from 200 μs up to 3000 μs depending on event complexity.

All digitizing electronics are located in dedicated CAMAC or VME crates which are read by specialized processors and the data are accumulated in a VME system crate which is connected to a host computer for data storage and monitoring.

STATUS AND OUTLOOK

All sub-detectors of the Crystal Barrel experiment, outlined above, are under construction at the various collaborating institutes or in industry. Prototypes of the two major parts, sectors of the jet drift chamber and of the CsI barrel have had exposure to test beams. In these runs (more or less) final readout and digitizing electronics were employed and satisfying first results, close to the expected performance, were achieved. A detailed analysis of the test beam runs is under way and the specifications for all detector parts will be finalized soon. The detector assembly at LEAR should start in spring '88 and we hope to take first physics data about one year later.

A few years of good physics runs at LEAR should enable us to clarify light quark spectroscopy, discover glueballs and hybrids (if they exist), study QCD dynamics and perhaps even find new paths beyond the standard model.

By the mid 90s it might be time to move on and install the Crystal Barrel detector at one of the newly build colliders. Due to its geometry and physics capabilities it would be well suited as a first stage detector at the Super LEAR Collider or an e^+e^- tau-charm-beauty factory.

SPECTROSCOPY OF NON-EXOTIC (qq̄) AND EXOTIC (gg, ggg, qq̄g, qq̄qq̄)

LIGHT MESONS WITH THE OBELIX DETECTOR AT LEAR

Ugo Gastaldi

CERN
Geneva
Switzerland

1. INTRODUCTION

The physics programme of the OBELIX experiment[1] includes spectroscopy of light conventional (qq̄) and exotic (glueball, hybrid, qqq̄q̄) mesons, dynamics of NN̄ interactions, and studies of annihilation of antinucleons with nuclei. This report is restricted to aspects concerning the first topic; it summarizes the relevant physics motivations, the experimental programme, and the principal characteristics of the detector.

OBELIX, together with the beams offered by the Antiproton Collector (ACOL), the Antiproton Accumulator (AA), and the Low-Energy Antiproton Ring (LEAR), enables a strategy of measurements that exploits physics possibilities that are unique and distinctive compared to other facilities (e.g. e^+e^- colliding rings) where spectroscopic work on light mesons has been carried out recently. These unique possibilities concern i) the change or selection, and the identification, of the initial state of the NN̄ system (whose annihilation produces light mesons), ii) the identification and measurement of the particles present in the final state of annihilation, and iii) the production, identification, and measurement of intermediate resonant states. These possibilities include:
  - i) Angular momentum polarization (AMP) and isospin polarization (IP): the distribution of quantum numbers of the initial state of NN̄ annihilations can be changed without modifying the other experimental conditions;
  - ii) extremely low momentum threshold for detecting, measuring, and identifying charged particles ($\pi^\pm$, $K^\pm$, p^+) present in the final state;
  - iii) very good resolution for reconstructing invariant masses of structures (produced in NN̄ annihilations) which decay into channels with charged pions and/or kaons;
  - iv) selection of large data samples with strangeness in the final state by triggering on charged and neutral kaons;
  - v) statistical accuracy limited more by the data-analysis time than by the data-acquisition time;
  - vi) KGB_L and KGB_I: *broad* structures can be identified by the comparison of identical decay modes in exclusive final states of the same type occurring from two sets of initial states with different angular momentum or isospin distributions.

The OBELIX experiment is in preparation. It is part of the diversified physics programme planned to take place at LEAR after ACOL completion[2]. The detector is a general-purpose large-acceptance high-resolution magnetic spectrometer which will use p̄ and n̄ beams. It features, for charged particles, good direction and momentum resolution, and complete identification (by dE/dx and time of flight) plus good direction measurement and identification for X-rays, gammas, and π^0's. The components include the OAFM magnet from the Axial Field Spectrometer (AFS) at the Intersecting Storage Rings[3] (ISR), a spiral projection chamber[4] and X-ray drift chamber[5] (SPC-XDC) central detector, the AFS jet drift chambers[6], a time-of-flight system (TOF), four high angular resolution gamma detector (HARGD) supermoduli, and two high density spiral projection chamber[7] moduli (HDSPC or BITURBO) in the end-caps. We emphasize in this written version of the oral presentation given at Erice the illustration of the physics possibilities which are new, and of the functional roles of the various components of the detector towards the selection, identification, and measurement of the $N\bar{N}$ initial state and of the annihilation final state.

This report is based on former work[8-15], on results of the ASTERIX experiment[16-19], and on discussions with and criticism from colleagues of the OBELIX Collaboration. For the topics of the physics programme of OBELIX not covered in this lecture, the reader is referred to the proposal[1] and to recent reviews and publications[20-22].

2. PHYSICS

2.1 Physics Motivations: Gluon to Gluon Coupling, Confinement, Structures

$N\bar{N}$ annihilations are a copious source of light mesons and mesonic resonances, and appear as a natural source of exotics such as glueballs (gg, ggg), hybrids ($q\bar{q}g$), and multiquark structures ($qq\bar{q}\bar{q}$, $qqq\bar{q}\bar{q}\bar{q}$), if they exist. The existence of these exotics is expected in the framework of Quantum Chromodynamics. They could be produced in $N\bar{N}$ annihilations with several types of ingredients such as, for example:
   i) quarks and antiquarks of the original $N\bar{N}$ pair which did not annihilate in the interaction;
   ii) pairs made of a new quark and a new antiquark created during the interaction;
   iii) gluons created by the annihilation of one, two, or three pairs of a constituent quark and a constituent antiquark of the original $N\bar{N}$ system.

In glueballs and hybrids the gluons play a double role: i) they mediate the interaction between constituent quarks and antiquarks as in $q\bar{q}$ mesons and in qqq baryons, and ii) they play the role of constituents of the hadronic structure. The constituent role can be played by a gluon because it can be coupled with the gluons which act as mediators of the strong interaction. This fundamental property of gluons — that they can be coupled directly also with another gluon — is at the basis of the existence of glueballs and hybrids, as well as being a building block of QCD. The unambiguous establishment of the existence of glueballs and/or hybrids would then give essential experimental support to the foundations of QCD. Notice that the photon (mediator of the electromagnetic force) does not couple directly with another photon. Notice also that the Z^0 (one mediator of the weak force) is expected to couple directly with the W^+ and W^- (the other two mediators of the weak force) and that, in order to observe the Z^0 coupling to W^+ and W^- via the decay of a virtual Z^0 to a W^+W^- pair, it will be necessary to upgrade and operate the LEP e^+e^- collider at ~ 200 GeV c.m. energy.

The basic conceptual importance of glueballs and hybrids motivates and justifies a major and diversified effort to pinpoint the observation of their existence. The energy region where they are expected to exist (1–3 GeV) is easily accessible. Instead, their identification is difficult and tricky as theorists are not in a position to make definite and complete predictions on energy, width, and decay properties of gluonic mesons, and candidates for glueballs and hybrids are generally also candidates for being conventional ($q\bar{q}$) or unconventional ($qq\bar{q}\bar{q}$, $qqq\bar{q}\bar{q}\bar{q}$) mesonic structures. However, glueballs and hybrids may have J^{PC} quantum numbers which are not accessible to $q\bar{q}$ mesons. This gives a chance of identifying some of them relatively quickly before the completion of the work of identifying and measuring all $q\bar{q}$ structures expected in the quark constituent model. However, to fully exploit this chance, it is necessary to be able to identify also broad structures in the mass spectra, as no solid theoretical argument forces a glueball or hybrid with exotic quantum numbers to have a narrow width and to show up as an outstanding peak. We will see, in the following, that $\bar{p}p$ and $\bar{p}d$ at rest in a gas target with a LEAR low-momentum beam, and a SPC–XDC initial-state dectector inserted in a general-purpose annihilation detector, offer chances to identify also broad structures. Notice that in 1963 already, a clear bump was observed in $\bar{p}p$ annihilations at rest[23-24] at the mass and in the channel ($K\bar{K}\pi$) where the best glueball candidate(s) has (have) been established recently in e^+e^- storage rings.

Besides offering exciting prospects in the spectroscopy of exotics, experimental work on $N\bar{N}$ annihilations has other outputs:

  i) basic information on $N\bar{N}$ interaction dynamics (this topic is not covered here);

  ii) new information on decay modes of known mesons obtained in production or in formation;

  iii) observation of conventional $q\bar{q}$ mesons not yet established;

  iv) possibility of observing $qq\bar{q}\bar{q}$ and $qqq\bar{q}\bar{q}\bar{q}$ structures.

The so-called 'conventional' $q\bar{q}$ meson spectroscopy is, in itself, of great interest and actuality[25]. Light mesons are more difficult to handle theoretically than heavy ones, but of great interest because the constituent light quark and antiquark move rapidly inside the meson volume and explore the periphery of the hadron, which is the confinement region.

2.2 Well-known Advantages of $N\bar{N}$ for Meson Spectroscopy

Some of the main advantages of $N\bar{N}$ annihilations for meson spectroscopy are well known:

  i) Extremely high statistics are possible with low $\bar{p}$ beam momenta: the annihilation cross-section increases with lowering beam momentum; at rest, all antiprotons can annihilate in the target, and only mesons and intermediate mesonic resonances are produced, as baryons cannot be made because of energy conservation.

  ii) The $\bar{p}$ beams available from sources where cooling is applied (at CERN and at Fermilab) are extremely cold ($\Delta p/p \leq 10^{-3}$) and permit high-resolution formation experiments.

  iii) The antiproton and the target nucleon are extended objects (unlike e^+ and e^-): annihilation can proceed via the many J^{PC} initial states which can be made with a non-pointlike fermion and antifermion pair (however, with the restrictions imposed by isospin conservation). While mesons with $J^{PC} \neq 1^{--}$ cannot be formed in e^+e^- machines at intermediate energies where two photon interactions are negligible, with $p\bar{p}$ collisions all $q\bar{q}$ mesons with mass higher than $2 m_p$ can be formed with, as a premium, extremely high energy resolution[26]. This basic idea has been nicely demonstrated in charmonium spectroscopy at the ISR[27], is at the origin of the E760 experiment at Fermilab and of JETSET PS202 at LEAR, and is obviously valid also near the $2m_p$ threshold where OBELIX will be active in $N\bar{N}$ formation measurements with $\bar{p}$ and $\bar{n}$ beams, and p^+ and d targets.

Notice that *in a $p\bar{p}$ formation experiment, only non-exotic quantum numbers are accessible.* To obtain structures with exotic quantum numbers starting from $p\bar{p}$ initial states, one must run production measurements and pick up combinations of particles in the final state that may be the decay products of an intermediate resonance with exotic quantum numbers.

  iv) High-statistics measurements of decay channels with kaons of resonances produced in low-energy $p\bar{p}$ annihilations are conceivable—if an adequate detector and trigger are operational—because of the very high rate of annihilations out of which the trigger can still extract a sizeable number of useful events per second. This feature is interesting since it is necessary to study and compare the properties of decay channels with and without kaons, in order to assess the nature of new states.

2.3 Newly Established Possibilities with $N\bar{N}$ Annihilations

Other major advantages offered by $N\bar{N}$ annihilations, especially for the identification of *broad* structures, are less well known, and some have had their physics and detector foundations proved only in recent years by the ASTERIX experiment at LEAR. They are related a) to the possibility of choosing, or selecting, or varying the distribution of angular momentum and isospin of the initial $N\bar{N}$ state without affecting the phase space open to the final state, and b) to the possibility of measuring precisely energy and momentum of the $N\bar{N}$ annihilating system. In the following part of this section we concentrate on the physics, beam, target, and detector aspects which allow the possibilities mentioned under points (a) and (b). We call these possibilities angular momentum polarization (AMP) and isospin polarization (IP). We elaborate then with the idea of exploiting AMP and IP in production experiments with a strategy of detector choices and measurement sequences that should allow the identification (and hopefully eventually also the measurement in detail) of broad structures. The approach is called — from the initials of colleagues of the Protonium, ASTERIX, and OBELIX Collaborations, who contributed indirectly to its elaboration — KGB_L when the distribution of the angular momenta is changed and KGB_I when the distribution of isospin is changed.

2.3.1 Angular Momentum Polarization

All the atomic levels of the $p\bar{p}$ atom are comprised in an energy window of 12 keV, which is the binding energy of the protonium ground state (1S level). The binding energy of an excited level with principal quantum numbers n is $n^{-2} \times 12$ keV. Figure 1 shows the energy levels of protonium

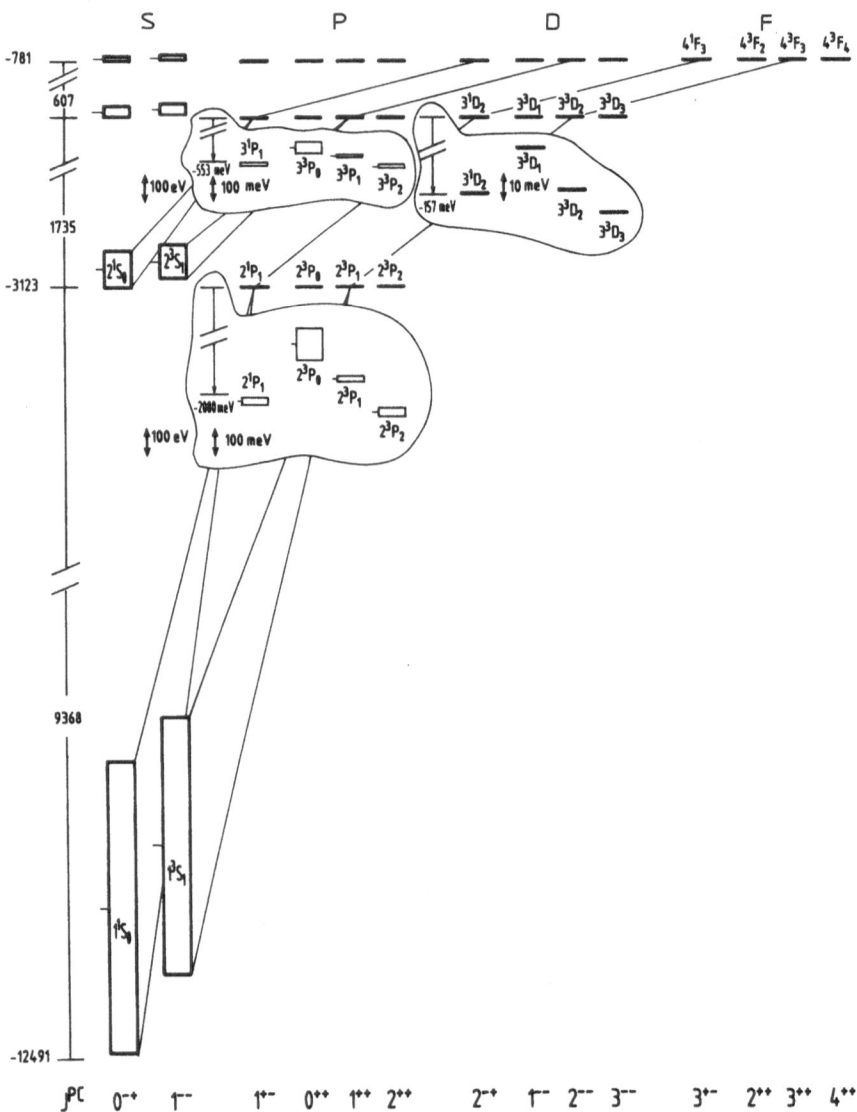

Fig. 1 Level scheme of the low levels of protonium, where X-ray emission spectroscopy is feasible (from ref. 15).

314

with indications of the effects due to strong interaction, spin, and relativity. Nearly all the antiprotons entering a H_2 target with momentum less than 50 MeV/c stop in the target, form $p\bar{p}$ atoms in relatively high n levels, undergo an atomic cascade, and eventually annihilate from nS or nP atomic levels (Fig. 2). A long time ago, Day, Snow and Sucher[28] had predicted that the fraction of annihilations in S- and P-wave should depend on the target density because of the combined effect of a) collisional Stark mixing of nearly degenerate atomic levels of the same principal quantum number n under the effect of the non-central electric field experienced by the neutral $p\bar{p}$ atom when crossing H_2 molecules of the target, b) radiative and Auger transitions to levels with lower n, c) annihilation in P and S levels. Their prediction has been confirmed indirectly by measurement of the yields of protonium X-rays with targets of various densities (liquid[29-30], four[31], eight[32], and one atmosphere at standard temperature[17-18], one atmosphere at various temperatures[33-34], and at several pressures below one atm[35]): lowering the target density the yields of transitions to the 2P and 1S levels increase; the ratio between the radiative width Γ^R_{2P} and the annihilation width Γ^A_{2P} of the 2P level is $\Gamma^R_{2P}/\Gamma^A_{2P} \approx 1\% $ [18].

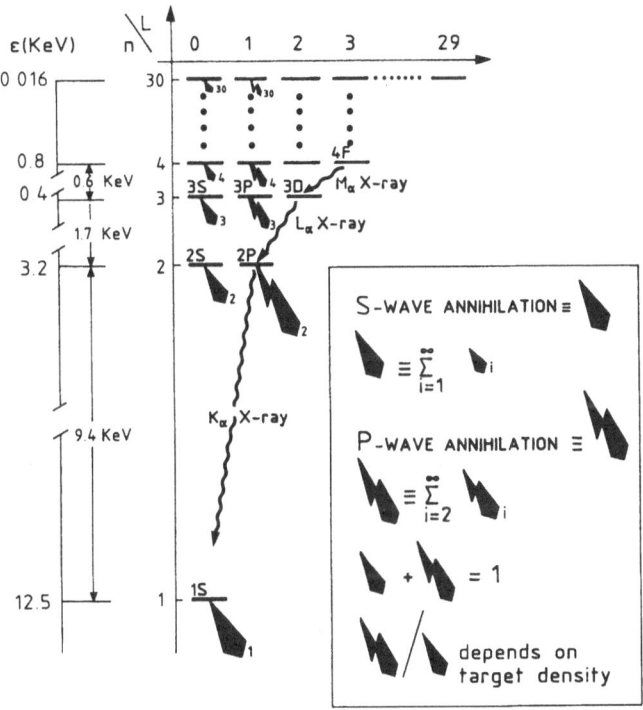

Fig. 2 Simplified scheme of protonium levels and $p\bar{p}$ annihilation at rest.

The Day–Snow–Sucher hypothesis has been confirmed directly by the striking differences in the values of branching ratios of $p\bar{p}$ exclusive final states measured in liquid H_2 by bubble-chamber experiments[36], and in H_2 gas at STP by the ASTERIX experiment at LEAR[16,19,37-38]. The data indicate that annihilation occurs dominantly from nS levels in liquid H_2 and from comparable amounts of nS and nP levels in H_2 gas at STP. Therefore, it is possible to change the distribution of initial states of $p\bar{p}$ annihilation at rest by switching from a liquid to a gas target, and by changing the density of the gas target. This possibility is not too practical in the first case because of the changes in hardware, and implies a thick container in the second case to stand under pressure and overpressure with respect to the ambient in which the final-state detector works.

The ASTERIX experiment has, however, also demonstrated that the branching ratios of some exclusive channels of $p\bar{p}$ annihilations, recorded in coincidence with X-rays in the energy region of the atomic transitions of protonium to the 2P level (nD–2P Lines: 1.7–3.1 keV), are markedly different from those measured ignoring the X-ray information[16,19]. This is due to two facts:
i) The fractions of S- and P-wave annihilation in H_2 gas at 1 atm are comparable.
ii) The request of X-rays in the energy region of protonium L lines selects preferentially P-wave annihilations from the 2P atomic level.
Point (i) is linked to the physics of the protonium cascade. Point (ii) is due in turn to two facts:
a) L X-rays from protonium have a much larger yield than the yield of X-rays from internal bremsstrahlung (IB) which may be emitted in protonium annihilations with charged particles in the final state from any nS or nP level in which annihilation occurred.
b) The L X-ray and also the internal bremsstrahlung emission overwhelmingly dominate the noise and background when using a SPC–XDC detector as proposed for[4,5] and developed by[39] the ASTERIX Collaboration. The SPC–XDC detector has extremely low mass and surrounds a cylindrical H_2-gas target. It is a gas detector equipped with sampling electronics based on flash analog-to-digital converters (FADCs). Its geometry allows the thinnest possible separation between the target volume and the detector's active volume, the maximum solid angle coverage, and the minimum amount of material causing multiple scattering of prongs and conversion and Compton scattering of gammas. The XDC has, furthermore, large granularity ($\sim 10^6$ virtual cells), constant segmentation in ϕ at all radii of the toroidal active volume, and high detection efficiency for L X-rays of protonium.

An essential physics point demonstrated experimentally by ASTERIX in recent years is that the branching ratio of a given exclusive channel of $p\bar{p}$ annihilation at rest depends on the distribution of initial states. That distribution is controlled by the target density and can be changed by requiring the coincidence of X-rays in the L lines energy region of protonium if one uses a gas target and a SPC–XDC X-ray detector. *We mean by angular momentum polarizability the fact that the distribution of nS and nP levels, from which $p\bar{p}$ annihilation at rest occur, can be changed in a H_2-gas target surrounded by a SPC–XDC.* This change (in the notation of Fig. 2, ⅄/⅄ varies from about 50/50 to about 90/10) can be done off line and does not affect the detection of the particles present in the final state. Moreover, it affects the ~ 2 GeV phase space open to the final state by less than 10^{-5} GeV. Conceptually, AMP is similar to spin polarization. Practically, it has the advantages that the target is pure H_2, has low density (thus featuring little absorption and very little multiple scattering), does not require a dedicated magnetic field to establish the 'polarization', the measurements with the two different polarizations are simultaneous and no time is required for polarization, repolarization, depolarization, and polarization inversion. Figure 3 summarizes the atomic and nuclear physics, and the accelerator and detector techniques that make AMP possible. Notice that, in order to detect efficiently L X-rays of protonium, a gas target is necessary, while LEAR is indispensable to stop antiprotons efficiently in a gas target. We will discuss later on, in the KGB_L subsection, the application of AMP to the search for broad new states. Angular momentum polarization is a by-product of a programme proposed long ago[8] to measure $p\bar{p}$ annihilations in P-wave and to produce mesons copiously starting with one more unit of angular momentum in the initial state. That programme requires LEAR, a 4π X-ray detector, and a complete final-state detector. It has been initiated with the ASTERIX experiment and is starting to bear fruit; it will be continued and should be brought to maturity in the ACOL era of LEAR.

2.3.2 Isospin Polarization

The change of distribution of isospin of the initial $N\bar{N}$ state requires changing the beam antinucleon or the target nucleon. $N\bar{N}$ initial states have the following isospins:

$\bar{p}n$ and $\bar{n}p$: $I = 1$
$\bar{p}p$ (and $\bar{n}n$): $I = 1$ and $I = 0$.

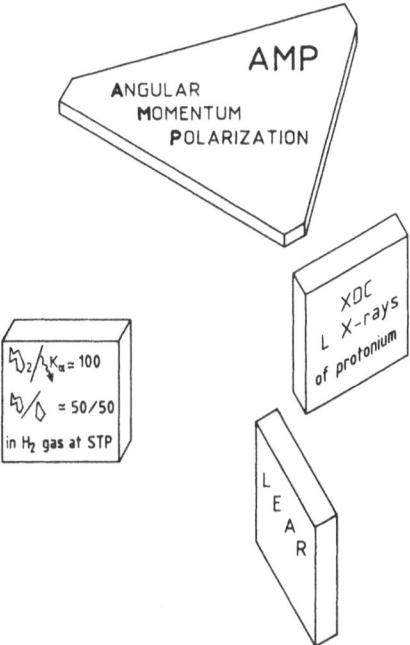

Fig. 3   Physics, accelerator, and detector foundations of angular momentum polarization.

In order to study directly pure $I = 1$ annihilations, an $\bar{n}$ beam has to be sent onto a H_2 target. This is part of the programme of OBELIX and it requires producing an $\bar{n}$ beam by charge exchange of an $\bar{p}$ beam on a liquid-H_2 $\bar{n}$ production target. About 10^3–10^4 antiprotons have to be consumed to get one antineutron in the $\bar{n}$ beam. A fraction of the antineutrons of the beam can interact in the target and it is convenient to have a dense target — hence liquid H_2 — in order to get appreciable interaction rates. The $\bar{n}$ beam has a large momentum band; however, the energy of each of the antineutrons which annihilates in the target may be determined by time-of-flight techniques[22]. Therefore the $\bar{n}$ beam can be used very conveniently to run formation experiments searching for $I = 1$ resonant states very near to the $N\bar{N}$ threshold, where with $\bar{p}$ beams there is the difficulty of the energy loss by ionization in the target. Change of the isospin of the initial state in the experimental configuration suitable for an $\bar{n}$ beam (liquid-H_2 experimental target in the centre of the detector and liquid-H_2 $\bar{n}$ production target upstream of the detector) is imaginable by shifting the production target off beam axis and drastically reducing the $\bar{p}$ beam intensity. However, the $p\bar{p}$ data with $I = 0$ and $I = 1$ would be taken with a much reduced performance of the final-state detector because of the limitations introduced by the liquid-H_2 experimental target (multiple scattering in the target and the container, reduced accuracy in vertex reconstruction, reduced capability in detecting and identifying — and of course also in triggering — neutral and charged kaons).

A second possibility of studying $I = 1$ reactions is to send an $\bar{p}$ beam onto a D_2 target and to measure $\bar{p}n$ annihilations in coincidence with the spectator proton. This has been done with limited statistics in liquid-D_2 bubble chambers, where spectator protons with momentum below 100 MeV/c were not detected because their track is too short[36].

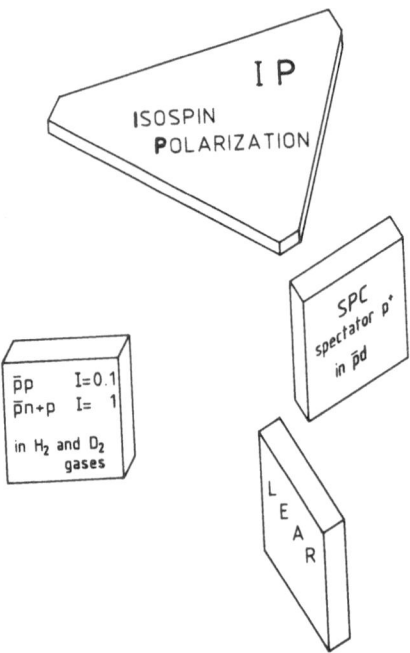

Fig. 4 Physics, accelerator, and detector foundations of isospin polarization.

With the LEAR beam it is possible to stop all antiprotons with a distribution width of stop points (corresponding to annihilation vertices) in the beam direction of less than 10 cm H_2 or D_2 gas at STP. It is easy to exchange H_2 and D_2 gases in a gas target, and it is possible to have in the same target volume a mixture of H_2 and D_2 gases. Prong counting allows one to distinguish $\bar{p}n$ annihilations from $\bar{p}p$ ones, and measurement of the momentum (for high-energy recoil protons) and of the range (for low-momentum protons) permits selection of those $\bar{p}n$ annihilations in which the proton was playing a mere spectator role (the energy distribution of these protons is due to their Fermi motion inside the deuteron). An SPC[4] surrounding a gas target can track the recoil proton in $\bar{p}d$ annihilations with a low-energy threshold well below that of deuterium bubble chambers. Recoil protons with momentum below 100 MeV/c have less than 3 mm range in liquid D_2 and could not be observed in the scanning process. Their range increases by a factor of about 10^3 in a gas target filled with D_2 gas or a D_2 and H_2 mixture, and the detection threshold in a surrounding SPC is essentially given by the transparency of the Mylar separation window. A SPC with a 6 μm thick Mylar cathode allows us to detect recoil protons with kinetic energy above 1 MeV, and to measure the direction and magnitude of their momentum.

The phase space open in p̄n annihilations with low-momentum recoil protons is little affected by the energy of the spectator proton; however, the determination of the p^+ momentum is essential to the complete kinematical reconstruction of the p̄n final states.

We mean by isospin polarizability the fact that the distribution of I = 0 and I = 1 states, from which NN̄ annihilations at rest occur, can be changed in a H_2–D_2 gas target. These changes in distribution can be done with measurements in H_2 and D_2 alternating in time, or with simultaneous measurement in a target filled with a mixture of both gases: the extremes are $(I = 0)/(I = 1) \approx 50/50$ in p̄p, with a pure H_2 gas target and $(I = 0)/(I = 1) = 0/100$ in p̄n with a pure D_2 gas target and selecting events with detected low-momentum protons. Notice that dE/dx and ranges of prongs in H_2 and D_2 are practically equal, so that for the measurement of the particles in the final state the target mixture is irrelevant. Again isospin polarization is conceptually similar to spin polarization. Practically, it has the advantage that the target has low density, does not require a dedicated magnetic field to establish the 'polarization', and the measurements with different isospin polarizations are either simultaneous, if a H_2/D_2 target mixture is employed, or require little time to change the 'polarization' by replacing the H_2 (or D_2) target gas with D_2 (or H_2). Figure 4 summarizes the detector, accelerator, and nuclear-physics ingredients which make IP possible. In the subsequent KGB_1 section we will discuss the envisaged application of IP to the search for broad structures.

2.4 Identification of Broad Structures with the KGB approach

The general idea of KGB for identifying broad structures is the following one: *compare with the same detector identical decay modes in exclusive final states of the same type, occurring from two sets of NN̄ initial states, with different angular momentum or isospin or spin polarization. The two sets of data are collected, without changes of the experimental set-up, simultaneously or by alternating the polarization.*

For the two sets of data the phase space available to the decay products is the same, as the total energy available to the two different types of initial states is the same (apart from minor differences that will be mentioned for each type of polarization).

Because there is no change in the detector configuration there is no variation of the acceptance for the detection of single particles (prong or γ), and moreover the global acceptance is the same when varying AMP and SP (while when varying IP the global event acceptance may vary because of the change of prong number by one unit).

If the detector has high enough segmentation, and the subdetectors used to select the initial state are not used in the trigger on the final state, the two event sets will also have the same trigger biases. By collecting the two sets of events with different polarization simultaneously, or with frequent changes of the polarization, the two data sets will be affected by the same detection and operation instabilities.

Finally, the event reconstruction and the data analysis will naturally be performed with the same software and by the same people. In the case of AMP, a set of dominant P-wave annihilation events can be selected off line as a subset of any set of reconstructed events by requiring the presence of an X-ray in the L energy region of protonium.

Under the circumstances mentioned above, the two event sets differ uniquely by their distribution of initial states. Differences in the features of the final state (e.g. shape of invariant mass spectra of combinations of decay particles, structures of Dalitz plots, shape of Dalitz plot projections, etc.) are due to dynamical effects or to selection rules. In order to construct the spectra corresponding to pure initial states, it will be necessary to perform linear combinations of the two sets of data. The proper multiplicative factors to be used depend on their degree of polarization. They are known (for IP) or can be measured independently (in AMP).

Two points are crucial:

i) as the initial states have different quantum numbers event by event, there are no interference effects;

ii) factorization occurs for all parameters linked with the apparatus.

Differences in the final-state spectra will show dynamical effects and can give evidence, in particular, of *broad signals* which may be produced with one type of initial state and absent with the other type.

An idealized situation for KGB application is summarized schematically in Table 1.

Table 1. KGB Ingredients

SAME	PHASE SPACE for decay products
,,	EXPERIMENTAL APPARATUS: hardware electronics trigger on final states
,,	ACCEPTANCES
,,	TRIGGER BIASES
,,	DATA COLLECTION TIME
,,	SOFTWARE
DIFFERENT AMP (or IP or SP) POLARIZATION	
Observe differences in final state of same type → Selection rule or dynamical effect.	

Dramatic variations in the final-state spectra have been observed with a change of AMP in the ASTERIX experiment at LEAR. Differences are expected in some two-body final states because of selection rules, but in addition annihilation dynamics enhances or depresses known resonances produced as intermediate states in $N\bar{N}$ annihilations.

Narrow resonances are, in any case, visible in mass or Dalitz plots, while broad ones may be overlooked or not noticed. Since glueballs and hybrids are not predicted as being particularly narrow, might be broad, and should not be a rare effect in $N\bar{N}$ annihilation, the KGB approach which we propose might contribute significantly to their observation and study.

Of course, it is not new to change one physical parameter in one experiment, keeping the rest unchanged, and performing a differential measurement. This has been done, for instance, at high energies by the NA3 experiment at CERN, switching from π^+ to π^- the beam shot onto p and n targets in the study of Drell–Yan $\mu^+\mu^-$ pairs[40] (here the candidate decay channel $\mu^+\mu^-$ was chosen among all final states containing a $\mu^+\mu^-$ pair of the type $\mu^+\mu^-$ + anything with 'anything' not measured). More recently, MARK III and DM2 have studied J/ψ decays with a low-bias trigger[41,42] and have compared final states of the types $J/\psi \rightarrow \gamma X$, ωX, ϕX with X fully identified (e.g. $X \rightarrow \pi^+\pi^-$, K^+K^-, $K_0\bar{K}_0$, $\eta\eta$, $K\bar{K}\pi,\eta\pi\pi$, $\phi\phi$, $\varrho\varrho$, $\omega\omega$). What is new in the $N\bar{N}$ case is that all the following features come in combination:

i) feasibility of changing the INITIAL STATE, as in NA3 Drell–Yan experiments,

ii) full identification of all particles present in the FINAL STATE, as in e^+e^- experiments,

320

iii) 'easy' experimental situation:
   a) c.m. of final state at rest, as in e^+e^- experiments,
   b) fixed-target experiment conditions (ease of access),
   c) low energies (easy particle identification + high resolution).

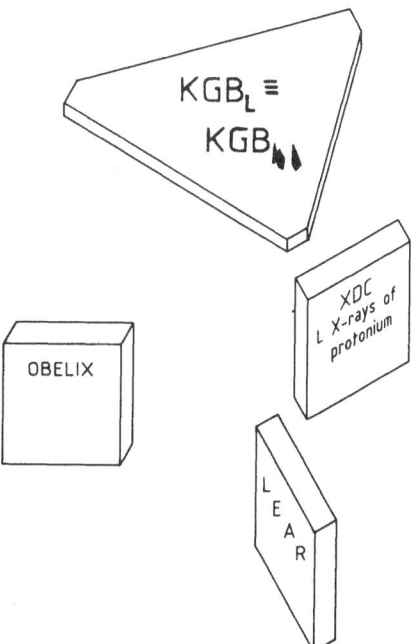

Fig. 5 Accelerator, initial-state detector, and final-state detector for differential measurements varying the distribution of angular momenta of the initial states (KGB$_L$).

One crucial point which distinguishes KGB from classical polarization experiments is the fact that the target is pure H_2 (in the D_2 case it is 'pure' n for all events with the recoil proton fully measured). Therefore, the final state of N$\bar{\text{N}}$ annihilation can be fully reconstructed and production experiments can be run. In classical N$\bar{\text{N}}$ polarization experiments the hydrogen of the target is mixed with nuclei of other elements chemically necessary for establishing the polarization, and only a few two-body channels such as $\pi^+\pi^-$, K^+K^-, p$\bar{\text{p}}$, that have little background because of their unique kinematics, can be uniquely identified as associated with p$\bar{\text{p}}$ interactions. Consequently, N$\bar{\text{N}}$ polarization experiments have been run so far only in formation by fixing the final state (e.g. $\pi^+\pi^-$) and studying its dependence on the energy of the N$\bar{\text{N}}$ system. As mentioned before, this approach could not give access to mesons with exotic quantum numbers.

2.4.1 KGB acting on the Angular Momentum Polarization: KGB$_L$

Figure 5 summarizes the beam, detector, and physics ingredients of KGB$_L$.

As a first simple example of the idea, let us consider the reaction $p\bar{p} \rightarrow K^+K^-\pi^0$. One can take one data set where the final state $K^+K^-\pi^0$ is selected off line, among the other final states with two prongs collected with a simple multiplicity trigger, and no further requirement made at the trigger level and in the off-line analysis on the initial state before annihilation. A second data set can be collected simultaneously by a parallel trigger which has the same requirements on the prong multiplicity and requires, in addition, the detection in coincidence of X-rays in the 1–4 keV energy region. The second data set contains events where the $K^+K^-\pi^0$ annihilation occurs dominantly in P-wave from sublevels of the 2P-level. One can measure, for instance, the spectrum of invariant masses of the K^+K^- system recoiling against the π^0 in the two sets of events and compare the two spectra, or linear combinations of them, to observe if a broad state X^0 decaying into K^+K^- is present. Notice that a state decaying into K^+K^- cannot have exotic quantum numbers (only $J^{PC} = 0^{++}, 1^{--}, 2^{++}, \ldots$ are allowed for a system decaying into K^+K^-).

As a second example of the idea, let us now consider the reaction $p\bar{p} \rightarrow K^+K^-\pi^0\pi^+\pi^-$. Again two data sets with and without the request of the presence of an X-ray in the L energy region are taken simultaneously with the same trigger requirements on the final state (e.g. trigger on four-prong multiplicity). One can study the Dalitz plots of the $K^+K^-\pi^0$ system in the two cases, and compare them — or their projections — or compare linear combinations of the two plots to observe if there are narrow and broad states X^0 decaying into $K^+K^-\pi^0$. A state decaying into $K^+K^-\pi^0$ can have exotic quantum numbers.

Notice that the KGB study requires, as a prerequisite, full understanding of the final-state event reconstruction and measurement, and good performances of the final-state detector. Notice also that we have picked up two examples of final states where no combinatorial background exists if the particle identification is complete. This is not fully the case with the ASTERIX detector. Nevertheless, analysis is in progress; in the $K^+K^-\pi^+\pi^-\pi^0$ final state a clear signal of a structure decaying into $K^+K^-\pi^0$ is visible in the E/i mass region, and we are curious to see if the signal depresses in the i/E mass region and enhances in the D region when X-rays are put in coincidence.

Let us now comment on various points following the guideline of Table 1.

The uncertainty on the total energy of the initial state is $\sim$ 12 keV for the set of events without selection based on the protonium X-rays. For 80%–90% of the events of the set with X-rays in the 1–4 keV energy region, the energy of the $p\bar{p}$ system in its c.m. is known to better than 1 keV, and the $p\bar{p}$ c.m. is at rest in the laboratory apart from the thermal motion of the $p\bar{p}$ atom.

In order to vary the AMP one has to detect the protonium X-rays. One can also think of changing the distribution of initial states by alternating a liquid-H_2 target with a H_2-gas target.

One could imagine optimizing one experiment around a liquid-H_2 target and then running also with H_2 gas in the same target container. This approach would be straightforward, with the proviso of having properly designed the entrance counter and windows, and could already have been applied by several experiments using liquid-H_2 targets at LEAR, but it has not been considered by anybody so far. One inconvenience for KGB would be that the annihilation vertex ($\sim \bar{p}$ stop point) distribution would be different in the two cases (acceptances would not factorize), and the momentum threshold for annihilation prong measurement would be higher with a liquid-H_2 target than the one for an experiment optimized around a gas target, because of the higher density of the target itself and the presence of the surrounding materials required by the cryogenics. Another drawback would be the relatively long time (a few hours) to get stable density conditions when going from liquid to gas: instabilities in detector performances and operation may not factorize under these circumstances. Angular momentum polarization would change in this scenario from dominant S-wave annihilation to an S/P ratio of about 50/50 at STP. The detector, the trigger biases, and the analysis software would be unchanged.

The approach of OBELIX to KGB_L is to use a H_2-gas target, to take *simultaneously* two sets of data with different AMP, and to change AMP by requiring in coincidence X-rays in the energy region of protonium L lines. The X-rays are detected by an XDC. The gas target, in conjunction with the surrounding SPC/XDC detector, allows the lowest possible momentum threshold and the best direction measurement for annihilation prongs. The X-ray request can also be put at the trigger level. In that case two types of master trigger have to be used alternatively with the same level of priority to avoid cross biases. The X-rays of protonium can enter only the central detector SPC/XDC, where they originate ionization clusters at their absorption point. Their presence (at most three per event) does not affect the final-state detection performances, since the segmentation and granularity of the central detector are high. Angular momentum polarization varies from a ratio of S/P annihilation of about 50/50 for the data set which ignores the X-rays, to about 90% P-wave annihilation for the data set with X-rays in the energy region of the protonium L lines in coincidence. Notice that the S-wave contamination in the data set with X-rays is low, but not the same for all final states since the X-ray internal bremsstrahlung activity depends on the annihilation channel. The design of the OBELIX experiment has been carried out with the explicit intention of realizing completely the unique possibility of KGB_L offered by $p\bar{p}$ at rest in H_2 gas, and to set up a final-state detector with the best performances achieved so far in the energy region $\sqrt{s} < 1900$ MeV.

The Crystal Barrel Collaboration is preparing a second detector which will be used to study $p\bar{p}$ at rest at LEAR. At the time of the Tignes Workshop they considered using a liquid and a gas target without X-ray detection[43]. Subsequently, in their proposal[44], they added the study of annihilation in a H_2-gas target surrounded by an XDC[5] to study $p\bar{p}$ annihilation at rest with L X-rays in coincidence. They plan to make dominant S-wave annihilation measurements with an assembly including a liquid target and two surrounding concentric cylindrical MWPCs; then to remove this assembly and to substitute for it a H_2-gas target and XDC assembly to measure dominant P-wave annihilation. This approach is motivated by selecting quickly dominant S-wave and dominant P-wave annihilations. It exploits AMP, but does not allow the thorough application of the KGB_L approach to search for broad structures. The use of KGB_L in their experiment would be unsafe for several reasons:
  i) the apparatus has to be opened and changed between two measurements (no more factorization on time-dependent figures such as detection efficiencies and stabilities);
  ii) the vertex distribution changes as well as the vertex reconstruction hardware and software;
  iii) the trigger biases are different for final states with prongs because of the change of central detector.

2.4.2 KGB acting on the Isospin Polarization: KGB_I

Figure 6 summarizes the ingredients of KGB_I. KGB_I exploits IP. As a first example of the idea, let us now consider the two reactions $p\bar{p} \rightarrow K^+K^-\pi^0$ and $\bar{p}n \rightarrow K^+K^-\pi^-$. One can measure, for example, the invariant mass of the K^+K^- system in the two reactions and compare the two spectra and linear combinations of them to observe if broad states X^0 decaying into K^+K^- are present.

Then let us consider two more complex final states of the same type as further examples: $p\bar{p} \rightarrow K^+K^-\pi^0\pi^+\pi^-$ and $\bar{p}n \rightarrow K^+K^-\pi^0\pi^0\pi^-$. One can study the Dalitz plot of the $K^+K^-\pi^0$ system in both cases and compare them, or linear combinations of them, to observe if there are broad states X^0 decaying into $K^+K^-\pi^0$. A state with such a decay may have exotic quantum numbers.

In both types of examples one should notice some difficulties in the application of KGB: i) the final states differ by one unit of charge, and the detection efficiencies for π^- and π^0 are not the same since different detectors concur in the π^- and π^0 measurements, and then detection efficiencies do not factorize, ii) the presence of the recoil proton in $\bar{p}n$ annihilation in d may alter, in principle (and in practice depending on the detector granularity), the detection efficiency of the other prongs, iii) a combinatorial background is present in the $K^+K^-\pi^0$ spectrum of the second example because of the presence of two π^0's in the final state. Nevertheless, the other parameters discussed in the general KGB section may remain unchanged by using a H_2- and a D_2-gas target or a mixture of the two gases as discussed in the section on IP, and KGB_I may relatively quickly indicate broad structures if they are produced with strong dependence on the isospin of the initial $N\bar{N}$ state.

Comparing $\bar{n}p$ with $\bar{p}p$ data for the purpose of KGB_I requires a liquid-H_2 target and would exhibit the following difficulties: the two sets of measurements would differ for the interaction vertex distribution, the $\bar{n}$ and the $\bar{p}$ momentum distribution at interaction would be different; the two data sets cannot be taken simultaneously and, because of the use of a liquid target, the detection threshold and direction error for measuring annihilation prongs may be relatively high.

In OBELIX, KGB_I will be applied with antiprotons at rest in H_2- and D_2-gas targets, and operation experience, both of the experimental apparatus and of the analysis chain, will tell us whether it is better to use a H_2–D_2 mixture to factorize operation instabilities or alternate the target gas to make the $\bar{p}p$ and $\bar{p}n$ event identification straightforward.

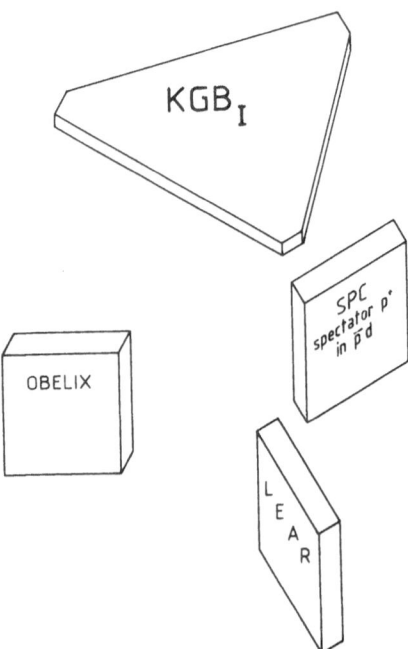

Fig. 6   Accelerator, initial-state detector, and final-state detector for differential measurements varying the distribution of isospin of the initial states (KGB_I).

2.4.3 KGB acting on the Spin Polarization: KGB_S?

Some years ago, we were considering the possibility of separating $\bar{p}p$ annihilations at rest in the singlet state from those in the triplet state, by detection of the K-line X-ray transition to the ground state, if the splitting of the 1^1S_0 and 1^3S_1 sublevels due to strong interactions should have turned out to be large[9]. In view of the experimental results obtained with ASTERIX[18], this hope appears to have been an optimistic dream because of the following facts: the K_α line is rather wide; we have not yet measured a still possible spin splitting of the ground state; the internal bremsstrahlung X-ray emission dominates the K-line emission, and therefore detection of an X-ray in the K-line energy region is not even a good signature of S-wave annihilation.

Hopes of producing polarized $\bar{p}$ beams[45] circulating in LEAR are instead still legitimate and a polarized-jet H_2 target[46] with a pure H_2 target material might be installed. If, in the long range, spin polarization is achieved at LEAR, then KGB_S will be feasible. KGB_S would require fixing the circulating beam momentum and studying $p\bar{p}$ annihilations changing the total spin from dominantly singlet to dominantly triplet, and studying and comparing sets of events with the same exclusive final state. These production experiments do not require record luminosities and would allow access to exotic final states if the detector would be able to measure completely complex final states. This type of spin experiment has not, so far, been considered to our knowledge, while groups interested in jet targets and polarization have focused their physics interest on two-body final states, which are the only ones that can be treated practically with existing polarized external targets, which contain other elements besides H_2.

In summary, KGB could be applied thoroughly with $p\bar{p}$ in the KGB_L and KGB_S cases. KGB_L is proved to be feasible. KGB_S needs many years to ascertain if SP is feasible. KGB_L has the further experimental advantage that a 6 μm thick Mylar 'beam pipe' can surround the interaction volume, while in the KGB_S case, the detector has to compromise with the beam pipe which must stand the ultrahigh vacuum of LEAR (this implies higher detection threshold and multiple scattering before direction measurement!). KGB_l has the advantage that the isospin polarization is completely known for each target, but has the drawback that the final states of the same type differ by one unit of charge.

3. THE OBELIX DETECTOR

3.1 General Structure

Figure 7 shows the general layout of the detector. The active elements of the detector components are disposed between and around the two polar expansions of the Open Axial Field

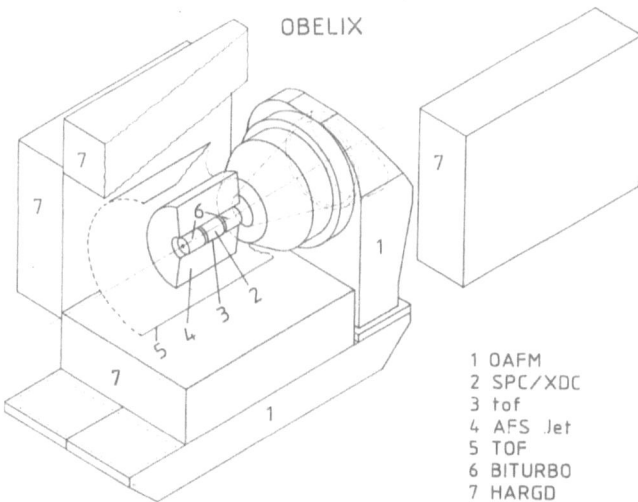

Fig. 7   OBELIX detector: 1) Open Axial Field Magnet, 2) Spiral projection chamber and X-ray drift chamber — SPC/XDC —, 3) Time-of-flight tof scintillators, 4) AFS jet drift chamber, 5) Time-Of-Flight TOF scintillators, 6) High density spiral projection chambers — BITURBO —, 7) High angular resolution gamma detector — HARGD —.

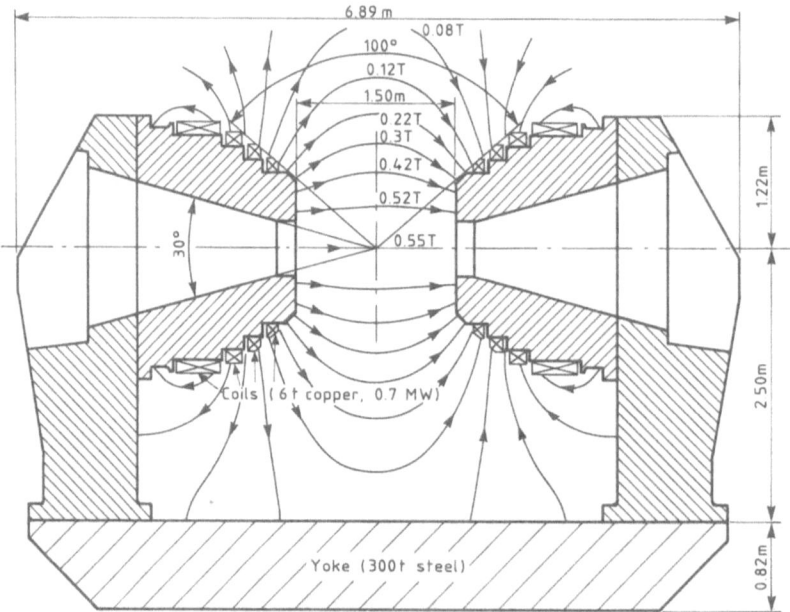

Fig. 8 Maquette of the OBELIX detector.

Fig. 9 Magnetic field lines in the OAFM.

Magnet (OAFM). The axis of the magnet (and of the experiment) is aligned with the straight section 1 of LEAR (LSS1) to enable the future IDEFIX experiment, which will need p$\bar{\text{p}}$ atoms produced in flight in LEAR with the $\bar{\text{p}}$H$^-$ co-rotating beams option[15,47].

Figure 8 shows a 1:5 scale model of the experiment with one HARGD supermodule and the south lateral wall of the TOF barrel removed. In the picture are visible: the OAFM magnet, the top and bottom HARGD supermoduli and — between the two conic poles of the magnet — the south AFS jet chamber. The tof barrel and the SPC plus BITURBO assembly are hidden by the AFS chamber.

The magnetic field of the OAFM has good axial rotation symmetry up to 1.5 m radius from the axis. The field lines in the vertical plane containing the beam axis are shown in Fig. 9.

Table 2. Identification and Measurement of the Initial-State Particles for Angular Momentum Polarization and Isospin Polarization

Function	Detector Component	
Protonium X-rays (and $\bar{\text{p}}$d X-rays): AMP		
Emission point	SPC (via vertex reconstruction in annihilations with prongs in the final state)	
Absorption point	SPC	
Energy	SPC/XDC	
Recoil proton in $\bar{\text{p}}$d $\rightarrow$ p + $\bar{\text{p}}$n annihilations in deuterium: IP		
	Low-Momentum 'Spectator' Proton	High-Momentum 'Actor' Proton
Emission point	SPC	SPC
Direction	SPC	SPC
Momentum	SPC	AFS
Energy loss	SPC	SPC, tof, AFS
Range	SPC	–
Velocity	–	tof + TOF

3.2 Initial- and Final-State Selection and Measurement

Figures 10 and 11 show the active volume of all the detector components in side and front cut views containing the detector centre. In the drawings are indicated the functional roles of the detector components with respect to detection, identification, and measurement of charged and neutral particles emitted before, together with, or after the N$\bar{\text{N}}$ annihilation process.

The functions of the detector components which participate in the identification of the initial and final states of annihilation are given in Tables 2 and 3.

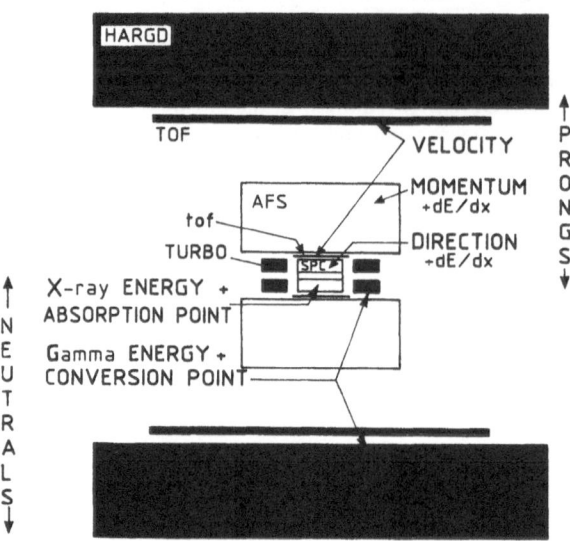

Fig. 10 Cut view of OBELIX detector components along a vertical plane containing the beam axis.

Table 3. Identification and Measurement of Particles Present in the Final State of p̄ Annihilations at Rest

Function	Detector Component	
	Charged Particles	
	High Momentum	Low Momentum
Emission point (vertex)	SPC	SPC
Direction	SPC	SPC
Momentum	AFS	SPC
Energy loss	SPC + AFS (+ tof)	SPC
Velocity	tof + TOF	–
Range	HARGD + TURBO	SPC
	Gammas and Bremsstrahlung X-rays	
	High Energy (MeV)	Very Low Energy (keV)
Emission point	SPC	SPC
Absorption point	–	XDC
Conversion point	HARGD + BITURBO	–
Energy	HARGD + BITURBO	XDC

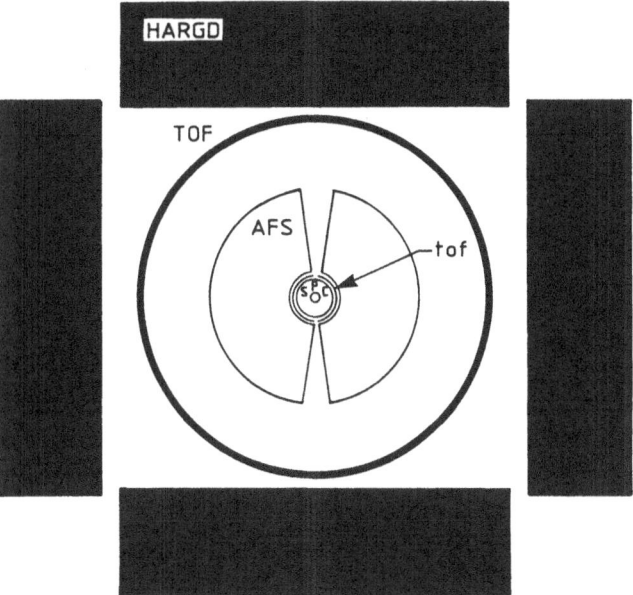

Fig. 11   Cut view of OBELIX detector components along a vertical plane orthogonal to the beam axis and containing the centre of the experiment.

3.3 Components of the Experiment

3.3.1 Beams

We plan to use $\bar{p}$ beams with momenta of 100 MeV/c, and 100–300 MeV/c. The 100 MeV/c beam will be used for stopping antiprotons in the centre of a gas target, with stopping rates varying between 10^3 and 10^5 depending on the trigger selected. Practically all antiprotons of the 100 MeV/c beam stop in the target.

Variable low momenta will be used for formation measurements near the $N\bar{N}$ threshold. With a 40 cm long H_2 target and an overall amount of 20 mg/cm^2 of material in the beam — including beam-pipe window, beam counters, target windows, window of the vacuum pipe to the beam dump — about 1% of the beam particles interact with the windows and 2‰ in the target. The longest drift-time in the central detector is $\sim$ 3 μs. Hence OBELIX can stand 10^5 interactions per second before experiencing pile-up problems. The data-acquisition system can record, on tape, typically 20–100 events per second. A beam intensity of up to 10^7 antiprotons per second could then be useful for measurements in flight of rare channels in those cases where the trigger is very selective and can reduce the rate of accepted events to the level of 20–100 per second that can be handled by the data-acquisition system, provided that the antiprotons which do not interact with the target and with the windows in measurements in flight can be dumped downstream of the experiment far away from the detectors, in a dump shielded from the detectors.

The beam comes into the experiment from the left side of the scale model of Fig. 8. For $\bar{n}$ measurements a liquid-H_2 target will be positioned in the upstream cone of the magnet to produce antineutrons by charge exchange. The $\bar{n}$ production target must be well shielded to protect the experiment against the p̄p annihilation particles. The flux of antineutrons in the wide-band n̄ beam going to the interaction target will be 10^3 to 10^4 times less intense than the p̄ flux on the production target. If a good enough shielding can be designed, p̄ beams with intensities up to 10^8 could be useful since the n̄ flux incident on the target would be 10^4–10^5 and only a fraction of the antineutrons would interact with the experimental target.

3.3.2 Targets

The gas-target assembly is a cylindrical unit plugged into the central detector (see Fig. 12). The target wall is a 6 μm thick (or thinner) aluminized Mylar tube, which acts also as the internal cathode of the central detector. The low density of the target, as well as the extreme thinness of the Mylar wall separating it from the active volume of the central detector, allows the measurement of the K, L, and M X-ray lines of protonium, the measurement of the recoil proton in p̄d annihilations down to proton momenta of about 40 MeV/c, the detection of charged pions and kaons with momenta down to 20 MeV/c, and the measurement of the direction of prongs emitted in the annihilation process without multiple scattering.

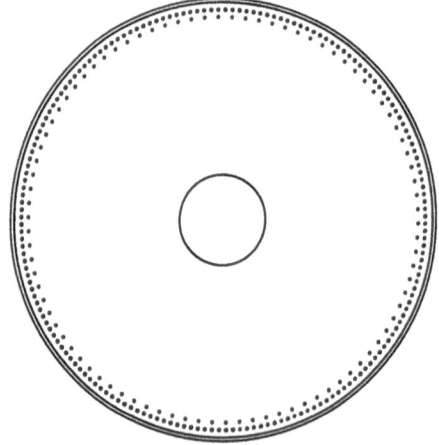

Fig. 12   Side and front cut views of the central detector SPC/XDC.

With an incident $\bar{p}$ beam of 100 MeV/c, the volume where antiprotons stop in H_2 and D_2 gas is contained within a cylinder 10 cm long and 6 cm in diameter for a target of 40 cm total length. The materials in the $\bar{p}$ beam line will be adjusted to have the stop volume positioned in the centre of the SPC. Nearly all the antiprotons of the beam annihilate in the stop volume since interactions in flight are negligible. The rate of interactions at rest in the centre of the detector can be changed by increasing the beam intensity from 10^3 to about 10^5.

In measurements in flight, annihilations occur uniformly all along the target, with higher densities at the target and beam windows.

The target plug can be removed and replaced by a plug with a larger diameter to give room for the liquid-H_2 target assembly foreseen for the $\bar{n}p$ measurements. The liquid-target cryogenics will be located upstream of the experiment together with the cryogenics of the $\bar{n}$ production target.

3.3.3 The Central Detector

Figure 13 shows the assembly of the SPC and the BITURBO end-cap calorimeters, which slides inside the tof barrel to be positioned in the centre of the experiment. The SPC (see also Fig. 12) covers 90% of the solid angle in order to detect prongs from annihilations in the target centre; it images tracks of charged particles over 80% of the solid angle with typically seven points per track, with a resolution of about 300 μm in the three dimensions. Ionization clusters drift to the sense wires along spiral trajectories, under the action of a radial drift field, and are reconstructed by hit wire numbers (ϕ), and by drift-time (r) and by charge division (z) measured by the electronics of the sense wires. Cathode strips, at an angle to the sense wires, cover the interior of the SPC container, and are connected at the downstream end of the SPC to FADC read-out electronics of the same type that equip each extremity of the sense wires. The cathode signals extend, by about one order of magnitude, the dE/dx range for pulses which saturate the electronics of the sense wires, and allow a spatial resolution in the beam direction (z) of the same quality as in the rϕ plane by centre of gravity between strips and by crossing with the sense wires.

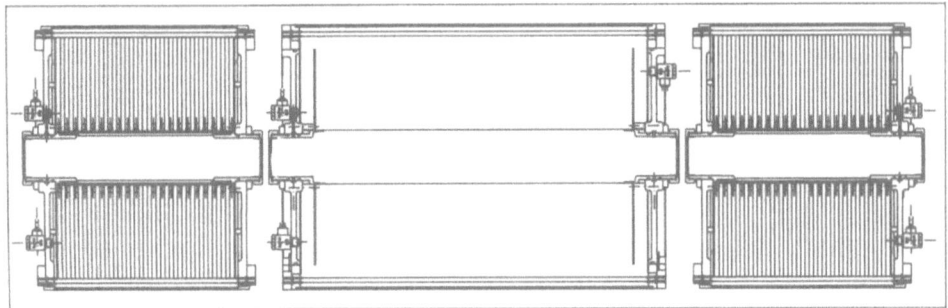

Fig. 13   SPC and BITURBO assembly.

Our SPC and target configuration provides a central detector with a record minimum of material between the active volume and the vertex: 6 μm Mylar to be compared with the beam pipe of e^+e^- colliding rings plus the entrance window of their detectors. This permits — besides detecting protons with transverse momentum in excess of 40 MeV/c, and measuring the momentum of soft pions and kaons which do not enter the AFS tracking detector — the precise measurement of the direction of the prongs which enter the AFS chambers before they undergo multiple scattering in the SPC container, in the tof scintillators, and in the AFS entrance wall. The AFS chambers measure, with great accuracy, the momentum of annihilation prongs and their direction after they have undergone multiple scattering. The energy lost by prongs, between the vertex and the AFS active volume, can be calculated and corrected for. The capability of the SPC to measure the direction of prongs before they suffer multiple scattering permits an accurate determination of the relative angles between charged particles emitted in the annihilation and, together with the momentum measurement of the AFS chambers, good resolution in the spectrum of invariant masses of hadronic resonances produced in the annihilation and decaying into charged particles.

The SPC detects protonium X-rays with good efficiency in the energy region 1–15 keV. Its overall efficiency for L X-rays of protonium (1.7–3.1 keV) is about 50%.

The SPC signals are sent to a FADC-based acquisition system and, in parallel, to a trigger system which allows the selection of events with one isolated cluster in the active volume of the chamber (X-rays absorbed in the SPC generate isolated ionization clusters) and permits triggering on the hit multiplicity at various radii inside the SPC active volume and on the hit multiplicity variation as a function of radius (K_S^0, Λ, and $\bar{\Lambda}$ decaying into $\pi^+\pi^-$, $p\pi^-$, and $\bar{p}\pi^+$ inside the SPC are associated with a hit multiplicity increase at larger radii). The possibility of triggering on isolated clusters permits the enrichment, by a factor of about 20, of the set of P-wave annihilation events collected in a given time (the absolute yield of L-line X-rays is $\sim 10\%$ and their detection efficiency is $\sim 50\%$).

a)

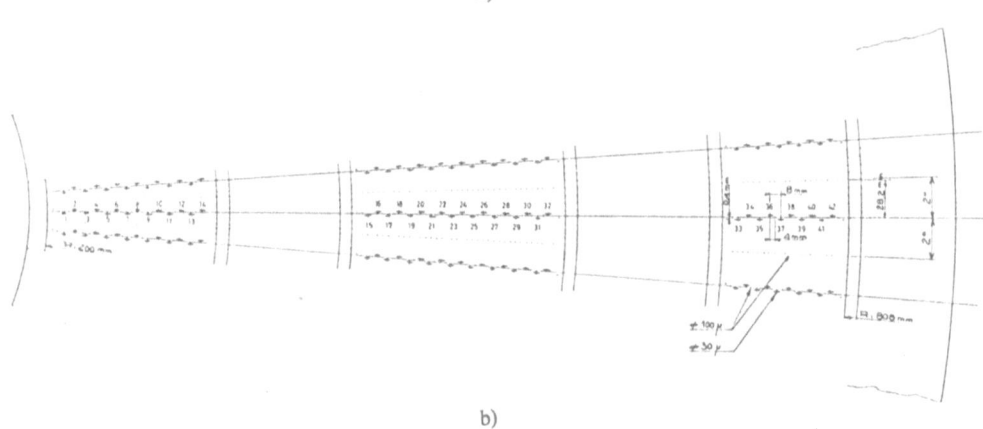

b)

Fig. 14 a) AFS jet chamber being completed and b) structure of one sector (from ref. 6).

3.3.4 The Tracking Chambers

The AFS jet chambers (see Fig. 14) are a well-known and reliable detector[3,6]. They will be equipped with FADC electronics for OBELIX. These chambers ensure a complete visualization, over a large solid angle ($\Omega/4\pi \approx 75\%$), of tracks (42 points in the radial direction, $4°$ azimuthal segmentation, $\sigma_{R\phi} \approx 200\ \mu$m), and a measurement with high resolution of the momentum of prongs ($\sigma \approx 200\ \mu$m, B ≈ 0.5 T, $R_{min} \approx 20$ cm, $R_{max} \approx 80$ cm, $\sigma_p \approx 2\%$ at 1 GeV/c) and of their energy loss ($\sigma_{dE/dx} \approx 11\%$).

3.3.5 The Time-of-Flight System

The time-of-flight system consists of a small barrel of 30 scintillators 1 cm thick at a radius of 18 cm and of a barrel of 90 scintillators, 3 m long and 4 cm thick, at a radius of 134 cm. It permits the direct measurement of the annihilation time for events with prongs in the final state. When antiprotons are stopped in gas the annihilation time jitters with respect to the time of entrance of the antiproton into the target (which is measured by the entrance beam counter) by several nanoseconds, because of the range straggling and the cascade-time. The time-of-flight system permits clean identification of charged kaons with momenta up to 1 GeV/c. It will also permit triggering on events with charged kaons in the momentum window 200–600 MeV/c.

3.3.6 The Gamma Detectors

Gammas are detected by the two TURBO moduli and the four HARGD supermoduli. These devices are gas-sampling calorimeters with large segmentation. The HARGD uses lead converter plates, limited streamer tubes with 1 cm^2 cross-section as active elements with strip and pad

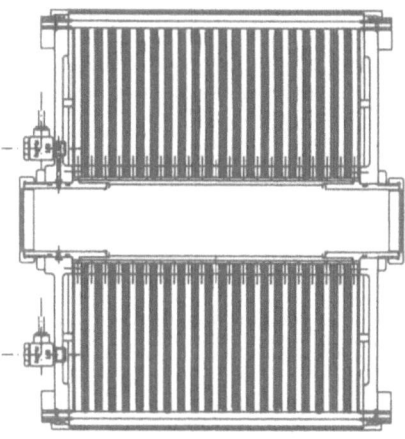

Fig. 15   Side and front cut views of a TURBO high density spiral projection chamber.

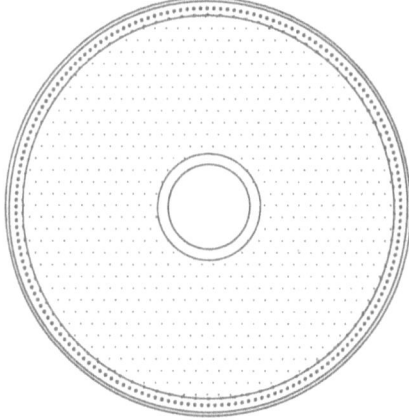

read-out. The TURBOs are SPCs instrumented with a radiator shaped somewhat like a turbine, which segments in the z direction the active volume of the SPC (see Fig. 15). The prongs produced by gammas converting in the disks of the radiator are tracked in the gas volumes between the radiator disks. These gas volumes are identified by charge division.

The BITURBO and HARGD permit the visualization of showers in three dimensions, and the measurement of γ energies with a resolution of 18% $E^{-1/2}$ for gammas whose shower is fully contained by the calorimeters. Their main advantage is in the reconstruction of a conversion point with high angular resolution, and therefore they permit a γ to be distinguished from a π^0 up to high energies ($\sigma_x \approx \sigma_y \approx 3$ mm in HARGD with a distance from the vertex larger than 1500 mm; $\sigma_\phi \approx 1.5°$, $\sigma_\theta \approx 10^{-3}$ rad for TURBO).

3.3.7 Trigger System

The system is designed to produce triggers at various levels by using signals from individual detector components and in combination.

The beam scintillators and the tof and TOF barrels provide fast strobe signals within a few tens of nanoseconds.

The prong multiplicity, at the radius of the sense wires of the SPC, and the overall prong and gamma multiplicity, counted by the pad towers in the HARGD, are available within about 400 ns for fast trigger decisions.

The primary ionization electrons take less than 0.5 μs in the AFS jet chambers, and less than 3 μs in the SPC and in the TURBOs, to drift to the sense wires. Spiral projection chamber signals at different time windows (corresponding to different radii in the SPC), and AFS signals at different radii (corresponding to different layers of sense wires), will be used to trigger on K^0, Λ, and $\bar{\Lambda}$ decaying into the SPC or AFS active volumes, and to trigger on gammas converting in the tof material and producing an e^+e^- pair entering the AFS.

The time-of-flight system will permit a fast selection of $K^\pm$ candidates. Subsequent use of SPC hits at minimum drift-time, tof, AFS, and TOF hits to identify track roads in a second-level trigger is foreseen in order to reduce the contamination in the first-level $K^\pm$ trigger produced by slow pions.

The SPC trigger electronics provides the X-ray flag within a delay of about 100 ns after the maximum drift-time in the SPC.

To give some orders of magnitude, the X-ray trigger has a rejection of about 20, the $K^\pm$ trigger has a rejection of between 20 and 100, and the multiplicity trigger rejection depends on the channel chosen and increases from 10 to 10^3 or more for events with increasing multiplicity in the final state.

3.3.8 Data-Acquisition System

The data-acquisition system collects the event information from all the detector components (and from the trigger system as a check) and writes onto tape the events selected by the trigger supervisor. Event records are rather long and complex because of the richness of information provided by the various detector components. The event length is proportional to the number of prongs and gammas of each annihilation. A typical event length is 10 Kbytes, and it will be possible to write from 20 to 100 events on tape per second. The beam intensity will be adjusted depending on the physical channels under investigation and the trigger rejection. In all cases it will be necessary to collect events with a minimum-bias trigger every fixed number n (e.g. n = 8) of events collected with a more selective trigger, in order to verify the stability of the apparatus and to measure the trigger biases.

Fig. 16 Views of the preparation in the LEAR experimental hall for OBELIX and of the installation of the OAFM magnet (Nov. 1986–Jan. 1987).

3.3.9 Off-Line Software

Data analysis requires a major software effort for the simulation of the detector and the event reconstruction.

Current activities include the simulation with GEANT III of the overall detector and of individual components, adaptation of the existing AFS reconstruction programs, and work on reconstruction and pattern recognition for the SPC.

For the development of the software of the SPC, real data from the ASTERIX SPC/XDC are used instead of simulated data. Two spin-off products of software work for the OBELIX SPC are the following[48]: i) the overall momentum resolution for individual prongs in the ASTERIX experiment has been improved by a factor of about two by introducing the SPC hits in the ASTERIX tracking; ii) the efficiency of the ASTERIX pattern-recognition programs for six prong events has been improved by about one order of magnitude by a pattern-recognition program which identifies tracks directly in the SPC.

3.4 Installation of the Experiment

The experiment is in the course of preparation and installation has started. Figure 16 shows a series of images taken during the preparation of the floor and the installation of components of the OAFM magnet in the LEAR South Hall. Figures 17 and 18 show views of the OAFM and of

Fig. 17 OAFM magnet in the LEAR hall and Frascati team in charge of its installation (Dec. 1987).

Fig. 18 OAFM magnet and the OBELIX experimental area seen from the beam side.

the experimental area at the end of 1987. Running in of the experiment with an $\bar{p}$ beam is foreseen at the end of 1988, and OBELIX could be ready to make calibrations of all the detector components in operation at the end of 1989.

4. EXPERIMENTAL PROGRAMME

As was mentioned in the Introduction, the OBELIX programme is much more extensive than the work on meson spectroscopy via $\bar{p}p$ and $\bar{p}d$ annihilations at rest, and it includes $\bar{p}p$ and $\bar{n}p$ formation experiments near the N$\bar{\text{N}}$ mass threshold and studies of $\bar{p}$ and $\bar{n}$ annihilations on nuclei. Here, we only consider production experiments and mention formation measurements. Proton–antiproton annihilation at rest offer an extremely convenient tool to: a) debug the detector, the trigger, and the data-acquisition system, b) calibrate the charged-particle detection via the reaction $\bar{p}p \rightarrow \pi^+\pi^-$ and $\bar{p}p \rightarrow K^+K^-$ and the gamma detectors via the reactions $\bar{p}p \rightarrow \pi^+\pi^-\pi^0$ and $\bar{p}p \rightarrow \pi^+\pi^-\pi^+\pi^-\pi^0$, c) exercise and test analysis software, and d) measure trigger biases. A good understanding of the experimental apparatus as well as a check of the achievement of the nominal detection performances is necessary before starting production runs.

After good understanding of the detector, meson-spectroscopy work could start with the priority being put on $\bar{p}p$ and $\bar{p}d$ at rest with a 100 MeV/c $\bar{p}$ beam. First data can be collected simply by triggering on an $\bar{p}$ entering the target. Successively, more and more selective triggers will be introduced to pick up interesting final states with small combinatorial background. Each final state trigger in p$\bar{p}$ at rest should be accompanied by a trigger running in parallel with the same requirements on the final state and, in addition, the request for an X-ray detected in the central detector, in order to have comparable statistics in these two sets of events with the same final state to be able to perform KGB$_L$. KGB$_l$ will be performed comparing $\bar{p}p$ data and $\bar{p}n$ annihilations in $\bar{p}d$ interactions with a low-momentum recoil proton.

Later on we envisage extending p$\bar{p}$ work in flight with runs at various momenta in the region 100–300 MeV/c and eventually running $\bar{p}p$ and $\bar{p}d$ at 1000 and 1800 MeV/c to extend the mass range for meson spectroscopy in production with the possibility of performing KGB$_l$ and to produce reference data for $\bar{p}A$ annihilations at these momenta.

Once the apparatus is working properly and reliably, the limit on the statistics which can be collected at rest seems to be given more by the event-analysis time than by the data-acquisition time. In fact, one hour of data taking at full speed requires about ten hours processing time at an IBM 3090 computer for DST production.

For $\bar{n}p$ formation measurements we plan to use an intense $\bar{p}$ primary beam with a momentum of about 300 MeV/c. The $\bar{n}$ runs will be our most expensive ones in terms of antiprotons from LEAR.

REFERENCES

1. R. Armenteros et al., OBELIX Collaboration, Proposal CERN/PSCC/86–4 (1986).
2. U. Gastaldi, Survey of the LEAR physics programme in ACOL time, Nucl. Phys. A 478:813 (1988).
3. H. Gordon et al., AFS Collaboration, Nucl. Instrum. Methods 196:303 (1982).
4. U. Gastaldi, Nucl. Instrum. Methods 188:459 (1981).
5. U. Gastaldi, Nucl. Instrum. Methods 157:441 (1978).
6. C. Fabjan et al., Nucl. Instrum. Methods 156:267 (1978);
   O. Botner et al., AFS Collaboration, Nucl. Instrum. Methods 196:315 (1982).
   D. Cockerill et al., Nucl. Instrum. Methods 176:159 (1980).
7. U. Gastaldi, OBELIX Note CERN–OX–01–86 (1986), p. 398;
   M. P. Bussa et al., Nucl. Instrum. Methods A252:321 (1986).
8. U. Gastaldi, 'Proc. 4th European Symposium on Antiproton Interactions, Barr (Strasbourg), 1978, A. Friedman, ed. CNRS, Paris (1979), vol. 2, p. 607.
9. R. Armenteros et al., ASTERIX Collaboration, 'Physics at LEAR with Low-Energy Cooled Antiprotons', Proc. 2nd LEAR Workshop, Erice, 1982, U. Gastaldi and R. Klapisch, eds. Plenum, New York (1984), p. 109.
10. U. Gastaldi, OBELIX Note CERN–OX–11–84 (1984).
11. U. Gastaldi, 'Atomic Physics 9', R. S. van Dick and E. N. Fortson, eds., World Scientific, Singapore (1984), p. 118.
12. R. Armenteros et al., OBELIX Study Group, 'Physics with Antiprotons at LEAR in the ACOL Era', Proc. 3rd LEAR Workshop, Tignes, 1985, U. Gastaldi, R. Klapisch, J. M. Richard and J. Tran Thanh Van, eds., Editions Frontières, Gif-sur-Yvette (1985), p. 369.
13. U. Gastaldi, OBELIX Note CERN–OX–01–86 (1986), p. 15.
14. U. Gastaldi, Lecture Notes in Physics 273:503 (1987).
15. U. Gastaldi, 'Fundamental Symmetries', P. Bloch, P. Pavlopoulos and R. Klapisch, eds., Plenum, New York (1987), p. 307.
16. S. Ahmad et al., ASTERIX Collaboration, 'Fundamental Interactions in Low Energy Systems', P. Dalpiaz, G. Fiorentini and G. Torelli, eds., Plenum, New York (1985), p. 279.
17. S. Ahmad et al., ASTERIX Collaboration, Phys. Lett. B157:33 (1985).

18. S. Ahmad et al., ASTERIX Collaboration, Measurements of the strong interaction shift and broadening of the ground state of the $p\bar{p}$ atom, CERN–EP/88–05 (1988), to be published in Phys. Lett. B.

19. M. Doser et al., ASTERIX Collaboration, CERN/EP 88–42, to be published in Nuclear Physics A.

20. G. Piragino, 'Hadronic Physics at Intermediate Energy', Folgaria Winter School, 1986, T. Bressani and R.A. Ricci, eds., North Holland, Amsterdam (1986), p. 293.

21. C. Guaraldo, $\bar{N}$-nucleus interactions at LEAR, to be published in Proc. 4th LEAR Workshop, Villars, Switzerland, 1987.

22. T. Bressani et al., Main performances of a low momentum tagged antineutron beam, to be published in same Proceedings as Ref. 21.

23. R. Armenteros et al., 'Proc. Int. Conf. on Elementary Particles, Sienna, 1963, G. Bernardini and G.P. Puppi, eds., SIF, Bologna (1963), vol. 1, p. 287.

24. P. Baillon et al., Nuovo Cimento 50A:393 (1967).

25. For recent reviews and lectures on exotic and light mesons see, for example:
   T. Barnes, in same Proceedings as Ref. 9, p. 201;
   T. Barnes, in these Proceedings;
   M. Chanowitz, 'Proc. Summer Inst. on Particle Physics', Stanford, 1981, A. Mosher, ed., SLAC, Stanford (1982), p. 41.
   M. Chanowitz, 'Proc. 2nd Int. Conf. on Hadron Spectroscopy', KEK, Tsukuba, Japan 1987, Y. Oyanagi, K. Takamatsu and T. Tsuru, eds., KEK Report 87-7 (1987);
   F. Close, in these Proceedings;
   N. Isgur, in these Proceedings;
   S. Narison, in same Proceedings as Ref. 12, p. 297;
   for an experimental review, see F. Couchot, in these Proceedings.

26. P. Dalpiaz, in these Proceedings and references therein.

27. M. Macri, in same Proceedings as Ref. 12, p. 423 and references therein.

28. T. B. Day, G. A. Snow and J. Sucher, Phys. Rev. 118:864 (1960).

29. M. Izycki et al., Z. Phys. A297:1 (1980).

30. J. R. Lindemuth et al., Phys. Rev. C30:1740 (1984).

31. E. Auld et al., 'Proc. 5th European Symposium on Nucleon–Antinucleon Interactions', Bressanone, 1980, M. Cresti, ed., CLEUP, Padua (1980), p. 437.

32. E. Auld et al., Phys. Lett. B77:454 (1978).

33. T. P. Gorringe et al., Phys. Lett. B162:71 (1985).

34. C. A. Baker et al., in same Proceedings as Ref. 15, p. 301.

35. R. Bacher et al., in same Proceedings as Ref. 15, p. 115.

36. For a review see:
   R. Armenteros and B. French, '$N\bar{N}$ Interactions in High-Energy Physics', E.H.S. Burhop, ed., Academic Press, New York (1969), vol. 4, p. 237.

37. S. Ahmad et al., ASTERIX Collaboration, in same Proceedings as Ref. 12, p. 353.

38. S. Ahmad et al., ASTERIX Collaboration, to be published in same Proceedings as Ref. 21.

39. U. Gastaldi et al., The central detector of the ASTERIX experiment at LEAR, to be submitted to Nucl. Instrum. Methods.

40. J. Badier et al., Phys. Lett. B86:98 (1979).

41. G. Eigen, in these Proceedings.

42. F. Couchot, in these Proceedings.

43. H. Koch, in same Proceedings as Ref. 12, p. 389.

44. E. Aker et al., Crystal Barrel Collaboration, Proposal CERN/PSCC/85–56 (1985).

45. For reviews, see the papers of A. Penzo and of W. Haeberli, to be published in same Proceedings as Ref. 21.

46. L. Dick et al., 'High-Energy Physics with Polarized Beams and Polarized Targets', C. Joseph and J. Suffer eds., Birkhäuser Verlag, Basel–Boston–Stuttgart (1981) and references therein.

47. U. Gastaldi and D. Möhl, in same Proceedings as Ref. 9, p. 649.

48. A. Coc and R. Landua, OBELIX Note CERN–OX/87–02 (1987);
   R. Landua, in ASTERIX News No. 8 (1987);
   A. Coc, OBELIX Note, in preparation.

SPECTROSCOPY AT LEAR WITH AN INTERNAL GAS JET TARGET

PS202 COLLABORATION

K. Kirsebom

Ist. Naz. di Fisica Nucleare (INFN)
via Dodecaneso 33
I-16146 Genova

ABSTRACT

At LEAR Experiment PS202 will study exotic states such as glueballs
(gg) and hybrids (gq$\bar{q}$) in $\bar{p}$p annihilation. This will be done using an
internal H$_2$ gas jet target. The set-up features high luminosity,
10^{31} cm^{-2} s^{-1}, a compact and large acceptance detector and a fast and
flexible trigger. Furthermore we will have precise tracking (silicon
strip detectors and straw chambers), full π/K/p identification (correlat-
ing data from the tracker and the RICH identifier) and a powerful electromagnetic calorimeter. This set-up, having excellent signal detection
efficiency and background rejection efficiency, will allow us to study a
variety of rare processes. Our sensitivity will be 480 nb^{-1}/day at
L = 10^{31} cm^{-2} s^{-1}. The first processes to be studied are p$\bar{p}$ → $\phi\phi$ →
K$^+$K$^-$K$^+$K$^-$ and $\bar{p}$p → $\bar{K}_s^0$K$_s^0$ → $\pi^+\pi^-\pi^+\pi^-$. Resonances are detected by monitoring the reaction rates as a function of the LEAR beam momentum in 0.6 <
p$_{\bar{p}}$ < 2.0 GeV/c. Later, reactions such as $\bar{p}$p → $\phi\omega$,$\phi\eta$,$\eta\eta$,$\phi\phi\pi^o$,4K (close
to threshold) and $\phi\phi$ (tagging on ϕ's) will be explored. Also the
future programme will include studies of $\bar{p}$p collisions with a polarized jet target.

INTRODUCTION

In this contribution we shall give a description of the PS202 experiment at LEAR. This experiment was approved on the 11 February 1987[1] and
is planned to start taking data in the spring on 1990. Essentially
JETSET is a large acceptance powerful detector surrounding an internal
gas jet target at the LEAR ring. Topics to be covered are:

1) The experimental technique/jet target.

2) The physics programme.

THE EXPERIMENTAL TECHNIQUE, THE JET TARGET

A supersonic H_2 jet is passed across the LEAR vacuum pipe. By optimizing a series of parameters (operating temperature and pressure, shape of
nozzle and skimmers) we will obtain a target density $\rho = 8 \cdot 10^{13}$ atoms/cm.
With a $\bar{p}$ revolution frequency of $f = 3.2 \cdot 10^6 \text{ s}^{-1}$ and $4 \cdot 10^{10} \bar{p}$ $(= N_{\bar{p}})$
injected into LEAR, the initial luminosity is $L = \rho N_{\bar{p}} f = 10^{31} \text{ cm}^{-2} \text{ s}^{-1}$.
The beam lifetime, given by $\tau = (\sigma_{tot} \rho f)^{-1}$, equals 11 hours ($\sigma_{tot} =$
100mb). So operating with two fills per day, each of $4 \cdot 10^{10} \bar{p}$ (about 6% of
the AAC capacity), the integrated luminosity is 480 nb^{-1}, assuming 100%
detection efficiency.

The experimental technique is illustrated by Fig. 1, and detailed in
Ref. 2. Consider the production of a resonance in the formation channel.
A suitable final state through which the resonance may be detected is
chosen and its rate (correctly normalized) is recorded as the $\bar{p}$ beam
momentum is varied.

Automatic ramping of the beam momentum
will soon become available, allowing us to
perform continuous and efficient scans in
the centre-of-mass range $1.9 \leq \sqrt{s} \leq 2.43$ GeV
(Ref. 3). This technique is seen to com-
bine several nice features:

• The accuracies by which the masses and
widths of resonances may be determined
essentially only depend on the beam para-
meters. For example, $\Delta M \leq 1 \text{ MeV/c}^2$ with
a momentum resolution $\Delta p/p = 1 \cdot 10^{-3}$ of the
coasting $\bar{p}$ beam.

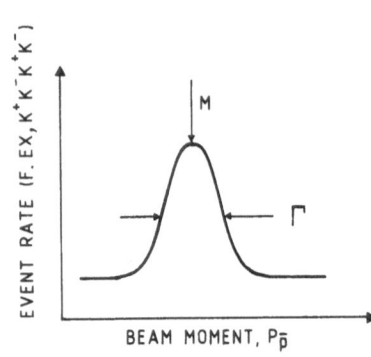

Fig. 1 Production of a
resonance in the formation
channel varying $p_{\bar{p}}$.

• High luminosity, corresponding to 480
nb^{-1}/day and efficient use of $\bar{p}$'s (60%

with two injections per day). Notice that the full luminosity may be
spent in a very narrow √s bin (Δ√s ~ 1 MeV/c²). This would not be
possible without the continuous cooling of the p̄ beam. The technique
therefore is well suited for establishing both narrow and wide resonances.

*   Considering the high luminosity the interaction volume is small,
typically $\sigma_x, \sigma_y, \sigma_z$ ~ 2, 1 and 4 mm. Also we have a 100% pure proton
target.

THE PHYSICS PROGRAMME

Initial physics

The first two processes to be measured are $\bar{p}p \to \phi\phi \to K^+K^-K^+K^-$ and $\bar{p}p$
$\to \bar{K}^0_s K^0_s \to \pi^+\pi^-\pi^+\pi^-$, see Figs. 2a-b. As the ϕ is pure $\bar{s}s$, the $\bar{p}p \to \phi\phi$
process is OZI suppressed and is mediated via a pure gluon intermediate
state. This means that the non-resonant background cross-sections are
low. Like other experimenters[4-8] we believe that the $\phi\phi$ channel is
the most promising in the search for exotic states (glueballs, hybrids,
etc.). All the cited experiments show data which indicate structures in
the mass range 2.0 ≤ M(φφ) ≤ 2.4 GeV/c². However, these data are inconclusive due to low statistics. In JETSET the picture will be much
cleaner, for several reasons:

*   Higher statistics. If we assume a detected cross section $\sigma(\bar{p}p \to \phi\phi \to$
    $4K^{\pm}) = 0.1$ μb, then we will record $5 \cdot 10^4$ φφ events per day. This is to
    be compared, for example, with the total statistics of $1.3 \cdot 10^4$ πBe
    → φφX events gathered by the WA67 Collaboration[5] .

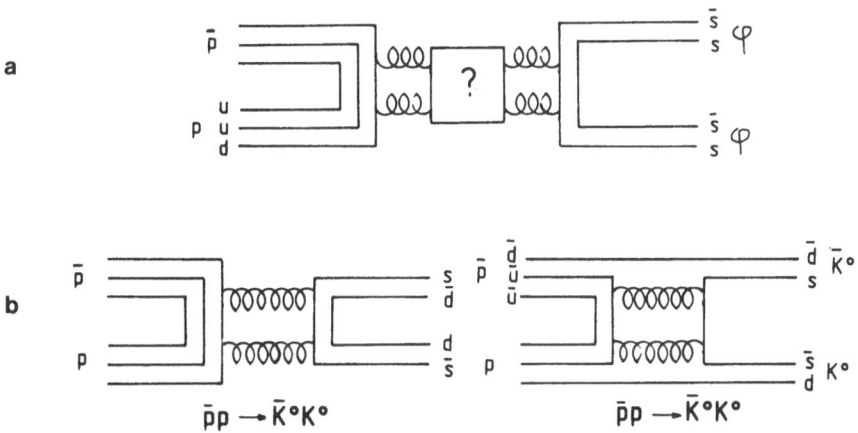

a

b

$\bar{p}p \to \bar{K}^0 K^0$ $\bar{p}p \to \bar{K}^0 K^0$

Figs. 2a-b Production mechanisms for φφ and $\bar{K}^0_s K^0_s$ final states.

343

- In addition, the experimental width of resonances will <u>not</u> be smeared
  by the mass resolution of the detector, $\sigma(M_{\phi\phi})$. In cases of limited
  statistics and a resonance width which is well below the experimental
  mass resolution this is particularly important. In JETSET $\sigma(M_{\phi\phi})$ =
  $\sigma(\sqrt{s})$ < 1 MeV), i.e. the mass resolution is determined only by the
  beam parameters and is far better than detector determined mass resolutions. Typically $\sigma(M_{\phi\phi}) \simeq \pm 25$ MeV/c^2, see ref. 5.

- The future use of a polarized target, which can clarify the partial
  wave analysis.

The annihilation of $\bar{p}p$ into $\bar{K}^0_s K^0_s$ is shown in Fig. 2b. We will investigate this process and measure the relative importance of the annihilation diagrams indicated.

Future possibilities

In the future our programme will be expanded to investigating other
channels such as $\bar{p}p \rightarrow \phi\omega$, $\phi\eta$, $\eta\eta$, $\phi\phi\pi^0$, $\pi^0\pi^0$ and 4K near threshold. This
will allow us to verify and extend the initial physics results. We will
also do inclusive physics. With a capability of tagging on ϕ's, we
can, e.g. measure $R \equiv BR[\phi \rightarrow \gamma\delta(980)]/BR[\phi \rightarrow \gamma S^*(975)]$, a ratio which is
very sensitive to the nature of the δ and S^* states[9]. Finally a wide
physics programme will open up when a polarized gas jet target becomes
available in 1991.

THE DETECTOR

The inner tracker is composed of three types of detectors to be
specified below. Further out one finds fast RICH counters for particle
identification, scintillation trigger counters and the electromagnetic
calorimeter (EC). Notice that the sequences of detectors are the same
in the barrel and forward regions. The geometric acceptances in polar
and azimuthal angles are: $13° < \theta < 130°$ and $0 < \phi < 2\pi$. The acceptance
loss due to the jet is very small; diameters of injection and recupera-
tion pipes being 16 and 32 mm, respectively.

Inner tracker

The three components of the inner tracker are: the OVAL chamber being
an instrumented vacuum tube, the silicon microstrip detectors and the
straw chambers. The role of the inner tracker is important to provide
multiplicity and tracking data in the second and third level triggers,
to be detailed in another section.

344

The OVAL chamber is a novelty which in a single structure (5.5 mm thick and 41 mm long) combines a mechanically strong, low mass vacuum pipe (2 × 0.3 mm of Al ≃ 0.007 X_o) with a position sensitive detector as close as possible to the vertex. There will be 64 cells, each measuring both ϕ and z (through charge division).

The silicon microstrip detector. There will be three layers in the barrel and forward directions. The silicon strips will measure the ϕ angle which together with r (distance to the beam axis) are the only relevant coordinates for the momentum determination. Silicon strips will be daisy-chained both in the barrel and on the forward planes in such a way that the readout electronics need only be mounted on the external rims of the active silicon surfaces. A front-end VLSI circuit is being developed. This is an Amplex based design[10] which will have both analog and digital readout. The amount of material is 4 10^{-3} X_o per Si layer (30% for support structures) and the expected r.m.s. position resolution is better than 20 μm. Main parameters when fixing this design figure are: the magnetic field strength, the lever arm available and the amount of multiple Coulomb scattering.

The straw chambers. These are individual cylindrical drift tubes. They are made from aluminized mylar (wall thickness = 75 μm) with a central tungsten anode wire. In the barrel region eight different tube diameters (6-8 mm) are needed to configure with maximum coverage two times four straw layers, in total 960 straws. Likewise in the forward region, there are two planes, each being made up of four layers of x- and y-configured straws, in total 320 straws. The longitudinal straw coordinate is measured through resistive charge division, a resolution of σ_z/z = 1-2% or for z = 30 cm, 3-6 mm is expected. The transversal coordinate measurement is based on the electron drift time, here the error will be ≃ 100 μm. Straw chambers have several nice features: they introduce very little material, one straw = 0.8 10^{-3} X_o, may be built to form a self supporting structure (thus reducing the amount of end-cap material needed) and, compared to conventional drift chambers, are fast allowing for their use in the second level trigger.

RICH identifier

The principles of the fast RICH are shown in Fig. 3. Both in the barrel and forward regions this type of detector will measure the speed of the particles, β. Two possible radiators are under investigation: a solid radiator[11], NaF with a refractive index n = 1.37 (E_γ = 6eV), and

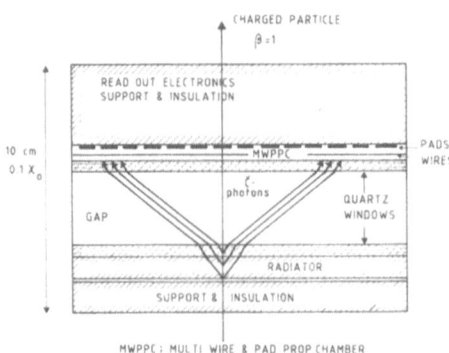

CHARGED PARTICLE
β≃1

READ OUT ELECTRONICS
SUPPORT & INSULATION

10 cm
0.1 X₀

MWPPC
PADS
WIRES

photons

GAP

QUARTZ
WINDOWS

RADIATOR

SUPPORT & INSULATION

MWPPC: MULTI WIRE & PAD PROP CHAMBER

Fig. 3 Principles of the fast
RICH identifier.

a liquid one, C_6F_{14} (liquid freon) with n = 1.28 (E_γ = 6eV). The advantage of the former is that of simpler mechanics while the latter has better particle identification properties. With 1 cm of freon we expect 21 photoelectrons in the MWPC for a β = 1 and normal incidence particle. The momentum thresholds for freon, respectively, are 180, 620 and 1180 MeV/c for π, K and protons. After the opening up of the Cherenkov cone in the "gap", the photons are converted with ~ 50% efficiency in the photosensitive gas (TMAE) of the MWPC.

Operated at 70°C the photon mean free path is 1.0 mm. The use of the RICH is threefold:

- By means of the anode wire information to perform fast counting of the number of Cherenkov photons. This allows us to reject in the second level trigger background processes such as, the $\bar{p}p \rightarrow 4\pi$, $2\pi 2K$ while retaining $\bar{p}p \rightarrow 4K$.

- With the pad data (induced signals coming from 5 by 5 mm pixels) we will measure β. Monte Carlo results predict a β resolution of 1%, $\Delta\beta/\beta$ = 0.01. We therefore expect to have good π/K separation in the range 0.17 → 2.0 GeV/c and K/p separation in 0.6 → 3.0 GeV/c.

- Charged particles (above and below the Cherenkov threshold) will produce "track" signals in a RICH. Therefore the RICH will provide a precise three dimensional point for each track, which will aid in the tracking.

Electromagnetic calorimeter (EC)

The ECs are built from 504 "supertowers", 324 in the barrel and 180 in the forward region. Each supertower (physical unit) consists of 4 towers, which are read out separately. The supertowers are by volume, made from 50% scintillating fibre and 50% lead. They are 16 X_o (17.6 cm) deep ensuring 98% containment for a 1 GeV γ shower and have a entrance

face of 3 x 3 cm². The towers point towards the vertex and the fibres
will be read out on the back side. Multi-channel PMs are used for this
purpose. These are low noise and fast (20 ns integration time) and contain 16 channels per PM. Monte Carlo studies show that the transverse
spatial resolution for a 1 GeV γ will be $\sigma_x = \sigma_y \simeq 1.5$ mm. First test
results of such a EC tower module shows excellent sensitivity ($\sim 100\%$
efficiency at $E_\gamma = 16.5$ MeV) and energy resolution ($\sigma/\sqrt{E} \simeq 10\%$). The
ECs will be used in a veto mode for the initial programme ($\phi\phi, \overline{K}^0_s K^0_s$) and
allow us to investigate channels containing neutrals in the future.

Trigger

Table 1 bears out the basic figures of our three level triggering
scheme. The trigger should take down the 1 MHz input rate (correspond-
ing to $L = 10^{31}$ cm^{-2} s^{-1} and $\sigma_{tot} = 100$ mb) to a reasonable rate without
discarding good events nor imposing a prohibitive deadtime on the data
acquisition system. The main features of the three triggering levels are
the following.

Table 1

Rates and deadtimes in the JETSET triggering scheme

	Input rate	Output rate	Deadtime per trigger	Relative deadtime
Level 1	1 MHz	30 kHz	200 ns	20%
Level 2	30 kHz	500 Hz	4 µs	12%
Level 3	500 Hz	50 Hz	100 µs	5%

1st level trigger: This is produced within 200 ns by the scintillation
counters (Fig. 4), the ECs and dedicated veto counters. Essentially at
this level triggers with wrong charged and/or neutral multiplicity are
discarded. The importance of the ECs in the 1st level trigger is evident;
for example, $\sigma(\overline{p}p \to 4\pi^\pm + n\pi^0) > 20$ mb.

2nd level trigger: Here we retain: i) triggers with many K's ($\gtrsim 3$)
and ii) those with a "high" content of K^0_s's. The former is done with
the fast RICH. The trigger counters and the RICH themselves indicate the
position of the track impact. Dedicated processors will then perform a

347

wire count (= Cherenkov photon count) in the vicinity of the impact. Only one out of four impact points will be allowed to have a Cherenkov count compatible with a $\beta > 0.8$ particle when we are triggering on $4K^{\pm}$. For the K^o_s trigger the selection is based on multiplicity counting in the OVAL chamber and all the straw layers. A $0\rightarrow4$ or $2\rightarrow4$ change in multiplicity would indicate a secondary vertex. Also fast processors will measure the impact parameters for vertices inside the vacuum pipe. The time needed for a 2nd level trigger decision is 4 μs.

3rd level trigger: This refines level 1 and 2 trigger decisions. Silicon strip and RICH pad information now enter in the trigger defini- tion. Data from the different detectors are cross-correlated and the kinematics of the events are checked. Processing will be in parallel (5 ms/event) and little additional deadtime will be introduced.

CONCLUSIONS

Starting in 1990 Experiment PS202 will study gluonic states produced in $\bar{p}p$ annihilation at LEAR. This is to be done with an internal gas jet target. First processes to be studied are $\bar{p}p \rightarrow \phi\phi \rightarrow 4K^{\pm}$ and $\bar{p}p \rightarrow K^o_s K^o_s \rightarrow 4\pi^{\pm}$ in the mass range $1.96 < \sqrt{s} < 2.43$ GeV. Later a rich physics programme is foreseen: triggering on final states including neu- trals (for example $\bar{p}p \rightarrow \phi\omega \rightarrow 2K^{\pm}2\pi^{\pm}\pi^o$) and employing a polarized jet target.

Acknowledgements

We would like to express our gratitude for the substantial contribu- tion of J. Seguinot and T. Ypsilantis in the design of the fast RICH.

REFERENCES

1. G. Bassompierre et al., JETSET: Physics at LEAR with an Internal Gas Jet Target and an Advanced General Purpose Detector, CERN/PSCC 86-23, 25 March 1986.

2. C. Baglin et al., CERN-EP/85-01 (1985).

3. See Contribution of P. Lefèvre to this Workshop.

4. A. Etkin et al., Phys. Rev. Lett. 49 (1982) 1620.

5. P.S.L. Booth et al., CERN-EP/85-138.

6. D.R. Green et al., Fermilab-Pub 85/120-E.

7. T.A. Armstrong et al., Phys. Lett 166B (1986) 245.

8. J.E. Augustin et al., LAL/85-27.

9. F. Close, private communication.

10. The UA2 Collaboration. Proposal for the installation of a Second Silicon Array in the UA2 Detector, CERN/SPSC 87-14.

11. R. Arnold et al., Proceedings of the London Conference on Position Sensitive Detectors (1987), CERN-EP/87-186.

12. D.W. Hertzog et al., Construction of a Generic Pb/SCIFI Calorimeter Segment, JETSET/87-23, 7 July 1987.

ANTIPROTON-PROTON FORMATION OF CHARMONIUM STATES

Gerald A. Smith*

Laboratory for Elementary Particle Science
Department of Physics
The Pennsylvania State University
University Park, PA 16802 USA

ABSTRACT

An approved experiment (E-760) to study charmonium states using a gas
jet in the Fermilab Antiproton Accumulator Ring is reviewed. Physics
goals and techniques are discussed and compared with ISR R-704, an
experiment which pioneered this method at CERN.

INTRODUCTION

In December, 1985, Fermilab approved E-760, "A Proposal to Investigate
the Formation of Charmonium States Using the Antiproton Accumulator Ring,"
a Fermilab-Ferrara-Genoa-Irvine-Northwestern-Penn State-Torino collaboration. We are presently engaged in construction of the detector, and
expect to start taking data in 1987 or 1988. This paper reviews the
physics goals of E-760, as well as the design and present status of the
construction of the detector. Since this is a second generation experiment using the $\bar{p}p$ formation technique, some discussion will be given to
comparisons with earlier results from CERN experiment R-704. The reader

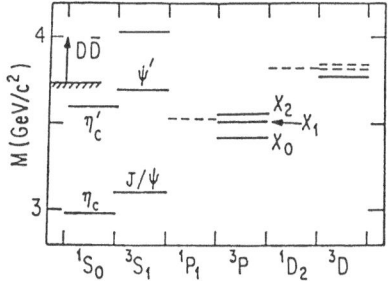

Fig. 1 - Low-lying charmonium states.

* Work supported in part by the U.S. National Science Foundation.

is referred to an earlier review [1] for additional discussion of R-704 and E-760.

PHYSICS GOALS

A great deal has been written about charmonium since its discovery in 1974, so it is reasonable to start this discussion by quoting from some theoretical experts on what else is to be learned from further studies of charmonium, and the prospects for a program of p$\bar{\text{p}}$ studies. For reference, I show in Figure 1 the spectrum of low-lying states [2]. Solid lines denote observed, and dashed lines predicted, levels. D$\bar{\text{D}}$ threshold is indicated by the shaded band.

1. Why p̄p?

(E.L. Berger, P.H. Damgaard and K. Tosokos, 1984 Durham Conference [3])

"As is well known, a very detailed and impressive amount of information about the different charmonium states has already been obtained by experimental studies in e$^+$e$^-$ colliders. Perhaps, then, the first question we should address in this talk is: why should we begin to search through the different charmonium channels again in p$\bar{\text{p}}$ collisions? There are several answers to this question. First, let us make a some- what simplified (but basically correct) order-of- magnitude argument. The most obvious charmonium reson- ance is the ψ, which has quantum numbers $J^{PC}=1^{--}$. This state is, of course, readily produced in e$^+$e$^-$ colli- sions. In terms of Feynman diagrams the process can proceed through e$^+$e$^-$ $\rightarrow$ γ^* $\rightarrow$ $\bar{\text{c}}$c. Only one intermediate off-shell photon is required. Thus, in the amplitude only one power of α_{EM} appears. In contrast, consider another charmonium state like the χ_2, which has quantum numbers $J^{PC}=2^{++}$. Since it is charge-conjugate even, at least two intermediate photons are required: e$^+$e$^-$ $\rightarrow$ $2\gamma^*$ $\rightarrow$ $\bar{\text{c}}$c. Therefore, even just at the amplitude level the process is (roughly) suppressed by a factor of 1/137. In the cross section this gives a suppression factor of ~ 10^{-4} as compared to the e$^+$e$^-$ $\rightarrow$ ψ cross section! Although this argument is rather sketchy (for one thing, it glosses over the differences between the wave functions of the ψ and the χ_2), the idea is right: there is no hope of producing the χ_2 state directly in e$^+$e$^-$ collisions. (Of course, it can be seen in radiative decays, but that is a different story.) What is the difference if we start invoking the strong interactions? Consider the process q$\bar{\text{q}}$ $\rightarrow$ ψ. Just from the J^{PC} quantum numbers alone this process should be able to proceed via one intermediate gluon (just as we in the QED case had one intermediate photon), but that Feynman diagram obviously has a van- ishing color factor. We are therefore forced to bring two extra off-shell gluons into the game, i.e. q$\bar{\text{q}}$ $\rightarrow$ $3g^*$ $\rightarrow$ $\bar{\text{c}}$c. This looks bad from the point of view of counting powers of the coupling constant: roughly speaking, a factor of α_s is involved in this amplitude. However, in this case it turns out to be rather conven- ient that the strong interactions indeed are quite strong; at the charmonium resonances we can expect to find a QCD running coupling constant of about $\alpha_s \approx 0.2$. Several powers of α_s are therefore quite harmless as far as cross section suppression is concerned. Moreover, in this case the production of the χ_2 resonance proceeds via two gluons: $q\bar{q} \to 2g^* \to c\bar{c}$. As compared to ψ production we would expect roughly the same order of magnitude for the cross section. Hence, if the ψ can be produced in $p\bar{p}$ collisions, then so should all other charmonium states which are allowed by the quantum numbers. This leads to the exciting experimental possibility that the whole of charmonium spectroscopy can be finished up in studies of $p\bar{p}$ collisions."

2. What New Physics Can We Expect From E-760?

(Highlights of comments made by R.L. Jaffee, opening speaker at the 1986 Fermilab Low Energy Antiproton Facility Workshop, in support of $\bar{p}p$ formation studies of charmonium [4])

. A New Era in Charmonium Spectroscopy

. Discover 3 previously unknown narrow states:
1) $^1P_1$ or 1^{++} (charmonium analog of B-meson): Search for

$$P \to J/\psi + \gamma \text{ (forbidden by C)}$$

$$P \to J/\psi + 2\pi \text{ (suppressed by P-wave phase space)}$$

$$P \to J/\psi + \pi^0 \text{ (isospin violating (S-wave)): (remember that}$$

$$\omega \to \pi^0\eta \text{ @ 8.7\%)}$$

2) $^1D_2$ or 2^{-+} (predicted by models): Note that $2M_D < M(^1D_2) < M_D + M_{D*}$ and decay to $D\bar{D}$ is forbidden by parity, so it's likely to be narrow, i.e. $D \to J/\psi + \gamma$ ($\Delta L = 2$ M1 transition)

$$\to J/\psi + \rho^0 \text{ (isospin violating (S-wave))}$$

3) $^3D_2$ or 2^{-+} ($c\bar{c}$ state mass close to $^1D_2$): Cascade decays to J/ψ will likely be strong, i.e.

$$^3D_2 \to \chi + \gamma \to J/\psi + \gamma\gamma \text{ (both allowed E1 transitions)}$$

$$^3D_2 \to J/\psi + \pi^0 \text{ (isospin violating (P-wave))}$$

. Confirm Weak η'_c

. Accurately Measure Masses of $^1P_1$, $^3D_2$, $^1D_2$: These are sensitive to spin, spin-orbit, tensor and relativistic terms in the charmonium potential [5-7].

. Accurately Measure Total Widths for All Narrow $c\bar{c}$ States: Presently total widths are poorly known (except for $^3S_1$). Total widths are predicted in QCD via 2 or 3 gluon annihilation diagrams [8,9] shown in Fig. 2.

. Measure Helicity Amplitudes in $\bar{p}p$ Production: In perturbative QCD, calculations of the diagrams in Fig. 2 coupled to $\bar{p}p$ with massless quarks give a helicity selection rule $\lambda_{c\bar{c}} = \lambda_p - \lambda_{\bar{p}} = \pm 1$ [10],

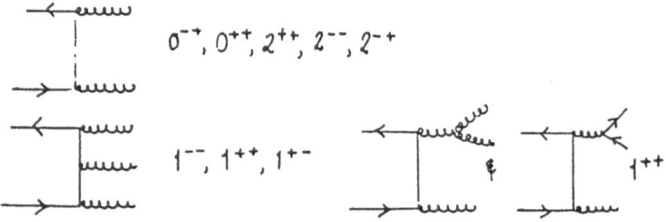

Fig. 2 - Two and three gluon diagrams responsible
for $c\bar{c}$ states.

which forbids states with $0^{-+}(\eta_c)$, $0^{++}(\chi_0)$ and $1^{+-}(^1P_1)$. However, it is known from studies of η_c decay that $BR(J/\psi \to \bar{p}p) \simeq 2\, BR(\eta_c \to \bar{p}p)$, so the rule is broken. Therefore, it is of interest to actually measure the ratios of all helicity amplitudes, 0 and ±1, to further probe spin effects in QCD.

<u>Unravel Multipoles</u>: Radiative decays of $c\bar{c}$ states often allow competing multipoles [11], e.g. $2^{++} \to 1^{--}$ via E1, M2 or E3 transitions, $2^{-+} \to 1^{--}$ via M1, E2 or M3 transitions. Multipoles are sensitive to high momentum components in the wavefunction, $(M_2/E_1) \sim v^2/c^2$, etc., and probe the regime where relativistic corrections are small.

The previous discussion suggests that $\bar{p}p$ studies hold a great deal of promise for discovery of significant new information about the $c\bar{c}$ system and a better understanding of QCD. Furthermore, the experimental technique has been proven practical by R704 at the ISR [12-16].

3. Comparison with R-704

The E-760 plan is to use the same basic technique as R704, namely insertion of a hydrogen gas jet into a stored and cooled $\bar{p}$ beam in the south straight section of the Fermilab $\bar{p}$ accumulator ring. Ultimately a luminosity of $10^{31} cm^{-2} s^{-1}$ is expected, assuming a stored beam of 1.5×10^{11} $\bar{p}$'s and a gas jet of density $1.2 \times 10^{14} cm^{-2}$ and a thickness of 0.9 cm, to be compared with typical luminosities for R704 of $\sim 10^{30}\, cm^{-2} s^{-1}$. In addition, the acceptance of the E-760 detector will be five times that of R704. The expected mass resolution in the $\bar{p}p$ system is 300-500 KeV (FWHM) in the 3-7 GeV/c region based on a momentum resolution of (2×10^{-4}).

In Fig. 3 we show (a) the R704 $\chi_{1,2}$ data [13] compared with (b) data from the Crystal Ball. A comparison of the widths of these states illustrates the power of this technique. With but a handful of events, $30(\chi_1)$ and $50(\chi_2)$, R704 extracted masses and total widths (Table 1) which are comparable to, or somewhat better than, those from the Crystal Ball. They also quote partial widths of $\Gamma(\chi_1 \to \bar{p}p) = 57^{+13}_{-11}$ eV and $\Gamma(\chi_2 \to \bar{p}p) = 233^{+51}_{-48}$ eV. With a factor of ~ 50 higher rate (x10 for luminosity and x5 for acceptance), a run of one week for E-760 will produce $\sim 3,600\ \chi_1$ and χ_2 events each. For reference and purposes of scaling for different states, we show in Table 2 effective cross sections expected in E-760, based on values determined in R704. The effective cross section is defined as the peak cross section including acceptance and resolution smearing. It is evident from Table 2 that even the most difficult states are statistically accessible (e.g. $^1P_1$ yields are ~ 150 events per one week run), provided backgrounds (typically 10^7 times larger than signal) can be handled.

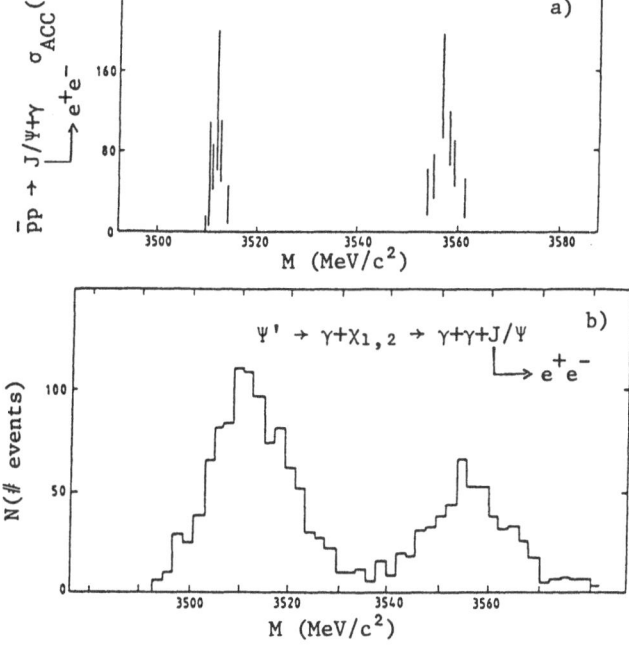

Fig. 3 - Results on $\chi_{1,2}$ states from (a) R704
and (b) Crystal Ball.

Table 1

Parameter	Experiment	χ_1	χ_2
M	R704	3511.3±0.4±0.4	3556.9±0.4±0.5
(MeV/c²)	Crystal Ball	3510.3±0.4±4.	3555.8±0.5±4.
Γ	R704	<1.3 (95% c.l.)	$2.6^{+1.4}_{-1.0}$
(MeV)	Crystal Ball	<3.8 (90% c.l.)	(0.85-4.9) (90% c.l.)

4. The E-760 Detector

As indicated earlier, at this time the detector is presently under construction. We show in Fig. 4 a schematic layout of the detector. A silicon detector array (Northwestern) which monitors luminosity via $\bar{p}p$ diffractive elastic scattering is not shown. Particles produced in the intersection of the $\bar{p}$ beam and the gas jet (Genoa) emerge through an inner tracking chamber system (Ferrara), the purpose of which is to define angles and multiplicities of charged tracks as they emerge into the outer layers of the detector.

The next layer is a gaseous threshold counter (Torino), comprised of two polar sectors, each with eight azimuthal counters. Angles from 70° to 15° are covered by this detector. Mirrors reflect light back onto one of

Table 2

Reaction	P_{lab} MeV/c	R704 σeff (pb)	R704 σeff$_{back}$ (pb)	E-760 σeff (pb)	E-760 σeff$_{back}$ (pb)	Acceptance Ratio
$pp \to \eta_c \to \gamma\gamma$	3698	22-42	9-30	90-170	36-120	4
$pp \to \chi_1 \to \psi + \gamma$ $\to e^+e^- + \gamma$	5553	130	< 5	650	25	5
$pp \to {}^1P_1 \to \psi + \dots$	5600-5615	3-11	< 8	15-55	40	5
$pp \to \chi_2 \to \psi + \gamma$ $\to e^+e^- + \gamma$	5727	150	< 5	750	25	5
$\to \gamma\gamma$		7-17	1-9	35-85	5-45	5
$pp \to \psi \to e^+e^-$	4068	3000	-	6x10	-	5

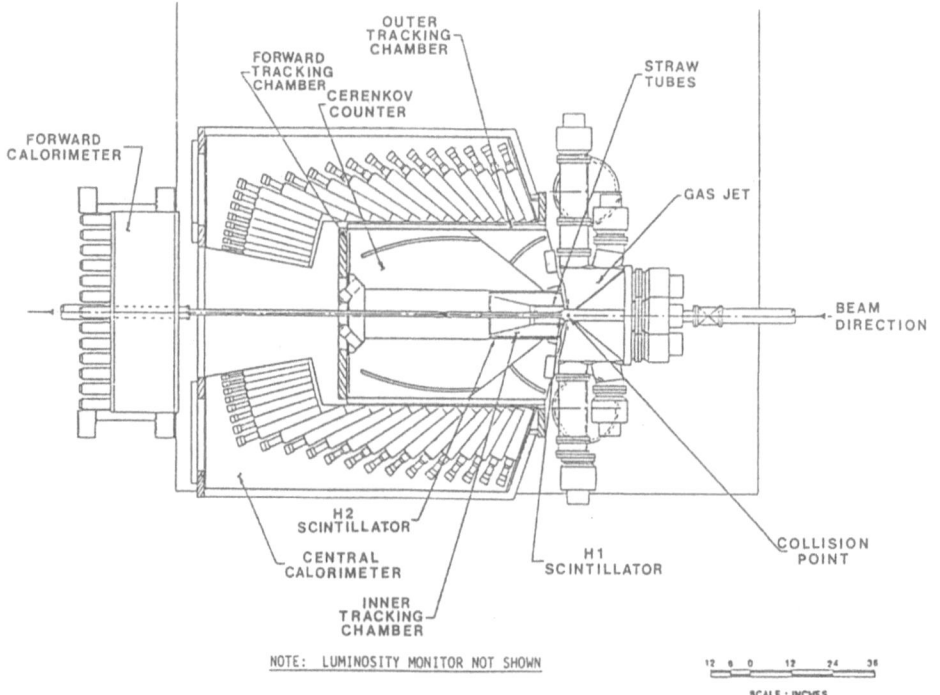

Fig. 4 - E760 Detector

sixteen phototubes. This detector identifies $e^\pm$ from J/ψ decay, with a rejection of ~ 10^{-3} per pion. Therefore, one can reduce backgrounds in triggers involving states which decay via $J/\psi \to e^+e^-$ by a factor of 10^6. This is sufficient to reduce the gross 700 kHz rate at the intersection to a comfortable rate of ~ 1 Hz for such events. One of the principal sources of background in identifying $J/\psi \to e^+e^-$ decays are Dalitz pairs and γ conversions resulting from π°'s, where the electrons trigger the Cerenkov counter. The inner tracking chamber includes dE/dx capability for the purpose of isolating two versus one electron on the same side of the beam pipe. This is followed by the outer and forward tracking systems.

The central electromagnetic calorimeter (Fermilab, Irvine, Northwestern, Penn State), is comprised of 1536 lead glass (SF2) modules with pointing geometry. The modules are organized into 64 azimuthal "wedges", each containing 24 modules. The number of radiation lengths per module varies from 16(70°) to 24(15°). Each module is read out by a photomultiplier tube coupled directly to the module. The RMS energy resolution of each module is expected to be ~ 5%/$\sqrt{E}$ over 50-3000 MeV, with an angular resolution of ~ 1-2°.

(a)

(b)
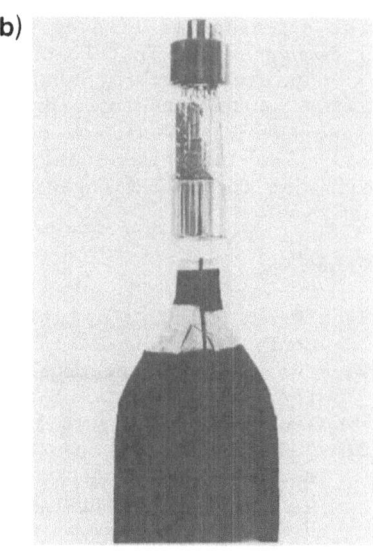

(c)

(d)

Fig. 5 - The forward calorimeter: (a) layers of scintillator and lead with wavelength shifter bar and lightguide (b) an Amperex XP2081B phototube coupled to the lightguide (c) several wrapped modules and (d) a stack of modules awaiting shipment to Fermilab.

The forward calorimeter (Penn State) closes the end-cap region down to ~ 2° around the beam pipe. It consists of 144 modules (non-pointing) of ~ 15 radiation lengths each. It is constructed in a lead-scintillator sandwich format, and read out through wavelength shifter bars to an Amperex XP2081B phototube. The detector has just been completed and shipped to Fermilab where it is presently being set up in a test beam. Photographs of it are shown in Fig. 5. We expect its resolution to approach that of the central calorimeter (~7%/$\sqrt{E}$, 1-2°).

The following activities are planned for the summer and fall of 1987: (1) the Fermilab group will decelerate and cool protons in the accumulator ring from injection at 8.9 GeV/c down through transition to 3 GeV/c, the lowest momentum of interest for charmonium studies. This would then allow first experimental studies with antiprotons in the 3-7 GeV/c range; and (2) debugging and installation of the gas jet, luminosity monitor, Cerenkov counters, tracking systems and the forward calorimeter. With the limited photon acceptance provided by this setup (~ 15% of the final setup) we should nonetheless be able to observe $\chi_{1,2} \rightarrow J/\psi + \gamma$ events this year. The lead-glass central calorimeter should be ready for 1988, at which time the full-fledged experimental program will begin.

REFERENCES

[1]   "Heavy Quarks in Antiproton-Proton Interactions," R. Cester, AIP Conf. Proc. No. 150, 1986, p. 468.

[2]   "Heavy Quark Spectroscopy," J.L. Rosner, review talk presented at the International Symposium on Lepton and Photon Interactions, Kyoto, Japan, August 19-24, 1985, EFI 85-63.

[3]   "QCD Prediction for Charmonium Production in $\bar{p}p$ Collisions," E.L. Berger, P.H. Damgaard and K. Tosokos, Antiproton 1984: Proceedings of the VII European Symposium on Antiproton Interactions, Durham, U.K., Inst. of Phys. Conf. Ser. No. 73, Adam Hilger Ltd., Bristol and Boston, p. 349.

[4]   "Antiproton-Proton Physics in the milli-Tev Region, R.L. Jaffee, review talk in the Proceedings of the Workshop on Antimatter Physics at Low Energy, Fermilab, April 10-12, 1986, p. 1.

[5]   P. Moxnay and J.L. Rosner, Phys. Rev. D28, 1132 (1983).

[6]   H. Grotch et al, Phys. Rev. D30, 1924 (1984).

[7]   S. Gupta et al, Phys. Rev. D31, 160 (1985).

[8]   "Production and Decays on Quarkonia in $\bar{p}p$ Collisions," W. Buchmüller, Physics with Antiprotons at LEAR in the ACOL ERA, LEAR Workshop, Tignes, France, Jan. 19-26, 1985, p. 327 and references cited therein.

[9]   A. Andrekopoulou, Z. Phys. C22, 63 (1984).

[10] S. Brodsky and G.P. Lepage, Phys. Rev. D24, 2848 (1981).

[11] M.G. Olsson et al, Phys. Rev. D31, 1759 (1985).

[12] C. Baglin et al, Phys. Lett. 171B, 135 (1986).

[13] C. Baglin et al, Phys. Lett. 172B, 455 (1986).

[14] C. Baglin et al, Phys. Lett. 187B, 191 (1987).

[15] C. Baglin et al, CERN-EP/87-30, 16 Feb. 1987.

[16] C. Baglin et al, CERN-EP/87-94, 21 May 1987.

LEAR, PAST, PRESENT AND NEAR FUTURE

Pierre Lefèvre

CERN
CH-1211 Geneva, Switzerland

INTRODUCTION

LEAR[1],[2] is a strong focusing synchrotron/storage ring 78.5 meters in circumference. The design range of operating momenta is 0.1 to 2.0 GeV/c and its present role is to provide intense, pure beams of low energy antiprotons. It has a separated function lattice BODFOFDOB and is a symetric four period machine with 4 bending magnets and 8 focusing doublets. The result is a very strong focusing machine, with a phase advance per period of $\approx$ 250° and a betatron wavelength of $\approx$ 30 meters. $\gamma = E/E_0$ at transition is imaginary for the standard working points. This leads to a large value for $|\eta|$,

$$\text{where } \frac{\Delta F}{F} \,/\, \frac{\Delta P}{P} = \eta = \frac{1}{\gamma_T^2} - \frac{1}{\gamma^2}$$

which favours good mixing for the stochastic cooling process and helps the ultraslow stochastic extraction. Since η is always negative, the machine operates below transition over all of its large momentum range, thus avoiding negative mass effects and the problems associated with operating close to the transition energy.

Chromaticity correction is performed using 18 sextupoles (14 normal and 4 skew). The closed orbit is controlled using 8 backleg windings and 4 dipoles in the horizontal plane and 6 dipoles in the vertical plane.

Present performance

The main mode of LEAR operation for experimental physics has been as a variable momentum antiproton beam stretcher ring. The essential features of the machine are phase space cooling to improve beam quality, ultra-slow extraction with spill times ranging from 15 minutes to 5 hours and a variable extraction momentum from 105 to 1700 MeV/c. Figure 1 shows the general layout of LEAR and its associated transfer lines.

The antiprotons for LEAR, typically a bunch of up to 3.0×10^9 particles, is unstacked from the AA and transfered to the PS, where

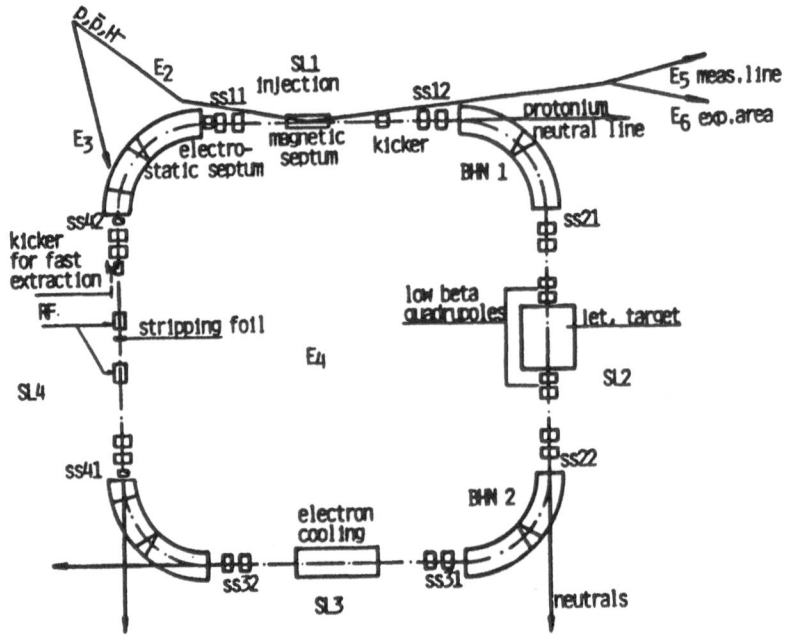

Figure 1 : LEAR : An overview.

the beam is decelerated from 3.5 GeV/c to 0.609 GeV/c, before being ejected towards LEAR. After about 3 minutes of stochastic cooling in LEAR at the injection momentum (609 MeV/c) the beam is either accelerated or decelerated to the required ejection momentum. At present, for ejection momenta above 609 MeV/c no further stochastic cooling is applied, however, for ejection momenta below 609 MeV/c, further beam cooling is necessary to maintain high beam quality, and to counteract the adiabatic emittance increase. During the deceleration cycle the beam is held on a series of fixed momentum "flat-tops" (309, 200 and 105 MeV/c) for several minutes of intermediate cooling. At each stage the emittances are restored to their previous levels. Under normal operating conditions, the minimum emittances obtained are around 10π mm.mrad transversely and $dp/p = \pm 1.0 \times 10^{-3}$ longitudinally (emittances containing 95% of the beam).

Figures 2, 3 and 4 show the result of 5 minutes stochastic cooling of 2.5×10^{9} particles at 609 MeV/c.

Figure 2 : Longitudinal Schottky scan, taken at the 20th harmonic of the revolution frequency (centre freq.= 40.56MHZ, freq. span = 300KHz). The curves show the square root of the particle density against momentum. The momentum spread is reduced from 5.5 o/oo to 2 o/oo.

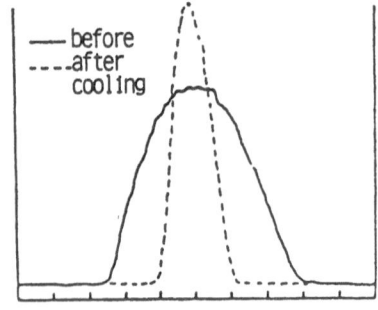

— before
----after
cooling

Figure 3 : Horizontal Schottky
scan, taken at the 100th revolution
harmonic (centre freq. = 207.8MHz,
freq. span = 1.87MHz). The height
of the betatron sideband is a measure of the transverse emittance.
Cooling decreases the emittance by
a factor of 3.5.

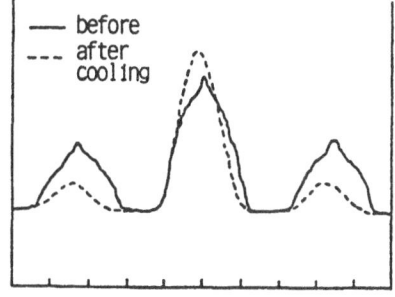

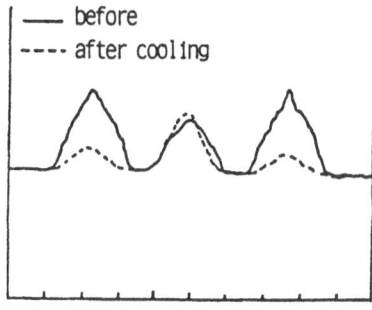

Figure 4 : Vertical Schottky scan,
taken at the 100th revolution harmonic (centre freq. = 207.8MHz,
freq. span = 2.078MHz). The vertical
emittance is reduced by a factor of
5 during the cooling.

The transverse stochastic cooling system consists of a limited
number of pick-ups and kickers. A set of coaxial relays, which
commute between different cable delays, are used to compensate for
the changing particle velocities over the range of operating
momenta. There are several sets of preadjusted delays, which allow
"fast" cooling on each of the fixed "flat-tops", with cooling time
constants of the order of one minute for 10^9 particles. In addition a
second series of relays and cables can be set to provide a slow
transverse cooling at any intermediate momenta. This slow cooling is
used to maintain beam quality during the long extraction process.

The longitudinal cooling system is again made in two parts, each
using the same pick-up. One, for use at 609 MeV/c, uses a line filter of
fixed electrical length, and the other, for all momenta below 609 MeV/c,
uses a line filter of variable electrical length from 70 to 700m. The
present LEAR stochastic cooling system is shown in Figure 5 and
described in detail in Ref. (3). Table 1 summarises the present "fast"
cooling system.

Table 1 : Present LEAR stochastic cooling system

	Longitudinal		Vertical	Horizontal
	609 MeV/c	309,200 MeV/c	609,309,200, 100 MeV/c	609,309,200, 100 MeV/c
Pick-ups	24 Gaps	24 Gaps	8 loops pairs	12 loop pairs
Noise figure amplifiers	2.5 dB	2.5 dB	2.0 dB	2.0 dB
Kicker	8 Gaps	8 Gaps	1 loop pair	1 loop pair
Bandwidth (MHz)	20-300	15-200	50-500	50-500

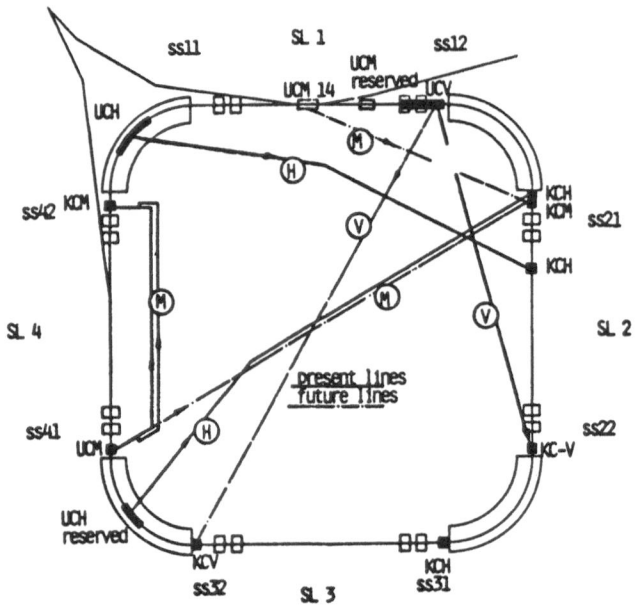

Figure 5 : LEAR Stochastic Cooling System

The ultra-slow stochastic extraction process is described in detail in Refs. (4,5).

Once the required momentum has been reached, the stochastic extraction process is initiated. Firstly, the beam is "shaped", i.e. the longitudinal distribution is made rectangular, Fig. 6, by applying an RF noise signal to the beam, with a well defined bandwidth around an harmonic of the revolution frequency between 15.5 and 17.5 MHz. This shaping takes 30 seconds and is done to ensure a constant spill rate. During acceleration or deceleration, the betatron tune and chromaticity (ξ = $\Delta Q/Q$ / $\Delta p/p$) are kept at or very close to Qh = 2.305, Qv = 2.725, ξh = ξv =0. After shaping, the machine quadrupoles and sextupoles are adjusted to give Qh = 2.325, Qv = 2.725, ξh = 0.6 and ξv = 0.0, and at the same time drive the extraction resonance 3Qh = 7. Figure 8 shows the various working points used during acceleration and extraction. Simultaneously, a local orbit bump is applied to move the beam close to the electrostatic and magnetic extraction septa. Finally, an RF noise signal is swept slowly into the beam, again at around 16 MHz, as shown in Figures 6 and 7, and the particles diffuse slowly towards the extraction resonance. The length of the spill is determined by the rate of sweeping of the extraction noise. Table 2 gives further details of the shaping and extraction process.

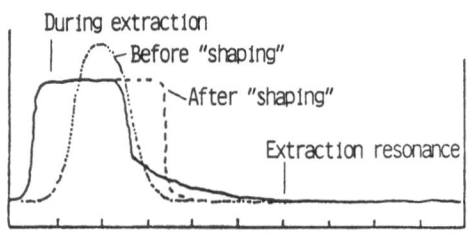

Figure 6 : Longitudinal particle distribution during "shaping" and extraction.

362

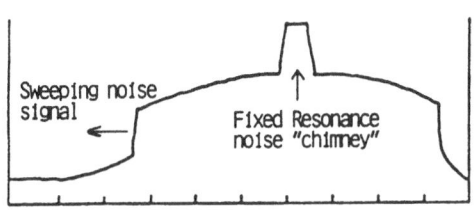

Figure 7 : Form of swept RF noise used for ultraslow extraction.

Table 2 : Parameters for ultraslow extraction at 350 MeV/c.

a) Shaping		
Bandwidth	60	kHz
Power	0.15	mW/Hz
Time	28.0	s
Centre frequency	17.340	MHz
b) Extraction		
Bandwidth	140	kHz
Power	0.02	mW/Hz
Time	3600	s
Centre frequency	17.458	MHz
Sweep	85	kHz
c) Resonance "chimney"		
Bandwidth	10	kHz
Resonance frequency	17.433	MHz
Resonance power	0.2	mW/Hz

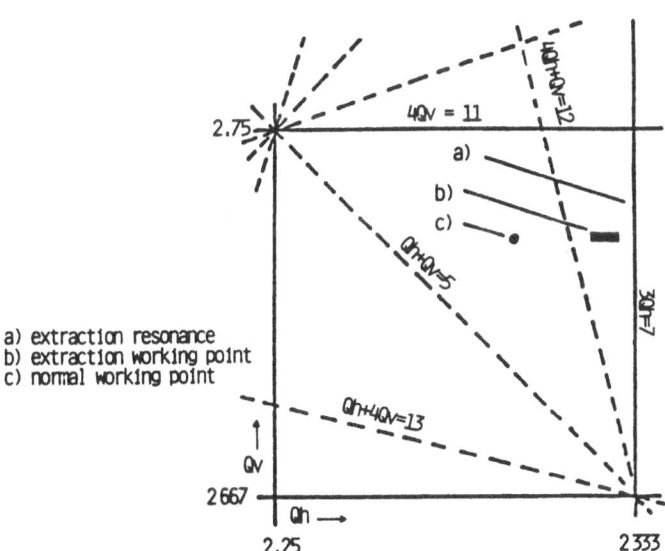

Figure 8 : Typical machine working points.

This extracted beam is then split into 3 different experimental lines, shown in Figure 1, using 2 beam splitter[6]. In this way, different experiments can receive particles simultaneously. Figure 9 shows the extracted beam intensity variation during a typical spill.

$\bar{p}$ intensity measured on the PS 172 scintillator

Horizontal scale : 1 mm/mm
Vertical scale : 100 000 $\bar{p}$/s per cm
Total intensity per spill :~4.5 E8 $\bar{p}$

Figure 9

In this way continuous spills with an 80% duty factor, intensities ranging from 2.0×10^6 antiprotons/sec to over 1.0×10^6 antiprotons/sec, and varying from 15 minutes to 5 hours in length, are routinely obtained.

In order to be capable of ejecting beam at any momentum within the LEAR operating range (at present 105 to 1700 MeV/c), an online procedure has been developed, which will calculate the power supply values and cycle timing necessary to create the new ejection momentum, by interpolation or "scanning" between existing ejection settings. All existing values for cycles with different extraction momenta are stored using an archive and retrieval system on the LEAR control computers[7],[8]. In this way it is possible to recall a previously used cycle or create an entirely new machine cycle in a few minutes.

A similar technique is used to "scan" the values for the transfer lines to the physics experiments, and to calculate new settings for the tranverse stochastic cooling. The time needed to change the extraction momentum is less than 1 hour, if recalling a previously used cycle, and about 2 hours for a completely new "scanned" ejection momentum. This scanning procedure is now a routine part of machine operation, but considerably more developement is needed both in the hardware and software domains to meet the demands of future experiments, as will be seen in the following sections.

Table 3 gives a breakdown of the antiproton running statistics from November 1983 to August 1986.

Table 3 : LEAR running statistics

Year	N° of different extraction momenta	Hours of operation		Antiproton pulses used
		exp. physics	total	
1983	2	352	712	233
1984	5	1320	2444	815
1985	24	2651	3837	1612
1986	36	2254	3370	1480

Post-Acol performance requirements

The addition of the Antiproton Collector (ACOL) to the CERN Antiproton Accumulator (AA) coincides with the installation of the second generation of LEAR experiments. These are, in general, bigger and more complex than the first generation, and more demanding in terms of machine performance[9,10]. Some of the problems posed by the new experimental demands on LEAR are outlined in this section. The proposed final layout of the new LEAR experimental area is shown in Figure 10.

The most obvious effect of ACOL on LEAR operation will be the increase in AA stack intensity. This increase in the antiproton stack density will mean a possible increase in stack emittance, therefore, in order to profit fully from the higher antiproton fluxes available, the acceptance of the PS-LEAR transfer line needs to be increased. The optics of this transfer line have been redesigned and two new quadrupoles added in order to reduce the beam dimensions in the transfer channel. At the same time seven horizontal and vertical beam position pick-ups are being installed at critical points down the line. These pick-ups are copies of the pick-ups developed for the ACOL machine, and they will cover a range of bunch intensities from $1.0 \ 10^9$ to $1.0 \ 10^{11}$ particles for bunch lengths ranging from 20 to 200 nsecs. The estimated position resolution is ± 1 mm.

Several approved LEAR experiments require ultra-low beam momenta, even below the 100 MeV/c design figure already attained. The different experimental methods impose varying constraints on the form of the LEAR extracted beam. Some need slow extraction of the type already performed at higher momenta, others need extraction times of between 100 msecs and a few seconds, and finally the "antiproton trapping" experiments need fast extraction of small antiproton bunches of around 200 nsecs in length.

One of the principle problems at low momenta has been the excitation of systematic machine resonances by the machine sextupoles. These sextupoles are primarily used for chromaticity adjustment and excitation of the extraction resonance, $3Qh = 7$. However in mid-1986 a series of machine experiments were started in order to compensate, using the existing sextupoles, the resonance $Qh + 2Qv = 8$ at 309 MeV/c. This compensation was found semi-empirically by placing the machine working point very close to the resonance in question and adjusting the compensation until the maximum beam intensity was injected[11]. The resulting optimum compensation was used at all momenta below 609 MeV/c, but only partially applied at momenta above 609 MeV/c, due to lack of available sextupole power.

Using the new sextupole configuration several improvements in beam behaviour have been observed, which are particularly important for ultra-low energy operation.

The dependance of Qh or Qv on the betatron amplitude was reduced effectively to zero, contributing to a large increase in the dynamic transverse acceptance of the machine. E.g. at 309 MeV/c the loss-free horizontal acceptance passed from around 40π mm.mrads to 100π mm.mrads. This is important both at injection and during stochastic extraction, when the particles excecute very large amplitude horizontal betatron oscillations on the last few turns before leaving the machine.

The beam lifetime at 105 MeV/c, with the stochastic cooling on, was increased from 13 to 50 minutes, which made spills of 1 hour feasible at 105 MeV/c and quadrupled the extraction efficiency. Indeed the first evidence of genuine transverse cooling at 105 MeV/c was seen after compensation of this coupling resonance. It was also possible to reduce the losses during deceleration from 200 to 105 MeV/c from 50% to 30%.

However, due to the non-symmetric distribution of sextupole power used to compensate the systematic resonance $Qh + 2Qv = 8$, and the location of the sextupoles in non-zero dispersion regions, a strong dependance of Qv on $(dp/p)^2$ is now observed. This dependance has been measured $dQv = 700(dp/p)^2$. Although relatively unimportant at high momenta, where dp/p is small, this effect becomes enormous during deceleration towards 60 MeV/c, where dp/p approaches 1%. Studies are underway to try and reduce this sextupolar asymmetry even more, but this task is complicated by the lack of space, and the fact that at least one sextupole has to be moved from its present postion to make way for a gas-jet target[12]. Any reduction in beam momentum spread at 100 MeV/c would significantly reduce the machine sensitivity to this sextupole asymmetry, and considerable effort is being invested into the developement of stochastic cooling pick-ups for use at or below 100 MeV/c.

In order to continue this work of resonance measurement and compensation a system for Q measurement by Fast Fourrier Transform analysis of the beam response to a transverse kick has been installed at LEAR[13]. It is now possible to measure accurately the amplitude and phase of a resonance and thus correctly compensate or excite it[14,15]. This method has been used during machine experiments to study resonances $Qh + Qv = 5$ and $3Qh = 7$.

The fast extraction process, which is required by two approved experiments, has already been tested and used for some initial trapping trials using a Penning trap, supplied with 200 MeV/c antiprotons, which are subsequently degraded down to the KeV energy range. This work culminated in the trapping and storing, for upto 10 minutes, of 3 KeV antiprotons[16]. The extraction method is a standard fast extraction of a bunched beam using a kicker module, similar to the injection kickers, placed a quarter of a betatron wavelength upstream of the magnetic septum. One of the important requirements of the trapping experiment was the possibility to set-up the experimental apparatus with slow extracted beam before receiving the fast extracted pulses. This meant that not only did the machine have to be capable of switching rapidly from one mode of ejection to the other, but also that the momentum of the fast extracted pulses had to be the same as that the slow spills. However, during stochastic extraction the particles, prior to ejection, are diffused across the machine aperture and their momentum increases, therefore the momentum of the slow ejected beam is $5 \ 10^{-3}$ higher than the coasting LEAR beam.

The solution to these problems was in fact relatively simple. Firstly the same magnetic machine cycle is used for both processes, making switching extraction modes very quick and simple. Secondly, instead of using a local radial orbit bump to move the circulating beam close to the extraction septum, as is done for stochastic extraction, the entire beam is accelerated to the momentum of the slow extracted particles, which corresponds to a fractional momentum increase of $5 \ 10^{-3}$, prior to fast extraction. This acceleration displaces the beam at the magnetic septum by almost the same amount as the normal dipole extraction bump, and ensures that the fast extracted bunches have exactly the same momentum as the slow extracted beam.

The whole fast extraction process lasts less than 1 sec, and can be performed at any time during the LEAR cycle. The beam is decelerated to the required momentum and the stochastic cooling switched on. When fast extracted beam is requested the coasting LEAR beam is bunched at the 4th RF harmonic, accelerated to the ejection orbit, and one bunch out of the four is ejected by sychronising the ejection kicker with one of the circulating bunches. Then the remaining 75% of the beam is decelerated back to the centre of the machine and debunched, ready for the next beam request. Between fast and slow extraction a small correction, at the magnetic septum, is needed, as the trajectories upon which the particles leave the machine are not the same in the two cases, but apart from this difference the settings for the experimental beam line are identical for either fast or slow extraction.

In 1988 this procedure will be reused at 105 MeV/c for the same experiment, and later a similar ejection will be required at 60 MeV/c for another trapping experiment, which uses an inverted RFQ to decelerate the extracted beam, rather than a simple energy degrader foil[17]. The expected characteristics of this rather specialised fast extracted beam are shown in Table 4.

The installation and operation of a gas-jet target[12] inside LEAR is also an integral part of the approved experimental program. The expected maximum luminosity is around $3.0\ 10^{30}\ cm^{-2}\ sec^{-1}$, which is determined by the antiproton production rate. However, the maximum available target density may limit this figure to $1.6\ 10^{30}\ cm^{-2}\ sec^{-1}$. As well as the obvious consequences for the machine vacuum, and severe space limitations for the experimental apparatus, this installation will pose several specific problems for LEAR[18].

Stochastic cooling will be necessary over the whole LEAR momentum range in order to combat the beam blow-up caused by repeated traversals of the gas-jet, as well as longitudinal cooling as a method for fine tuning of the final beam momentum. At present, stochastic cooling is only available at or below 609 MeV/c.

For an internal target it is advantageous to increase the acceptance angle of the machine by reducing the b values at the interaction point. This is especially true for momenta below 800 MeV/c, where the loss rate due to single Coulomb scattering in the gas-jet, without a reduced β at the interaction point becomes comparable with the strong interaction rate. For this reason a low β insertion has been designed for operation with the gas-jet[19]. The experiment requires continuously variable beam momentum, in order to scan a wide mass range in the proton/antiproton formation channel. With this in mind a considerable revision of the control system hardware and software has been undertaken, which is dealt with in the following section.

Table 4 : Fast extracted beam for RFQ deceleration.

Momentum = 61.3 MeV/c, Kinetic energy = 2 MeV
Eh = Ev = 5 to 10π μm.mrad.
Dp/p = 2 to 4 x 10^{-3} , Energy spread = 8 to 16 KeV
Bunch length = 250 to 500 nsec.

367

Machine upgrade for the new experiments

In order to be able to satisfy the needs of the new generation of experiments, the LEAR consolidation program has been approved and started.

The effective range of the stochastic cooling system is being extended. In one direction, the system must be capable of operation up to 2 GeV/c, for the gas-jet operation, and in the other direction cooling at and below 100 MeV/c is essential for efficient deceleration towards 60 and even 20 Mev/c! For high momemtum cooling the problem is one of space for new kickers, as the increasing particle velocity means that the cooling signals have to take bigger and bigger short-cuts with respect to the beam. The proposed layout of these new lines is shown in Figure 10. At low momenta the small beam intensities and the low revolution frequency mean that a very poor signal- to-noise ratio is the major difficulty, and a range of new pick-ups are being developed to try and overcome this problem. Two existing transverse pick-ups have been connected in "travelling wave" mode in order to coherently sum the beam signals. They have also been equiped with cryogenically cooled pre-amplifiers and terminating resistors. In this way the thermal background noise is reduced by a factor of four. This system has been used with some success at 105 MeV/c.

The same longitudinal cooling system used at 200 MeV/c has been tried at 105 MeV/c, but very little effective cooling has been observed. Two new designs of travelling pick-ups for stochastic cooling of 60 MeV/c beams are at present being tested. The pick-ups must have a high coupling impedance up to 50 Mhz, above which the Schottky bands overlap. For betatron cooling a pair of 20 cm long "meander couplers" have been built, the meander line and the copper ground plane are electroplated onto the surface of ceramic plates. The phase velocity of the induced signal is designed to be as close as possible to the beam velocity, about 6.5% of the velocity of light. Proton beams of 100 and 200 MeV/c have been used to estimate the coupling impedance of one coupler. The measured impedance is about one half of the theoratical prediction at 100 MeV/c. This reduction in coupling impedance is thought to be caused by unwanted signals induced at the ends of the pick-up. The available beam current at 60 MeV/c in LEAR is not yet high enough at present to permit measurements of the coupling impedance at the pick-up design momentum.

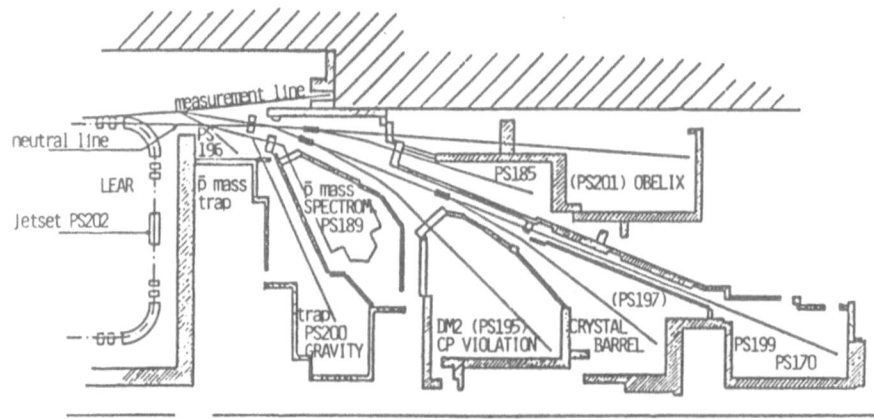

Figure 10 : The new LEAR experimental zone

For momentum cooling, a helix coupler is being fabricated, which will be installed in LEAR in May 1987. The structure is similar to that used in TARN[20,21] The pick-up will be located in the straight section used for injection and has an aperture 32 cm wide by 6 cm high. The coupler is divided into three pieces, which are connected in series for 60 MeV/c operation, and give a theoretical coupling impedance of 500Ω at 100 Mhz. With 100 MeV/c beam the couplers can be connected in parallel, and in this way a reasonable coupling impedance at 100 MeV/c can be obtained. The expected maximum is about 200Ω at 70 Mhz.

As part of the overall improvement in low energy performance a series of new more stable power supplies are being installed for the LEAR start-up in June 1987. All of these supplies will be bipolar, i.e. able to change polarity during a machine cycle. Upto now this has not been the case, which has, for example, proved to be a limiting factor in the optimum use of the horizontal dipoles for closed orbit correction and in the sextupolar excitation of the extraction resonance.

As was mentioned earlier, it will be necessary to scan the beam momentum for gas-jet target operation. Until now the momentum scanning process has been restricted to producing one new momentum flat-top in any one machine cycle, but, in this case, momentum scanning has to be possible, whenever requested by the users, and without the injection of a new beam. With the old control system this would not be possible, as the Function Generators, which controled the power supplies could only be updated at the end of a complete cycle. A new series of Digital Function Generators are being developed, which will be updatable on-line. This will make the continuous momentum tuning possible, as well as considerably speeding up the present momentum scanning procedures, since the results of requested changes will be immediately measurable, without the necessity of recycling and injecting a new antiproton pulse.

Continued software effort is required to make effective use of the new hardware, and new programs for the momentum scanning procedure as well as the new Digital Function Generators are under development. Software and hardware development is also continuing in the area of machine surveyance and fault-finding. A new CAMAC unit has been developed to memorise all the cycle timing events as they occur, and make timing system faults easier and quicker to diagnose. It is hoped to simplify the control of the stochastic cooling system, by the use of local intelligence, in the form of several 6809 uprocessors, installed in the local G64 interface.

Work on the machine vacuum system is also underway, to try and push the overall vacuum towards $1\ 10^{-12}$ Torr, by the installation of several new NEG pumps to isolate the machine vacuum from that of the transfer lines. In addition the installation of the gas-jet and the electron-cooler both pose very particular vacuum problems, which have to be solved in order to fully integrate these devices into LEAR. This vacuum improvement is an essential part of the quest for lower and lower momenta, and has made possible some of the tests, which have been performed with H⁻ ions and will be discussed in the following section.

Studies for future options

The installation of an electron cooler[22], developed from the original ICE test device, has started and a series of experiments aimed at running-in the device with proton beams are scheduled to take place this summer. In this way we will be able to study the electron cooling itself and the possible combination of electron and stochastic cooling[23] for use with internal targets in the low momentum range (<300 MeV/c).

The availability of H⁻ ions, from Linac 1, has made possible a series of measurements of H⁻ beam lifetimes and stripping mechanisms. A variety of stripping processes have been studied, including intra-beam stripping, beam-gas stripping and photon-induced stripping. For the first time it has been possible to make accurate measurements of the cross-sections for these processes at LEAR energies, and the results show the stripping cross-sections to be somewhat lower than previously estimated[24,25]. This could open the way to the storage and use of beams of H⁻ ions for experimental physics purposes, e.g. the storage of co-rotating H⁻ and antiproton beams.

CONCLUSIONS

The first generation of LEAR experiments have already pushed the machine performance beyond the goals set in the initial machine design, and the second generation promise to extend still further the possibilities of the machine. This extension is possible due to the "consolidation program", which is underway, and the first part of which should be finished in time for the restart of the machine in June 1987. The years to come will see LEAR far beyond its initial role as a low momentum beam stretcher, and rather come of age as a multi-purpose storage and decelerator ring.

This presentation results from the work performed by the LEAR team : D. Allen, E. Asséo, S. Baird, J. Bengtsson, M. Chanel, J. Chevallier, R. Giannini, P. Lefèvre, F. Lenardon, R. Ley, D. Manglunki, E. Martensson, J.L. Mary, C. Mazeline, D. Möhl, G. Molinari, J.C. Perrier, T. Pettersson, P. Smith, N. Tokuda, G. Tranquille, H. Vestergård.

REFERENCES

1. Design study of a facility for experiments with Low Energy Antiprotons (LEAR), ed. G. Plass, CERN/PS/DL 80-07.
2. LEAR, P. Lefèvre, CERN/PS 84-07 (LEA).
3. G. Carron et al., Status and future possibilities of the stochastic cooling system for LEAR, in Proceedings of the Third LEAR Workshop, Tignes, January 1985. (Editions Frontières, Gif-sur-Yvette, 1985, U. Gastaldi et al., eds.) pp. 121-128.
4. R. Cappi et al., Ultraslow extraction (Status report), in Proceedings of the Second LEAR Workshop, Erice, May 1982, (Plenum Press, New York 1984, U. Gastaldi and R. Klapisch, eds.) pp. 49-54.
5. The LEAR Team, Performance of LEAR, IEEE Trans. NS32, pp. 2652-56 (1985).
6. LEAR machine and experimental areas : M. Chanel et al., Invited paper at the VIIth European Symposium on Antiproton Interactions, Durham, July 1984; also CERN/PS 84-26 (LEA).
7. M. Chanel and T. Pettersson, The LEAR control system phase II, Nucl. Inst. Meth. A247, pp. 146-152, 1986.
8. P. Lienard and T. Pettersson, Development of the LEAR control system, in Proceedings of the Third LEAR Workshop, Tignes, January 1985. (Editions frontières, Gif-sur-Yvette 1985, U. Gastaldi et al., eds.) pp. 121-128.
9. P. Lefèvre, D. Möhl, D.J. Simon, Future machine improvements in LEAR, in Proceedings of the First Workshop on Antimatter Physics at Low Energy, Fermilab. April 1986, pp. 69-82.
10. R. Landua, The future physics at LEAR, ibidem, pp. 35-69.

11. M.   Chanel et al., Resonance compensation in LEAR, CERN/PS/LEA Note 87-08 (1987).

12. G. Bassompierre et al., Jetset: Physics at LEAR with an internal gas-jet target, CERN PSCC 86-23.

13. See a series of papers by E. Asseo CERN/PS/LEA Note 87-01. CERN/PS/LEA Note 86-14, CERN/PS/LEA Note 85-9, CERN/PS/LEA Note 85-3.

14. J.   Bengtsson, Programs for symbolic solving of differential equations, CERN/PS/LEA Note 86-12 (1986).

15. J. Bengtsson and M. Chanel, Resonance measurements using Fourier spectrum analysis of beam oscillations, CERN/PS/LEA Note 86-15 (1986).

16. G. Gabrielse et al, First capture of antiprotons in a Penning trap: a Kev source, Phys. Rev. Letters 57 (1986) pp. 2504-2507.

17. N. Jarmie, A measurement of the gravitational acceleration of the antiproton, presented at the 8th Conference on the Application of Accelerators in Research Industry, Denton November 1986.

18. S. Baird, Some numbers concerning a gas-jet in LEAR, CERN/PS/LEA Note 87-06 (1987).

19. R. Giannini, Straight section 2 of LEAR as a low-beta insertion, CERN/PS/LEA Note 87-07 (1987).

20. N. Tokuda, A helix coupler for a pick-up of a low velocity beam, CERN/PS/LEA Note 86-05 (1986).

21. N. Tokuda, Formulation of a flat-helix coupler for a pick-up for a low velocity beam, CERN PS/LEA Note 87-05 (1987).

22. A. Wolf et al., Status and perspectives of the electron cooling device under construction at CERN, in Proceedings of the Third LEAR Workshop, Tignes January 1985, (Editions frontières, Gif-sur-Yvette 1985, U. Gastaldi et al., eds.) pp. 121-128.

23. D. Möhl, Phase-space cooling techniques and their combination in LEAR, in Proceedings of the Second LEAR Workshop, Erice 1982 (Plenum Press, New York 1984, U. Gastaldi and R. Klapisch, eds.) pp. 27-48.

24. M. Chanel et al., Measurement of H$^-$ intra-beam stripping cross-section by observing a stored beam in LEAR, CERN/PS/LEA 87-12 (1987).

25. The LEAR Team, Epluchage des H$^-$ dans LEAR par le gas residuel, la lumière des jauges par un faisceau laser ou une lampe spot, CERN/PS/LEA Note 87-2 (1987).

SLEAR, A HIGH LUMINOSITY 10 GEV PROTON-ANTIPROTON STORAGE RING WITH SUPERCONDUCTING MAGNETS

R. Giannini, P. Lefèvre, and D. Möhl

CERN
1211 Geneva 23
Switzerland

ABSTRACT

The design of a 10 GeV antiproton storage- and cooling ring (SLEAR) desired for precision studies with internal targets and $\bar{p}p$ colliding beams is revisited. Possible geometries of a compact, high luminosity machine with superconducting magnets are discussed. Apart from being a powerful tool for particle physics, SLEAR could be an attractive "testbed" of new accelerator technology. Complementary possibilities for a $\bar{b}b$ "pre-experiment" at CERN are mentioned.

INTRODUCTION

Several groups[1],[6] have expressed interest in an antiproton storage- and cooling ring covering the range of, say, 2 to 10 GeV/c of circulating beam momenta. Working with internal targets and/or $\bar{p}$-p colliding beams, such a ring should permit precision measurements in a domain of centre of mass energies where a number of new particles and states containing c and b quarks have been found or are to be expected. This range (Figure 1) is currently not accessible with $\bar{p}$-p at CERN.

In the present note, we summarize the tentative machine aspects of a compact high luminosity storage ring presented before[7],[8] and discuss alternative machine designs. Performance limitations will be discussed in a companion paper[9].

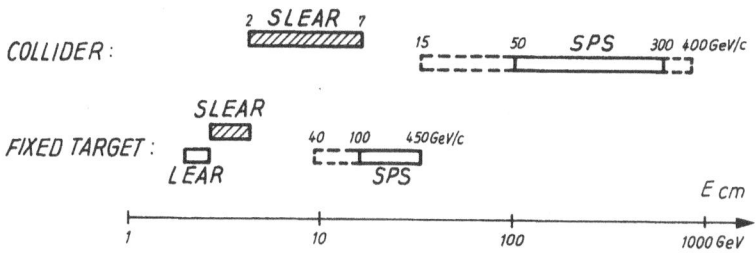

Figure 1 : Centre of mass energies available with $\bar{p}$+p.

Table 1 : SLEAR basic parameters and typical performance at 7 GeV/c.

Momentum range	1.5 - 7 GeV/c		
Injection momentum	3.5 GeV/c		
Circumference	120 m		
Intensity	$\leqslant 10^{12}$ $\bar{p}$		
Possible extension of momentum range	1 - 10 GeV/c		
PERFORMANCES	INTERNAL TARGET MODE	HIGH LUMINOSITY $\bar{p}$-p collider, $(10^{12}\bar{p})$	HIGH RESOLUTION $\bar{p}$-p collider, $(10^{11}\bar{p})$
Luminosity $(cm^{-2}sec^{-1})$	10^{32} *	$0.3\text{-}3 \times 10^{31}$ **	$0.05\text{-}0.5 \times 10^{30}$
r.m.s beam size $(\sigma_h \& \sigma_v)$ at interact. p.	0.5 mm	2.5 mm	0.5
r.m.s. momentum spread $(\sigma_{\Delta p/p})$	3×10^{-4}	3×10^{-3}	1×10^{-3}
bunch length (σ_1)	120 m (coasting beam)	0.5 - 1m	0.25 - 0.5

* limited by antiproton flux ($\dot{N} = 10^7$/sec) available
** limited by space-charge effects.

LAYOUT

    Basic parameters assumed in Ref. 7 and 9 are compiled in Table 1. A "high resolution collider case" is included for convenience. It works with only $10^{11}\bar{p}$ and has smaller beam size and shorter bunches at the expense of a reduced luminosity.

    The tentative layout, presented in Ref. 7-9 uses 8 superconducting 45° bending magnets and 16 normal conducting quadrupole doublets (Figure 2). The doublet structure has been chosen as it allows us to push transition energy above the working range.

    In Ref. 7 the medium straight sections have small beta (small beam size - large angular acceptance) and zero dispersion for the installation of the collision region and/or relatively thick targets. The short straights have large beta and large dispersion for targets or other applications requiring parallel beam and orbit separation by momentum.

    Some characteristics of the bending magnets are recalled in Table 2. As an example, a 5m long dipole requiring 5 Tesla at 10 GeV/c is considered. This field level is well within the realm of present techno- logy. With high field superconductors under development for the LHC, fields of, say, 8 Tesla corresponding to 15 GeV/c in SLEAR could be aimed at. To work up to this momentum in a future extension, the quadrupoles also would have to be converted to a superconducting design (gradients of 30 Tesla/m instead of 15 Tesla/m assumed in the "normal" layout).

    Ramping of the fields from injection to the final momentum can be done very slowly (dB/dt < 1 Tesla/min), so that essentially "d.c." magnets can be used.

374

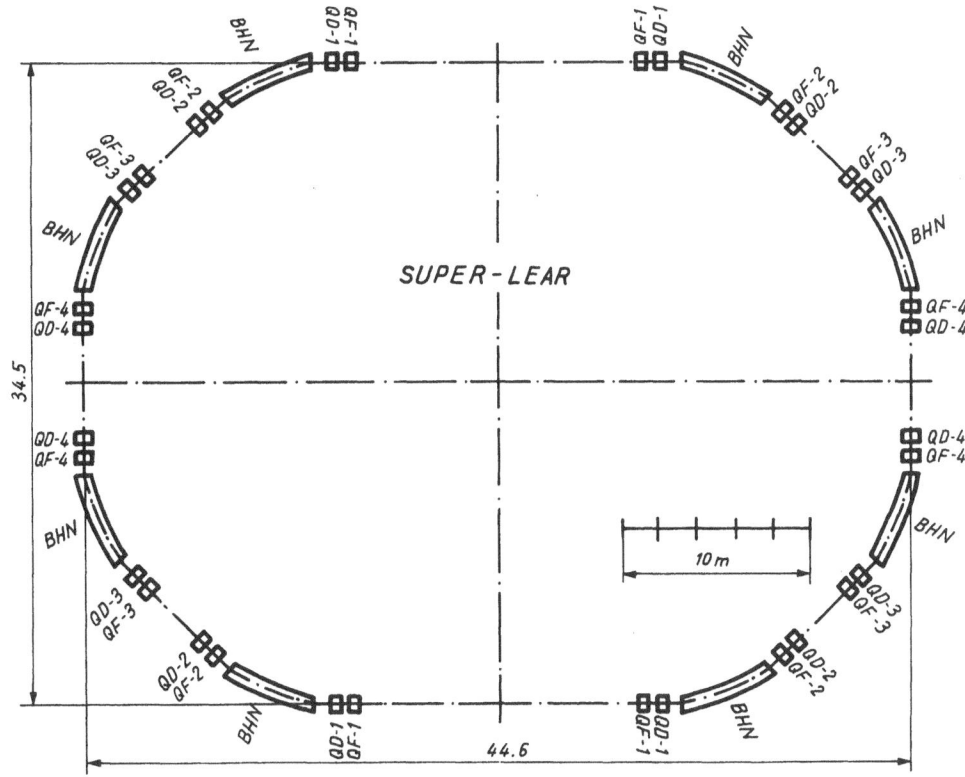

Figure 2 : SLEAR tentative layout (SLEAR 1)

It is convenient to transfer antiprotons directly from ACOL-AA at 3.5 GeV/c except if a location for SLEAR is chosen, where it is easier to inject the p̄-beam via the PS. The protons for tests and collisions have to come from the PS in any case.

The final geometry of SLEAR will depend on the location chosen. To save cost, construction in an existing hall (West Hall ?, ISR service building 181 ?, East Hall?, South Hall?) should be envisaged. In the following we wish to discuss alternatives to the geometry of Figure 2.

Table 2 : Superconducting bending magnets for SLEAR 1-3.

Superconducting magnets		
Length (cent. orbit)	[m]	5.0
Bending angle	[deg]	45
Bending radius	[m]	6.37
Sagitta	[m]	0.5
Field at 10 GeV/c	[T]	5.3
Bore diameter	[mm]	100
Gradient (nominal)		0
Tolerances DB/B		1E-4

A number of other geometries has been studied. The following extreme cases (see Table 3) will be discussed here :

SLEAR 1 : the "octagonal racetrack" (Figure 2) discussed above.

SLEAR 2 : a "symmetric octagonal" similar to SLEAR 1 but with 8 straight sections of (almost) equal length.

SLEAR 3 : a "square shaped machine" with 4 very long and 4 shorter straight sections (Figure 3).

SLEAR 5 : a "hexagonal racetrack" with 2 long and 4 shorter straight sections (Figure 4).

SLEAR 8 : a racetrack with 2 long and 10 short straight sections (Figure 5).

The number and the tentative length of the straight sections is summarized in Table 3. Obviously machines with less focusing elements (like SLEAR 3) have more free straight section space. On the other hand the beam tends to become larger in these long sections, a rule of thumb being that the focusing and dispersion functions increase (i.e. the acceptance for betatron oscillations and momentum spread decrease) linearly with the length of the long straight sectors. Thus larger aperture elements are required.

A condition imposed on all lattices was to have transition much larger than 10 GeV, preferably imaginary, and to be able to tune to small dispersion and small beam size (low beta) in at least one medium or long straight section. Doublet focusing gives enough flexibility to meet these requirements, at least in the cases considered.

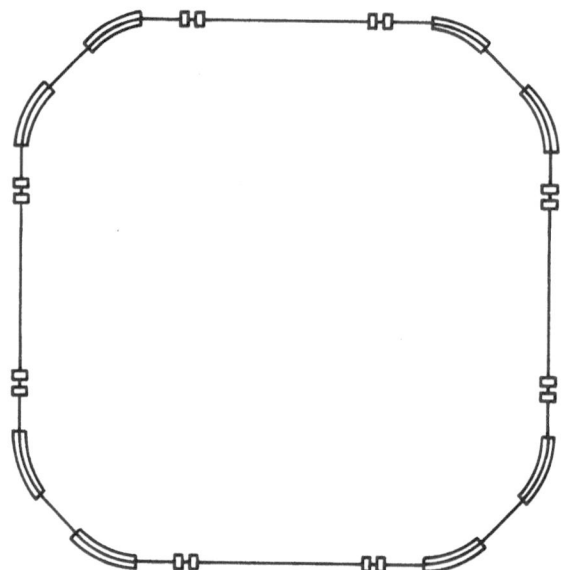

Figure 3 : Square configuration (SLEAR 3).

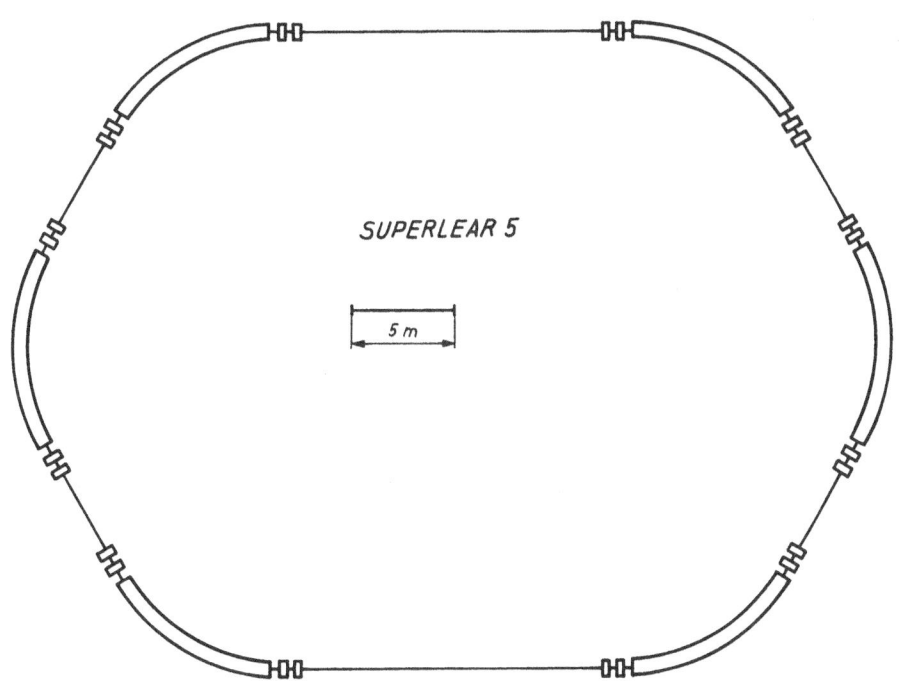

SUPERLEAR 5

5 m

Figure 4 : "Hexagonal Racetrack", SLEAR 5.

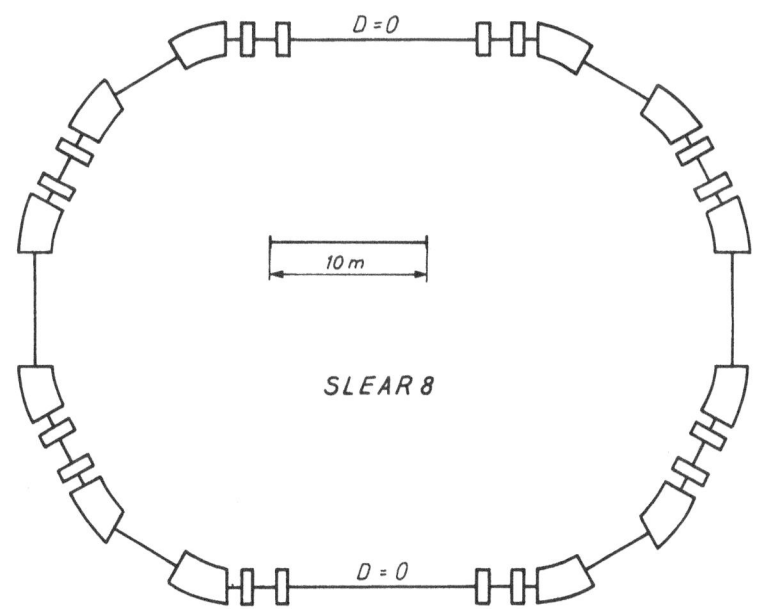

$D = 0$

10 m

SLEAR 8

$D = 0$

Figure 5 : "Twelve sided Racetrack", SLEAR 8.

Table 3 : Some machine configurations considered.

Machine Configuration	Bending magnets N°./length	Doublets N°.	Straight sections N°./free length long	medium	short
SLEAR 1	8/5 m	16	2/12 m	2/5 m	4/4 m
SLEAR 2	8/5 m	16	-	8/6 m	
SLEAR 3	8/5 m	8	4/11	4/4 m	-
SLEAR 5	6/7 m	12	2/15 m	4/6 m	-
SLEAR 8	12/3.3m	8	2/14 m	6/4 m	4/2 m

After a number of "cross-checks" including aperture and beam density limitations we have come to the conclusion that the structures discussed are viable alternatives to SLEAR 1 but that none has any clear-cut advantage. The SLEAR 1 geometry can be used as a convenient model, without too much loss of generality to work out performance limits, as it has been done in Ref. 7-9 (see also Table 1 above for a summary of performances).

LUMINOSITY ESTIMATES FOR A BOTTONIUM "PRE-EXPERIMENT" IN THE SPS

Following a suggestion of P. Dalpiaz, L. Dick, R. Klapisch, L. Montanet and others[1] a preliminary study has been made[1] of the luminosity obtainable with a $\bar{p}$-beam circulating in the SPS at about 40 GeV and the jet target of the UA6 collaboration[1]. Results assuming beam conditions similar to those of the 1986 collider runs are summarized in table 4 below, which indicates a luminosity of 5×10^{29} cm^{-2}sec^{-1}.

Improvements seem possible using the higher intensity available with the new antiproton source (factor 6), trying to obtain a better matching between beam- and target size (factor 2-3) and by pushing the target density. In this way a luminosity of say 3×10^{31} cm^{-2}sec^{-1} could be obtained after a certain amount of development. This is only a factor of 3 below the "production limit" of 10^{32} cm^{-2}sec^{-1} which is the best obtainable assuming that the flux of 10^7 $\bar{p}$/sec from AA-ACOL is completely consumed in the interaction with a total cross-section of 100 mbarn[9].

Table 4 : Performance estimate for a $\bar{b}$-b pre-experiment in the SPS assuming beam and target characteristics comparable to those of the 1986 collider runs.

Number of circulating antiprotons	10^{11}
Transverse beam size at target at $\approx$ 45 GeV. ($\sigma_h \times \sigma_v$)	3×2 mm^2
Momentum spread (bunched beam) $\sigma_{\Delta p/p}$	10^{-3}
Horizontal diameter (total) of jet.	3 mm
Effective density (taking account of imperfect overlap with beam).	10^{14} p/cm^2
Luminosity	5×10^{29} cm^{-2}sec^{-1}

As the antiprotons can in principle be distributed into many bunches or even a coasting beam can be used, some of the very stringent limitations pertaining to the collider mode may be absent or alleviated. Clearly a more detailed feasibility study of running the SPS in internal target mode in the bottonium range (beam energies of say 40-100 GeV) is necessary to arrive at more precise performance estimates.

CONCLUSIONS

A compact high luminosity antiproton ring working in the 1.5 to 10 GeV/c range looks feasible. Small circumference and hence efficient use of antiprotons become possible by use of superconducting magnets. Ultimate luminosities of 10^{32} cm^{-2} sec^{-1} in the internal target mode and 3×10^{30} cm^{-2} sec^{-1} with $\bar{p}$-p colliding beams at 7 GeV/c can be expected for a single ring scheme with head-on collisions. Strong phase-space cooling, a powerful RF system with superconducting cavities to make very short bunches and/or perhaps - a two ring scheme may allow us, also for the collider, to aim at $L \to 10^{32}$ cm^{-2} sec^{-1} which is the best obtainable with an antiproton production rate of 10^7/s.

A staged approach starting with a simple internal target machine looks then quite feasible. A strong RF-system advanced phase-space cooling and eventually a second ring could be added step by step to arrive at a powerful $\bar{p}$-p collider.

Sufficient flexibility in the lattice design exists to find a geometry that fits into an existing hall like the East Hall or the ISR service building 181.

Apart from being a valuable tool for particle physics, SLEAR could be an attractive "test bed" of new accelerator technology.

A bottonium pre-experiment running the SPS as a $\bar{p}$-storage ring at about 45 to 100 GeV and using the target of UAG looks feasible on first sight but needs detailed study to arrive at precise estimates of the performance obtainable with ACOL intensities.

REFERENCES

1. P. Dalpiaz, Charmonium and other onia at minimum energy, Proc. of 1st LEAR Workshop, Karlsruhe 1979, p. 111 (KfK report 2836, ed. H. Poth).
   P. Dalpiaz, Experimental possibilities of charmonium and bottonium spectroscopy, Proc. 2nd LEAR Workshop, Erice 1982, p. 725 (Plenum Press, New York & London, 1984 ed. U. Gastaldi and R. Klapisch).
   P. Dalpiaz, Possibilities of experimentation at Super-LEAR, Proc. 3rd LEAR Workshop, Tignes 1985, p. 441 (éditions Frontières, Gif-sur-Yvette 1985 ed. U. Gastaldi, et al.).
2. A. Martin, Some aspects of Quarkonium spectroscopy, Proc. 3rd LEAR-Workshop, p 321.
3. W. Buchmüller, Production and decays of quarkonium in $\bar{p}$p-collisions, Proc. 3rd LEAR Workshop, p. 327.
4. M. Macri, Formation of charmonium states in proton-antiproton annihilations, Proc. 3rd LEAR Workshop, p. 423.
5. C. Baglin et al, Plaidoyer for charmonium spectroscopy at Super-LEAR, Proc. 3rd LEAR Workshop, p. 441.
6. L. Montanet, Heavy flavour physics with a low energy $\bar{p}$p collider, Proc. 4th LEAR Workshop, Villars, Switzerland, September 1987, to be published.

7. E. Gianfelice, P. Lefèvre, D. Möhl, A Superconducting Low Energy Antiproton Ring (Super-LEAR), Proc. 3rd LEAR Workshop, p. 65.

8. P. Dalpiaz et al. : Physics at Super-LEAR, CERN/EP internal report 87-27.

9. R. Giannini, P. Lefèvre, D. Möhl, SLEAR, a superconducting anti-proton storage ring for p̄-p physics in the centre of mass energy range of 2 to 20 GeV., Proc. 4th LEAR workshop.

10. R. Billinge, Future possibilities and limitations of the CERN antiproton source, Proc. 4th LEAR Workshop.
    E. Jones, Initial performance of AAC, Proc. 4th LEAR Workshop.

11. A. Bernasconi et al. (UA6 collaboration), Proposal for the study of e^+e^-, γ, π^0 and hyperon production in p̄p reactions using an internal jet target at the SPS; Addendum CERN-SPSC internal report 86-35, Dec. 1986.

12. M. Macri (R 704 collaboration), A clustered H_2 beam, Proc. 2nd LEAR Workshop, p. 691.

13. See e.g. M. Sands, The Physics of electron storage rings, SLAC report 121 (1970).
    E. Keil, Beam-beam interaction in p-p storage rings, in: Theoretical aspects of the behaviour of beams, CERN report 77-13 (1977), p.314.
    S. Kheifets, Experimental observations and theoretical models for beam-beam phenomena, SLAC report 2700 (1981).
    B. Zotter, Experimental investigation of the beam-beam limit of proton beams, in Proc. Xth Conf. on High Energy Accel., Protvino, USSR (1977), p. 23.

14. E. Keil, Handy formulae for p-p and p-p̄ colliders, CERN/LEP/TH internal report 83-25.

15. L. Tecchio, Electron cooling at intermediate energies, Proc. 3rd LEAR Worskhop, p. 135.

16. U. Bizzarri et al., LEAR, Double-LEAR, Super-LEAR as collider, Proc. 2nd LEAR Workshop, p. 729.

17. J. Gareyte, W. Kubischta, P. Lefèvre, D. Möhl, A bottonium pre-expe-riment in the SPS using a stored p̄-beam and the cluster jet target of UA6; Rough luminosity estimates, CERN internal note, to be published.

PHYSICS WITH MEDIUM-ENERGY ANTIPROTON BEAMS—SUPERLEAR

Pietro Dalpiaz

Dipartimento di Fisica, Università di Ferrara and
INFN Gruppo di Ferrara, Italy

1. INTRODUCTION

As is well known, the CERN antiproton source[1] [the Antiproton Accumulator (AA)] is used for experiments at maximum energy with the CERN $p\bar{p}$ Collider[2] and at minimum energy with the Low-Energy Antiproton Ring (LEAR)[3].

From 1977, it was proposed to use the antiproton beams for the spectroscopy of charmonium states, in formation experiments of $c\bar{c}$ states, from $p\bar{p}$ annihilation, with the aim of determining the widths of all the $c\bar{c}$ states, tuning the antiproton beam energy[4]. This kind of experiment, which is important to test quantitatively QCD[5], is not feasible with e^+e^- colliders because in this case only the $J^P = 1^-$ particles are directly formed, since the e^+e^- interactions are dominated by one-photon exchange.

In $p\bar{p}$ interactions all J^P states are allowed in formation, but the background is much larger than that in e^+e^- interactions. To perform such experiments, it is necessary to have a high luminosity coupled with a narrow beam-momentum distribution and to use thin targets which preserve the intrinsic $\bar{p}$ beam-momentum definition given by the cooling techniques. These conditions were met[6] in the experiment R704 at the Intersecting Storage Rings (ISR); in fact the $\bar{p}$ beam was coasting in a ring of the ISR and interacted with a H_2 gas-jet target of a suitable density. This experiment has produced very interesting results[7] on the properties of $\chi_1(^3P_1)$ and $\chi_2(^3P_2)$, and gave some hints on the $^1P_1$ question. Unfortunately, data was taken for only 5 days before the ISR was closed, but from this experiment we learned that charmonium physics is accessible through $p\bar{p}$ annihilation.

A similar experiment, but with larger acceptance, is in construction at the Fermilab $\bar{p}$ Accumulator[8]. The data taking is expected to start in 1989.

At the third LEAR workshop, held in Tignes in 1985, Möhl presented[9] a project called SUPERLEAR (SL). It is a superconducting antiproton ring, which has a momentum range of 2 to 10 GeV/c. Such a machine can be used with a H_2 gas-jet target on a coasting $\bar{p}$ beam or as a collider with two coasting beams—one of protons and the other of antiprotons; with $p\bar{p}$ annihilation it can reach a centre-of-mass energy of 2.5 GeV $< E_{cm} < 20$ GeV, so, in principle, charm and beauty spectroscopy appear to be accessible.

2. SUPERLEAR

The SL[9] machine has a circumference of 120 m with 4 bending superconducting magnets of 4 T, an injection momentum of 3.5 GeV/c, and a momentum range of 2 to 10 GeV/c.

Three versions of SL will be considered:

SL1: one antiproton ring, of 2 to 10 GeV/c, with a H_2 gas-jet target which can reach a luminosity of $L = 10^{32}$ cm^{-2} s^{-1} with $\Delta p/p = 10^{-3}-10^{-4}$;

SL2: a proton–antiproton collider, up to 10 + 10 GeV/c, with one ring which can reach a luminosity of $L = 3 \times 10^{30}$ cm^{-2} s^{-1};

SL3: a proton–antiproton collider (2 separated rings) with powerful cooling, where a luminosity of 1.5×10^{31} cm^{-2} s^{-1} can be reached.

3. PHYSICS AT SUPERLEAR

We should like to show that SL could provide original opportunities to test fundamental questions of nature[10]. In particular, we discuss:

i) the possibility of performing new tests on CP and CPT;

ii) the possibility of studying (c$\bar{c}$ and b$\bar{b}$) not directly accessible to e$^+$e$^-$ colliders, and to measure their widths and decay angular distributions;

iii) the possibility of studying the production and decay properties of hadrons in the new quark environment (heavy flavours, charmed hyperons);

iv) the possibility of searching for exotic states predicted by QCD (gg, gq$\bar{q}$).

3.1 CP and CPT

With SL2 one could envisage studying CP and CPT violation in a new system, namely for the $\Lambda\bar{\Lambda}$ produced in exclusive reactions. With the luminosity of SL2, 3×10^6 $\Lambda\bar{\Lambda}$ per day could be produced in a symmetric way. With such $\Lambda\bar{\Lambda}$ production we can study the following: CPT, by measuring the Λ and $\bar{\Lambda}$ lifetimes; CP, by measuring for reactions $\Lambda \to \pi^-$p and $\bar{\Lambda} \to \pi^+\bar{p}$ the branching ratios[11] and their decay symmetry in the decay parameters[12].

3.2 Heavy quarkonia (c$\bar{c}$, b$\bar{b}$)

With SL1 and a suitable large-acceptance magnetic spectrometer equipped with a high-resolution γ detector and able to identify events with e$^+$e$^-$ in the final state it is possible:

i) To discover c$\bar{c}$ states not directly accessible to e$^+$e$^-$ colliders: $^1P_1$, $^1D_2$, $^3D_2$, and η'_c.

ii) To measure accurately the widths of the c$\bar{c}$ states, with a resolution of $\Delta M \sim 50$ keV, which are calculable in QCD[5] and constitute a unique quantitative test of QCD[13].

iii) To measure the production and angular distribution of reactions such as p$\bar{p} \to \chi \to J/\psi + \gamma \to$ e$^+$e$^-$ + γ, where the angular distribution can be related via the single-quark radiation model to the anomalous magnetic momentum of the quarks[14].

iv) To repeat part of the c$\bar{c}$ physics on the b$\bar{b}$ region with SL3. However, even with SL3, the feasibility of these experiments may become marginal for a b$\bar{b}$ system, because[15] $\sigma(\text{p}\bar{\text{p}} \to \text{b}\bar{\text{b}})/\sigma(\text{p}\bar{\text{p}} \to \text{c}\bar{\text{c}}) \approx 10^{-3}-10^{-4}$. Progress in electron cooling could improve the SL3 luminosity and make possible this experimentation.

3.3 SUPERLEAR as a heavy-flavour hadron factory

The J/ψ was discovered in 1974 but so far only one charmed baryon, the Λ_c, is known. The decay properties of the other multiplet members ($J^{PC} = 1/2^+, 3/2^+$) with cud, cuu, cdd, csd, csu, etc. quarks would shed light on the behaviour of quarks and diquarks in a heavy-quark environment.

The QCD predictions for exclusive reactions of p$\bar{p}$ annihilation are:

$$\sigma(\Lambda_s\bar{\Lambda}_s) \approx 10\ \mu b, \quad \sigma(\Lambda_c\bar{\Lambda}_c) \approx 1\ \mu b \quad \text{and} \quad \sigma(\Lambda_b\bar{\Lambda}_b) \approx 10\ \mu b.$$

This cross-section gives very high rates with SL2 and SL3. With collider operation it is easier to detect large-p_T events in the presence of leptons than it is in fixed-target experiments.

3.4 Glueball, hybrid, and exotic states

Quantum Chromodynamics predicts several glueball (gg) and some exotic states such as $c\bar{c}g$ ($4.3 < m_{c\bar{c}g} < 5.8$ GeV), and $b\bar{b}g$ ($10.7 < m_{b\bar{b}g} < 11.6$ GeV).

The mass of glueballs is probably in the range of LEAR. With SL1, SL2, and SL3, we cover the range of exotic states. The difficulty encountered in the search for $q\bar{q}g$ states is to obtain a clear signature in order to identify the events. One solution could be the search for states such as $J^{PC} = 1^{-+}$ not permitted in the $q\bar{q}$ system. As trigger we can look for reactions $c\bar{c}g \to J/\psi + \ldots \to e^+e^- \ldots$.

The detector must have a large acceptance and be able to perform a magnetic analysis of the momentum of the particles produced in $p\bar{p}$ annihilation.

4. CONCLUSIONS

Important aspects of CP, CPT, and charmonium spectroscopy could be examined in a new way with a medium-energy $\bar{p}$ machine such as SUPERLEAR. With such an experimental facility, the way may be open for bottonium spectroscopy.

REFERENCES

1. E. Jones, 'Physics with Antiprotons at LEAR in the ACOL Era', Proc. 3rd LEAR Workshop, Tignes, 1985, U. Gastaldi, R. Klapisch, J.M. Richard and J. Tran Thanh Van, eds., Editions Frontières, Gif-sur-Yvette (1985), p. 25, and references therein.
2. C. Rubbia, *Rev. Mod. Phys.* 3: 57 (1985) and references therein.
3. 'Proc. Joint CERN–KfK Workshop on Physics with Cooled Low Energy Antiprotons', H. Poth, ed., KfK 2836, Karlsruhe (1979).
   P. Dalpiaz, *Nucl. Phys.* A446: 219 (1985).
4. P. Dalpiaz, 'Electromagnetic Annihilation in a Low Energy $p\bar{p}$ Colliding beam', CERN–$p\bar{p}$ note 06, $p\bar{p}$ First Study Week, CERN (1977).
   P. Dalpiaz, 'Charmonium and other Onia at Minimum Energy', KfK 2836, Karlsruhe (1979), p. 111.
   P. Dalpiaz, 'Proc. 5th European Symposium on Nucleon–Antinucleon Interactions', CLEUP, Padova (1980), p. 711.
   P. Dalpiaz, 'Research Programme at LEAR', LA 8775C, Los Alamos (1981), p. 300.
5. R. Barbieri, R. Gatto and E. Remiddi, *Phys. Lett.* 106B: 497 (1981).
   E. Remiddi, *in* 'Physics at LEAR with Low Energy Cooled Antiproton beams', U. Gastaldi and R. Klapisch, eds., Plenum, NY and London (1984), p. 711.
   A. Martin, same Proc. as Ref. 1, p. 321 and references therein.
6. P. Dalpiaz, V. Gracco and M. Macri, CERN/ISRCI 79–23 (1979).
   M. Macri, same Proc. as Ref. 1, p. 423.
7. C. Baglin et al., *Phys. Lett.* B172: 455 (1986).
   C. Baglin et al., *Phys. Lett.* B171: 135 (1986).
8. V. Bharadway, J. Griffin, S. Holms, W. Kells, J. MacCarthy, J. Peaples, P. Rapidis, D. Yang, R. Calabrese, P. Dalpiaz, P. Ferretti, E. Luppi, F. Petrucci, M. Savriè, M. Marinelli, M. Mattera, F. Tomassini, V. Valbusa, G. Borreani, R. Cester, E. Menichetti, S. Palestrini, N. Pastrone, G. Rinaudo and L. Tecchio, Fermilab proposal 760 (1985).
9. D. Möhl et al., same Proc. as Ref. 1, p. 83.
10. P. Dalpiaz, same Proc. as Ref. 1, p. 441.
    P. Dalpiaz, R. Klapisch, P. Lefèvre, M. Macri, L. Montanet, D. Möhl, A. Martin,

J.M. Richard, H.J. Pirner and L. Tecchio, preprint CERN/EP 87–27 (1987), to be published *in* 'The Elementary Structure of Matter', E. Aslanides, N. Boccara and J.M. Richard, eds., Springer, Heildelberg (1988).

11. Ling Lee Chau, *Phys. Rep.* C95: 62 (1983).

12. J. Donoghue, *Phys. Lett.* B179: 319 (1986).

13. W. Büchmuller, Quarkonium spectroscopy, *in* 'Fundamental Interactions in Low-Energy Systems' P. Dalpiaz, G. Fiorentini and G. Torelli, eds., Plenum, NY (1985), p. 233, and references therein.

14. M.G. Olsson, same Proc. as Ref. 1, p. 341.

15. W. Büchmuller, same Proc. as Ref. 1, p. 327.

e^+e^- HEAVY FLAVOUR FACTORIES BASED ON NEW IDEAS

Guy Coignet

LAPP

BP 909

74019 Annecy–Le-Vieux, Cedex

France

INTRODUCTION

Today, High Energy Physics is interpreted in terms of the Standard Model (electroweak interaction) and of the Quantum Chromodynamics (strong interaction). Until now, the Standard Model (S.M.) has been very successful and no deviation from its predictions have been experimentally observed. On the other hand, the S.M. is unsatisfactory from a theoretical point of view since it needs 21 input parameters, it does not explain the mass generation problem and it does not explain the origin of CP and T violation in K^O meson decay.

Two different alternatives can then be considered to improve our knowledge: either go to higher energy to produce directly more massive particles or use machines with improved luminosity and detectors to perform more precise measurements at previously reached energies.

We will consider here the second approach for an improved study of charm (c) and bottom (b) quarks, as well as tau (τ) lepton. The main physics result already obtained or which could be obtained in the near future will first be surveyed. At the same time, we will also try to estimate the physics potantial of a new e^+e^- collider, with a luminosity increased by typically two orders of magnitude. Two new ways to possibly reach this luminosity improvement will then be presented : circular colliders with multi-bunch operation and independent e^+ and e^- rings, or linear colliders with special emphasis on an accelerator complex which could be used either as a $c\bar{c} - \tau\tau$ factory or as a $b\bar{b}$ factory.

MOTIVATION FOR e^+e^- HEAVY QUARK FACTORIES

The arguments in favour of such high luminosity factories have been advocated by many people and has been reviewed recently in various workshops[1-5]. We will only recall here what are the main reasons for such a need in luminosity increase.

Figure 1 shows the total cross section for $e^+e^- \rightarrow$ hadrons as a function of $\sqrt{s}$, the center of mass energy. On top of the $1/s$ general trend of the electromagnetic interaction, one observes a few anomalies. First, there are the two $c\bar{c}$ bound state resonances $J/\psi, \psi'$ followed by a cross section increase around $\sqrt{s} = 3.6$ and 3.8 GeV corresponding to the crossing of the

385

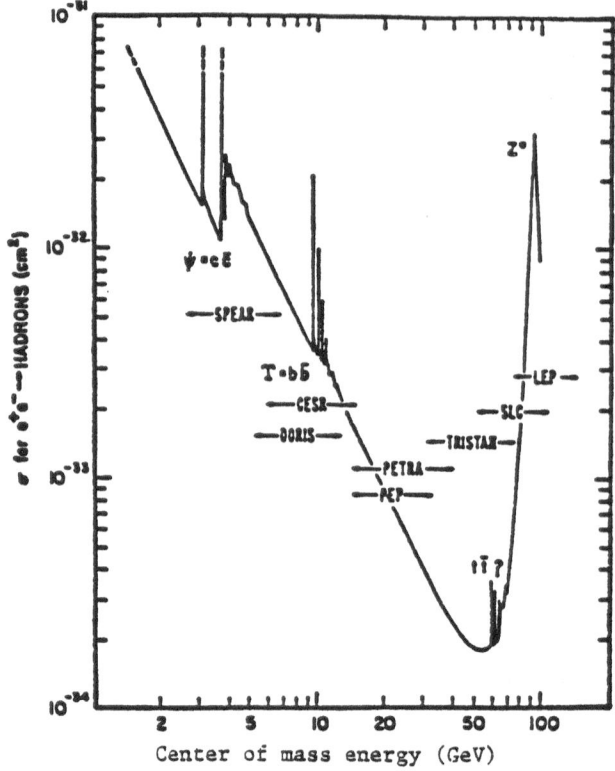

Fig.1. The cross section for $e^+e^- \rightarrow$ hadrons in the
2 to 100 GeV center of mass energy domain

$\tau\bar{\tau}$ and $c\bar{c}$ production thresholds. The storage rings, used until now including SPEAR (SLAC), the only one presently in exploitation in this energy
domain, have been working with luminosities in the 10^{30} cm^{-2}s^{-1} range : a
few 10^6 J/ψ and a few 10^5 D$\bar{\text{D}}$ and $\tau\bar{\tau}$ events per experiment have then
been recorded.

The spectroscopy of the ψ system has been well studied, but a lot of
interesting questions are still unanswered. Among them, the search for glueballs in radiative and hadronic decays of the J/ψ has been inconclusive,
mainly because of statistics limitation. General properties of the lowest
lying $c\bar{q}$ (D$^+$,D^0,D$_s$ and D^{*+},D^{*0} and D$_s^*$) are known, but the $c\bar{q}$ 1P states
are still not found. The baryons containing c quarks are very poorly known
and measurement of their lifetimes and their partial decay rates are very
much needed for QCD computations in weak decays. Measurements of the V_{dc}
and V_{sc} Cabibbo-Kobayaski-Maskawa matrix elements, rare decays such as
Cabibbo suppressed decays, family number violation decays, and D^0-$\bar{\text{D}}^0$
mixing limits would also gain from a large increase in statistics.

Concerning tau lepton physics, the case was independently discussed[6]
at this school and will not be repeated here. I just would like to mention
a few topics for which a significant increase in statistics would be mandatory : precise determination of the τ weak decay parameters ; improved
limit on the ν_τ mass ; search for lepton flavour violation ; search for
second class currents and for any unexpected effect in rare decays.

It has to be noted that a new machine, BEPC (Bejing) covering this
energy domain with a luminosity L $\simeq 1.5\times10^{31}$ cm^{-1}s^{-1}, is expected to come
into operation at the end of 1989.

A pattern similar to the charm one is reproduced Figure 1, near $\sqrt{s} \simeq 9\text{-}12$ GeV, with the crossing of the $b\bar{b}$ production threshold. Figure 2 shows in more details the measured hadronic cross section for the production of the first four $T(b\bar{b})$ bound states. Two other resonances, $T(5S)$ and $T(6S)$ have also been observed. These measurements, performed at DORIS (Hamburg) and CESR (Cornell) working with typical luminosity of $L = 10^{31} cm^{-2} s^{-1}$, led to some 0.5×10^6 $T(1S)$ and 10^5 $B\bar{B}$ events per experiment. The measurements of the T states, and of the X_b and X'_b states have provided very important constraints for potential models used to describe the quarkonium system. Nevertheless, the discovery of the still missing η_b and h_b states would require much higher luminosities. The search for new particles [Higgs,axion...] in T decays, as well as better determination of the strong coupling constant α_s at low Q^2, would also gain from a luminosity increase.

The $T(4S)$ resonance is much wider than the three first states (see Figure 2 and Table 1), since it decays into a pair of B mesons $B_u^+(b\bar{u})B_u^-(b\bar{u})$ or $B_d^0(b\bar{d})\bar{B}_d^0(b\bar{d})$ produced nearly at rest. The $B_s(b\bar{s})$ meson, for which there are indications for threshold production crossing just below the $T(5S)$, has till not been directly seen. The $B_c^+(b\bar{c})$ meson and the b flavoured (bqq) baryons are still to be observed and studied.

The study of B_u and B_d mesons and of their weak decays has been started. The B^0, B^+ mass have been determined with precision, owing to the good energy resolution (4-10 MeV) of the machines and the fact that only mono-energetic B and $\bar{B}$ mesons are produced in the $T(4S)$ decay. Only 10% of all the exclusive channels, mainly $B \to Dn\Pi$, $D^*n\Pi$, ψK, have been reconstructed. Some inclusive decays and semi-leptonic decays have also been measured. Together with the large B meson lifetime measurements performed at PEP and PETRA, they allowed the first estimates of the Cabibbo-Kobayaski-Maskawa matrix elements V_{cb} and V_{ub} which represent the coupling of the b quark to charged weak current. A precise determination of these quantities needs separate lifetime measurements for $B^\pm$ and $B^0, \bar{B}^0$ mesons as well as separate branching ratio measurements. This emplies that a future high luminosity $b\bar{b}$ factory must have the possibility to work at $\sqrt{s} \geq 13\text{-}15$ GeV

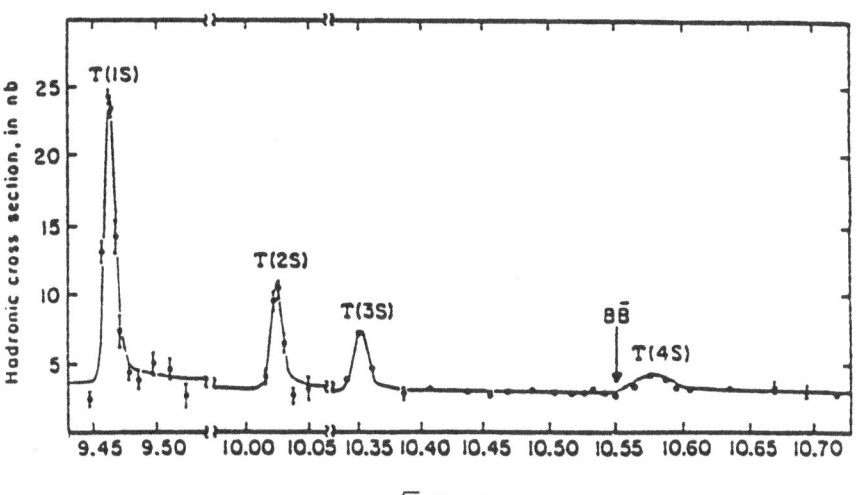

Fig.2 The cross section for $e^+-e^- \to$ hadrons
in the upsilon energy range.

Table 1. Masses of the bound $b\bar{b}$ and open b systems [from "Review of Particles Properties", Phys. Lett. 107B, 1986]

	Mass (GeV)	Full Width (MeV)		Mass (GeV)
T(1S)	9460.0 ± 0.2	0.043 ± 0.003	B^+_u, B^-_u	5271.2 ± 3.0
T(2S)	10023.4 ± 0.3	0.030 ± 0.007	B^o_d, $\bar{B}^o_d$	5275.2 ± 2.8
T(3S)	10355.5 ± 0.05	$0.012 \begin{smallmatrix}+0.010\\-0.004\end{smallmatrix}$	B^o_s, $\bar{B}^o_s$	≈ 5400
T(4S)	10577 ± 4	24 ± 2		
T(5S)	10865 ± 8	110 ± 12	B^+_c, B^-_c	≈ 6600 (estimate)
T(6S)	11019 ± 9	79 ± 16		

such that B meson path length before decay would be large enough to be detected in a special vertex detector and then the reconstruction efficiency
increased. Rare decays allowed by b → u transitions, with typical
10^{-3}–10^{-4} branching ratios, and possible flavour changing neutral current
decays, with typical 10^{-4}–10^{-6} branching ratios, which are sensitive to
m_t, the top quark mass, or to new mass scales, could also be investigated
within these conditions.

Another topic, B^o–$\bar{B}^o$ maxing, has recently obtained considerable attention. It may proceed via two W exchange (box diagrams), and substantial
mixing has been predicted in the B^o_s–$\bar{B}^o_s$ system. Mixing was first experimentally observed by the UAl group[7] at the SppS (CERN) as an excess of like-
sign dimuons attributed to primary semi-leptonic decays of B. The measure-
ment can be expressed as $\chi = \dfrac{BR(b \to \bar{B}^o \to B^o \to \mu^+X)}{BR(b \to \bar{B}^o \to \mu^{\pm}X)} = 0.121 \pm 0.047$ for B_d
or B_s without distinction. A surprising result came from the ARGUS experi-
ment at DORIS which observed[8] an unambigous mixing evidence in the B^o_d–$\bar{B}^o_d$
system. In the framework of the Standard Model the value obtained for
$\chi_d = 0.17 \pm 0.05$ is fully consistent with other results of e^+e^- experiments
and of UAl. This result imply that m_t is large ($m_t \geq 50$ GeV) or that the
ratio V_{ub}/V_{cb} is larger than presently believed. Better precision on χ_d
and χ_s would then be obtained with a high luminosity $b\bar{b}$ factory allowing
to study B_d separatly from B_s, by varying the energy from T(4S) to T(5S).

Another important issue concerns the possible detection of CP violation in B decays, which can be due to mixing, final state interactions or
both. The straightforward way to look for CP violation would be to look
for the like-sign lepton asymmetry in double semi-leptonic decays. In the
Standard Model, a 10^{-3}–10^{-4} asymmetry is expected and more than 10^9 B^o–$\bar{B}^o$
events would be required to observe it. The search for CP violation looks
more promising[9] when looking at specific channels with small branching
ratios (10^{-3}–10^{-4}) but where the asymmetry due to CP violation could be
large (1-10 %). One has to distinguish the cases where B and $\bar{B}$ decay into
two final states with the same CP eigenvalues (i.e. $\pi^+\pi^-$, $D\bar{D}$) for which a
simultaneous double tagging is necessary, from the case where B and $\bar{B}$
decay into different CP eigenvalues (i.e. $B^o \to D^+\pi^-$ and $\bar{B}^o \to D^-\pi^+$) for
which double tagging is in principle not necessary. The possibility to run
at $\sqrt{s} \leq 13$-15 GeV is also worth considering for this type of CP violation
study. Taking into account realistic tagging and reconstruction efficiencies, which are very much dependent upon the accelerator working conditions and the associate detector, it has been estimated[10] that a minimum
of 10^7–10^8 events would be required to detect CP violation.

Table 2. Rough estimates of B$\bar{\text{B}}$ production rates at various machines.

Accelerat.	$\sqrt{s}$ * (GeV)	$L(10^{31}$ cm^{-2}s^{-1})	L/day (pb^{-1})	$\sigma_{b\bar{b}}$ (nb)	Signal / Backgr	# B$\bar{\text{B}}$/year (200 full days)
Machines in operation or nearly completion						
DORIS	10.6	2	1	1	0.25	1.5×10^5
CESR now	10.6	3.5	1.5	1	0.25	3×10^5
1988	"	18	7	"	"	1.5×10^6
VEPP-4 now	10.6	2	0.9	1	0.25	1.5×10^5
impr.	"	30	12.5	"	"	2.5×10^6
PEP	25	6	2.5	0.05	0.09	2×10^4
SLC	93	$0.02 \to 0.5$	$0.01 \to 0.20$	6.5	0.22	$0.1 \to 2 \times 10^5$
LEP	93	1	0.5	6.5	0.22	5×10^5
Sp$\bar{\text{p}}$S [TeVI]	540 [2000]	0.5	0.2	1×10^4	1.5×10^{-4}	4×10^8 $\downarrow_{\to 10^6}$ after cuts
Machines under discussion						
KEK	10.6	10	4	1	0.25	8×10^5
NPEP	10.6	50	20	1	0.25	4×10^5
	14	200	80	0.1	0.09	1.5×10^6
SIN	10.6	$50 \to 500$	$20 \to 200$	1	0.25	$4 \to 10 \times 10^6$
	13.5	$5 \to 100$	$2 \to 40$	0.12	0.09	$0.5 \to 10 \times 10^5$
Lin.Collid. (U.A+G.C)	10.6	100	40	1	0.25	8×10^6
	15	10^3	400	0.1	0.09	8×10^6
Lin.Collid. (D.C et al)	10.6	10^3	400	1	0.25	8×10^7
SSC [LHC]	40×10^3 [20]	10	4	2×10^5	1.5×10^{-3}	1.5×10^{11} $\downarrow_{\to 10^8}$ after cuts

*For the e$^+$e$^-$ machines quoted here, typical fractional energy spread is $\sigma_{\sqrt{s}}/\sqrt{s} \simeq 0.4\text{-}1\times10^{-4}$ except for SLC ($\sigma_{\sqrt{s}}/\sqrt{s} \simeq 2\times10^{-3}$) and the Linear Collider (U.A+G.C) working at $\sqrt{s} = 15$ GeV ($\sigma_{\sqrt{s}}/\sqrt{s} \simeq 1.3\times10^{-2}$).

Experiments at PEP and PETRA (29 and 35 GeV respectively) have provided complementary information on D mesons, τ leptons and B mesons, taking mainly advantage of the large boost given to these particles. When increasing $\sqrt{s}$ up to TRISTAN (KEK) energy, i.e. $\sqrt{s} \simeq 50$ GeV, the total cross section as well as heavy flavour cross sections reach a minimum (Figure 1) unless the top-antitop production threshold is crossed. Taking into account the top mass limit[11] $m_t > 44$ GeV, obtained by UA1, we will disregard this eventuality.

At the Z^0 pole, which will be reached soon by SLC (SLAC) and LEP (CERN), the cross section increases by more than two orders of magnitude; it will then be a nice source of heavy flavours[12] since it decays into b$\bar{\text{b}}$, c$\bar{\text{c}}$ and τ$\bar{\text{τ}}$ with typical branching ratios of 15 %, 11,5 % and 3 % respectively. If $m_t < m_{Z^0}/2$ it will also decay to t$\bar{\text{t}}$ pairs in roughly 1 or few %

of the cases, depending strongly on the top mass value. If $m_t > m_{z^0}/2$, e^+e^- collisions would still be a clean way to study t quark physics[13]. SLC and LEP will contribute to improve our knowledge in τ,c,b and possibly t sectors, especially for lifetime measurements, B^0-$\bar{B}^0$ mixing and sensitive tests of the electroweak theory by forward-backward asymmetry measurements. However, a few 10^6 Z^0 per year will be produced at LEP, and a Z^0 machine would only compete with the heavy flavour factories discussed here if its luminosity was in the range of $\simeq 10^{33} cm^{-2} s^{-1}$.

This is seen in Table 2 which gives rough estimates of $B\bar{B}$ production rates at the various machines previously mentioned and at the ones to be discussed next. High energy hadron-hadron colliders are also quoted for comparison. One has to remember that the signal to background ratio is an important parameter and that the final number of useful $B\bar{B}$ events will also be detector dependent.

NEW CIRCULAR COLLIDERS AS B-$\bar{B}$ FACTORY

For an e^+e^- storage ring, the luminosity L can be written[14] :

$$L = \frac{n\ I_b^2}{4\pi\ e^2 f_o\ \sigma_x^*\sigma_y^*}$$

with   n = number of bunches per beam, I_b = bunch current,
       e = electron charge, f_o = revolution frequency, σ_x^* and σ_y^*
       standard deviation of the horizontal and vertical beam
       dimensions at the interaction point.

The horizontal (x) or vertical (y) emittance is related to the $\beta_{x,y}^*$ function by

$$\varepsilon_{x,y} = (\sigma_{x,y}^*)^2/\beta_{x,y}^*.$$

Taking into account the space charge effects of opposite bunches which causes a tune shift ΔQ, the maximum luminosity is obtained for

$$L \propto \frac{n\ \varepsilon_x}{\beta_y^*}\ \Delta Q_{max}^2.$$

The maximum horizontal emittance is limited to $< 10^{-6}$ m, the two remaining parameters are then n and β_y^*, since $\Delta Q_{max} \simeq 0.025$ in classical storage rings.

At CESR, a luminosity upgrade programme[15] has been undertaken and first improvements have already been obtained. By going to seven bunches per beam crossing at only one interaction, and by reducing β_z^* as well as the bunch length to 1 cm, a luminosity of 1.5 to $2 \times 10^{32} cm^{-2} s^{-1}$ may be expected sometime in 1988. No improvement programme is forseen for DORIS. Let's just mention here the possible transformation of two existing machines which have been contemplated : The use of PEP as a One Interaction Region Collider[16], with low $\beta_y^* = 4$ cm, working at $\sqrt{s} = 14.5$ GeV with $L = 10^{32} cm^{-2} s^{-1}$; the Tristan accumulator ring upgrade[17] for B physics which, in multibunch operation and with micro β would be able to reach $L \simeq 10^{32} cm^{-2} s^{-1}$.

A much more interesting proposal for a really new $B\bar{B}$ meson factory has been made[14,18] for SIN (Zurich). In this design, multibunches (up to n = 10) of positrons and electrons circulate into two separate rings with combined insertions in the two interaction regions (Figure 3). The main advantage of the double ring rests in the independent control of the two beams, with possibly different n, and in the beam-beam effects localized only at the crossing point. The basic difficulty of this scheme is the

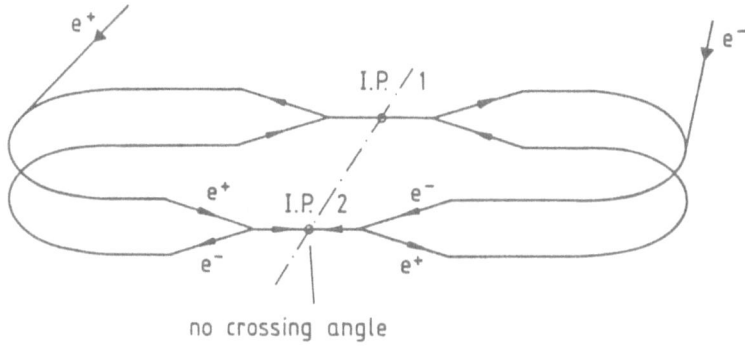

no crossing angle

Fig.3 Sketch of the double ring with head on collision.

vertical beam separation for which an electrostatic solution has been chosen. The arcs consist mainly of periodic FODO cells containing two bending
magnets and two quadropoles each, and the emittance is increased to the
required value with wiggler magnets. Eight 5-cell RF cavities per ring,
working at an accelerating frequency f_{rf} = 500 MHz, are used. With 485 mA
current per beam, an rf power of P_{rf} = 1.6 MW per ring is required.

The layout of the proposed B-Meson Factory is shown in Figure 4.
Electrons and positrons are accelerated up to 200 MeV in two linac structures. They are then accumulated and compressed in a small storage ring to
provide a very intense single bunch. The two bunches are next accelerated

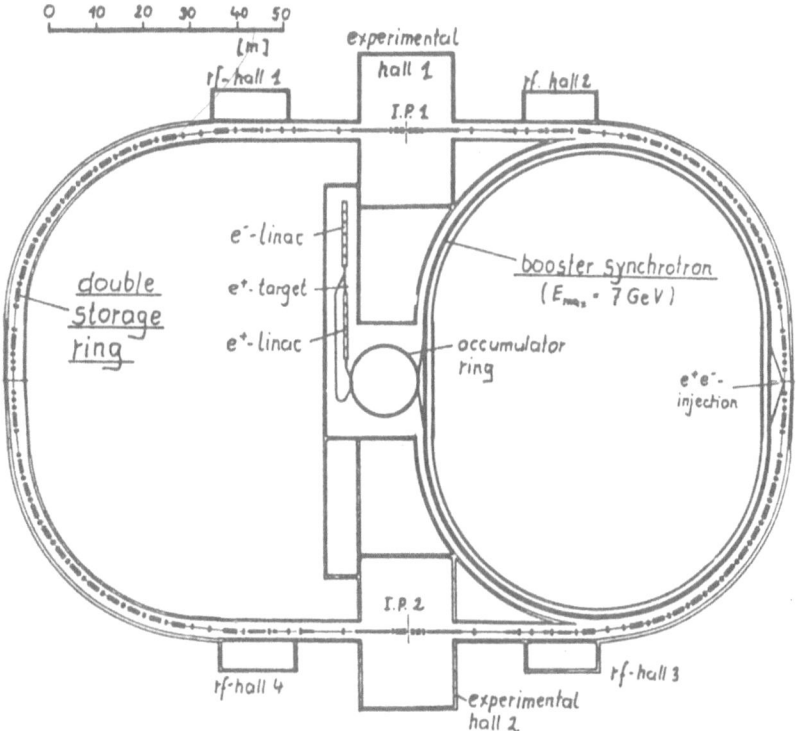

Fig.4 The B-Meson Factory with injector and booster synchrotron.

in a booster synchrotron up to the operating energy before injection in the two main rings. The variation of the collider luminosity as a function of the beam energy $E = \sqrt{s}/2$ is shown in Figure 5 (version 1). One sees that the luminosity reaches $L = 5 \times 10^{32} \text{cm}^{-2} \text{s}^{-1}$ at T(4S) and is larger than $10^{32} \text{cm}^{-2} \text{s}^{-1}$ up to $\sqrt{s} \simeq 13$ GeV. More recently, K.Wille has considered[19] further improvements to upgrade the luminosity : by increasing the tune shift from $\Delta Q = 0.025$ to $\Delta Q = 0.050$, by reducing β_y^* from 3 cm to 1 cm and by increasing the RF power from 1.6 MW to 2.5 MW per ring, a peak luminosity of the order of $L = 5 \times 10^{33} \text{cm}^{-2} \text{s}^{-1}$ seems possible to achieve. This luminosity improvement would probably be done in steps, as shown in Figure 5.

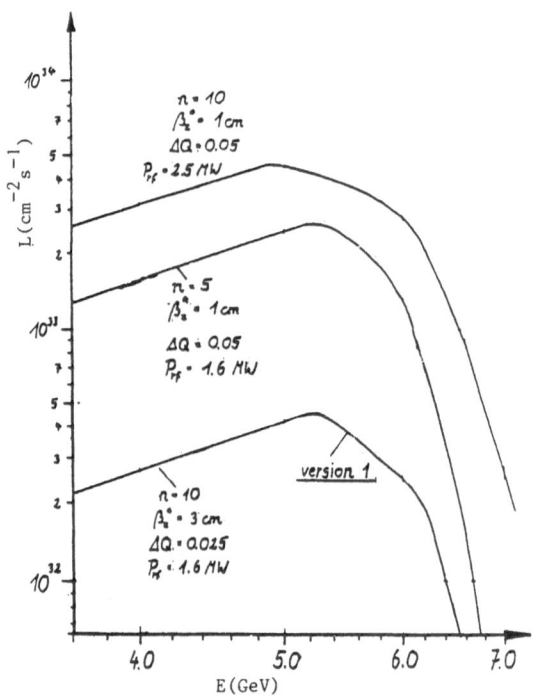

Fig.5 Luminosity as a function of beam energy for different operating conditions of the proposed SIN B-Meson Factory.

Following the same approach, the transformation of PEP in NPEP, a two ring multibunch colliding beam facility, working at $L = 2 \times 10^{33} \text{cm}^{-2} \text{s}^{-1}$ over a wide energy range, $4.5 < E < 15$ GeV, has also been considered[20].

In a linear collider, at the interaction point of two round beams with energy $E = \gamma \, m_e \, c^2$, the luminosity is[21]

$$L = \frac{N^2 \, n \, f \, H}{4\pi \, \sigma_t^2} = \frac{N^2 \, n \, f \, H}{4\pi \, \sigma_t^2 \, \varepsilon_n \, \beta^*}$$

where N = number of particle per bunches ; n = number of bunches per pulse ; f = average pulse repetition rate ; $\sigma_t (\sigma_z)$ = transverse (longitudinal) standard deviation of gaussian bunches ; H = "pinch factor" (which varies from 1 to 6); ε_n = transverse invariant emittance given by $\varepsilon_n = \gamma \, \varepsilon_t = \gamma \, \sigma_t^2/\beta^* n$; β^* = β-value at the crossing point.

The power of each beam is then $P = N \, n \, f \, E$.

The focusing effect produced by one bunch acting on the particles in the other bunch depends on the disruption parameter D. When the particles pass through the opposite bunch they emit synchrotron radiation characterized by the beamstrahlung parameter δ. Typically, the energy spectrum of the electron-positron collisions has a peak close to $\sqrt{s} = 2E$ and a tail extending down to zero. Three parameters can be used to describe the average electron (positron) energy spectrum : $\langle \varepsilon \rangle = \langle E' - E/E \rangle$, the average fractional energy loss ; N_γ the average number of radiated photons ;

$\sigma_{\sqrt{s}} = [(\sqrt{s} - \langle \sqrt{s} \rangle)^2]^{1/2}$ the r.m.s. C.M. energy spread. They can be expressed as function of δ and Y, where $Y = E_c/E$, and E_c is the average photon critical energy. When $Y \ll 1$, quantum effects are not important and classical formulae can be used. In a linear collider, once five parameters have been chosen, all other quantities are fixed.

A conceptual design of a multipurpose high luminosity heavy flavour factory has been proposed[22] by U.Amaldi and myself. The accelerator complex, shown in Figure 6, is based on a superconducting linear collider with recirculators and damping rings. Its energy could be varied over a wide range. After production, electron and positron bunches are accelerated in two linear structures by S.C. radiofrequency cavities with accelerating gradient $G = 7$ MV/m, quality factor $Q = 5 \times 10^9$ and $f_{rf} = 350$ MHz (LEP 200 type). In the first stage, after three recirculations, they can reach up to $\simeq 2.5$ GeV before injection into optimized damping rings in which the required invariant emittance, 2×10^{-6} m, will be produced. It has been computed that this may be achieved by putting, for instance, five bunches at a distance $\lambda_{rf} = 85$cm separated by about 10m in four damping rings. Each ring contains 8×5 bunches and a refined feed-back system as well as fast kickers are needed. After extraction, electrons and positrons can collide at $\sqrt{s} = 3$ to 5 GeV, with a luminosity roughly a hundred times larger than previously obtained in $\tau\bar{\tau}$ and $c\bar{c}$ factories. A possible list of parameters is given in Table 4. The bunch dimensions and intensity are similar to the nominal SLC ones, while the luminosity gain is mainly achieved through the higher values of f and D (and then H), and the lower ε_n value. The quoted energy resolution is due to beamstrahlung only and finally, the energy resolution will be dominated by the damping ring energy spread, i.e. $\sigma_{\sqrt{s}} = 3$ MeV.

This assumes that positrons and electrons are produced with sufficient flux ($> 10^{14}$s^{-1}). Present electron sources would be able to provide the necessary current, but recent developments[23] at Stanford University and Los Alamos are very encouraging. By irradiation of a photocathode by a laser, it seems that it would be possible to produce $N \geq 10^{10}$ electrons/bunch of a few MeV with $\varepsilon_n \simeq 3 \times 10^{-6}$m, and thus to avoid, at

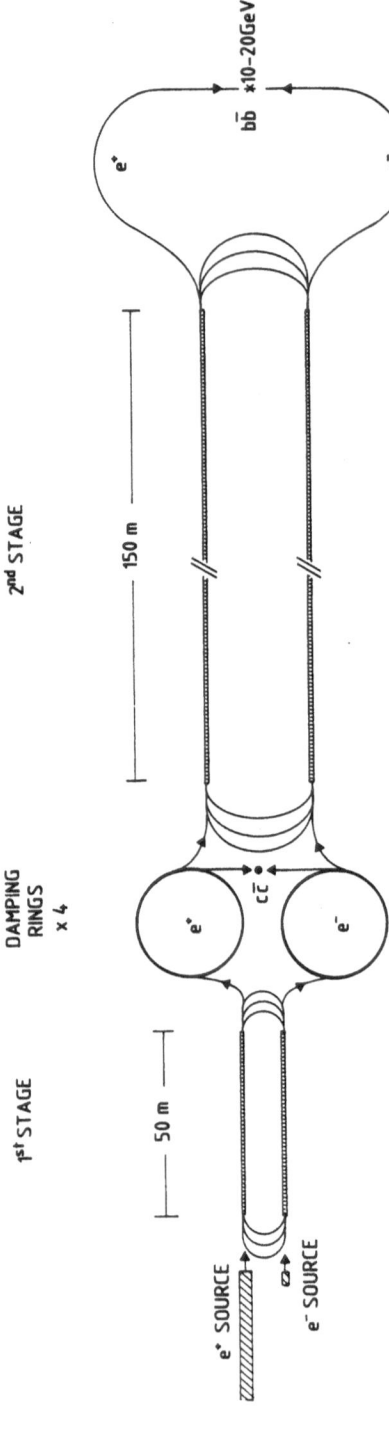

Fig. 6. Schematic drawing of the Superconducting accelerator complex which can be used as a $c\bar{c}$–$\tau\bar{\tau}$ factory or as a $b\bar{b}$ factory with adjustable energy.

Table 4. Parameters for a tau and charm factory around (2+2)GeV.

Symbol/Unit	Values	Symbol/Unit	Values
E/GeV	$1.5 - 2.5^{*}$	β^{*}/mm	5
L/cm^{-2}s^{-1}	3×10^{32}	σ_t/μm	1.6
P/MW	0.13	E_c/MeV	0.17
f/kHz	10	Y	8.5×10^{-5}
ε_n/m	2×10^{-6}	$<\varepsilon>$	4.7×10^{-5}
D	17	N_γ	1.4
H	6	$\sqrt{s}$/GeV	4.0
N	$4 \quad 10^{10}$	$\sigma_{\sqrt{s}}$/$\sqrt{s}$	4.5×10^{-5}
σ_z/mm	1.5	$\sigma\sqrt{s}$/MeV (due to beamstrahlung)	0.1

*The other parameters are given for $E = 2.0$ GeV.

least for electrons, the expensive damping rings. Positron production is a serious problem because of the low energy of the collider. One can re-sort (Figure 6) to a 200 MeV high current SC specialized electron linac to produce e^{-} hitting an e^{+} production target. A more elegant solution would make use of the high energy electrons available since the e^{+} production rate is directly proportional to the incident beam power. In any case, roughly 1/3 of the average e^{-} beam power has to be dissipated in the target. Different types of positron sources are worthwhile studying.

After the damping rings, trains of five e^{-} and five e^{+} bunches, with a frequency of 2.4 kHz are available. They can be accelerated in the second racetrack linac, (second stage in Figure 6) up to the necessary energy and brought to collision through the arcs and final focus system very similar to that of the SLC.

Two working conditions have been considered for a b$\bar{\text{b}}$ factory :

a) the "high resolution mode", $<\varepsilon> \simeq 5 \times 10^{-4}$, at $\sqrt{s} = T(4S)$, obtained by one recirculation of the beams in the second stage.

b) the "low resolution mode", $<\varepsilon> \simeq 10^{-2}$, at $\sqrt{s} = 15$ GeV (two recirculations) or at $\sqrt{s} = 20$ GeV (three recirculations), in which the luminosity is increased to compensate for the b$\bar{\text{b}}$ cross section decrease.

The main parameters for both running modes are given is Table 5. In the high resolution mode, L = 10^{33}cm^{-2}s^{-1} is achieved with an average current of 200 μA and a peak current of 200 A, the beam dimensions and the emittance being similar to the SLC nominal values. In the low resolution mode, the emittance and the frequency are similar to the previous ones, while the other parameters are somewhat more demanding to achieve L = 10^{34}cm^{-2}s^{-1}.

Table 5. Parameters for the high resolution and low resolution modes of the beauty factory.

Quantity	Symbol/Unit	High resolution mode	Low resolution mode
Bunch energy	E/Gev	5–6[*]	6–10[*]
Luminosity	L/cm^{-2}s^{-1}	10^{33}	10^{34}
Power/beam	P/MW	0.5	1.5
Bunch average frequency	f/kHz	12	12
Invariant emittance	ε_n/m	2×10^{-6}	2×10^{-6}
Disruption parameter	D	16	13
Pinch factor	H	6	6
Particles/bunch	N	5×10^{10}	8×10^{10}
Bunch r.m.s. length	σ_z/mm	1.3	0.4
β-value at IP	β^*/mm	6.0	4.0
Bunch r.m.s. radius	σ_t/μm	1.1	0.60
Av.critical energy	E_c/MeV	2.3	85
E_c/E	Y	4.5×10^{-4}	8.5×10^{-3}
Fractional energy loss	$<\varepsilon>$	4.5×10^{-4}	2.5×10^{-2}
Av.number of photons	N_γ	2.5	7.5
Collision energy	$\sqrt{s}$/GeV	10–12	12–20
Fract.r.m.s. of $\sqrt{s}$	$\sigma_{\sqrt{s}}/\sqrt{s}$	3.5×10^{-4}	1.3×10^{-2}
r.m.s. of $\sqrt{s}$	$\sigma_{\sqrt{s}}$/MeV	3.4	250

[*]The other parameters are given for E = 5 GeV and E = 10 GeV respectively.

Finally, this accelerator complex which can be built in stages, could offer other possibilities for heavy flavour study :

- A Toponium factory : when the mass of the top quark is known, high energy resolution implying low r.f. frequency (i.e. $f \simeq$ 350 MHz) will be required, and a fully superconducting linear collider, without recirculator, would be a natural extension of the $b\bar{b}$ factory. To make it not too long, higher gradient S.C. cavities are needed. Assuming 100 GeV Toponium mass, a possible set of parameters have been computed[22] providing $L = 2 \times 10^{32}$cm^{-2}s^{-1} for $\sigma_{\sqrt{s}}$ = 20 MeV.

- A Z^0 factory : The bunches extracted from the damping rings have the emittance and the repetition frequency required by the "low energy S.C. driving beam" used in the CLIC scheme[24] to power the $\simeq$ 30 GHz copper structure to large field gradients ($\simeq$ 100 MV/m) which in turn accelerates "the main beam" to high energy. Reusing for instance the 350 MHz cavities of the second stage to power the driving beam, the main beams would reach 50 GeV, and collisions at $\sqrt{s} = m_{Z0} \pm 1$ GeV with $L = 2.5 \times 10^{33}$cm^{-2}s^{-1} could be achieved[22], producing $\simeq 10^9$ Z^0, i.e. > 10^8 $b\bar{b}$ per year.

The case for a high energy resolution $L \geq 10^{34} \text{cm}^{-2}\text{s}^{-1}$ linear collider $b\bar{b}$ factory has also been advocated by D.Cline[25]. This linear collider would be based on new particle sources, new accelerating structures and new final focus systems for which a research and development programme could be started soon. In line with this approach, two 2-beam accelerator solutions with new high frequency (11.5 GHz) power sources to produce high field gradients (50-250 MV/m) have been contemplated : The Free Electron Laser[26] and the Relativistic Klystron[27] driven linear $b\bar{b}$ factories. As in the design[22] described before, high repetition rates are used, but induction units requiring more power are used to recover the energy extracted from the driving beam. Sets of parameters to achieve $L = 10^{33}\text{cm}^{-2}\text{s}^{-1}$ have been computed for both cases[26,27].

One interesting feature of the linear colliders is that asymmetric beam energies can be used[10].This could offer special interest for detectors with improved final state tagging and reconstruction efficiencies. As an example, a 2.5 GeV (e-) × 12.5 GeV (e+), with $N_{e^+} = 0.02\ N_{e^-}$, $B\bar{B}$ factory has also been considered[26].

SUMMARY

In this contribution I have reviewed the physics motivations for dedicated high luminosity e^+e^- heavy flavour factories. In particular, a $B\bar{B}$ factory with luminosity $L = 10^{33}\text{cm}^{-2}\text{s}^{-1}$ would provide important informations on Cabibbo-Kobayaski-Maskawa matrix, on symmetry breaking mechanism and on new mass scales, while with $L = 10^{34}\text{cm}^{-2}\text{s}^{-1}$ some evidence for CP violation could also be attainable.

Two main schemes to reach the forseen luminosity have been presented: The multibunch double ring approach and the linear collider approach. The first one, which is already at the proposal stage looks attractive for a $b\bar{b}$ factory since $L = 5 \times 10^{32}\text{cm}^{-2}\text{s}^{-1}$ would certainly be achieved and improvement with time can be expected. The linear collider approach, still at the conceptual level needs probably more research and development : it can be built in steps, at least for the SC accelerator complex, going from a $\tau\bar{\tau} - c\bar{c}$ factory to a $b\bar{b}$ factory and possibly to a $t\bar{t}$ factory ; it would allow to use special detectors going very close to the interaction point ; it offers the best potentiality to reach very high luminosity ; around it, new techniques needed for future high luminosity TeV linear colliders would be developed.

ACKNOWLEDGEMENTS

I wish to thank the organizers of this school and in particular R.Klapish, the Director of the School, for inviting me to present this contribution.

It is a pleasure for me to mention that the work on linear colliders has been done in a very friendly collaboration with U.Amaldi. I would also like to thank E.D.Bloom, D.B.Cline, A.Fridman, K.R.Schubert and K.Wille for many useful informations and discussions. Finally, I acknowledge Mrs C.Le Marec who typed this manuscript.

REFERENCES

1) Proceedings of the Workshop on e^+e^- Physics at High Luminosities. SLAC, Stanford, November 30 - December 1, 1984 and April 5-6, 1985. SLAC-283, UC-34D-1985.

2) Proceedings of the Moriond Workshop on Flavour Mixing and CP viola-
    tion, La Plagne, Jan.13-19,1985, Ed.Frontières (J.Tran Than Van Ed.).

3) Proceedings of the Int. Symposium on Production and Decay of Heavy
    Hadrons, Heidelberg, May 20-23,1986 (K.R.Schubert and R.Waldi Eds).

4) AIP Conference Proceedings 156 of Advanced Accelerator Concepts
    Madison, August 21-29,1986 (F.E.Mills Ed.).

5) Proceedings of Linear Collider $B\bar{B}$ Factory Conceptual Design Workshop,
    UCLA Los Angeles, January 26-30,1987 (to be published).

6) J.Kirkby, contribution to these Proceedings.

7) C.Albijar et al., UA1 Collaboration, Phys. Lett. B186, 247(1987).

8) H.Albrecht et al., ARGUS Collaboration, Phys. Lett. B192, 245(1987).

9) See for instance : I.Bigi, Reference 3, p.344 ; L.L.Chau, Reference
    3, p.352 ; D.B.Cline and A.Soni, ULCA/87/TEP6 and Ref. therein.

10) See for instance P.Oddone et al., Reference 5.

11) M.Della Negra, Proceedings of the 2nd Topical Seminar on Heavy Fla-
    vours, San Miniato, May 25-29,1987 (to be published).
    C.Albijar et al., UA1 Collaboration, Search for New Heavy Quarks at
    the CERN Proton-Antiproton Collider, CERN EP 87-XXX (to be submitted
    to Z.Phys.C.)

12) See for instance W.Buchmuller et al., Vol.1, p.203 and A.Ali, Vol.2,
    p.290 of Physics at LEP (J.ELlis and R.Peccei Eds), CERN Yellow
    Report 86.02.

13) P.Igo-Kemenes et al., Proceedings of ECFA Workshop on LEP 200, Aachen
    Sept. 29 - Oct. 1, 1986, (A.Böhm and W.Hoogland Eds), CERN 87-08,
    ECFA 87/108, Vol. II, p.251.

14) K.Wille, Reference 3, p.473.

15) D.Rubin, Reference 5.

16) M.Sullivan, Reference 5.

17) H.Aihara, Reference 5.

18) R.Eichler, T.Nakada, K.R.Schubert, S.Weseler and K.Wille, Motivation
    and Design Study of a B-Meson Factory with High Luminosity,
    SIN-Preprint, SIN-PR-86-13 (1986).

19) K.Wille, SIN-Preprint, SIN-BFP/87-3 (1987).

20) E.D. Bloom, Reference 5.

21) See for instance, U.Amaldi, G.Coignet, R.Evans, J.Le Duff, S.Tazzari
    and T.Weiland, Vade-Mecum of Concepts and Basic Formulae on e^+e^-
    Linear Accelerator, ECFA Report, ECFA/87/AP/NTPA/15 and Ref. therein.

22) Original version : U.Amaldi and G.Coignet, An Electron-Positron Linear Collider as a B-B̄ Factory, LAPP-EXP-86-71 and Ref.4. More complete version : U.Amaldi and G.Coignet, Conceptual Design of a Multi-purpose Beauty Factory Based on Superconducting cavities, CERN-EP/86-211 and Nucl. Instr. and Methods (in print) ; also Ref.5. See also U.Amaldi, CERN-EP/87-104.

23) 100 A of peak current with $\varepsilon_n \simeq 10^{-5}$ m have been produced by irradiation of a Cs_3Sb photocathode by a neodymium laser ; See for instance T.I.Smith, Nucl. Instr. and Methods A250, 64(1986) and also R.Sheffield, Photocathodes in Accelerator Applications, Proceedings of the Workshop on Low Emittance Beams, Brookhaven National Lab., March 20-25, 1987 (to be published).

24) Report from the Advirosy Panel on the Prospects for e^+e^- Linear Colliders in the TeV range, submitted the CERN Long Range Planning Committee chaired by C.Rubbia, CERN-CLIC note 38 (April 1987).

25) D.B.Cline, WICS-86-276 and WISC-86-285, also in Reference 4, p.435, and private communication.

26) J.S.Wurtele and A.M.Sessler, Reference 4, p.322.

27) S.Yu and W.Barletta, Reference 5.

A τ-CHARM FACTORY AT CERN

Jasper Kirkby

CERN
1211 Geneva 23
Switzerland

ABSTRACT

We advocate a new experiment at CERN which will perform an extensive programme of precise measurements to confront the Standard Model. The experiment will probe the fundamental constituents ν_τ, τ and c with a sensitivity which exceeds present data by two-to-three orders of magnitude and, in addition, will make a definitive search for the gluonic matter of QCD. The experiment involves a compact high-resolution detector of advanced technology, integrated with an intense e^+e^- storage ring. The storage ring has a luminosity $L = 10^{33}$ cm^{-2} s^{-1} at $\sqrt{s} = 5$ GeV and delivers data samples per year of 10^7 $\tau^+\tau^-$, 10^7 $c\bar{c}$ or 10^{10} J/ψ events, with very low backgrounds. The experiment makes extensive use of present CERN infra-structure, such as the LEP e^+e^- injector (LIL, EPA and PS), the ISR transfer tunnels and the ISR experimental area. The experiment can be constructed within a three year period at an estimated cost of 44 MSF for the storage ring and 29 MSF for the detector.

1. OVERVIEW

Today, particle physics finds itself in the remarkable situation where all established experimental observations can be interpreted within a single theoretical framework: the Standard Model. In order to make progress we must now perform experiments of the following type:

* Precise measurements of the parameters and properties of the fundamental constituents of the Standard Model.

* Searches, at higher masses or at higher sensitivities, for new phenomena which go beyond the Standard Model.

Here we advocate an experiment of this nature, which focuses on the properties and interactions of the τ and ν_τ leptons, of the c quark, and of the fundamental carrier (g) of the strong force. The experiment involves a compact high-resolution detector, together with an intense e^+e^- storage ring, operating in the range $3 < \sqrt{s} < 5$ GeV. The machine and detector are designed together into a single integrated and optimized experiment: the τ-charm factory. The high luminosity of the machine,

Table 1.   Design goals of the τ-charm factory

Item

Item	
Beam particles	e^+e^-
Energy range, $\sqrt{s}$	3–5 GeV
Luminosity	10^{33} (cm^{-2} s^{-1})
Integrated luminosity per year[1].	2500 pb^{-1}
Number of $\tau^+\tau^-$ events per year	10^7
Number of $c\bar{c}$ events per year	10^7
Number of J/ψ events per year	10^{10}

1. One operational year is taken to be 10^7 s (100 days) at a mean luminosity, <L> = 0.25 × L_{max}.

together with the optimized (monochromator) optics, provides extremely
large data samples (Table 1), with the low background conditions characteristic of e^+e^- colliders. In the study of τ, charm and J/ψ decays, this
experimental environment is unmatched by any present or planned machine.

We can illustrate the experimental possibilities with the DELCO measurement, at SPEAR, of the hadronic cross-section ratio R (Fig. 1). The
collision energy, $\sqrt{s}$, is a sensitive experimental 'knob' which may be
used to tune into different physics. Several key operating points are
readily identifiable in Fig. 1, as follows:

1. J/ψ (3.10). Study of gluonic particles [glueballs (gg) and hybrids
    ($q\bar{q}g$)] and light quark (u,d,s) spectroscopy. The monochromator optics
    provide a peak cross-section ~ 5 times larger than the value at SPEAR
    (R_{PK}= 250). The J/ψ production rate is ~ 6 KHz, with a non-ψ back-
    ground of 2 × 10^{-3}.

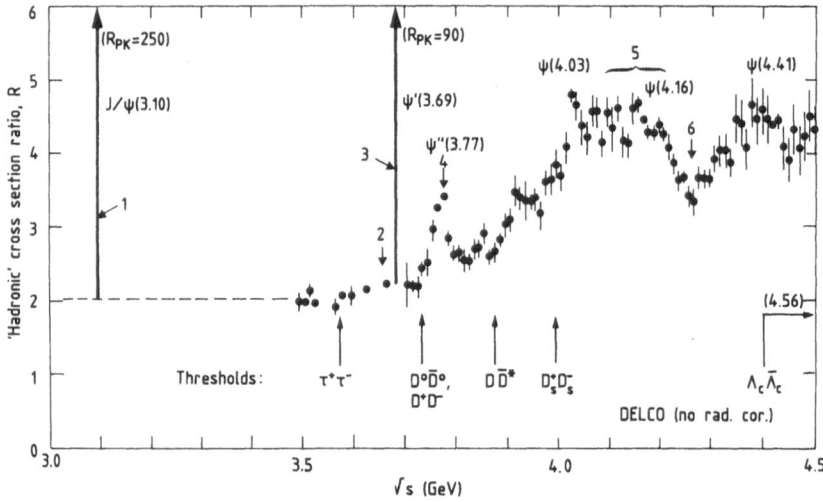

Fig. 1.   The 'hadronic' cross-section ratio, R = σ(e^+e^- → hadrons)/
          σ(e^+e^- → $\mu^+\mu^-$) vs. centre-of-mass energy, $\sqrt{s}$, where the annihila-
          tion proceeds via one photon, as measured by DELCO at SPEAR. Cor-
          rections have been made for the radiative tails of the J/ψ and ψ'.
          (The peak R measurements indicate the experimentally observed
          values.) No other radiative corrections or subtractions (such as
          $\tau^+\tau^-$ production) have been applied. The numbered regions (1–6)
          indicate machine operating points where specific experiments are
          optimized, as discussed in the text.

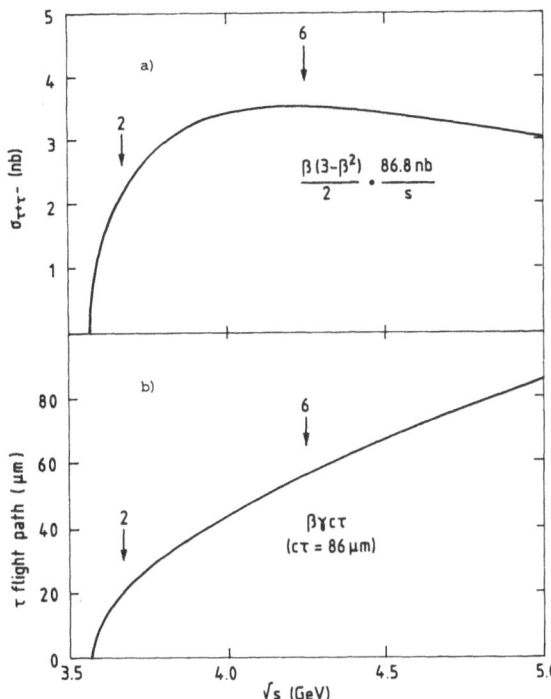

Fig. 2. a) The $\tau^+\tau^-$ cross-section vs. $\sqrt{s}$. No radiative corrections are
included. b) The mean τ flight path vs. $\sqrt{s}$. The indicated operating points (2,6) correspond to those shown in Fig. 1.

2. $\sqrt{s}$ = 3.67 GeV. Study of $\tau^+\tau^-$ decays, essentially at rest (β_τ = 0.23)
    and with zero charm background. Since the $\tau^+\tau^-$ cross-section increases
    rapidly above threshold (Fig. 2a), it has reached 50% of the maximum
    value at this energy, which is only 50 MeV above threshold. The $\tau^+\tau^-$
    production rate is 2 Hz.

3. ψ' (3.69). Study of charmonium states produced in radiative ψ' decays.

4. ψ'' (3.77). Study of tagged $D^0\bar{D}^0$ and D^+D^- events, with zero back-
    ground from higher-mass charmed particles or from mixed $D\bar{D}$ states
    (D^0D^-,$D^+\bar{D}^0$). The $D\bar{D}$ production rate is 8 Hz and the $D\bar{D}$ cross-section
    is 30% of the total hadronic cross-section.

5. $\sqrt{s}$ = 4.05-4.20 GeV. Study of tagged $D_s^+D_s^-$ events. The optimum energy
    for the study of D_s^{+-} is not known from present experiments; it will
    therefore be necessary to map out the R plot, with higher statistical
    accuracy, at the outset of this programme.

6. $\sqrt{s}$ = 4.25 GeV. Study of $\tau^+\tau^-$ events in a region of high cross-section
    and low charm background and, in addition, where the τ flight-path and
    decay impact-parameter are appreciable (Fig. 2b). The $\tau^+\tau^-$ production
    rate is 4 Hz.

7. $\sqrt{s} \sim$ 5 GeV. Study of $\Lambda_c\bar{\Lambda}_c$ charmed baryons.

    It is clear that this e^+e^- experimental region has an excellent physics
potential. We point out that Fig. 1 corresponds to $\sim$ 100 K hadronic
events accumulated at SPEAR over a period of $\sim$ 12 weeks; the τ-charm
factory would achieve a similar statistical accuracy after $\sim$ 3 hours!

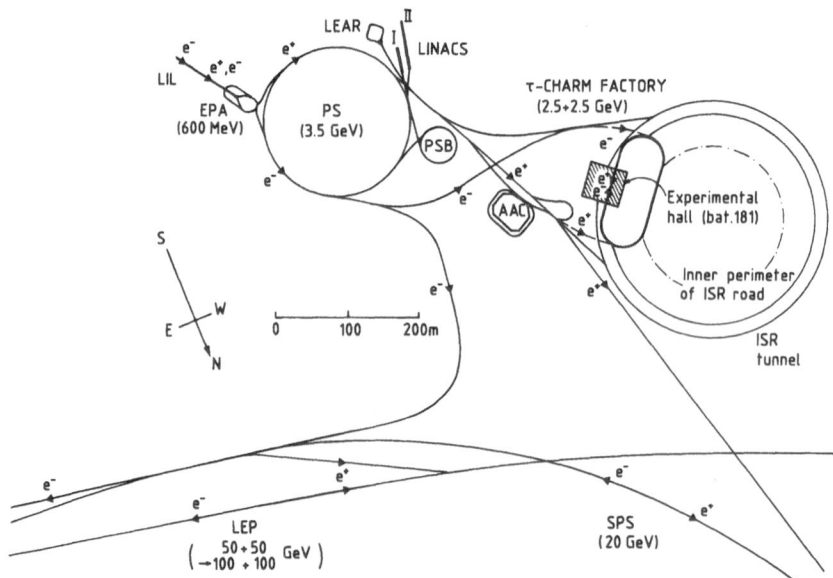

Fig. 3.   The e^+e^- injector chain for LEP and the proposed location of the
τ-charm factory. The factory makes use of the LEP Injector
Linacs (LIL), Electron-Positron Accumulator (EPA) and Proton
Synchrotron (PS) (operated at 2.5 GeV), as well as the ISR transfer tunnels and ISR experiment hall (bât. 181). The new tunnels
are indicated by dashed lines. The area occupied by the factory
does not require further excavation.

We argue that the fastest and least expensive route to this physics is
to perform the experiment at CERN. The current performance of the LEP
e^+e^- injector chain [LEP Injector Linacs (LIL), Electron-Positron
Accumulator (EPA) and Proton Synchrotron (PS)] exceeds the requirements of
the τ-charm factory. The storage ring can be filled, at collision
energy, in a few cycles of PS operation (< 2 minutes). This implies
straightforward operation of the storage ring in parallel with LEP and,
indeed, compatibility with essentially any mode of operation of the CERN
accelerator complex. Furthermore, a natural site exists at CERN for this
experiment, which involves minimal civil engineering. The proposed location (Fig. 3) uses an existing experimental hall (bât. 181, located at the
ISR) for the detector. The racetrack-shaped ring is fully accommodated on
the existing floor of bâtiment 181 and the adjoining ISR tunnel, together
with the present surface of the adjacent ISR perimeter road. The civil engineering merely involves ~ 150 m of transfer tunnel connections and a
SPEAR-like concrete block housing for the (50%) fraction of the ring which
extends onto the ISR perimeter road.

For the same reasons of speed and minimum cost, we further argue that
the e^+e^- collider is a storage ring and not a linear collider. We now
have twenty years of experience with e^+e^- storage rings and a feasible design (Ref. 1 and Fig. 4) can be made, based on quantities measured
in existing rings, which indicates a τ-charm factory can be built with a
peak luminosity of 10^{33} cm^{-2} s^{-1}. The same is not true for linear colliders, where we have yet to see even the first demonstration of such a
machine -- the SLC. A case is made, in Refs. 2 and 3, for the development
of a linear e^+e^- supercollider which proceeds through τ-charm and bottom
factories. However, it is clear that the performance figures[2,3] required
for such a linear collider will take several more years of intensive machine
R & D before we will know whether or not they are feasible. Moreover, the

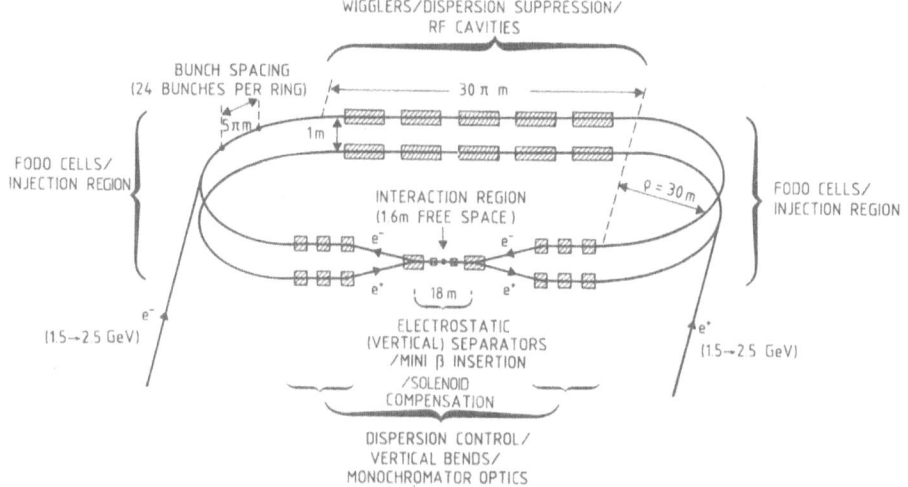

WIGGLERS/DISPERSION SUPPRESSION/
RF CAVITIES

BUNCH SPACING
(24 BUNCHES PER RING)

30 π m

5 π m

1m

FODO CELLS/
INJECTION REGION

INTERACTION REGION
(1.6m FREE SPACE)

ρ = 30 m

FODO CELLS/
INJECTION REGION

e⁻
(1.5→2.5 GeV)

e⁻

e⁻

e⁺

e⁺

e⁺
(1.5→2.5 GeV)

18 m

ELECTROSTATIC
(VERTICAL) SEPARATORS
/MINI β INSERTION
/SOLENOID
COMPENSATION

DISPERSION CONTROL/
VERTICAL BENDS/
MONOCHROMATOR OPTICS

Fig. 4.   The τ-charm factory storage ring.

well-known scaling laws for the costs of circular colliders ($\propto E^2$) and
linear colliders ($\propto E$), with equal costs expected to occur at E ~ 300 GeV,
strongly favour a storage ring for the τ-charm factory.

The improved sensitivity of this experiment results from both an intense machine **and** a new generation of detector. The detector (Fig. 5) is
based on a precise compact tracker (Si microstrips and straw chambers) in

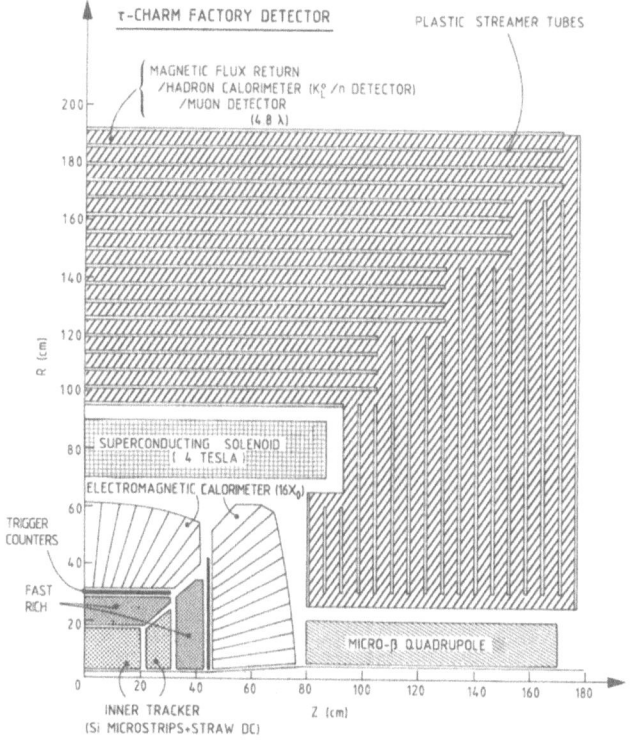

Fig. 5.   The τ-charm factory detector.

an ultra high magnetic field (4 T). This follows the concept of the JETSET
detector at LEAR. Particle (e,π,K,p) identification is made in a fast
RICH, using a solid (NaF) radiator. Electromagnetic energy is measured in
an array of crystals of a high-Z scintillator (BGO or BaF_2) which are read
out with photodiodes. The outer detector comprises a fine-grained muon
detector/hadron calorimeter (K_L^0/n identifier), of thickness 4.8 λ, which
also serves as a flux return. This high-resolution and high-rate detector
represents a new generation of instrumentation for low energy e^+e^-
storage rings, with substantial improvement in performance relative to
previous detectors.

We estimate the experiment would commence data-taking three years after
approval of the proposal, assuming sufficient resources are applied. The
estimated cost of the machine is 44 MSF and of the detector is 29 MSF.

In the remainder of this paper we will briefly describe the physics,
machine and detector aspects of this experiment, after first raising some
general arguments in favour of low energy factories.

2. THE CASE FOR LOW ENERGY FACTORIES

There are two basic approaches to the next generation of particle
accelerators:

1. New energy frontier.

2. New luminosity frontier ('factory') at present energies.

The first approach is vital and properly receives the major share of
our effort and resources. The machines in this group include LEP, SLC,
HERA, SSC, LHC, etc. However, it is important not to overlook the merits
of the second approach -- low-energy factories --which we summarize as
follows:

• Precise low-energy measurements fundamentally influence the Standard
Model and constrain high-energy phenomena (through virtual propagator
effects, etc.). Examples are mixing and CP violation in the K^0 system,
the absence of flavour - changing neutral currents in K decays, upper
limits on rare K decays, the Weinberg angle in ν interactions and measurements of the KM matrix elements.

• Our present experimental knowledge of the heavier fundamental constituents -- ν_τ, τ, c and b -- is much more superficial than that of the
lighter constituents -- ν_e, e, ν_μ, μ, u, d and s.

• The key missing element of the Standard Model -- the Higgs particle
-- may be light and yet have escaped detection at the present level of
experimental sensitivity.

• Interpretation of the experimental results from high energy colliders
repeatedly relies on a precise knowledge of the properties of the lighter
particles. An example is the interpretation of the dimuon data at the SPS
Collider; this requires a detailed understanding of the semileptonic
branching ratios and spectra of the c and b quarks.

• In general, the optimal energy for the study of a fundamental particle
is close to its threshold, at an e^+e^- collider; this region provides the
largest production rate, lowest background, highest detector resolution and
least demanding requirements on the detector.

• Improvements in accelerator technology and design now make factories feasible with luminosities 100–1000 times higher than those of previous colliders.

• Moreover, advances in detector and data acquisition technologies, since the initial round of experiments, extend the physics-reach well beyond simply the relative statistical improvement.

• Low-energy factories are very cost effective; the cost of the τ–charm factory (machine and detector) is less than the cost of a single contemporary high-energy collider detector.

• A low-energy factory serves to maintain a broad experimental base at CERN; it complements and diversifies the supercollider programme. In particular it helps avoid the danger of relying completely on the 'monolithic' supercollider programme. Historically, we have seen many examples of a democracy amongst machines, in which new discoveries are not made only at the highest energies.

• Finally, and of particular importance, low-energy factories provide an excellent and attractive environment for the introduction and training of young physicists in experimental particle physics. Here a physicist can work in a small group, understand and participate in the complete experiment, perform a precise physics measurement and be clearly identified with the result and, finally, complete a thesis on a reasonable timescale. (The τ–charm factory could easily be a thesis factory as well!) It need hardly be emphasized how vital it is for the future of our field that we continue to attract and train young physicists.

3. PHYSICS HIGHLIGHTS

3.1 General comments

We can readily appreciate the outstanding physics prospects of a factory devoted to ν_τ, τ and c. The need to study the properties of these fundamental constituents, and their relationships with other known (and perhaps unknown) particles, is evident. Our present knowledge, especially in the case of the enigmatic τ lepton, is based on experiments with relatively low statistics (see Table 2). The Mark III detector at SPEAR, for

Table 2. Present world statistics for J/ψ, τ and c

				Statistics of parent sample	
Experiment	Production	$\sqrt{s}$ (GeV)	J/ψ	$\tau^+\tau^-$	$c\bar{c}$
DM2/Orsay	e^+e^-	3	$9\ 10^6$	–	–
Mk III/SPEAR	e^+e^-	3–4	$6\ 10^6$	$4\ 10^4$	10^5 1.
ARGUS/DORIS II	e^+e^-	10	–	$2\ 10^5$	$3\ 10^5$
Mk II/PEP	e^+e^-	29	–	$3\ 10^4$	$4\ 10^4$
E691/FNAL	γN	18	–	–	$5\ 10^5$ 1.
[τ–c factory/CERN	e^+e^-	3–5	$1.5\ 10^{10}$	10^7	$2\ 10^7$] 2.

1. Mk III and E691 have ~ equal samples (~ 10 K events) of reconstructed charm decays, despite the difference in parent $c\bar{c}$ samples.
2. Statistics per operational year (100 days).

example, has collected approximately 8 pb^{-1} (40 K c$\bar{c}$ events) at the ψ''
(3.77). We also note the great success of Mark III despite its modest
(X 3) statistical improvement over previous experiments, which underscores the richness of the τ-charm physics still to be discovered. Here,
we propose an experiment which collects each year more than 100 times the
total Mark III statistics. It is <u>inevitable</u> that this experiment will be
highly productive; it is <u>possible</u> that this experiment will make a spectacular discovery.

There are several reasons why this physics is best explored by an e$^+$e$^-$
factory which operates near a maximum energy in the τ-charm threshold
region, as follows:

• <u>Peak luminosity</u> at the energy of interest ($\sqrt{s}$ = 4-5 GeV).

• <u>Highest cross-section</u> ($\sigma \sim 1/E^2$, away from resonances).

• <u>Lowest background</u>. In this region, τ and c are the only heavy particles. Furthermore, $\tau^+\tau^-$ pairs can be generated with zero charm
background in the region 3.57 < $\sqrt{s}$ < 3.73 GeV. In the case of charm,
pure final states are produced without accompanying jet fragmentation
particles.

• <u>Tagged charm states</u>. At the ψ'' (3.77), the charm events involve
<u>either</u> D$^0\bar{D}^0$ <u>or</u> D$^+$D$^-$, with no extra particles. An equivalent region,
above $\sqrt{s}$ = 4 GeV, can be selected for tagged D$_s^+$D$_s^-$ events.

• <u>Best resolution</u>. The presence of a precise $\sqrt{s}$ constraint, exclusive
production, low-energy final-state particles and absence of jets lead to
mass resolutions $\sim$ 1 MeV/c^2 compared, for example, with $\sim$ 100 MeV/c^2
for charmed particles at the Z^0.

We illustrate the fundamental quality of the physics programme by the
following brief (and incomplete) summary of the highlights.

3.2 τ/ν_τ physics

• ν_τ mass. The possibility of non-zero neutrino masses is of central
importance to physics. Finite neutrino masses would profoundly affect the
Standard Model -- allowing mixing and oscillations between the families --
and also provide answers to two of the major puzzles of cosmological
physics -- the deficiency of solar neutrinos and the composition of the
dark matter of the universe. There are two experimental methods which are
used to measure m(ν_τ):

a) Measurement of the end-point of the <u>energy spectrum</u> in the decays
$\tau^- \to \nu_\tau$X, where X = π^-, ρ^- or $\pi^-\pi^+\pi^-$. Sensitivity to m(ν_τ) occurs at
small values of the neutrino energy, E(ν_τ) = E$_{beam}$-E$_X$.

b) Measurement of the end-point of the <u>mass spectrum</u> in the decays
$\tau^- \to \nu_\tau$X, where X = (4π)$^-$, (5π)$^-$ or K$^-$K$^+\pi^-$. At the end-point,
m(ν_τ) = m$_\tau$ - m$_X$.

Although the first method has much higher statistical accuracy, it can
only be used effectively at low $\sqrt{s}$, and so most measurements (made at
DORIS II, PEP and PETRA) have relied on the second method. The latter requires a careful understanding of backgrounds.

The <u>optimum machine energy for the measurement of m(ν_τ) is close to
$\tau^+\tau^-$ threshold</u>; specifically, the τ-charm factory operating at
$\sqrt{s}$ = 3.67 GeV (region 2 in Figs. 1 and 2). The reasons are as follows:

a) Highest production rate (see Section 4.3).

b) ΔE_{beam} spread is only ~ 0.5 MeV, due to the monochromator optics.

c) Radiative corrections are small.

d) ΔE_X errors are minimized. The low particle momenta imply small measurement errors, (precise) mass values contribute to the determination of E_X, and the nearby $\psi'(3.69)$ resonance provides an ideal, stable calibration point for the detector energy and momentum measurements.

e) m_τ is measured, in the same detector, with a precision ~ 0.1 MeV/c².

f) Zero background from heavy flavour (c,b,t) production.

g) τ velocity is small ($\beta = 0.23$) and so also is the Lorentz smearing. This will maximize the sensitivity of a given statistical sample to $m(\nu_\tau)$.

h) Longitudinal beam polarization may be applied to enhance the sensitivity and to provide experimental cross-checks on the data.

The present mass limit is $m(\nu_\tau) < 50$ MeV/c² with 95% CL (from the ARGUS analysis of the mass spectrum $\tau \to \nu_\tau 5\pi$); here we are sensitive to $m(\nu_\tau)$ ~ 1 MeV/c². In several models which involve neutrinos with finite masses, they follow the same mass hierarchy as the charged leptons. One such model predicts the mass relationship, $m(\nu_\tau) = (m_\tau^2/m_e^2)m(\nu_e)$, in which case a sensitivity of 1 MeV/c² for $m(\nu_\tau)$ is equivalent to 1 eV/c² for $m(\nu_e)$.

• **Nature of the τ-W-ν_τ current**. The V,A, etc. nature of the τ-W-ν_τ current is determined from the lepton spectrum in $\tau^- \to \ell^- \bar\nu_\ell \nu_\tau$. This is best-measured near to threshold where the lab spectrum is close to the (ideal) spectrum in the τ rest frame. At present, the only τ decay parameter which has been measured is the Michel parameter: $\rho = 0.73 \pm 0.07$, averaged over all e, μ measurements. Although this is consistent with the pure V–A spectrum ($\rho = 0.75$) expected in the Standard Model, it barely restricts a possible right-handed, V+A, component ($\gtrsim 47\%$ V+A excluded at 95% CL). Here, the large statistics would allow effective tagging of the polarization state of one τ by its partner, and each of the four decay parameters -- ρ, η, ξ and δ -- could be measured. In these studies, the option of longitudinally-polarized beams at the τ-charm factory is of great importance, since the τ polarization would be exactly known. The experimental error on the Michel parameter would be approximately ± 0.01.

• **Cabibbo angle in τ decays**. This is measured by the relative branching ratios,

$$b(\tau^- \to K^- \nu_\tau)/b(\tau^- \to \pi^- \nu_\tau) = (0.57 \pm 0.15)\%/(10.3 \pm 1.2)\% .$$

The precision of $\tan^2 \theta_c$ in τ decays is therefore $\sim 25\%$, compared with 1.5% in other processes. A precise comparison of the τ Cabibbo angle with the conventional value will be a sensitive test of the Standard Model of τ decays.

• **Anomalous τ decays**. Present measurements indicate that the sum of the exclusive one-prong τ decays do not add up to the inclusive one-prong branching ratio $B_1 = (86.6 \pm 0.3)\%$. The sum of the individual one-prong channels, excluding 'multiple-neutral' modes, is $(70.2 \pm 1.6)\%$. Although the multiple-neutral channels, i.e. $\tau^- \to \nu_\tau \pi^- X$, where $X = n\pi^0 (n > 2)$, $\pi^0 \eta$,

$2\pi^0 + \eta$, etc., are poorly measured in present data, they appear to have a total branching ratio $\lesssim 11\%$. The result is that $\sim 5\%$ of the single-prong τ decays are not accounted for. The recent reports of a large branching ratio for $\tau^- \to \nu_\tau \pi^- \eta$, which involves second-class currents (hadronic system $J^P = 1^+$, $G = +$), have not been confirmed in other experiments. These observations raise the possibility that the τ lepton may not be as 'well-understood' as is widely assumed, and underscores the need for a study of τ decays in a detector which has good performance for both charged <u>and</u> neutral particles.

• <u>Lepton-flavour violation in τ decays</u>. The search for rare τ decays which involve lepton-number violation is a sensitive experimental tool with which to search for new physics. These modes involve neutrino-less decays such as $\tau^- \to e^-\gamma$, $e^-\mu^+\mu^-$, $e^-\gamma\gamma$, $e^-\pi^0$, $e^-\rho^0$, $\pi^-e^-\mu^+$, $K^-e^-\mu^+$, etc. (or similar modes with e and μ interchanged). The equivalent experimental limits on lepton-number violation in μ decays are 10^{-12}-10^{-13}. These are compared with the current τ limits of $\sim 5 \times 10^{-5}$, and future sensitivities of 10^{-7} in the τ-charm factory. In general, models suggest the largest violations of lepton-flavour conservation may occur in the τ sector; a sensitivity of 10^{-7} in lepton-flavour violating τ decays could be equivalent to 10^{-15} in μ decays.

3.3 Charm Physics

• <u>$D^0\overline{D}^0$ mixing</u>. In analogy with the neutral kaon system, we expect D^0-$\overline{D}^0$ oscillations and B^0-$\overline{B}^0$ oscillations to occur. The latter have now been observed, by UA1 and ARGUS, whereas the (ARGUS) upper limit on $D^0 \to \overline{D}^0$ is 1.4%, at 90% CL. The Standard Model expectation for $D^0 \to \overline{D}^0$ transitions, including long-distance effects, is $\lesssim 10^{-4}$; this is well within the experimental sensitivity of the τ-charm factory. The best signatures involve semileptonic D^0 decays, in order to avoid the problem of doubly-Cabibbo suppressed hadronic decays (which occur with branching ratios $\sim 10^{-3}$). The D^0 is tagged either at the $\psi''(3.77)$, using the opposite $\overline{D}^0$, or at $\sqrt{s} = 4.03$ GeV, using the soft π^+ signature in $D^{*+} \to \pi^+D^0$. An observation of an anomalous rate for $D^0 \to \overline{D}^0$ oscillations would be a clear indicator of new physics, e.g. charm-changing neutral currents.

• <u>KM matrix elements, V_{cs}, V_{cd}</u>. These charm KM matrix elements have not yet been directly measured. If we do not make the a priori assumption of only three generations, we merely know from present experimental data that $|V_{cs}| > 0.66$ (90% CL). The charm KM matrix elements are best-measured by comparing the semileptonic branching ratios, $D \to Ke\nu$ vs. $D \to \pi e\nu$, or $D \to K^*e\nu$ vs. $D \to \rho e\nu$. These measurements have not yet been made because of the small branching ratio of the Cabibbo-suppressed semileptonic decays (0.1-0.5%).

• <u>D^+, D_s^+ purely leptonic decays</u>. These decays allow an unambiguous measurement of the decay constant f_D, which is of importance both in calculating the magnitude expected for $D^0\overline{D}^0$ mixing and in calculating the magnitudes of weak flavour annihilation and Pauli interference, which have been invoked to explain the difference in D^+ and D^0 lifetimes. Such decays have not yet been observed; Mark III measures the upper limit, $b(D^+ \to \mu^+\nu_\mu) < 0.8 \times 10^{-3}$, at 90% CL. The expected branching ratios are, for example, $b(D^+ \to \mu^+\nu_\mu) \sim 2 \times 10^{-4}$, $b(D^+ \to \tau^+\nu_\tau) \sim 5 \times 10^{-4}$ and $b(D_s^+ \to \tau^+\nu_\tau) \sim 2\%$. The last decay is of substantial interest since this process is the major source of ν_τ in possible future beam dump experiments which seek to observe ν_τ directly.

• <u>Flavour-changing neutral currents and lepton-flavour violation in D decays</u>. The absence of these processes is implicit in the Standard Model. Such decays include $D^0 \to \mu^+\mu^-$, μ^+e^- and $D^+ \to \pi^+\nu\overline{\nu}$, $\pi^+e^+\mu^-$, $\pi^+\mu^+\mu^-$, etc. Observation of a decay of this type would signify new physics such as leptoquarks, horizontal gauge bosons, contact interactions or lepton/quark substructure, as in the Rishon Model. (Some of the apparent neutral current decays are, in fact, allowed in the Standard Model via second-order weak interactions, with a very low branching ratio.) The decays are sensitive to very heavy propagators and therefore serve to constrain certain high-energy processes. The interesting upper limits start at $< 10^{-4}$, which translate to cut-off masses, $\Lambda \gtrsim 1$ TeV/c^2. In view of the intense theoretical interest in these processes, there is a major effort now devoted to the equivalent decays in the K system. These are measured presently to branching ratio upper limits of 10^{-6}-10^{-8}; the new experiments aim for sensitivities of 10^{-10}-10^{-12}. However, the non-observation of these processes in K decays may not suppress the equivalent D decay. For example, $D^0 \rightarrow \mu^+e^-$ could be caused by massive leptoquarks which couple up-type quarks to charged leptons and down-type quarks to neutral leptons; hence $K^0(d\bar{s}) \rightarrow \bar{\nu}_\mu\nu_e$, but not μ^+e^-. In the D system, the current experimental limits on these processes are much less stringent: $b(D^0 \rightarrow \mu^+\mu^-) < 3 \times 10^{-4}$ and $b(D^0 \rightarrow \mu^+e^-) < 1.5 \times 10^{-4}$. The τ-charm factory is sensitive to branching ratios near to 10^{-7}.

3.4 Spectroscopy

• $J/\psi \rightarrow \gamma$ + glueballs/hybrids. Radiative J/ψ decays are the best experimental tool to search for gluonic matter. The key reasons are as follows:

a) $J/\psi \rightarrow \gamma gg \rightarrow \gamma X$ involves a pure two-gluon intermediate state, with a mass m(gg) < 3.1 GeV/c^2, i.e. in the expected mass region of the gluonic spectrum.

b) The initial and final states have well-defined quantum numbers: $J^{PC} = 1^{--}$.

c) The nature of the state X can be tested experimentally by comparing $J/\psi \rightarrow \gamma X$, ωX and ϕX. If X is gluonic, then the decays ωX and ϕX will be suppressed; in contrast, ωX is enhanced if X has u,d quark content, and ϕX is enhanced if X has s quark content.

d) The high statistics of J/ψ decays. Current experiments have $\sim 10^7$ decays; the τ-charm factory will have a parent J/ψ sample of 1.5×10^{10} decays per year.

It is evident that many interesting results are emerging from the Mark III and DM2 analyses which would benefit from even a factor of two or three times more data, let alone the factor 10^3 of the τ-charm factory. It is also clear that these studies again call for a detector with good π/K particle identification, and with high resolution on both charged and neutral particles, so that final states such as $\gamma\pi^0\pi^0$, $\gamma\eta\eta$, $\gamma\phi\pi^0$, $\gamma KK\pi$, $\gamma\eta\pi\pi$ etc. can be analysed.

• $J/\psi \rightarrow$ u,d,s mesons. J/ψ decays are also the ideal laboratory in which to study light quark ($q\bar{q}$) spectroscopy, and four-quark ($q\bar{q}q\bar{q}$) or molecular states. In the 1-2 GeV/c^2 mass region there are still $\sim$ 20 'missing' $q\bar{q}$ states. These should be experimentally measured, both to test QCD and to identify strangers, which would indicate new (possibly gluonic) matter. The τ-charm factory could make the definitive searches for these missing light quark states.

• $\tau \rightarrow$ u,d,s mesons. The τ decays via $\tau^- \rightarrow \nu_\tau W^-$ (virtual), followed by W^- (virtual) $\rightarrow \bar{u}d, \bar{u}s$. We therefore have the novel possibility of exploring light quark spectroscopy (m < m$_\tau$) produced from the charged weak current. As in e^+e^- annihilation, this may prove to be a far cleaner process than hadronic production for creating certain mesons, e.g. the 'missing' radial pseudoscalar (J^{PC} = 0^{-+}), π^{*-}, may appear as $\tau^- \rightarrow \nu_\tau \pi^{*-} \rightarrow \nu_\tau (\rho\pi)^-$. In these studies, the option of polarized τ^- production would provide important information .

• <u>Charmonium</u>. Our experimental and theoretical understanding of $c\bar{c}$ spectroscopy below threshold is now rather good. However, above threshold the ψ states and their e^+e^- partial widths, Γ_{ee}, fail to behave monatonically. These states are therefore the cleanest systems for studying how the presence of open decay channels modifies the naïve potential model. Strategically, we need to study the R plot (Fig. 1) in much finer detail than present experiments and measure the 3S, 4S, ...; 1D, 2D, ... etc. charmonium resonances. This study can also potentially reveal new types of state, e.g. hybrid charmonium, $c\bar{c}g$.

3.5 New light particles

• <u>Higgs</u>. There are <u>no effective experimental constraints on the mass of the neutral Higgs boson</u> (H$^\circ$) of the Standard Model above $\sim$ 30 MeV/c^2. A sensitive search for a low mass H$^\circ$ is therefore of the highest importance. Higgs with a mass below 3 GeV/c^2 are best-produced at low energy in J/ψ decays, with a branching ratio b(J/$\psi \rightarrow \gamma$H$^\circ$) = 2 X 10^{-5}. Since this machine will generate $\sim$ 1.5 X 10^{10} J/ψ decays per year, with essentially zero background, there is clearly the potential for an extremely sensitive search for the Higgs. In comparison with other machines, we see (Table 3) that the H$^\circ$ production rate at this machine is 100 times higher than elsewhere. More-

Table 3. Comparison of light Higgs H$^\circ$ production rates (m_H < 3 GeV/c^2)

	τ-c factory	B factory	LEP	
Reaction	J/$\psi \rightarrow$ H$^\circ\gamma$	Υ(1S) $\rightarrow$ H$^\circ\gamma$	Z$^\circ \rightarrow$ H$^\circ\gamma$	Z$^\circ \rightarrow$ H$^\circ\mu^+\mu^-$
Integrated L per year[1.] (pb^{-1})	1250 [2.]	2500	50	50
σ_{peak}(e$^+$e$^- \rightarrow$ V) (nb)	12000 [3.]	20	35	35
V production rate per year	1.5 10^{10}	5 10^7	2 10^6	2 10^6
Branching ratio (V $\rightarrow$ H$^\circ$X)	2 10^5	10^{-4}	2 10^{-6}	2 10^{-4}
H$^\circ$ production rate per year	3 10^5	$\lesssim$ 10^3	4	4 10^2

1. See Table 5.
2. This is reduced by a factor of two to account for the lower luminosity at √s = 3.1 GeV.
3. This assumes a factor of five increase above the SPEAR value, due to the monochromator optics.

over the detection of a low-mass H° is optimal (signal-to-background, detector resolution, etc.) in J/ψ decays. The cleanest signals involve $H^\circ \rightarrow \mu^+\mu^-$ or, for $m(H^\circ) < 2\, m_\mu$, $H^\circ \rightarrow e^+e^-$.

• **Axion.** By 'axion', we imply a generic title for a light, spinless, long-lived and non-interacting particle. The axion, a, can be produced in the decays, $J/\psi \rightarrow \gamma a$, which result in the distinct experimental signature involving, γ (monochromatic) + nothing. The present (Crystal Ball) upper limit is $b(J/\psi \rightarrow \gamma a) < 1.4 \times 10^{-5}$ (90% CL).

4. STORAGE RING

4.1 Overview

Following the discussion in Section 1, we have based the collider design on a circular, rather than linear, machine. Jowett[1] has made an initial study of the τ-charm factory collider using an approach -- a multi-bunch double storage ring -- which is similar to that of the SIN B factory design study[4]. Since the details of Jowett's design can be found in Ref. 1, we will restrict the present discussion to a summary of the main features.

The τ-charm factory embodies several notable features which contribute to its unique physics potential. These are as follows:

• Very high luminosity: $L = 10^{33}$ cm^{-2} s^{-1}.

• High operational efficiency. The beams can be dumped and the two rings refilled in about two minutes, reflecting both the excellent (achieved) performance of the LIL/EPA/PS e^+e^- injector chain and the optimized choice of parameters for the storage ring. This will result in good compatibility of the factory with other operations of the CERN accelerator complex, and also ensure a mean (operational) luminosity which can be maintained close to the peak value.

• Monochromator optics. The effective collision energy spread, $\sigma \sim 0.5$ MeV, is a factor of five less than the 'natural' energy spread. This translates into a factor of up to five gain in the counting rate on narrow resonances, such as J/ψ and ψ'. It also increases the experimental sensitivity to sharp structures which may be present above the naked charm threshhold, e.g. exotic charmonium states. Finally, a narrow beam-energy spread will improve the precision of the measurements of the τ and ν_τ masses.

• Longitudinal beam polarization. The feasibility of this option is under study and the initial results are encouraging. The polarization times can be made ≤ 20 minutes (near the top energy) and longitudinal polarization of the beams can be generated by means of a system of spin rotators or Siberian snakes, located in the (intentionally) long straight-sections. This option would open up entirely new experimental possibilities in the studies of τ and charm, production and decay. For example, a precise study of the nature of the τ decay current could be made by the combination of a known τ polarization with 'self-analyzing' decays such as $\tau^- \rightarrow \pi^- \nu_\tau$.

4.2 Design

The design parameters and performance of the storage ring are summarized in Table 4. The main features (see Fig. 4) are as follows:

• Two rings with a single interaction point. The maximum possible luminosity is concentrated in one detector.

413

Table 4.   Parameters and performance of the τ–charm factory

Top energy	E	2.5	GeV
Circumference	C	376.99	m
Arc radius	R_a	30.0	m
Bending radius	ρ	$\simeq 12$	m
β function at IP	β_x^*	1.0	m
	β_y^*	0.01	m
Betatron coupling	κ^2	0.01	
Betatron tunes	Q_x	$\simeq 6$	
	Q_y	$\simeq 6$	
Natural emittance	ε_{x0}	40 π	nm
Energy loss per turn	U_o	$\leqslant 0.4$	MeV
RF frequencies	f_{RF1}	496.2	MHz
	f_{RF2}	1.489	GHz
RF voltage	V_{RF1}	$\leqslant 1$	MV
	V_{RF2}	$\leqslant 2$	MV
Number of bunches	k_b	24	
r.m.s. bunch length	σ_z	$\leqslant 8$	mm
Total beam current	I	92	mA
Beam–beam parameter	ξ_y	0.04	
Luminosity	L	10^{33}	$cm^{-2}\ s^{-1}$

• Many (24) bunches. Each bunch makes a single collision per revolution. The separation between bunches is 5π m (resulting in a time of
52.4 ns between collisions).

• Zero crossing-angle between the colliding beams. This avoids the strong
transverse space-charge forces which were found to limit significantly the
luminosity of DORIS I.

• Superconducting wigglers in the long straight sections. These maintain an emittance and beam size which are constant with beam energy, E_B.
This, in turn, will result in the slowest ($\propto E_B^2$) fall-off in luminosity
from the peak value (L = 10^{33} cm^{-2} s^{-1} at E_B = 2.5 GeV). They are also
important for reducing the polarization time of the beam.

• Very small β function at the interaction point (β_y^* = 1 cm). This is
achieved by a superconducting mini-β insertion, with only 1.6 m free-
space in the interaction region.

• Small bunch length ($\sigma_z \simeq 8$ mm). This matches the small β_y^* and avoids the loss of luminosity which would occur if the beta function at the interaction point were to change significantly along the length of the bunch. The small bunch-length is achieved by the use of two RF frequencies: 500 MHz and 1.5 GHz. In order to avoid bunch-lengthening, special attention must be paid to reducing the impedance (smoothing) of the beam pipe.

• Long straight sections. These are sufficient both to accommodate the present beam elements, such as RF systems, wigglers, monochromators and spin rotators, and also to allow for future additions and upgrades to the storage ring.

4.3 Comparison with other machines

We compare, in Table 5, the τ-charm factory with other e^+e^- machines. Table 5 clearly shows the potential for a dedicated τ-charm factory to play a unique and dominant rôle in the future study of ν_τ, τ and c. Hadron colliders and fixed target accelerators are not included in this list since they are essentially insensitive to τ decays and, in the case of J/ψ decays, they cannot complete with the favourable rate and cleanliness of an e^+e^- collider. Fixed target charm physics is, however, competitive with current e^+e^- colliders for certain studies, as demonstrated by the highly successful charm photoproduction experiment, E691 at FNAL. This experiment has reconstructed ~ 10 K charm decays from a total data sample of 10^8 triggers (containing an estimated 5×10^5 charm events). The reconstructed charm samples of this experiment and Mark III are comparable. The disadvantages of charm photoproduction in future experiments, relative to production in an e^+e^- factory, are as follows:

• Limited statistics.

Table 5. Comparison with other e^+e^- machines

	τ-c factory	SPEAR/DORISI	BEPC[1.]	SIN B factory	LEP
E_{cm} (GeV)	3–5	3–7	3–6	9–11	93 (Z^0)
L_{max} (cm^{-2} s^{-1})	10^{33}	$2 \cdot 10^{30}$	$2 \cdot 10^{31}$	10^{33}	$2 \cdot 10^{31}$
Integrated L per year[2.] (pb^{-1})	2500	5	50	2500	50
$\sigma_{\tau\tau}$ (nb)	4.	4.	4.	0.8	1.
Number $\tau^+\tau^-$ events per year	10^7	$2 \cdot 10^4$	$2 \cdot 10^5$	$2 \cdot 10^6$	$5 \cdot 10^4$
σ_{cc} (nb)	8.	8.	8.	1.	4.
Number $c\bar{c}$ events per year	$2 \cdot 10^7$	$4 \cdot 10^4$	$4 \cdot 10^5$	$3 \cdot 10^6$	$2 \cdot 10^5$

1. Beijing Electron-Positron Collider.
2. One year is taken to be 10^7 s (100 days) at a mean luminosity, <L>
   $\simeq 0.25 \times L_{max}$.

- High multiplicity of non-charm jet fragments. E691 measures $<n_{ch}>$ ~ 14 in the charm events, i.e. ~ 10 non-charm secondaries.

- Relatively low detection efficiency (typically ~ 5%).

- Relatively high backgrounds (typically ~ 30%).

- Inability to handle 0-prong or 1-prong decays.

- Inability, in general, to handle decays involving neutrals.

Finally, we address the following question: "Why not build a B factory and run it at a lower energy?" There are two strong arguments against this approach. The first is that the luminosity would be poor, since it falls off at least as $(E/E_{max})^2$. Therefore the luminosity of a machine with $E_{max} = 11$ GeV, which is run at 4 GeV, is only $(4/11)^2$ ~ 1/8 of the peak value. The second reason is simply the long running time which would be necessary for B physics. The B region has several distinct operating points, e.g. Υ (1S), Υ(4S) [$B_d\bar{B_d}$, B^+B^-], > $B_s^0\bar{B_s^0}$ threshold etc., each of which requires a minimum of one or two years data-taking. Moreover, the important measurements of CP violation in the $B_d\bar{B_d}$ and $B_s^0\bar{B_s^0}$ systems will probably need even larger data samples. It is clear that there would be a <u>minimum</u> of five years, and more probably ten years, before the B program could be interrupted.

5. DETECTOR

5.1 <u>Concept</u>

The strength of the τ-charm factory is not only due to the improvement in machine luminosity but also to the substantial progress in detector technologies since the original experiments. Even the Mark III detector at SPEAR -- which represents the state-of-the-art in 4π detectors at this energy -- was designed 10 years ago. We can readily appreciate the potential improvement in the physics sensitivity of a new detector by consideration of the following points:

- There has never been a 4π detector at these energies which combines a) precise momentum measurements of charged tracks and, b) precise photon (γ,π^0,η,ω,etc) measurements in a high-Z crystal calorimeter.

- Similarly, no 4π detector has so far included a hadron calorimeter (K_L^0/n identifier), for ν tagging by missing E and E_T.

- The key measurement of $\pi/K/p$ identification has been carried out by time-of-flight techniques, which only operate efficiently and cleanly over a narrow momentum range.

We base the design of our detector (Fig. 5) on a compact geometry for the following reasons:

- In order to fit within the restricted free space (1.60 m) between the quads of the mini β insertion.

- In order to maximize the detection efficiency for soft $K^{\pm}$. In the τ-charm factory detector, the Cerenkov radiator is located at a mean distance of only 25 cm from the interaction point. This results in good $K^{\pm}$ detection efficiency, e.g. the loss is only 15% at 200 MeV/c.

• In order to minimize the false μ signatures from π → μν and K → μν decays.

• In order to minimize the volume and cost of the electromagnetic calorimeter.

Beyond the overall requirement of a compact geometry, there are several additional design criteria, as follows:

• Full solid angle (4π) coverage, since we are concerned with exclusive analyses. This will maximize the detection efficiencies and minimize backgrounds involving missing particles.

• Precise measurement of both charged and neutral particles (γ). In particular, the photons must be well-measured at low energies (≤ 100 MeV).

• Precise secondary vertex measurement capability.

• π/K/p identification up to ~ 2.5 GeV/c.

• Good e and μ identification.

• Hermeticity, i.e. hadron calorimetry to identify and measure the presence of neutral hadrons, K_L^0 and n. (The charged hadrons are better-measured by magnetic analysis.) This will allow, for the first time at a low-energy e^+e^- collider, the possibility to identify the presence of a missing ν or ν-like object.

• Fast detector elements, and an advanced trigger and data acquisition system. The latter involves full event processing in real time, with a system of parallel microprocessors, to provide a versatile and sophisticated level 3 trigger. A system of this nature will be vital to select the signals of interest in an environment of high data rates of 'good' events, e.g. 6 KHz at the J/ψ.

Guided by these criteria, we have prepared a preliminary design for the τ-charm factory detector, which we briefly describe in the following section.

5.2 Design and performance

The detector is based on the JETSET concept of a compact high resolution tracker in a strong magnetic field. The features are as follows:

• An inner tracker of silicon microstrips and straw chambers. The straw chambers are used primarily for track finding, both off-line and in the level 2 hardware trigger. The microstrips provide precise track points for the momentum measurement and for the identification of secondary vertices. The tracker system involves 3400 straws and 1.7 m² Si microstrips.

• Magnetic analysis provided by a high field superconducting solenoid. With a 4 T field, the momentum precision is,

$$(\sigma_p/p)^2 = [0.4\% \ p \ (GeV/c)]^2 + [1.1\%]^2 \ .$$

The clear-bore diameter of the solenoid is 1.40 m and the length is 1.60 m. The stored energy is 20 MJ.

417

• A <u>fast RICH</u> for e/π/K/p identification. This involves a solid (NaF) radiator together with a photosensitive MWPC with pad readout. The RICH can also participate in the level 2 and 3 triggers. The particle identification limits (corresponding to 1% misidentification) are: e/π, 0.9 GeV/c; π/K, 3 GeV/c; and K/p, 5.2 GeV/c. The RICH detector involves 700 wires, and 56 K pads with VLSI readout.

• An array of <u>trigger scintillation counters</u>. These are used in the fast level 1 hardware trigger and to provide a precise measurement of the event time. There are 48 counters in the barrel region and 48 pie-shaped counters in each end cap.

• A <u>precise electromagnetic (em) calorimeter</u> based on crystals of a high Z scintillator with photodiode readout. The material is either BaF_2 or BGO. The main advantage of BaF_2 is its superior radiation resistance, although it seems that BGO is also radiation hard when made with high purity. BGO has the advantages of faster response (τ = 300 ns cf. 620 ns for BaF_2), smaller radiation length (1.12 cm cf. 2.05 cm) and smaller Molière radius (2.3 cm cf. 3.4 cm). These characteristics would result in a more compact BGO calorimeter and consequently the mini-β quads could be located closer to the interaction point (±65 cm cf. ±80 cm for BaF_2). Furthermore, a BGO calorimeter would lead to smaller transverse shower sizes and consequently better shower separation, which is especially important in multi-photon final states. Finally, the cost difference is quite small (BGO/BaF_2 calorimeter cost ratio = 1.2) when account is taken of the required volumes (500 ℓ of BGO cf. 1340 ℓ of BaF_2). Overall, BGO is probably the better choice, providing it can be readily fabricated with sufficient purity. (The dimensions in Fig. 5, however, reflect the size of the larger calorimeter, BaF_2.) The two materials are equivalent in other aspects of the performance. The L3 BGO measurements indicate an energy resolution $(\sigma_E/E)^2 \sim [1.5\%/\sqrt{E(GeV)}]^2 + (0.5\%)^2$, which gives $\sigma_E/E \sim 5\%$ at E = 100 MeV. The second term reflects the excellent stability, uniformity and dispersion of the calibrated system. The L3 tests show σ(position) = 3 mm at E = 2 GeV, which is equivalent to $\sigma_\theta = \sigma_\phi \sim$ 3 mm/300 mm = 10 mr in the τ-charm factory detector. The measured πe rejection is 2 × 10^{-3} at E = 2 GeV. We anticipate an improved πe rejection in our calorimeter by measuring also the longitudinal energy profile. This is achieved by mounting two photodiodes at the entrance to each crystal, in addition to the two photodiodes at the exit. There are approximately 2800 projective towers in the calorimeter, each having an entrance face ~ 3 × 3 cm², and a depth of 16 X_0.

• A combined <u>hadron calorimeter (K_L^0/n detector)/muon detector</u> which also serves as the magnetic flux return. This is constructed from 5 cm Fe plates and is instrumented with 1 × 1 cm² plastic streamer tubes which have strip (tracking) and pad (tower energy) readout, following the ALEPH design. The functions of this fine-grained system are as follows:

    a) To detect and measure K_L^0/n. The angular measurements will be good ($\sigma_\theta \sim \sigma_\phi \sim$ 3 mm/1000 mm = 3 mr), whereas only an approximate energy measurement will be available [$\sigma_E/E \sim 80\%/\sqrt{E(GeV)}$]. The hadron detection efficiency is excellent; the probability of not interacting in the combined 6 λ_{int} material (1.1 λ_{int} em calorimeter + 0.35 λ_{int} solenoid + 4.5 λ_{int} Fe) is 0.25%. The detection of K_L^0/n is important in order to study more completely the decays of J/ψ, τ, D, D_s and Λ_c and also, of particular importance, in order to tag and veto K_L^0/n events in the study of the production of neutrinos or neutrino-like objects.

    b) To identify muons. A muon is signified by a non-interacting track which penetrates to a depth in the iron which corresponds to the measured

momentum of the incident track. Muons below 1.2 GeV/c will range out inside the calorimeter volume. The misidentification probability due to
punch-through will typically be less than 1%. The hadron decay probabilities are ~ 1.4%/p (GeV/c) for π, and ~ 11%/p (GeV/c) for K. The resulting
μ background from K decays is effectively reduced [$\lesssim$ 3%/p (GeV/c)] by the
precise inner tracker and RICH, which identify the K presence after a
flight path of only 25 cm.

   c) To reduce the πe background in which there is hadronic leakage
out of the back face of the em calorimeter.

   This system will be used to tag the presence of ν_τ in the following
way. First, in the case of 'hadronic' (no e,μ) final states, the requirement of no K^0_L/n and a large missing E and E_T will signal the presence of
(2) ν_τ. (We comment that the precision on the measurement of missing E and
E_T will be excellent, e.g. the error on missing energy is ~ 40 MeV/4000 MeV
= 1.0%.) Charm can only contribute a background to this signature via the
presence of an unidentified semi-leptonic decay. Second, in the case of
'leptonic' (e or μ present with hadrons) final states, charm and τ can
be distinguished since the missing p_T + lepton + hadrons will reconstruct
to a D mass in the former case (single ν) but not in the latter case
(which involves 3ν).

   The total Fe mass is 210 t. The instrumentation involves 23 K streamer
tubes, which cover a total area of 600 m^2.

• An <u>advanced trigger + data acquisition system</u>. In order to facilitate
the trigger, the detector components are chosen to have a fast response and
are assembled in highly granular systems. The short interval (52 ns) between successive bunch crossings implies the need for front-end electronics
involving digital and analogue pipelining, in order to avoid large deadtime losses. Sufficient pipeline depth (~ 1 μs) is required to allow time
for formation of the level 1 trigger decision. This pipeline is also used
as a simple time-tagger which associates data, e.g. em calorimeter pulse
heights, to the appropriate bunch crossing. The level 1 hardware trigger
involves the trigger counters and em calorimeter. The level 2 hardware
trigger executes track finding and RICH wire counting, as well as em
cluster energy measurements, within 10 μs. Providing the level 1 output rate is below 10 KHz there will be tolerable (< 10%) dead time losses
at level 2. Finally, in level 3, the events are fully processed in an
array of parallel microprocessors, such as M68K's. Here, sophisticated
software filtering of the data is applied before recording events on tape.
The level 3 system will have an input rate of a few X 100 Hz and an output rate of several 10's Hz.

6. COST ESTIMATE

   The cost estimate of the τ-charm factory storage ring is summarized
in Table 6. We point out that a considerable saving (approximately 50 MSF)
has been achieved by the presence at CERN of the high quality LEP e$^+$e$^-$
injection complex (LIL, EPA and PS), and by the existing ISR transfer
tunnels and ISR experimental hall (bât. 181).

   The cost estimate of the detector is given in Table 7. The most expensive item is the electromagnetic calorimeter, which represents ~ 40% of
the total.

   The estimated cost of the storage ring is 44 MSF and the detector is
29 MSF, to give a total estimated cost for the τ-charm factory of 73 MSF.

Table 6. Cost estimate of the τ-charm factory storage ring

Item	Cost (MSF)
1. RF system	10
2. Vacuum system	8
3. Warm magnets	5
4. Superconducting mini β quads	2
5. Superconducting wigglers	5
6. Electrostatic separators	1
7. Power supplies	4
8. Controls	4
9. Civil engineering (tunnels + ring housing)	5
TOTAL	44 MSF

Table 7. Cost estimate of the τ-charm factory detector

Item	Cost (MSF)
1. Si microstrip tracker	2.4
2. Straw tracker	2.0
3. Fast RICH	1.2
4. Trigger counters	0.4
5. Electromagnetic calorimeter[*]	12.0
6. Superconducting solenoid + cryostat + liquid He plant	2.5
7. Instrumented flux return (hadron calorimeter/muon detector)	3.5
8. Trigger + data acquisition system	3.0
9. Luminosity monitor	0.5
10. Detector infrastructure	1.5
TOTAL	29 MSF

[*] This figure refers to a BaF_2 calorimeter; for BGO, the cost is 14.0 MSF.

7. CONCLUSION

The possibility of achieving extremely high luminosities in low energy e^+e^- colliders, combined with recent developments in detector technologies, indicate the potential exists for a precise experimental confrontation of the fundamental constituents ν_τ, τ and c with the Standard Model. The proposed τ-charm factory storage ring and detector both make such a large step beyond their predecessors that a rich physics programme is guaranteed -- and the capacity exists for a spectacular discovery. The τ-charm factory could be built over a short timescale at a relatively low cost and would complement the high-energy collider programme of CERN during the 90's. We invite the collaboration of interested physicists to explore more fully the physics potential of the τ-charm factory and to prepare a detailed experimental proposal to CERN for an integrated machine and detector.

8. ACKNOWLEDGEMENTS

I would like to express my appreciation to J.M. Jowett for his study of the machine design. I have been guided by the advice of J.D. Bjorken, F. Close, C. Heusch, M. Perl, K. Wille and, in particular, C. Rubbia. Finally, I would like to thank the directors of this school -- U. Gastaldi and R. Klapisch -- for having organized an extremely enjoyable and stimulating meeting.

REFERENCES

1. J.M. Jowett, Initial Design of the CERN τ-charm Factory, CERN LEP-TH/87-56 (November 1987).
2. U. Amaldi and G. Coignet, Conceptual Design of a Multipurpose Beauty Factory Based on Superconducting Cavities, CERN-EP/86-211, Nucl. Instrum. Methods A260, 7 (1987).
3. U. Amaldi, A Superconducting Radiofrequency Complex for Molecular, Nuclear and Particle Physics, Proc. Topical Seminar on Heavy Flavours, San Miniato, CERN-EP/87-104 (June 1987).
4. R. Eichler, T. Nakada, K. Schubert, S. Weseler and K. Wille, Motivation and Design Study for a B Meson Factory with High Luminosity, SIN PR-86-13 (November 1986).

EUROPEAN HADRON FACILITY[*]

F. Scheck[1], F. Bradamante[2], and J.M. Richard[3]

[1]Physics Institute, Mainz University, Postfach 39800
D - 6500 Mainz, Germany

[2]University fo Trieste, I - 34127 Trieste, Italy

[3]Institut des Sciences Nucléaires, F 38026 Grenoble-Cedex
France

(Presented by J.M. Richard)

ABSTRACT

In this contribution we survey the physics potential of a facility such as EHF in somewhat general terms. In other words, we outline some fundamental questions in nuclear physics and low energy particle physics to whose advancement EHF can and will make substantial contributions, without going into specific experiments needed to answer them. It is the role of the case studies, presented in the EHF proposal(1), to illustrate the kind of experimental effort (typical beam requirements, characteristic detectors, size of experiments, etc.) needed at EHF for the physics one wishes to clarify.

1. THE EHF INITIATIVE

The EHF Study Group was created in March 1983 to analyse the actual situation in intermediate energy particle physics and high energy nuclear physics, and to develop new perspectives and options for the years 1990 and beyond. Today, the group consists of 23 members from seven European countries who meet at regular intervals to discuss physics and accelerator options for this field. Table 1 lists its present and former members.

Over the last three years members of the EHF Study Group have organized a series of topical seminars, workshops, and conference aimed at a careful and detailed review of the basic physics issues for the next decade and beyond, viz :

---

[*] This written contribution is a very slightly modified version of the first chapter of the EHF proposal(1).

1)	The Workshop on the Future of Intermediate Energy Physics. Freiburg im Breisgau (2)	10 - 13 April	1984
2)	The Workshop on Nuclear and Particle Physics at Intermediate Energies with Hadrons, Trieste (3)	1 - 3 April	1985
3)	A series of topical meeting held at the Max-Planck Institute, Heidelberg on:		

Hyperon-nucleon interactions (4)	4 - 5 June	1985
Polarization phenomena	2 - 3 October	1985
Quark model and hadron spectrometry	3 - 4 December	1985
Muon Physics	4 - 5 February	1986
Coherent production	8 - 9 April	1986
Hyperon physics	3 - 4 June	1986
Rare Meson Decays	2 - 3 December	1986
Heavy Quarks	3 - 4 February	1987

4)	The Meeting on the Future of Medium and High Energy Physics in Switzerland, Les Rasses (5)	17 - 18 May	1985
5)	The Winter School on Hadronic Physics at Intermediate Energy, Folgaria (6)	17 - 22 February 1986	
6)	The International Conference on a European Hadron Facility, Mainz (7)	10 - 14 March	1986

Very many prominent colleagues from particle physics and nuclear physics have contributed to these events — the list being too long to be reproduced here — and have helped to shape the final objectives through many fruitful suggestions and ideas. Thus, the proposal for a 30 GeV high intensity proton synchroton, rather than any of the other possible options that were raised during the discussions, is the result of a very detailed and thorough analysis and discussion of the physics. It is our belief that EHF offers both the necessary quality and the breadth for a varied spectrum of first-rate experiments in a broad area of subnuclear physics.

In particular, the Mainz Conference represented a very important step in the formulation of the present proposal. Its main purpose was to review the physics programme both in low energy particle physics and in modern nuclear physics, with special emphasis on strong interactions in the confinement regime of QCD. These topics were thoroughly spelled out at the Conference, and properly summarized in the proceedings (7), which we consider to be an integral part of this proposal. The Mainz conference also gave a vivid impression of the composition, size and strength of the community of future users of EHF in Europe.

2. GENERAL REMARKS

Regarding the physics and basic issues of strong interactions, nuclear and particle physicists speak very much more the same language today than they did, say, ten years ago. Indeed, the investigation of subnuclear quark-gluon degrees of freedom has become a research goal common to the resonance spectroscopist, the quark-gluon plasma hunter, the lepton-nucleus scatterer and the antiproton annihilator. EHF as a facility for low energy particle physics and advanced nuclear physics will provide a bridge between these fields and will strengthen the common research goals, to the benefit of both. At the same time, given the size of a collaboration and the time scale for a typical experiment at EHF, EHF will be rather close to the universities and to their role as training institutions for the younger generation.

Thus, EHF not only opens up a rich novel spectrum of experimental possibilities in nuclear physics of the future and in specific domains of elementary particle physics, it also provides the meeting ground of the two disciplines and constitutes the complement to the very large accelerator projects necessary for the advancement of subnuclear physics as a whole.

In the letter of intent, we divided the physics for which EHF is the optimal tool into three major themes, each of which will cover a long-range research programme of considerable depth, viz.

(A) Electroweak interactions and signals of new physics

(B) Quantum chromodynamics in the quark confinement regime

(C) Nuclear substructure as revealed by nonnucleonic probes

Here we follow essentially the same subdivisions, bearing in mind, however, that there are many cross-relations and overlaps between these topics. For instance, the investigation of a rare, but existing, semi-leptonic decay of a hadron is relevant for testing QCD in the regime of confinement and chiral symmetry. Conversely, probing hadronic weak currents may reflect the limits of the standard model, i.e. of flavour physics.

3. ELECTROWEAK INTERACTIONS AND BEYOND

The point of reference in the discussion of electroweak interactions, at all energies, is the standard SU(3) x SU(2) x U(1) model of unified local gauge interactions. Among the basic questions to which EHF can provide partial answers, important hints or clues are the following :
- The nature and the level of parity violation
- Signals of unification beyond the minimal SU(3) x SU(2) x U(1) gauge structure.
- Radiative corrections, i.e. quantum effects which test and establish the standard model as a renormalization quantum field theory through the finite quantum effects it predicts.
- Nature of leptonic quantum numbers and structure of the leptonic mass sector.
- Quark masses and quark state mixing.
- Lepton-quark symmetry.
- Signals of "horizontal" symmetry, technicolor and / or technicolor symmetries.
- Signals of compositeness.

There are many reasons why experiments at EHF energies and intensities can provide important information, complementary to the pioneering investigations at higher energies :
(i) Experiments at low energies with high quality beams can be made selective and dedicated in the sense that complete information on final states can be obtained and selection rules can be utilized as "filters" for specific aspects of interest.
(ii) Many of these experiments can reach high precision. This is particularly important, for instance, in precision tests of the standard model, and in the search for signals of new physics in the TeV range. For example, the recent series of precision experiments on muon decay have shown that charged current weak interactions are sensitive to virtual mass scales in the range of 0.5 to 1 TeV. Similarly, muon number changing decays, at the level of 10^{-11} to 10^{-12} in branching ratio, test mass scales up to 100 TeV.
(iii) Rare and ultrarare decay processes are, and will continue to be, sources of information for a variety of fundamental questions. These decays can only be studied, mainly for reasons of backgrounds, with the dedicated, high intensity beams of EHF.

The possibilities and options that EHF will open up, include the following:

3.1 Neutrino physics

A dedicated and most challenging experiment on neutrino oscillations is described in the EHF proposal (1), as one of the cases studied : by directing the high intensity neutrino beam of EHF towards the ICARUS detector in the Gran Sasso tunnel, the mass-squared vs. mixing-matrix-element limits on $\nu_\mu - \nu_e$ oscillations can be improved by at least two orders of magnitude. This is a significant step forward and is particularly interesting since the experiment is highly relevant to the solar neutrino puzzle. It will help to distinguish different solar matter effects on the oscillations even where the Gallium experiment is insensitive. In contrast to the solar neutrino experiment of Davies et al., which is a disappearance experiment, the EHF oscillation experiment is of the appearance type.

Similarly, the neutrino beams of EHF can be used for a dedicated experiment measuring the weak mixing angle $\sin^2 \theta_W$ to very high precision. This is not spelled out in the case studies, but is obviously a challenging possibility at EHF (8).

3.2 Rare and ultrarare decays

The rare and ultrarare processes, when seen in the light of the standard model, can be divided into three classes :

(i) processes which are allowed in the standard model, but in which the basic electro-weak vertices are affected by the strong interactions. These processes give access to hadronic matrix elements of currents which couple to the various gauge bosons and, therefore, provide an important "laboratory" and testing ground for chiral dynamics, QCD in the confinement regime, and nonperturbative analysis (such as QCD sum rules and lattice calculations).

(ii) processes which are not forbidden in the standard model, but which depend on unknown parameters of the model, such as fermionic mass scales and mixing matrix elements.

(iii) processes which are forbidden in the standard model and whose existence would provide unambiguous hints to "new physics" beyond the model.

For example, Buchmüller and Wyler have made a systematic analysis of low energy effective interactions (9), signalling new interactions at higher mass scales for the exchanged particles. They have concluded that there are many open windows for various channels which can be investigated by means of specific and well-chosen experiments at low energies. Similarly, as has been discussed by J. Ellis et al., signals of superstring physics (10) may well show up in precision studies at low energies, depending on the specific pattern with which the superstring symmetries are broken.

Regarding ultra-rare decays, the purely leptonic processes such as $\mu \to e\gamma$, $\mu \to 3e$ and $\mu - e$ conversion on nuclei have now reached the level of 10^{-11} to 10^{-12} in branching ratio. Although these may be improved by about two orders of magnitude, the main interest would be with semileptonic decays, where the achieved level in branching ratio is now 10^{-8} at best. Here an improvement by several orders of magnitude is possible and is of vital interest for almost any extension of the standard model that has been proposed, (such as left-right symmetric theories, extended technicolor models, composite models, and supersymmetric extensions of the standard model).

Finally, it is likely that the physics of CP violation will continue to be of great topical interest when EHF comes into operation. Clearly, there will then be a number of specific aspects of the CP complex which one will be able to study only by means of the dedicated beams at EHF. CP violation can and will be studied with kaon beams at EHF. If an antiproton-proton collider at low energy were added to the complex, the intensities at EHF would be sufficient to study CP violation and other fundamental laws for charm and beauty particles.

3.3. Matter-antimatter symmetry, gravitation experiments

Present technology allows us to trap bunches of antiprotons and to cool them to very low temperature ($\sim 10°$ K) (11). Further progress can be expected in the near future in trapping efficiency, the cooling rate and in the maximum number of $\bar{p}$s that can be trapped at a time (12).

These facts open up the possibility of comparing the fundamental properties of p and $\bar{p}$ to a high degree of accuracy, using modern versions of classical experiments on both particles, either at the same time or alternately.

Obviously, the physics of these measurements is mainly related to two fundamental problems :

- CPT invariance
- gravity

The inertial masses (or actually the c/m ratios) can be compared with a relative accuracy of $10^{-9} - 10^{-10}$ (13). A measurement of the matter-antimatter gravitational interaction ($\bar{p}$s against the earth's gravitational field) will be very selective in the framework of gravitation theories (supergravity), that deduce from general principles a vector and a scalar potential of the Yukawa type. This measurement is in fact the only one able to see the sum and not the difference of these to Yukawa terms. The measurement of the magnetic moment has not been studied in detail so far but the accuracy could be of the order of the one obtained in the geonium experiment (14).

Both measurements (inertial mass and gravitational acceleration) have been approved at CERN (15). The proposed experiment to form $H_0 (\bar{p}e^+)$ in the LEAR ring (16) (quoted rate $\sim 10^{-4}$ H_0 s per second at a few hundred MeV/c) would probably not reach the statistics needed to perform a spectroscopic measurement on the 2P - 1S and 2P - 2S lines at the 10^{-3} level. Similar experiments using trapped and cooled $\bar{p}$s (17) would produce $\bar{H}_0$ s at roughly the same rate in the form of a low velocity atomic (antiatomic) beam, but the statistics would be probably too small in this case, too.

The beautiful test of CPT in electomagnetic interactions needs development of the cooling techniques in the ring and in the trap, and a substantial improvement in the $\bar{p}$ production rate, but the key problem is how to confine the $\bar{H}_0$ s. Overcoming this last problem also needs high production rates of $\bar{H}_0$s.

Turning now to the antideuteron $\bar{d}$, EHF would be the first accelerator able to produce $\bar{d}$s at a reasonable rate of 5×10^3/sec (18). Using a long collecting time (~ 2000 s), one could obtain bunches of 10^7 $\bar{d}$s, that could be handled, cooled and trapped using present-daty technology. This offers the possibility of increasing the significance of the measurement on gravity, but more essentially the possibility of cheching CPT invariance very precisely by measuring the mass defect $2m_{\bar{p}} - m_{\bar{d}}$ to a relative accuracy of 10^{-7}.

4. QUANTUM CHROMODYNAMICS IN THE QUARK-GLUON CONFINEMENT REGIME

Particle physics and nuclear physics aspects

With reference to the standard model, one may divide the basic issues in our present understanding of elementary particle physics into questions of flavour dynamics and, as far as strong interactions are concerned, problems of colour dynamics. Although this subdivision is not sharp, one may say that the highest energy machines address primarily flavour physics, whereas intermediate energy facilities like EHF allow the study of colour physics, i.e. strong interaction physics in a regime where it exhibits its most interesting and most characteristic features. This applies equally well to experiments with elementary particle targets and to experiments with nuclear targets. As we remarked earlier, the distinction between these is somewhat artificial for the purposes of QCD in the confinement regime, in so far as an experiment in the nuclear medium can be as informative as, or complementary to, experiments on elementary targets. In the next section, we discuss first a few general aspects of QCD in the confinement regime. We then illustrate our remarks with a couple of specific topics in strong interaction physics for which EHF provides the optimal tools of investigation.

4.1 Colour physics : general aspects and fundamental questions

Local colour symmetry, formulated quantitatively by QCD, appears today as the key to understanding the filigree of strong interaction physics revealed by the accelerators of the sixties : AGS, PS, SLAC, etc. This "revelation", however, has not become a full and deep understanding for a number of reasons, both experimental and theoretical. On the theoretical side, QCD has not really developed into a complete and unfailing calculational tool, as compared to QED for instance, due to the considerable problems posed by a full-fledged quantum field theory with a complex ground state structure and essential nonperturbative features. Some of the basis issues in QCD have not been solved in a satisfactory manner, such as the closeness of our hadronic world to the chiral limit, or the correlations in the QCD ground state as manifested by the appearance of quark and gluon condensates. Although these problems are difficult, and although nonperturbative methods to solve them are still in their infancy, we expect important progress towards understanding a quantum field theory such as QCD in its nonperturbative regime (the physics of confinement) by testing our ideas and lines of attack with subtle and diversified sets of data.

On the experimental side, the initial phase of enthusiasm throughout the sixties was followed by a period of disenchantement with hadronic physics, and a concentration of efforts on "asymptotic" physics at the highest energies. It would be unwise, however, to put aside this whole domain of physics as a kind of "chemistry" of strong interactions, too complex and too difficult to be tractable. On the contrary, we believe that real progress is possible in the future, due to very much improved experimental techniques (as compared to the sixties), and thanks to the interplay between theory and experiment which helps in understanding and spelling out the subtleties of the hadronic world at low energies.

In the assessment of the standard model of electroweak and colour interactions, the hard problems of QCD are in fact thought to be best approached through large-scale computer calculations, i.e. by means of Monte Carlo simulations of local gauge theories on lattices. Therefore, one might be tempted to claim that the machine needed to scan the "confinement frontier" is a dedicated supercomputer, rather than an experimental facility such as EHF. Putting aside the fact that these calculations are still in their infancy, we believe that this view is likely to be false. The structure of the QCD ground state and the hadronic spectrum at intermediate energies are too rich and too complex to be explorable by computer simulations. (In fact, there is an intriguing controversy about the nature of the ground state of QCD and its possible instability which is not settled satisfactorily (19)). Instead, we need important hints from Nature, through experiments at a dedicated facility such as EHF, if we wish to explore the confinement frontier of colour physics.

To dispose in the nineties of a tool capable of producing a quantity of hadronic data is seen as a crucial step towards finally understanding the key problems of hadronic matter.

4.2 Spectroscopy

The quark model has achieved considerable success. The aim now is to refine the study of the quark dynamics and to search for new states with constituent gluons.

Ordinary hadrons

One should complete a detailed scanning of radial and orbital excitations of mesons to see whether there is room left for new states like multiquarks, hybrids or gluonia. For baryons too there is the possibility of multiquarks or hybrids, and also some open questions concerning the three-quark states. Most observed baryon resonances, for instance, are compatible with a quark-diquark picture where one relative distance is frozen. States where both oscillators (corresponding to the two Jacobi relative coordinates) are excited should be identified to confirm the three-body nature of baryons, i.e. to count the number of degrees of freedom in hadronic matter.

Gluonia

Gluonia have been studied mostly in J/ψ decay. Other entrances channels should be considered : $N \bar{N}$ annihilation, production in meson-baryon or baryon-baryon scattering, etc. with a variety of final orbits. The couplings to strange and non-strange particles are necessary to unravel the gluonium nature of a state.

Theoretically, the gluonium gives rise to interesting calculations: specifically, in lattice QCD one open question is whether the ground state in JPC is 0^{++} or 2^{++}.

Hybrids

As soon as the gluon is acknowledged as a coloured constituent, as in gluonium, one expects hybrid mesons or baryons : $q\bar{q}g$, $qqqg$. These states are likely to decay into orbitally excited mesons or baryons. This means a rather high total multiplicity, requiring sophisticated detectors. Hybrids might have escaped detection in previous spectroscopy experiments.

Heavy flavours (hidden)

Quarkonia have been studied with e^+e^- machines, with a great deal of success. However, this allows only for the formation of JPC = 1^{--} states and the observation of some C = 1 states via γ emission. The $q\bar{q}$ - spectrum is thus incomplete and one cannot determine the spin structure of the $q\bar{q}$ potential. In $p\bar{p}$ collisions, all $q\bar{q}$ states are, in principle, accessible. The feasibility of $p\bar{p} \to c\bar{c}$ has been illustrated by the R 704 experiment at ISR. Apart from the mass spectrum, $p\bar{p}$ experiments will allow for a systematic study of the decay of the χ - states (similar to the very rich information obtained from the study of J/ψ decay).

Heavy flavours (open)

The spectroscopy of open charm and beauty is rather poor so far. We know only some ground states, like D, D^*, D_s, D_s^*, Λ_c, Σ_c, B, etc., with one exception, a candidate for an orbital excitation of the D. These states are however extremely interesting, since the light quarks, when associated with heavy quarks, experience more relativistic corrections than in ordinary hadrons (remember than pe^- is more relativistic than e^+e^-).

Multiquarks

Exotic multiquark states have been predicted (22) by extrapolating reasonable quark models which reproduce the properties of ordinary mesons and baryons. On the other hand, some "unexpected" candidates have been tentatively seen (23). This fascinating sector of hadron spectroscopy urgently needs clarification.

4.3 The case for spin physics in the light of QCD

At the "confinement frontier", spin observables provide very powerful tools to penetrate the intricacies of hadronic phenomena at intermediate energies. At the hadron level, spin emerges as a combination of spins and angular momenta of quarks. Thus the observation of spin effects will shed light on the dynamics that confines spin 1/2 quarks within well-defined regions of space-time (bags). In particular, as already documented, the presence of sizeable spin effects at large momentum transfer runs counter to the expectations of perturbative QCD, although according to general belief, QCD should be directly applicable in this region. Thus, in the experimental programme of EHF, the investigation of polarization phenomena at large angles and transverse momenta is of primary importance. The results will be crucial for obtaining information on how quark confinement works at short distances in space-time.

4.4 Quark confinement and nuclear physics

Much of what we said above applies also to the investigation of nuclear structure at the scale of quark and gluon degrees of freedom. For instance, one of the oldest, and yet to a large extent unresolved, basic problems of nuclear physics is the nature of the hard core observed in nucleon-nucleon scattering. In the light of the quark model, the hard core seems to be linked to the inhibiting effects of the Pauli principle between constituent quarks. If the simple explanations that were put forward have some truth in them, what would one find in Λ - N or Σ - N scattering ?

QCD as a non-Abelian local gauge theory has much more structure than one would expect on the basis of perturbative quantum field theory. Just as in the case of electrodynamics, it may well be that QCD in an extended medium like nuclear matter exhibits new phenomena of many-body collective type. Such effects may show up in processes like

(i) Production and propagation of resonances in nuclei
(ii) Antiproton annihilation in nuclei
(iii) Coherent production

As very little is known to date, this field is in a stage of exploration and speculation. Quite obviously, many experiments are needed in order to clarify the basic mechanisms.

Even at a more modest and less speculative level there are basic issues and questions in nuclear physics that must be clarified, such as : Where and how does the quark-gluon substructure of nuclei show up ? How can one reconcile the description of nuclear dynamics in terms of meson and baryon degrees of freedom with the QCD description of nuclei in terms of quarks and gluons ?

Some examples of experimental projects in this area are the following :

(i) Production and scattering of baryonic resonances in nuclei. We already have some information on Δ, $\Lambda(1116)$ and Σ interactions with nuclei, from which a very simple pattern seems to emerge, in agreement with naive model pictures but still a long way from real understanding. Very little is known, however, about the formation of higher resonances in nuclei and their interactions in the nuclear medium. For instance, the states Λ (1405) and Λ (1520), which would be produced via the (K - p) channel, are of particular interest. In the constituent quark model, both are simple orbital excitations of the Λ (1116), and their interaction with nuclei should be accurately calculable. Investigating their spatial structure and their interactions would give clues for additional structures if the data does not follow the naive expectations.

(ii) The spin-orbit couplings of N, Δ, Λ, Σ and Ξ show simple regularities and may help to discriminate between models typical of QCD and conventional one-boson-exchange picture.

(iii) K^+ - nucleus interactions are of great interest because they should be calculable with high accuracy, and yet the comparison of experimental results with calculations shows serious discrepancies (20).

(iv) Production in nuclei of $(q\bar{q})$ meson states such as η' (958) which can have a glueball admixture may turn out to be the best way to reveal these gluonic components, typical for QCD (20).

(v) Interactions of antiprotons with nuclei, under various experimental conditions, will continue to provide an important source of information on hadron dynamics in the nuclear medium.

(vi) Drell-Yan-like processes in nuclei are likely to shed light on the dynamics which is reponsible for the EMC effect.

(vii) Production of hyperons or antiprotons below and above the nominal threshold for individual production on nucleons will give information on coherent and collective mechanisms as well as on the high momentum components of the nucleus wave-function.

In this field even more than the others mentioned above, close collaboration between theory and experiment is vital. The quark-gluon structure of nuclear matter raises challenging problems of non-linear dynamics which require new concepts. New theoretical ideas must then be checked against experiment before one can proceed. This requires a high-quality programme of dedicated experiments which shed light on specific facets of hadron dynamics. Thus, close coordination of experimental and theoretical research will enable decisive progress in this field. Again, this ambition goal calls for a dedicated facility such as EHF.

5. FURTHER REMARKS AND ADDITIONAL OPTIONS

5.1 Further remarks

The case studies contained in the following chapters serve to illustrate some of the topics discussed above. They are selected according to what seems most interesting today and according to the taste and preferences of their authors, but, as emphasized in the preface, they do not exhaust the potential of EHF. We give them in consecutive order, without names on the title page. In Table 1 of the Preface the reader will find a list of all the topics that were discussed, the names of all contributors, and the names of the conveners of the various groups.

5.2 Additional options

Among the many extension programmes one might think of for EHF, a most appealing possibility would be a kind of SUPERLEAR (21) facility. This is an extra option for EHF, realizable at extra cost but with great scientific potential. Chapter 11 of the EHF proposal (1) is devoted to the antiproton physics concerned with this option.

ACKNOWLEDGEMENTS

I would like to thank the organizers, R. Klapisch, F. Close and U. Gastaldi for the stimulating atmosphere of this school and all participants for enjoyable discussions.

Table 1 : EHF Study Group

R. BERTINI	Saclay	C. KLEINKNECHT	Mainz
R. BEURTEY	Saclay	H. KOCH	Karlsruhe
F. BRADAMANTE	Trieste	G.C. MORRISON	Birmingham
T. BRESSANI	Torino	B. POVH	Heidelberg
W. BREUNLICH	Wien	J.M. RICHARD	Grenoble
D. BUGG	QMC, London	M. ROY-STEPHAN	Orsay
A. CITRON	Karlsruhe	R. RÜCKL	DESY
P. DALPIAZ	Ferrara	F. SCHECK	Mainz
J. DOMINGO	SIN	J. SPETH	Jülich
R. ENGFER	Zürich	T. WALCHER	Mainz
S. GALSTER	Karlsruhe	W. WEISE	Regensburg
W. JOHO	SIN	B. ZEITNITZ	Karlsruhe
B. JONSON	Göteborg		

REFERENCES

1. The EHF study group, "Letter of intent for a European Hadron Facili-
    ty", August 1986, edited by F. Scheck, Mainz, Germany
    F. Scheck et al., "Proposal for a European Hadron Facility", ed.
    J.F. Crawford (available from the authors)

2. Proceedings of the "Workshop on the future of intermediate energy
    physics in Europe", 10-13 April 1984, ed. by S. Galster, KfK Karls-
    ruhe, Germany

3. Proceedings of the "Workshop on nuclear and particle physics at in-
    termediate energies with hadrons", Miramare, Trieste, Italy, 1-3 /
    April, ed. by T. Bressani and G. Pauli

4. Organized by B. Povh and C. Wiedner, Max Planck Institute für Kern-
    physik, Heidelberg, Germany

5. Proceedings of the Meeting on "The future of medium and high energy
    physics in Switzerland", Les Rasses, Switzerland, 17-18 May 1985,
    ed. by the SIN Documentation Group, 1 October 1985

6. Proceedings of the Winter School on Hadronic physics at intermediate
    energy, Folgaria, Italy, 17-22 February, 1986, ed. by T. Bressani
    and R. A. Ricci, North-Holland

7. Proceedings of the "International conference on a European hadron
    facility", Mainz, Germany, March 10-14, 1986, ed. by Th. Walcher,
    North-Holland Publishing Co., reprinted from Nucl. Phys. B279 (1987)
    Nos., 1,2.

8. J.L. Vuillemier, ASTOR proposal, SIN, December 1986

9. W. Buchmuller and D. Wyler, Nucl. Phys. B268 (1986) 621

10. J. Ellis et al., CERN preprint, to be published

11. G. Gabrielse et al., Phys. Rev. Lett. 57 (1986) 2504,
    D.J. Wineland et al., Adv. At. and Mol. Phys. 19 (1984) 135

12. Research on these subjects is in progress at Los Alamos Lab., Rice
    Univ., Texas A & M University, Seattle University in USA. Universi-
    tà di Pisa and Università di Genova in Italy.

13. R.S. van Dyck Jr. et al., Phys. Rev. Lett. 47 (1981) 395,
    G. Graff et al., Z. Phys. A297 (1980) 35,
    G. Gartner and E. Klempt, Z. Phys. A287 (1978) 1

14. J. Scherk in "Unification of fundamental particle interactions", eds. S. Ferrara, J. Ellis and P. van Nieuwenhuizen, Plenum Press, New-York, 1989, p. 381

15. G. Gabrielse et al. "Precision comparison of antiproton and proton masses in a penning trap", CERN proposal PSCC/85-21/P83-/22, March 1985, N. Beverini et al., "A measurement of the gravitational acceleration of the antiproton", CERN proposal PSCC/86-2/p94/16, January 1986

16. A. Wolf, "Production of fast antihydrogen and the first electron cooling for low energy antiprotons", Antiproton 86, Thessanoliniki, Sept. 1-5 1986, to be published, J. Berger et al. "Feasibly study for antihydrogen production at LEAR", CERN proposal PSCC/85-46/S86, H. Poth, private communication.

17. N. Beverini and G. Torelli "Antiproton traps and related experiments", Workshop on intermediate energy physics, Trieste 1-3.4.1985, "Nuclear and particle physics at intermediate energy with hadrons", Eds. T. Bressani and G. Pauli, Conf. Proc. SIF Vol. 3 p. 111, Kells in "Proceedings of the workshop on the design of a low energy antimatter facility", University of Wisconsin, Madison, Wisconsin, USA (October 1985), to be published

18. H. Koch et al., "Antideuterons at LEAR", Third LEAR workshop, Tignes, France, January 19-26, 1985, p. 877

19. L. Cosmai and G. Preparata, Phys. Rev. Lett. 57 (1982) 2613

20. F. Lenz in : Proceedings of international conference on a European hadron facility, Mainz 1986, Nucl. Phys. B279 (1987) Nos. 1.2, F. Cannata, "Hadronic Physics at EHF", Bologna preprint

21. P. Dalpiaz et al., Physics at SUPERLEAR, CERN internal report CERN/EP87-27, 12 February 1987, to appear in the Elementary Structure of Matter", Proc. les Houches Workshop, 1987, ed. E. Aslanides et al., Springer-Verlag (in press). See also the contribution by P. Dalpiaz in this volume.

22. See, e.g. the contributions by H.J. Lipkin, L. Heller and J.M. Richard, in "The Elementary Structure of Matter", loc. cit.

23. M. Bourquin et al., Phys. Lett. 172B, (1986), 113

Participants of the second course of the International School of Physics
with Low Energy Antiprotons on Spectroscopy of Light and Heavy Quarks,
held May 23-31, 1987, in Erice, Sicily, Italy

PARTICIPANTS

T. Barnes   Department of Theoretical Physics
                                        Toronto University
                                        Toronto
                                        Ontario M5S 1A7
                                        Canada

F. Binon   Institut interuniversitaire
                                         des Sciences nucléaires
                                        5, rue d'Egmont
                                        B-1050 Bruxelles
                                        Belgium

M. Botlo   Institut für Radiumforschung
                                        Universität Wien
                                        Boltzmanngasse 3
                                        A-1090 Wien
                                        Austria

M. Boutemeur   L A P P
                                        B.P. 909
                                        F-74019 Annecy-le-Vieux
                                        France

K. Braune   CERN - EP
                                        CH-1211 Geneva 23
                                        Switzerland

J.M. Butler   CERN - EP
                                        CH-1211 Geneva 23
                                        Switzerland

A. Coc   CERN - EP
                                        CH-1211 Geneva 23
                                        Switzerland

F. Close   Rutherford Appleton Laboratory
                                        Chilton near Didcot
                                        Oxfordshire OX11 0QX
                                        England

G. Coignet   L A P P
                                        B.P. 909
                                        F-74019 Annecy-le-Vieux
                                        France

F. Couchot   L A L
                               Centre d'Orsay
                               F-91405 Orsay
                               France

P. Dalpiaz   Istituto di fisica
                               Università di Ferrara
                               Via Paradiso 12
                               I-44100 Ferrara
                               Italy

G. Eigen   C A L T E C H
                               Pasadena
                               California 91125
                               USA

F. Feld-Dahme   Sektion Physik
                               Universität München
                               Am Coulombwall 1
                               D-8046 Garching
                               FRG

G. Folger   Sektion Physik
                               Universität München
                               Am Coulombwall 1
                               D-8046 Garching
                               FRG

U. Gastaldi   CERN - EP
                               CH-1211 Geneva 23
                               Switzerland

D. Gromes   Institut für Theoretische Physik
                               Universität Heidelberg
                               Philosophenweg 16
                               D-6900 Heidelberg
                               FRG

N. Isgur   Department of Theoretical Physics
                               Oxford University
                               Keble Road
                               Oxford OX1 3NP
                               England

J. Kirkby   CERN - EP
                               CH-1211 Geneva 23
                               Switzerland

R. Klapisch   CERN - DG
                               CH-1211 Geneva 23
                               Switzerland

R. Le Gac   Laboratoire René Bernas
                               Bâtiment 108
                               F-91406 Orsay
                               Fance

P. Lefèvre   CERN - PS
                               CH-1211 Geneva 23
                               Switzerland

M. Macri

Dipartimento di fisica
Università di Genova
Via Dodecaneso 33
I-16146 Genova
Italy

D. Möhl

CERN - PS
CH-1211 Geneva 23
Switzerland

L. Montanet

CERN - EP
CH-1211 Geneva 23
Switzerland

N. Nägele

Austrian Academy of Science
Boltzmanngasse 3
A-1090 Wien
Austria

A. Palano

Dipartimento di fisica
Università di Bari
Via Amendola 173
I-70126 Bari
Italy

J.-M. Richard

Laboratoire de physique théorique
Université Pierre et Marie Curie
F-75230 Paris cedex 05
France

M. de Saint Simon

Laboratoire René Bernas
Bâtiment 108
F-91406 Orsay
France

G.A. Smith

Pennsylvania State University
303 Osmond Laboratory
University Park
Pennsylvania 16802
USA

C. Thibault

Laboratoire René Bernas
Bâtiment 108
F-91406 Orsay
France

F. Touchard

Laboratoire René Bernas
Bâtiment 108
F-91406 Orsay
France

LEAR (continued)
emittances, 360
experimental area, 365, 368
low β insertion, 367
momentum scanning, 369
phase-space cooling, 359-361
pick-ups, 368
running statistics, 364
stochastic cooling system, 361, 368
stochastic extraction, 362
ultra-slow extraction, 339
vacuum system, 369
working point, 363
LEP, 312, 389
Leptoquarks, 411
LIL LEP Injector Linacs, 401, 404, 413
Lorentz invariance, 80
Luminosity, 341, 384, 390, 392, 393, 402, 404, 408, 413
L X-rays of protonium, 315, 316

MARK I, 167
MARK II, 30, 167, 170, 220, 222, 225, 407
MARK III, 30, 167, 169, 172, 174, 175, 177, 183, 193, 194, 195, 197, 198, 201, 204, 212, 217, 218, 220, 221, 223, 225, 229, 233, 239, 241, 260, 320, 407, 410, 411, 415, 416
Mass
of bound $b\bar{b}$ and open b systems, 388
effective, 50
generation problem, 385
limit for ν_τ, 409
prediction for glueballs and hybrids, 188, 189
splitting, 49
Matrix,
S, T and K, 99
K matrix formalism, 107
M matrix formalism, 107
Meson
B, 387
$q^2\bar{q}^2$, 29
resonances, 85-158
spectroscopy, 85-158
Michel parameter, 409
Microstrip detectors, 283, 341, 417
MIS, 150
Mixing, 124, 129, 193, 406, 425
$D^0\bar{D}^0$, 410
angle, 242
effects in η_c, 220
Molecules, 25, 88, 156, 241
MPS, 151

Monochromator, 413, 415
M1 transition, 185
Multibunch double ring, 397
Multichannel spin-parity analysis, 210
Multipoles, 354
Multiquark, 429
Muon decay, 425

$\bar{n}$ beam, 317
NA3 experiment at CERN, 320
NA12 experiment at CERN, 271 (see also GAMS)
Natural parity mesons, 87
Neutrino physics, 426
Neutrino tau ν_τ, 386, 401
Nuclear spectroscopy notation, 85

OAFM Open Axial Field Magnet, 312, 324, 325, 326, 336, 337
OBELIX, 46, 62, 311-339
Oddball, 186
OMEGA - WA76, 277
Oval chamber, 345

Parity violation, 425
Partial wave analysis, 241, 258, 265
Penning trap, 366
PEP, 389, 390, 408
Perturbation theory, 44, 67
PETRA, 389, 408
Phase shift, 101
$\bar{p}H^-$ co-rotating beams, 327
Pinch factor H, 393
PLUTO, 33, 167
Polarization, 321 (see also AMP, IP, spin)
Polarized jet target, 325, 341
Polarized targets, 325
Polarized $\bar{p}$ beams, 325
Pole of S matrix, 95
Pomeron, 111, 278
double pomeron reaction, 151, 277
Positronium, 139
$p\bar{p}$ colliding beams, 373 (see Super LEAR)
Potential
effective scalar, 69
interhadron, 25
interkaon, 32
intermeson, 38
interpion, 28
kaon-antikaon, 28
long-range, 67
quark-antiquark, 67
short-range, 67
static quark-antiquark, 75

442